Socio-Affective Computing

Volume 3

Series Editor
Amir Hussain, University of Stirling, Stirling, UK

Co-Editor
Erik Cambria, Nanyang Technological University, Singapore

This exciting Book Series aims to publish state-of-the-art research on socially intelligent, affective and multimodal human-machine interaction and systems. It will emphasize the role of affect in social interactions and the humanistic side of affective computing by promoting publications at the cross-roads between engineering and human sciences (including biological, social and cultural aspects of human life). Three broad domains of social and affective computing will be covered by the book series: (1) social computing, (2) affective computing, and (3) interplay of the first two domains (for example, augmenting social interaction through affective computing). Examples of the first domain will include but not limited to: all types of social interactions that contribute to the meaning, interest and richness of our daily life, for example, information produced by a group of people used to provide or enhance the functioning of a system. Examples of the second domain will include, but not limited to: computational and psychological models of emotions, bodily manifestations of affect (facial expressions, posture, behavior, physiology), and affective interfaces and applications (dialogue systems, games, learning etc.). This series will publish works of the highest quality that advance the understanding and practical application of social and affective computing techniques. Research monographs, introductory and advanced level textbooks, volume editions and proceedings will be considered.

More information about this series at http://www.springer.com/series/13199

Seng-Beng Ho

Principles of Noology

Toward a Theory and Science of Intelligence

Seng-Beng Ho
Social and Cognitive Computing,
　Institute of High Performance Computing
Agency for Science, Technology
　and Research (A*STAR)
Singapore, Singapore

2009–2014
Temasek Laboratories
National University of Singapore
Singapore, Singapore

ISSN 2509-5706
Socio-Affective Computing
ISBN 978-3-319-32111-0   ISBN 978-3-319-32113-4   (eBook)
DOI 10.1007/978-3-319-32113-4

Library of Congress Control Number: 2016943131

© Springer International Publishing Switzerland 2016
This work is subject to copyright. All rights are reserved by the Publisher, whether the whole or part of the material is concerned, specifically the rights of translation, reprinting, reuse of illustrations, recitation, broadcasting, reproduction on microfilms or in any other physical way, and transmission or information storage and retrieval, electronic adaptation, computer software, or by similar or dissimilar methodology now known or hereafter developed.
The use of general descriptive names, registered names, trademarks, service marks, etc. in this publication does not imply, even in the absence of a specific statement, that such names are exempt from the relevant protective laws and regulations and therefore free for general use.
The publisher, the authors and the editors are safe to assume that the advice and information in this book are believed to be true and accurate at the date of publication. Neither the publisher nor the authors or the editors give a warranty, express or implied, with respect to the material contained herein or for any errors or omissions that may have been made.

Printed on acid-free paper

This Springer imprint is published by Springer Nature
The registered company is Springer International Publishing AG Switzerland

To Leonard Uhr

Preface

Despite the tremendous progress made in the past many years in cognitive science, which includes sub-disciplines such as neuroscience, psychology, artificial intelligence (AI), linguistics, and philosophy, there has not been an attempt to articulate a principled and fundamental theoretical framework for understanding and building intelligent systems. A comparison can be made with physics, which is a scientific discipline that led to the understanding of the physical universe and the construction of various human artifacts. The feats we have achieved through physics are indeed incredible: from mapping the cosmos to the end of space and time to the construction of towering skyscrapers and rockets that took human beings to the moon. Physics provides the principles and theoretical framework for these incredible feats to be possible. Is it possible to construct a similar framework for the field of cognitive science?

Whether intelligent systems are those that exist in nature, such as animals of all kinds, or the various kinds, of robots that human beings are trying to construct, they are all autonomous intelligent agents. Moreover, animals are *adaptive* autonomous intelligent agents (AAIAs), and the robots that human beings construct are also intended to be adaptive, though we have been falling short of the achievement of nature in this regard so far.

Interestingly, neuroscientists and psychologists do not seem to construe their respective disciplines as attempting to uncover the nature and principles of intelligence per se nor do they often characterize the systems they study as AAIAs. Neuroscientists are primarily concerned with uncovering the neural mechanisms in various human and animal brain subsystems such as the perceptual systems, affective systems, and motor systems, and how these various subsystems generate certain behaviors. Psychologists also attempt to understand human and animal behaviors through behavioral experiments and fMRI. But it is not just behavior per se but *intelligent* behavior that the various animals and humans exhibit that improve their chances of survival, allow them to satisfy certain internal needs, etc. They are also *adaptive* and *autonomous* intelligent agents. Hence, the numerous experimental works conducted in the fields of neuroscience and psychology so far have not benefitted from or been guided by a principled theoretical framework that characterizes adequately the systems that they are studying.

On the other hand, AI has been concerned with constructing AAIAs (also called "robots") right from the beginning. However, the shortcoming of AI at its current state of development is that the major "successes" are in creating *specialized* intelligent systems – systems such as the Deep Blue chess playing system that can beat human chess masters, the Watson questioning-answering system that can outperform human opponents, the upcoming autonomous vehicles (such as the Google self-driving car) that can drive safely on the roads, etc. But some researchers in the AI community do attempt to strive toward constructing *general* AAIAs in the long run. This is reflected in the emergence of a field called artificial general intelligence (AGI), though ironically, AI, in its very inception, was meant to be AGI to begin with.

An interesting question arises concerning human-constructed intelligent systems. Can a system that is not mobile in itself but that has remote sensors and actuators benefit from the principles guiding adaptive *autonomous* intelligent systems? The answer is yes, and we can think of the system as a kind of "static" robot. Because, with remote sensors and actuators, it can achieve the same effect as in the case of an AAIA as it learns about the environment through observation and interaction, and enriches its knowledge and changes its future behavior accordingly.

AGI can certainly learn from the rest of cognitive science. For example, in traditional AI research, the issues of *motivation* and *emotion* for an adaptive intelligent system, which provide the major driving force behind the system and are hence critical in its adaptive behavior, are never discussed (a scan of the major current textbooks in AI would reveal that these terms do not even exist in the index), while these are often studied extensively in neuroscience and psychology. The issues of affective processes, however, are gaining some attention recently in the field of AGI/AI.

In this book we therefore set out toward a more systematic and comprehensive understanding of the phenomenon of intelligence, thus providing a principled theoretical framework for AAIAs. The methodology is similar to that of AGI/AI, in which representational mechanisms and computational processes are laid out clearly to elucidate the concepts and principles involved. This is akin to the quantitative and mathematical formulation of the basic principles of physics that embodies rigorous understanding and characterization of the phenomena involved. There are two advantageous to this approach: on the one hand, these mechanisms translate to directly implementable programs to construct artificial systems; on the other hand, these mechanisms would direct neuroscientists and psychologists to look for corresponding mechanisms in natural systems. Because of the relatively detailed specifications of the representational mechanisms and computational processes involved, they may guide neuroscientists and psychologists to understand brain and mental processes at a much higher resolution, and also understand them in the context of AAIAs, which is a paradigm that is currently lacking in these fields.

The representational mechanisms and computational processes employed in this book are not strange to people in the field of AI: predicate logic representations, search mechanisms, heuristics, learning mechanisms, etc. are used. However, they are put together in a new framework that addresses the issues of *general* intelligent systems. Some novel computational devices are introduced, notably the idea of rapid effective causal learning (which provides a rapid kind of learning subserving critical intelligent processes), learning of scripts (which provides a foundation for knowledge chunking and rapid problem solving), learning of heuristics (which enhances traditional AI's methodology in this regard which often employs heuristics that are built-in and not learned in a typical problem solving situation), semantic grounding (which lies at the heart of providing the mechanisms for a machine to "really understand" the meaning of the concepts that it employs in various thinking and reasoning tasks), and last but not least, the computational characterizations of motivational and affective processes that provide purposes and drives for an AAIA and that are critical components in its adaptive behavior.

From the point of view of identifying fundamental entities for the phenomenon of intelligence (much in the same spirit in physics of identifying fundamental particles and interactions from which all other phenomena emerge), two ideas stand out. One is *atomic spatiotemporal conceptual representations* and their associated processes, which provide the ultimate semantic grounding mechanisms for meaning, and the other is the *script*, which encodes goal, start state, and solution steps in one fundamental unit for learning and rapid problem solving operations.

We think it necessary to introduce the term "noology" (pronounced \nō-ˈä-lə-jē\, in the same vein as "zoology") to designate the principled theoretical framework we are attempting to construct. Noology is derived from the Greek word "nous" which means "intelligence." The Merriam-Webster dictionary defines noology as "the study of mind: the science of phenomena regarded as purely mental in origin." The function of noology – a theoretical framework for and the science of intelligence – is like the function of physics. It provides the principled theoretical framework for, on the one hand, the understanding of natural phenomena (namely all the adaptive autonomous intelligent systems (AAISs) that cognitive scientists are studying), and on the other, the construction of artificial systems (i.e., robots, all kinds of autonomous agents, "intelligent systems," etc., that are the concerns of AGI/AI).

In cognitive science, it has been quite a tradition to use "cognitive systems" to refer to the "brain" systems that underpin the intelligent behavior of various kinds of animals. However, as has been emphasized in a number of works by prominent neuroscientists such as Antonio Damasio and Edmund Roll, cognition and emotion are inseparable and together they drive intelligent behaviors as exhibited by AAISs. Therefore, "noological systems" would be a more appropriate characterization of systems such as these.

We do not pretend to have all the answers to noology. Therefore, the subtitle of this book is *"toward* a theory and science of intelligence." But we believe it sets a new stage for this new, and at the same time old, and exciting endeavor.

For readers familiar with the computational paradigm (e.g., AI researchers), it is recommended that they jump ahead to take a look at Chaps. 6 and 7 and perhaps also 8 and 9, where our paradigm is applied to solve some problems that would typically be encountered by AI systems, before returning to start from the beginning of the book. Our method derives "noologically realistic" solutions that use very different processes than that of traditional AI methods. After having a rough idea of what we are trying to achieve, the purpose of the relatively extensive groundwork covered in the first few chapters would then be clearer.

Singapore, Singapore   Seng-Beng Ho

A Celebration of Concavity

Shelf: Designed by Seng-Beng Ho
Design Registration No. D2014/888/A

"Science and art in the pursuit of beauty, and hence, truth"

Acknowledgements

In a sense the beginning of this book is lost in the author's personal mist of time. My formal journey on the path to understand intelligence began in the Fall of 1981 when I started my graduate school program at the University of Wisconsin-Madison. Because the effort culminating in this book was an incessant effort from more than 30 years ago, some acknowledgements are due to some of the people from that time.

My Ph.D. supervisor was the late Professor Leonard Uhr who was one of the pioneers in artificial intelligence, as reflected by the fact that one of his important works is collected in the book, *Computers and Thought*, published in 1963, 8 years after John McCarthy coined the term "artificial intelligence (AI)." One of the editors of the book was Edward Feigenbaum of Stanford University, a well-known researcher in AI. Leonard Uhr's research was a little off "mainstream AI" and hence he was not as well-known as, say, Edward Feigenbaum or Marvin Minsky of MIT in the field of AI.

However, my thanks are due to Leonard Uhr for allowing me the freedom to think outside the mainstream research trend in AI of that time, which allowed me to address critical issues ignored by others. This constituted my Ph.D. thesis work of that time and it set the stage for my subsequent pursuit in understanding intelligence in the same spirit, and which then led to this book of today. Thanks are also due to Chuck Dyer, Josh Chover, Gregg Oden, Lola Lopes, Robin Cooper, and Berent Enc, who were professors of that time who interacted with me in one way or another and who were equally encouraging and supportive of my unique focus on issues related to intelligent systems.

The intellectual origin of this book can be traced back to more than 30 years ago but the bulk of the work contained herein was carried out in the Temasek Laboratories of the National University of Singapore in the period 2011–2014. I wish to thank the Head of the Cognitive Science Programme at Temasek Laboratories, Dr. Kenneth Kwok, who, like the professors before him, fully supported my research that addressed critical and important issues not attended to by other AI researchers.

All the computer simulations in the book and the associated graphical renderings were done by Fiona Liausvia. My sincerest thanks to her for a fantastic job done.

Erik Cambria's work on sentic computing was an inspiration. My own approach to computational affective processes started on a different track, but knowing about Erik's work and exchanging ideas with him provided me with the confidence that we are moving in the right direction with regards to a more complete computational characterization of intelligent systems.

My immediate family has always provided great intellectual companionship. Thanks are absolutely due to them for keeping me intellectually alive all through the years.

Singapore, Singapore Seng-Beng Ho
January 1, 2016

Contents

1	**Introduction**		1
	1.1	The Core Functional Blocks of a Noological System	6
	1.2	Basic Considerations Motivating a Noological Processing Architecture	8
	1.3	A Basic Noological Processing Architecture	15
	1.4	The Bio-noo Boundary and Internal Grounding	20
	1.5	Motivation/Goal Competition and Adaptation, and Affective Learning	25
	1.6	External and Internal Grounding: Breaking the Symbolic Circularity	26
	1.7	Perception, Conception, and Problem Solving	29
	1.8	Addressing a Thin and Deep Micro-environment for Noological Processing	33
	1.9	Summary of the Basic Noological Principles	36
	Problem		37
	References		37
2	**Rapid Unsupervised Effective Causal Learning**		41
	2.1	Bayesian Causal Inference and Rapid Causal Learning	43
	2.2	On Effective Causality	44
	2.3	Opportunistic Situation and Rapid Cause Recovery	48
	2.4	Diachronic and Synchronic Causal Factors in Rapid Causal Learning	51
	2.5	Desperation, Generalization, and Rule Application	59
	2.6	Application: Causal Learning Augmented Problem Solving Search Process with Learning of Heuristics	61
		2.6.1 The Spatial-Movement-to-Goal (SMG) Problem	61
		2.6.2 Encoding Problem Solution in a Script: Chunking	67
		2.6.3 Elemental Actions and Consequences: Counterfactual Script	75

		2.6.4	Maximum Distance Heuristic for Avoidance of Anti-goal..............................	79
		2.6.5	Issues Related to the SMG Problem and the Causal Learning Augmented Search Paradigm............	80
	2.7	Toward a General Noological Forward Search Framework..		83
	Problem..			86
	References..			87
3	**A General Noological Framework**.........................			89
	3.1	Spatial Movement, Effort Optimization, and Need Satisfaction..............................		90
		3.1.1	MOVEMENT-SCRIPT with Counterfactual Information..................................	92
	3.2	Causal Connection Between Movement and Energy Depletion................................		97
	3.3	Need and Anxiousness Competition.....................		99
		3.3.1	Affective Learning.............................	103
	3.4	Issues of Changing Goal and Anti-goal..................		104
	3.5	Basic Idea of Incremental Knowledge Chunking...........		107
		3.5.1	Computer Simulations of Incremental Chunking.....	111
	3.6	Motivational Learning and Problem Solving..............		119
	3.7	A General Noological Processing Framework.............		122
	3.8	Neuroscience Review.................................		124
		3.8.1	The Basal-Ganglionic-Cortical Loops..............	125
		3.8.2	The Cerebellar-Cortical Loops...................	129
		3.8.3	The Houk Distributed Processing Modules (DPMs)..	129
		3.8.4	The Computational Power of Recurrent Neural Networks...............................	131
		3.8.5	Overall General Brain Architecture................	134
	3.9	Mapping Computational Model to Neuroscience Model......		136
	3.10	Further Notes on Neuroscience........................		138
		3.10.1	The Locus of Memory Is Not the Synapse..........	139
		3.10.2	Function of the Prefrontal Cortex and the Entire Brain...........................	140
	Problems...			141
	References..			141
4	**Conceptual Grounding and Operational Representation**.........			145
	4.1	Ground Level Knowledge Representation and Meaning......		146
	4.2	An Operational Representational Scheme for Noological Systems..		152
	4.3	Operational Representation............................		153
		4.3.1	Basic Considerations...........................	153
		4.3.2	Existential Atomic Operators....................	158

	4.3.3	Movement-Related Atomic Operators	160
	4.3.4	Generation of Novel Instances of Concepts	162
	4.3.5	Example of 2D Movement Operations	162
	4.3.6	Characteristics of Operational Representation	163
	4.3.7	Further Issues on Movement Operators	164
	4.3.8	Atomic Operational Representations of Scalar and General Parameters	171
	4.3.9	Representing and Reasoning About Time	173
	4.3.10	Representation of Interactions	174
4.4		Issues on Learning	183
4.5		Convergence with Cognitive Linguistics	185
4.6		Discussion	187
		Problems	188
		References	188

5 Causal Rules, Problem Solving, and Operational Representation ... 191

5.1	Representation of Causal Rules		191
	5.1.1	Materialization, Force, and Movement	192
	5.1.2	Reflection on Immobile Object, Obstruction, and Penetration	196
	5.1.3	Attach, Detach, Push, and Pull	198
5.2	Reasoning and Problem Solving		198
	5.2.1	Examples of Problem Solving Processes	201
	5.2.2	Incremental Chunking	205
	5.2.3	A More Complex Problem	208
5.3	Elemental Objects in 2D: Representation and Problem Solving		210
5.4	Natural Language, Semantic Grounding, and Learning to Solve Problem Through Language		213
5.5	Discussion		215
	Problem		218
	References		218

6 The Causal Role of Sensory Information ... 221

6.1	Information on Material Points in an Environment		222
6.2	Visual Information Provides Preconditions for Causal Rules		226
6.3	Inductive Competition for Rule Generalization		230
6.4	Meta-level Inductive Heuristic Generalization		234
6.5	Application to the Spatial Movement to Goal with Obstacle (SMGO) Problem		238
	6.5.1	Script and Thwarting of Script	241
	6.5.2	Recovery from Script Thwarting Through Causal Reasoning	244

	6.6 A Deep Thinking and Quick Learning Paradigm............	280
	Problem...	282
	References...	282
7	**Application to the StarCraft Game Environment**..............	**283**
	7.1 The StarCraft Game Environment.......................	284
	7.1.1 The Basic Scripts of the StarCraft Environment......	286
	7.2 Counterfactual Scripts and Correlation Graphs of Parameters..	291
	7.3 Desperation and the Exhaustiveness of Observations and Experiments......................................	299
	7.4 Rapid Learning and Problem Solving in StarCraft..........	300
	7.4.1 Causal Learning to Engage/Attack Individual Enemy Agents.................................	301
	7.4.2 Affective Competition and Control: Anxiousness Driven Processes..............................	313
	7.4.3 Causal Learning of Battle Strategies..............	317
	7.5 Learning to Solve Problem Through Language............	340
	7.6 Summary...	341
	References...	342
8	**A Grand Challenge for Noology and Computational Intelligence**...	**343**
	8.1 The Shield-and-Shelter (SAS) Micro-environment..........	344
	8.1.1 Basic Considerations...........................	344
	8.1.2 Activities in the Micro-environment..............	345
	8.1.3 Further Activities and Concepts..................	350
	8.2 The Generality of the SAS Micro-environment............	351
	8.3 The Specifications of the SAS Micro-environment Benchmark...	354
	8.4 Conclusion...	356
	Problem...	356
	References...	356
9	**Affect Driven Noological Processes**.........................	**359**
	9.1 Learning and Encoding Knowledge on Pain-Causing Activities...	360
	9.2 Anxiousness Driven Noological Processes.................	364
	9.3 Solutions to Avoid Future Negative Outcome.............	367
	9.3.1 Causal Reasoning to Identify the Cause of Negative Outcome..........................	368
	9.3.2 Identifying a Method to Remove the Cause of Negative Outcome..........................	370
	9.3.3 A Second Method to Remove the Cause of Negative Outcome..........................	377
	9.4 Further Projectile Avoidance Situations and Methods........	383
	9.4.1 Structure Construction and Neuroticism Driven Processes.....................................	386

		9.5	Summary and Future Investigations	387

      Problems 389
      References 389

10 Summary and Beyond 391
      10.1 Scaling Up to the Complex External Environment 394
      10.2 Scaling Up to the Complex Internal Environment 399
      10.3 Explicit Representations of Learnable Mental Processes 402
      10.4 Perception, Semantic Networks, and Symbolic Inference 406
      10.5 Personality, Culture, and Social Situations 408
      10.6 Comparisons with Methods in AI 408
            10.6.1 Connectionism vs Symbolic Processing 409
            10.6.2 Effective Causal Learning vs Reinforcement
                    and Other Learning Mechanisms 411
            10.6.3 Deep Learning vs Deep Thinking
                    and Quick Learning 413
      10.7 A Note on Biology 413
      References 415

Appendices 419
      Appendix A: Causal vs Reinforcement Learning 419
      Appendix B: Rapid Effective Causal Learning Algorithm 421

Index 425

Chapter 1
Introduction

Abstract This introductory chapter provides an overview of the principled and fundamental theoretical framework developed in this book to characterize intelligent systems, or "noological systems." Firstly, a set of fundamental principles is stated that lies at the core of noological systems. Then, a simple but illustrative "micro-world" is used to address all critical processes of a noological system – from perception to detailed action execution, including motivational and affective processes. A critical concept, the bio-noo boundary, is introduced, which constitutes the internal ground level of a noological system that defines its ultimate motivations. Lastly, the basic idea of semantic grounding is discussed. This constitutes the breaking of "symbolic circularity" that enables a noological system to "truly understand" the knowledge that it represents internally for intelligent functions. The discussions culminate in the design of a fundamental architecture for noological systems.

Keywords Noological systems • Intelligent systems • Principles of noology • Principles of intelligent systems • Architecture of noological systems • Architecture of intelligent systems • Bio-noo boundary • Internal grounding • Semantic grounding • Motivation • Affordance • Affective computing • Affective learning

As mentioned in the preface, a principled and fundamental theoretical framework is needed to provide a sound foundation for understanding the phenomenon of intelligence as well as for building intelligent systems. This would benefit both the engineering discipline of AGI/AI, which aims to construct truly intelligent machines that match human intelligent behavior and performance, as well as the various cognitive sciences – neuroscience, psychology, philosophy, linguistics, anthropology, sociology, ethology, etc. – that attempt to understand the natural phenomenon of intelligence as embodied in various kinds of animals.[1] Moreover, there is a recent surge of research in biology in which cells are being seen as

---

[1] AI is usually considered a sub-discipline of cognitive science. In this chapter when we refer to cognitive science we will sometimes include AI and sometimes distinguish the more "engineering" discipline, namely AI, from the more "scientific" disciplines. Philosophy, strictly speaking, is neither science nor engineering.

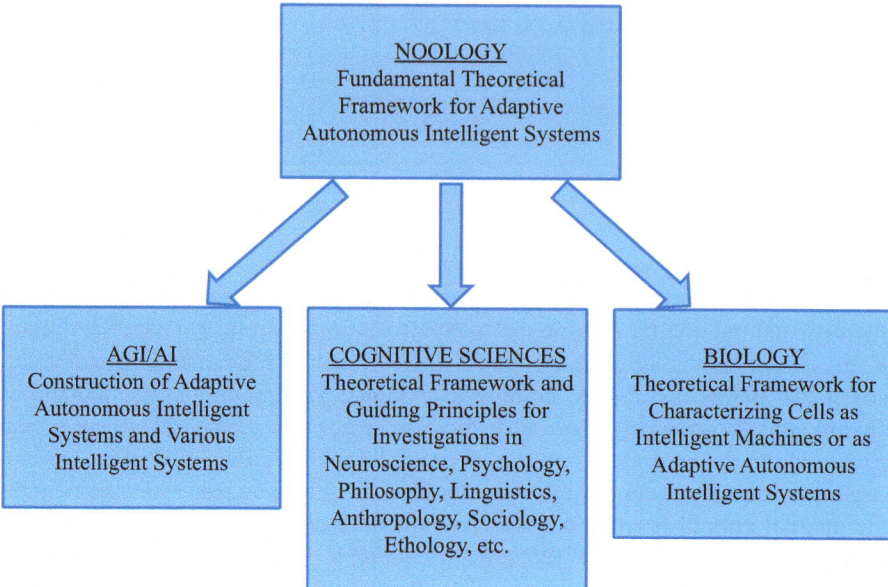

Fig. 1.1 The relationships between noology, AGI/AI, the cognitive sciences, and biology. Noology provides the theoretical framework to unify and support the research in these other areas

machines (Alberts et al. 2014; Fages 2014) and possibly intelligent machines (Albrecht-Buehler 1985, 2013; Ford 2009; Hameroff 1987), and therefore the principles uncovered here could be applicable at the cellular levels as well. We see the basic intelligent systems involved in AGI/AI (e.g., robotic systems), in the cognitive sciences (i.e., the naturally occurring animals), and possibly in biology (i.e., cells) as adaptive autonomous intelligent agents (AAIAs). We coined a new term, "noology," to refer to a discipline that provides a principled theoretical framework for adaptive autonomous intelligent systems (AAISs), much in the same way as physics provides the theoretical framework for the characterization of the physical world. Figure 1.1 illustrates the connections between noology, AGI/AI, the cognitive sciences, and biology. Basically, noology is the theoretical framework for both the engineering domain on the one hand, and the scientific and philosophical domains on the other.

Our approach in formulating a set of general principles underpinning noology is similar to the approach taken in AGI/AI, and that is to elucidate the representational mechanisms and computational processes involved. As mentioned in the preface, this provides a rigorous understanding and characterization of the noological issues involved, and at the same time these mechanisms can translate directly into implementable computer programs for AGI/AI systems. This also provides guiding principles for cognitive scientists and biologists to characterize naturally occurring intelligent systems in higher resolution.

1 Introduction   3

And as has been discussed in the Preface, "noological systems" would be a more appropriate characterization of AAISs instead of "cognitive systems" as cognition and emotion are inseparable in driving the behaviors exhibited by the systems.[2] As an example, it will be shown in Chap. 2 that when a novel principle called "rapid effective causal learning," which we purport lies at the foundation of a noological system's rapid learning process, is being developed, we will see that the functioning of something as fundamental as that cannot be satisfactorily formulated and understood without bringing in an important affective state – the state of *desperation*. That is precisely why many previous attempts at understanding intelligence, especially at the computational level such as the various efforts in traditional AI, failed. What we are going to show in this book is that causal reasoning and other mechanisms in terms of computationally "precise" algorithms can be formulated for an intelligent system, but they are not merely mathematically sound or cognitive, they are "noological."

Some of the representational and computational devices used in the ensuing discussions, such as predicate logic representations and search mechanisms, are derived from traditional AI and are hence familiar to the researchers in that field. However, one of the major intended contributions of the current effort is to look at the various issues involving intelligent processing of information in a new light and place them in a new framework. Therefore, similar representational and computational devices such as these will be used throughout the book, but in a different manner from that in traditional AI.

It is thought that in traditional AI, sometimes the basic aims and purposes of individual "intelligent processing functions" are not properly understood. For example, a lot of effort has been devoted to research addressing the issues surrounding object recognition, pattern recognition, and scene description in computer vision. It is always thought that these are important "intelligent functions" that are desired in an intelligent system, as most natural intelligent systems are able to perform these functions. Commercially, object or pattern recognition systems by themselves are useful as standalone systems to augment human activities. However, not enough thought has been given to the ultimate purpose behind object or pattern recognition specifically and sensory systems in general as part of an adaptive, autonomous, and general intelligent system.[3] Sensory systems, after all, perform a "service function" to the entire AAIS. In a complete AAIS, they are not ends in

---

[2] There are projects such as the Cognitive and Affect Project (Aaron 2014), short-formed CogAff, that also recognize the importance of both cognitive and affect in the total functioning of an intelligent system but the term "noological system" is obviously more succinct than "cognitive and affective system" in referring to a system such as this.

[3] Recently, attempts have been made in computer vision to move from vision to cognition (www.visionmeetscognition.org). This is good development in the correct direction. However, as will be seen in this book, there are further critical issues to be considered and included, and an integrated approach is necessary if noological systems are to be properly characterized and understood. The vision-cognition approach is complementary to the paradigm to be discussed in this book and will greatly enhance the "width" direction of our framework as will be explained in Sect. 1.8.

themselves but means to other ends. They support the functioning of the system in a certain way. Understanding this is critical for the understanding of how to build toward a general AAIS. The premise in this book is, of course, that without considering and characterizing the entire AAIS instead of just particular subsystems, intelligence as exhibited in natural systems and as desired in artificial systems cannot be properly understood and realized respectively. This is again what distinguishes a complete noological system from an object recognition or a scene description subsystem. Therefore, in Chap. 6 we discuss the "causal role of sensory systems" to address this issue at the fundamental level.

In a nutshell, this book propounds the view that a noological system's basic functioning is executing problem solving processes to achieve some built-in primary goals. This constitutes the fundamental "backbone" of noological processing. Affective processes serve to motivate and drive the intelligent system toward certain problem solving priorities. Some of the knowledge for problem solving is built-in, but most of it is learned from observing and interacting with the environment. The primary kind of knowledge learned is causal knowledge that enables effective problem solving and this is learned primarily through an unsupervised rapid learning process with a small number of training examples. For intelligent systems which have developed a symbolic system for communication (e.g., natural language in human beings), learning of knowledge can be carried out in a rapid supervised symbolic manner. Higher level conceptual knowledge acquired by the system is linked to a basic level of epistemically grounded knowledge and grounded knowledge supports various levels of problem solving. The system chunks the knowledge it acquires in the form of scripts and chunked knowledge accelerates subsequent problem solving processes. In problem solving processes, heuristics are learned along the way to further accelerate future problem solving. Heuristics encode inductive biases that are the essence of the "learning to learn" (Braun et al. 2010) or transfer learning (Pan and Yang 2010) process. Internal conceptual processes of generalization, categorization, and semantic network organization further assist problem solving.

This book introduces novel learning methods, representational structures, and computational mechanisms to support the above noological processes. In summary, the set of principles that lies at the core of a noological system is:

- **A noological system is characterized as primarily consisting of a processing backbone that executes problem solving to achieve a set of built-in primary goals which must be explicitly defined and represented. The primary goals or needs constitute the bio-noo boundary** (to be explained in Sect. 1.4. Traditional AI and the cognitive sciences have not articulated a coherent and definitive characterization of an AAIS).
- **Motivational and affective processes lie at the core of noological processing and must be adequately computationally characterized.** (Traditional AI ignores the issues on emotion and affect (Russell and Norvig 2010) but there is a recent emergence of research work in this regard (e.g., Cambria and Hussain

2015; Cambria et al. 2010). Neuroscience and psychology do not provide computational accounts for these processes.)
- **Rapid effective causal learning provides the core learning mechanism for various critical noological processes**. (Existing methods of learning investigated in AI and the cognitive sciences such as reinforcement learning, supervised learning, unsupervised learning, etc. are not sufficient in accounting for the adaptive speed and power of noological systems.)
- **The perceptual and conceptual processes perform a service function to the problem solving processes** – they generalize and organize knowledge learned (using causal learning) from the noological system's observation of and interaction with the environment to assist with problem solving. (This fact has not been clearly articulated in AI and the cognitive sciences.)
- **Learning of scripts (consisting of start state, action steps, outcome/goal, and counterfactual information) from direct experience with the environment enables knowledge chunking and rapid problem solving**. This is part of the perceptual/conceptual processes. Scripts are noologically efficacious fundamental units of intelligence that can be composed to create further noologically efficacious units of intelligence that improve problem solving efficiency, in the same vein that atoms are composed into molecules that can perform more complex functions. (Traditional AI investigated the concept of scripts but did not propose means to learn these representations from direct experience with the environment (Schank and Abelson 1977). Recently there has been some efforts devoted to "script" learning: e.g., Chambers and Jurafsky (2008), Manshadi et al. (2008), Regneri et al. (2010), and Tu et al. (2014). However, these efforts focus on using scripts for question-answering whereas our effort focuses on using scripts for problem solving – see the first principle above – and hence our script's structure is more complex: an example would be the inclusion of counterfactual information in a script to greatly enhance problem solving. There is scant discussion on scripts in neuroscience and psychology.)
- **Learning of heuristics further accelerates problem solving.** Similarly, this derives from the perceptual/conceptual processes. (Traditional AI methodology often employs heuristics that are built-in and not learned in a typical problem solving situation (Russell and Norvig 2010). Natural noological systems do have built-in heuristics and artificial noological systems' intelligent behaviors can certainly be jump-started with some built-in heuristics, but the learning of heuristics is certainly an important component of a noological system.[4] There is scant discussion on heuristics in neuroscience and psychology.)
- **All knowledge and concepts represented within the noological system must be semantically grounded** – this lies at the heart of providing the mechanisms for a machine to "really understand" the meaning of the knowledge and concepts that it employs in various thinking and reasoning tasks. There exists a set of

---
[4] This is similar to the recent interests in "learning to learn" (Braun et al. 2010). Basically, learning of heuristics is the learning of inductive biases that facilitate future learning.

ground level atomic concepts that function as fundamental units for the characterization of arbitrarily complex activities in reality. (Traditional AI does not address the issue of grounding (Russell and Norvig 2010). Linguistics (e.g. Langacker 2008) and psychology (e.g., Miller and Johnson-Laird 1976) do address this issue but no computational mechanisms are forthcoming.)

We wish to highlight the two constructs above that provide a sort of "fundamental units" that subsequent complex constructs are built on to engender intelligent behaviors, much like complex molecules and structures are built from simpler atoms and molecules. They are the atomic spatiotemporal conceptual representations that provide the mechanisms for semantic grounding (to be discussed in depth in Chap. 4) and the scripts (learned through experience), which consist of start state, action steps, outcome/goal, and counterfactual information in packaged, noologically efficacious units that provide the mechanisms for rapid problem solving, which will be discussed first in Chap. 2 in detail and then throughout the book as well.

Based on the above principles, we define *intelligence* as the ability to identify causality and effect generalization of causal rules for effective problem solving to achieve some goals of the system involved.

Before we delve into the detailed learning methods, representational structures, and computational mechanisms in the subsequent chapters, we would like to begin in this chapter with an overall consideration of the issues involved in AAIS to set the stage for subsequent discussions. Some of the issues that will be discussed in this chapter are:

- The core functional blocks and overall processing architecture of an AAIS.
- Issues motivating a novel and important learning mechanism – the rapid effective causal learning mechanism.
- Issues on need, motivation, and affect and how they drive the overall functioning of a noological system.
- Issues on external and internal grounding of knowledge and concepts.

1.1 The Core Functional Blocks of a Noological System

In the current view of cognitive science (e.g., Gleitman et al. 2010; Gazzaniga et al. 2013), an intelligent system typically consists of a number of different aspects of noological (cognitive, affective, and motor) processing such as those shown in Fig. 1.2. These processes range from Perceptual Processes to Attention Processes, various Memory Processes, Motivational and Affective Processes, Conceptual Processes, Goal Formation Processes, Action Planning Processes, Detailed Action Execution Processes, etc. Learning takes place in all aspects of processing. The vertical linear arrangement of the blocks in Fig. 1.2 is not meant to suggest a certain order of processing, but it is generally agreed that some information enters or is

1.1 The Core Functional Blocks of a Noological System

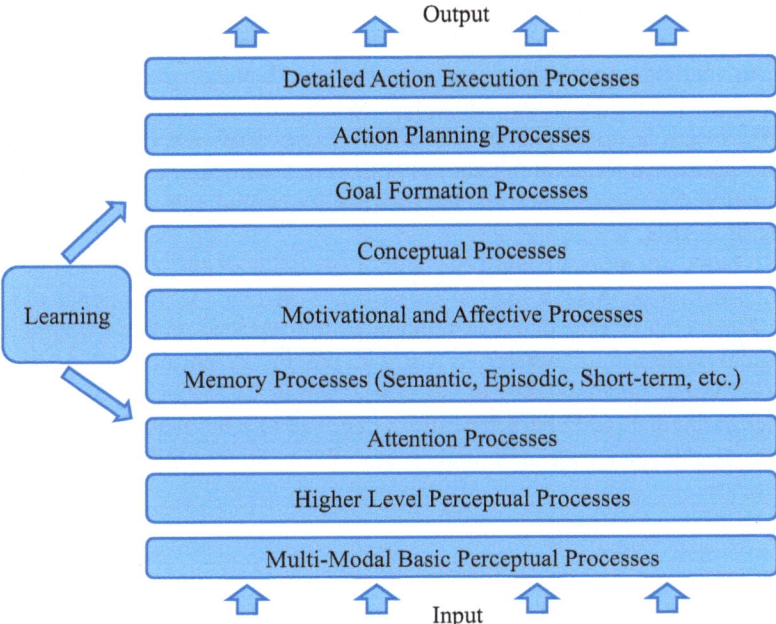

Fig. 1.2 Aspects of noological processing. (©2013 IEEE. Reprinted, with permission, from Ho, S.-B., "A Grand Challenge for Computational Intelligence: A Micro-Environment Benchmark for Adaptive Autonomous Intelligent Agents," Proceedings of the IEEE symposium series on computational intelligence – Intelligent Agents, Page 45, Figure 1)

picked up by the intelligent system from the outside world through its sensory/perceptual systems (the bottom blocks) and then after some internal processing, some actions are emitted (the top blocks).

The emphases on the importance of the various functional blocks vary among the various sub-disciplines of cognitive science. Suppose we use a typical layout and emphasis of treatment of the various issues in a current textbook as a good gauge of the way these issues are viewed in the respective disciplines, then whereas psychology and neuroscience textbooks typically consider motivational processes as being an important aspect of noological processing and would discuss it at the beginning of the texts (e.g. Gleitman et al. 2010), AI textbooks typically ignore the issue entirely (e.g., Russell and Norvig 2010). To the effect that the processes are deemed important to be discussed, the order of discussion of the processes involved which reflects a ranking of importance or a logical sequence that allows the concepts involved to be built-up step-by-step also vary among the sub-disciplines. Thus, the discussion on perceptual mechanisms is typically the concerns of the last few chapters in an AI textbook (e.g., Russell and Norvig 2010) while psychology textbooks typically emphasize the fact that perception is the source of empirical knowledge and its discussion takes place relatively early in the texts (e.g., Gleitman et al. 2010).

While the approach taken by this book is concerned with the detailing of computational mechanisms to elucidate the functioning of intelligent systems, which in methodology is more akin to that of AI, the importance it places on some of the issues concerning intelligent systems is more akin to that of psychology. For example, motivational and affective processes are considered to be of central importance to an intelligent system's functioning and attempts are made to computationalize these processes to elucidate its central role in the overall functioning of AAISs.

This book attempts to construct a representational and computational level characterization of AAISs that is "noological" in that the functional blocks of Fig. 1.2 are clearly elucidated in relation to the purpose and functioning of the noological system as a whole. As we will see in the ensuing discussions, this "purpose-based" understanding of the various sub-processes of a noological system and their computational characterizations can lead to a unification of the disparate attempts at characterizing these processes in the various sub-disciplines of cognitive science.

1.2 Basic Considerations Motivating a Noological Processing Architecture

We begin by considering a simple scenario depicted in Fig. 1.3 in which an Agent (represented by a square) has a need to increase its internal energy level in order to survive (assuming that its internal energy level keeps depleting whether it is stationary or moving). This need/primary goal is represented as Increase(Agent, Energy). An energy or "Food" source is represented by a hexagon that can help to increase the Agent's internal energy level. The Agent will die if its internal energy is depleted. (Assuming that the Agent does not have an explicit knowledge of this "death" condition but it just has the incessant drive to look for food to increase its energy.) Without this need or drive, the Agent can stay stationary for all of eternity and would not even move itself, let alone exhibiting any purposeful behavior. Hence this is the simplest and most fundamental characterization of a noological system – it beings with a built-in internal need and its ensuing purposeful behavior is to satisfy this need. This example, though simple, serves to elucidate the purposes of the various functional blocks of a noological system as depicted in Fig. 1.2 and how they fit together in serving the overall functioning of the system.

Let us next consider how this "nano" noological system goes about satisfying its internal need (in this simple scenario, there is only one need). In this simple nano-world that the Agent inhabits, there is just itself, a few pieces of Food (assuming that they will be replenished at a random location in the nano-world once the earlier pieces are consumed) and some space in between them. The dumbest thing that can happen is that the Agent has no sensory organs and all it can do is to just blindly move about and hope to find the Food fortuitously and also in time – before it "dies"

1.2 Basic Considerations Motivating a Noological Processing Architecture 9

Fig. 1.3 Agent and food

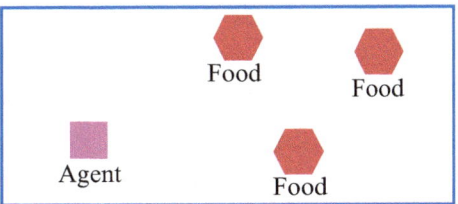

as a result of total energy depletion. Suppose in its random wandering, the Agent fortuitously finds a Food source and consumes it and its energy is recharged. (In this simple scenario, since the simple Agent has no "mouth," we assume that by simply touching the food the Agent would have its energy recharged.) However, its energy would then immediately begin discharging so that it would be beneficial for the Agent to find another piece of Food soon. Because the Agent has no sensory system to identify another piece of Food, it can only continue these blind searches until it gets unlucky – not being able to find any Food before it dies.

How can the Agent improve its chances of survival?

It is conventional wisdom that sensory information about the outside world is necessary to improve the survivability of an Agent in a situation such as the one depicted in Fig. 1.3. While that is basically true, it is our thesis that an even more important issue is how the Agent can identify, given the sensory information about the various entities, which objects or entities in the world can satisfy its internal needs. Hence the establishment of a causal relationship between certain entities in the world and the increase in the Agent's internal energy is the critical mechanism for the survival of the Agent.

Generally, the idea of "food" – energy source – does not have to be a visible object. One can imagine that in the nano-world of Fig. 1.3 there is an "energy grid" and the Agent can increase its internal energy by being at certain locations. However, to take advantage of the situation, the Agent still needs to have "sensory" information on location. We put "sensory" in quotes because this information on location can either be something that is perceived in the outside world – e.g., there is a coordinate marking at every possible location in the world – or it can be an internal signal (much like a "GPS" signal) that informs the Agent of its location – such as one that is derived from the proprioceptive sense.

Consider a simpler situation in which the Agent only has the location sense but not the visual sense. Suppose, as shown in Fig. 1.4a, there are energy points in the environment at which the Agent's internal energy starts to increase as soon as it is at those locations and these energy points can charge the energy of the Agent to any level needed. Now suppose the Agent will charge its energy to a "full" level, after which it will move away from the charging point and wanders about, and it has a satiation threshold above which it does not seek energy increase unless it is already at an energy charging point. Suppose also that the Agent has the ability to execute a straight-line (shortest distance) motion to any location of interest to minimize energy consumption and time of travel (learning to execute a straight-line motion as a means to minimize energy consumption and time of travel is a topic to be

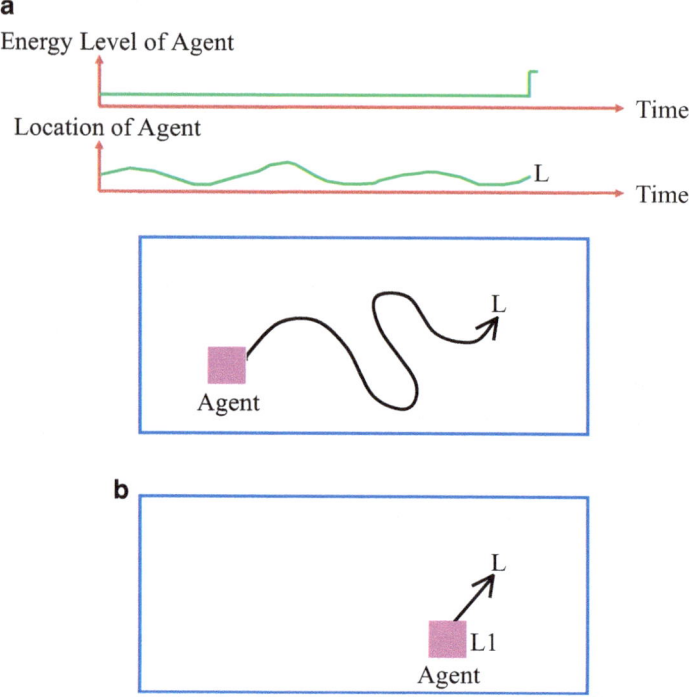

Fig. 1.4 (**a**) Agent wanders around and stumbles upon an "energy-point" location L at which its energy is being charged up. (**b**) Agent, starting from location L1, heading straight to the location L to charge up its energy

discussed in Chap. 2. For the current discussion, we assume this is already learned earlier).

Figure 1.4a shows that initially the Agent wanders around and fortuitously finds that its energy begins to increase when it is at a certain location L. In order to exploit this experience for the benefit of the future satisfaction of its internal need, the Agent needs to establish a causal relationship between being at a certain location and the increase in its internal energy.

The top of Fig. 1.4a shows the changes of both the location (Location) as well as the internal energy (Energy) values of the Agent over time – it shows at the precise time that the Agent is at location L, the energy begins to increase. A causal learning mechanism is needed for the learning of this causality: Location(Agent, L) → Increase(Agent, Energy). Using an effective causal learning method (Ho 2014), the full detail of which will be discussed in Chap. 2, the Agent encodes the temporal correlation between L and energy increase as a causal relationship: Location(Agent, L) → Increase(Agent, Energy).

Consider now in a future situation – Fig. 1.4b – in which the Agent is somewhere else (location L1, say) and its energy has dropped below the threshold of satiation. Its built-in goal of increasing energy represented as Increase(Agent, Energy) matches the right hand side of the above causal rule. It then retrieves the causal rule learned above and uses a backward chaining process to guide its action to satisfy its need/goal of increasing the energy – it moves, in a straight-line and purposeful manner from its current location of Location(Agent, L1), to a location that satisfies the precondition of the causal rule above – Location(Agent, L) – at which point it should experience an increase in energy and fulfil its need.[5]

Suppose now a visual sense is made available to the Agent in addition to the location sense and this situation is illustrated in Fig. 1.5. Suppose also now that the Agent has visual recognition ability that allows it to distinguish the shapes of the objects in the world and a simple touch sense that allows it to sense its contact with any object. Figure 1.5a shows initially the Agent randomly explores the environment and accidentally touches the hexagonal shape which is Food (assuming the other shapes are not food) and its internal energy consequently increases. A predicate Touch(Agent, Hexagon) is used to describe the state.

The situation is now more complicated as both the Touch and the Location parameter changes are correlated with the Energy value change. There are two conjunctive pre-conditions identified here: Location(Agent, L1) and Touch(Agent, Hexagon) → Increase(Agent, Energy). At this stage the Agent has the belief that it needs to be at the location L1 and touching the Hexagonal shape in order to gain energy. Suppose in further exploration as shown in Fig. 1.5b the Agent touches another Hexagonal shape fortuitously at a different location L2 and it also experiences an energy increase. Based on an effective causal learning algorithm to be described in more detail in Chap. 2 (Ho 2014), the Agent now relaxes the location condition (i.e., considers the location not being relevant) for the consequential increase in energy. This results in a more general rule of Touch(Agent, Hexagon) → Increase(Agent, Energy). After learning this more general rule, the Agent is now able to head straight to any Hexagonal shape to charge up its energy should there be a need to do so as shown in Fig. 1.5c. This is of course provided that the Agent has the visual sense to identify the hexagon at a certain location and is now able to exploit that visual sense. The general rule thus encodes a conceptual generalization of a method of increasing the Agent's energy. The earlier rule that is more specific can also help the Agent find an energy source but it is less "helpful" in that the rule dictates that the Food must be at a specific location. (There could be an alternative world in which the Food has to be at a specific location and be touched by the Agent at the same time for the energy increase to take place, in which case the more general rule will not be established.)

The events of Fig. 1.5a, b take place at different times, T1 and T2, respectively. We have simplified the discussion above by ignoring the time parameter. Had the time parameter been also included in the consideration, it would also have been generalized away like the location parameter for the energy increase rule.

---

[5] This is assuming the situation is non-probabilistic – that every time the Agent is able to get the energy at the location L. In Chap. 2 and subsequent chapters we illustrate how the statistics from the various instances of experience can be stored as part of the scripts encoding these experiences.

Fig. 1.5 (**a**) Agent wanders around and accidentally discovers that touching a hexagonal shape at location L1 results in its increase in internal energy. (**b**) Agent has a second fortuitous experience of touching another hexagon at location L2 that also results in its increase in internal energy. (**c**) After experiences in **a** and **b**, agent generalizes that the location is not important but the shape of the object is important in causing its internal energy to increase. Suppose now it is in need of energy, using its visual sense, it heads, in a straight line, purposive manner, toward a hexagonal shape to increase its energy. [*Square shape* = Agent; *Hexagonal shape* = Food]

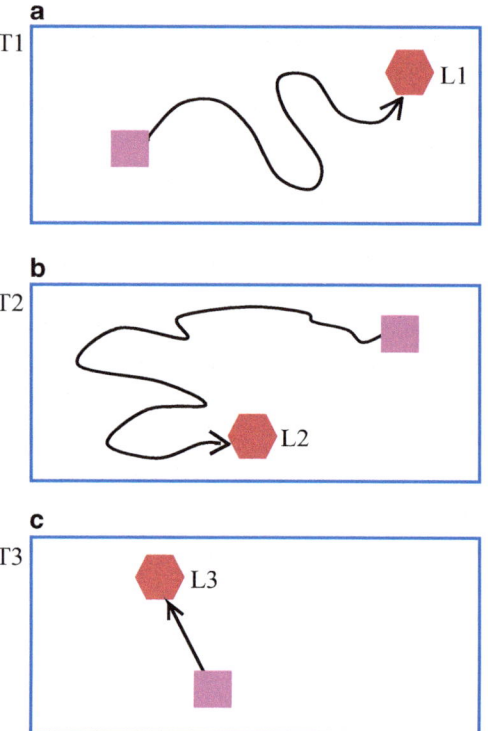

As can be seen in the simple example above, the knowledge of what is available in the environment that allows the Agent to satisfy its internal need(s) is of course not something that is built into the Agent and the Agent has to discover it by exploring and interacting with the environment. As shown above and as will be shown repeatedly in this book, the mechanism of effective causal learning is necessary to rapidly learn the useful rules to satisfy the internal need(s) in an unsupervised manner. A detailed treatment of effective causal learning will be given in Chap. 2.

In Appendix A we compare a popular method of learning, reinforcement learning (Sutton and Barto 1998) with the causal learning process described above using a simplified nano-world similar to that employed above.

We would like to add further descriptions to Fig. 1.5c to summarize the processes discussed above and also illustrate some other related issues. This is shown in Fig. 1.6. In Fig. 1.5c the Agent has learned a useful rule through causal learning for increasing its energy and now it exploits this when it has the need to increase its energy. Figure 1.6 summarizes a number of components of the process involved.

The process begins with an Energy Need. This need is for the Agent to maintain a level of energy (Desired Level) which could be higher than a Current Level. This need is a Primary Goal which we will discuss in more detail in Sect. 1.4. This need

1.2 Basic Considerations Motivating a Noological Processing Architecture

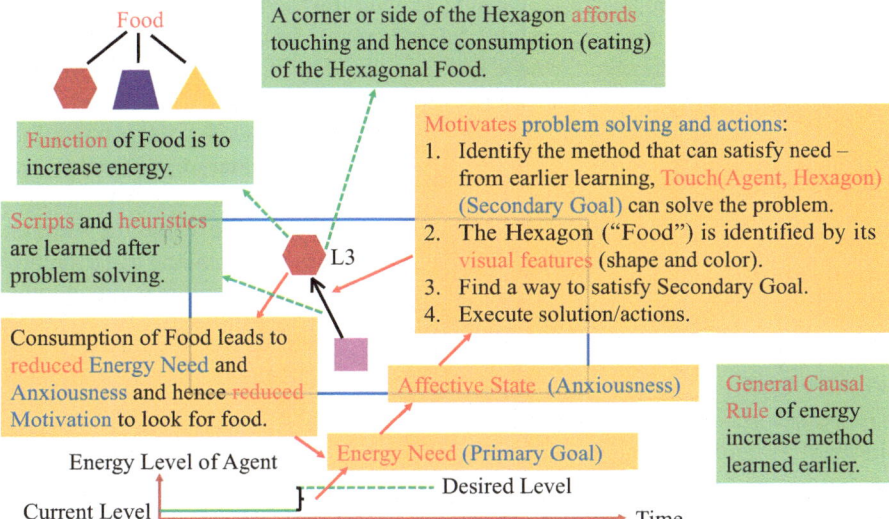

Fig. 1.6 Using Fig. 1.5c to illustrate the various concepts such as Need, Affective State, Motivation, Food, Primary and Secondary Goals, Scripts, Heuristics, Affordance, and Function (This is after the general causal rule for energy increase has already been learned in the situations of Fig. 1.5**a**, **b**, and now the Agent has a need for energy increase)

creates an Affective State of Anxiousness in the Agent. Anxiousness arises when there is a possibility of a future negative consequence for an agent given its current situation – in this case, the depletion of energy can lead to death. This link between the Energy Need and Anxiousness is built in – in Chaps. 8 and 9, we will encounter a situation in which the link between anxiousness and a certain situation's future possibility leading to the hurting of an agent is learned.

The Energy Need and Anxiousness create a Motivation for the Agent to identify a means to satisfy the need. In the process the Agent has to carry out problem solving and should a solution be found, it executes the solution to satisfy the need. The state of Anxiousness is dependent on two aspects of noological processing: (1) Can a solution be found from this problem solving process? (2) If a solution is found, can it be executed in time to prevent the future negative consequence? If there is less certainty to the positive answers to these two questions, there will be a higher level of anxiousness for the agent involved, and it will try harder to increase the certainty of positive answers.

The figure shows that there are a few sub-steps involved in the Motivational step. In Step 1, the Agent identifies a potential solution – since the need is to increase the energy level, it searches its knowledge base to see if there is any method to do so. The general causal rule learned earlier as discussed in connection with Fig. 1.5a, b prescribes just such a method: Touch(Agent, Hexagon) → Increase(Agent, Energy) (no matter where the Hexagon is located). Using backward chaining, Touch(Agent, Hexagon) becomes a Secondary Goal. In this step, if no usable causal rules are found, the Agent would execute the action of wandering like in Fig. 1.5a, b (in fact, this was how the Agent learned the useful causal rule to start with). In Step 2, the Hexagon, whose function would be "food" if Food is defined as something that can increase an agent's energy, is identified through the visual process by its shape (and color, if its color has been included as a feature). Step 3 would be to carry out problem solving to satisfy the Secondary Goal (in this case, the solution of the problem solving process is for the Agent to move along a straight-line path to touch the Hexagon). Step 4 would be to execute the solution. Once the Energy Need is reduced by the consumption of this piece of Hexagonal Food, the Anxiousness and hence Motivation for looking for and reaching another piece of Food is reduced.

The experience above is recorded as a *script* consisting of a goal (in this case the Primary Goal), start state, and action steps. In the future, should the same Primary Goal arises (i.e., Energy Need), and the Agent is in the same start state (i.e., same location as before), the script can be retrieved right away with the recommended action steps to solve the problem without having to carry out the earlier problem solving process. This script may not be very useful as it is a specialized script that requires that the start states of the Agent and Food be at specific locations. If the Agent has another experience of a start state consisting of itself and the Food at different locations and it carries out the same process as above and finally plots a straight-line path for itself from that other location to achieve the goal, a generalization could be made to create a more general script that relaxes the specific condition of the start state. This script will be useful in future similar problem situations as that in Fig. 1.6 or more complex ones than could recruit the script as part of a problem solution. Thus, scripts, a form of knowledge chunking, can greatly assist in problem solving processes.

If the Agent is executing problem solving the first time to discover a path toward the Secondary Goal (touching the Hexagonal Food), or for that matter toward any goal that is a physical location, it will also discover a shortest distance *heuristic* which results in a straight-line path toward the physical location goal. This process is detailed in Chap. 2, Sect. 2.6.1. Suffice it here to note that the learning of heuristic(s) is part of the problem solving process.

Figure 1.6 also shows that suppose the Agent discovers that a number of objects in the environment with other visual features (such as purple trapezoid and yellow triangle) can also provide energy increase, then *functionally* they are all Food. Therefore, the concept of food does not prescribe any shape for the Food object involved. Hence, the functional concept of Food admits a disjunctive list of objects with different shapes. The simple category hierarchy of Food in Fig. 1.6, much like what is normally represented in a semantic network, will greatly assist in problem solving processes to satisfy the Agent's internal needs. In Sect. 1.7 there will be a more detailed discussion on functional vs visual characterization of a concept and its connection with needs and motivations using a more complex example.

Figure 1.6 also introduces a concept called "affordance" (Gibson 1979; Norman 1988). Affordance is the kind of action that an object allows an organism or agent to interact with it. And this action usually leads to utilizing the object involved to realize certain function. Therefore, certain surface parts of the Hexagonal Food afford the touching of the Food by the Agent that would lead to the consumption of the Food and the subsequent energy increase of the Agent involved.

We can see that given the consideration of this simple scenario of an agent learning about and looking for food, all the various aspects of noological processing of Fig. 1.2 are engaged. The learning of the causal rule Touch(Agent, Hexagon) → Increase(Agent, Energy) requires the Episodic Memory – difference episodes of experience are compared to arrive at the general causal rule. Semantic Memory is involved in organizing the various kinds of food (the different shapes in Fig. 1.6) in a category hierarchy. Another aspect of Semantic Memory involves "meaning" and the meaning of food (an external entity) for the agent is something that can increase its energy. These processes – generalization of causal rules and construction of Semantic Memory – also constitute the Conceptual Processes, together with other processes such as the script construction processes. In the process of learning about and identifying an external object having certain visual attributes that can function as food, Perceptual Processes are engaged. Attentional Process directs the agent's processing resources toward items in the environment that have the highest possibilities of satisfying its needs. The internal Energy Need (Primary Goal) of the agent generates the Motivational and Affective (in this case Anxiousness) Processes for it to formulate a Secondary Goal. This is followed by problem solving and executing the attendant solution to satisfy the Secondary Goal and hence consequently the Primary Goal – the need. Goal Formation, Action Planning, and Action Execution are all part of these processes. Last but not least, Learning takes place in these various aspects of processing.

Given this simple example and scenario, it can also be seen why it is important that the built-in primary goal of a noological system be explicitly defined and represented, as stated in the first principle of a noological system. Explicit representation allows the learning of the rule that encodes the causal connection between an external entity, in this case a hexagonally shaped object that functions as Food (Fig. 1.5), and the internal primary goal, in this case the Energy Level of the Agent. This allows the Agent to be maximally adaptive to the environment.

1.3 A Basic Noological Processing Architecture

The foregoing discussion establishes a few basic principles for a noological system. Firstly, a noological system must consist of explicit motivational specifications that represent needs and primary goals (See Fig. 1.6). Primary goals are goals that are "built-in" such as what a biological agent is born with (e.g., the need to alleviate hunger when a "hungry" signal is activated). For an artificial agent, it would be something the builder builds in (e.g., the need to charge up energy level when the level falls below a certain threshold). There are also secondary goals that are derived from the primary goals that are learned in the process of experiencing the world (i.e., learn that money can be used to exchange for food that can in turn be used to alleviate hunger, or that an electrical charging point enables the charging up of energy level). The various goals will also need to compete for priorities of being attended to.

Secondly, a noological system's actions are channeled to achieving these primary or secondary goals unless the goals have been temporarily achieved.[6] To do this, it carries out problem solving to find a solution – a sequence of actions that can achieve the goals. This may involve forward or backward reasoning using the knowledge available.

Thirdly, the roles of perception and conception are "service" in nature – they serve to assist in speeding up the problem solving process in satisfying an agent's needs. Without them, the agent will still emit actions to try and satisfy the needs, but the chances of finding a sequence of actions – a solution – for the needs to be satisfied is low. (The simple example of the previous section illustrates this – *perception* of the Food and the *conceptual generalization* that Food is something of a specific shape but not necessary at any location serves to drastically reduce the amount of effort needed to find the Food. Otherwise, random exploration is needed.) Perception and conception work hand-in-hand – perception provides the information necessary for the conceptual processes to find the best (and hence most "powerful") generalizations and (causal) rules to make problem solving more effective. There is also the learning of scripts – also a kind of conceptual generalization – for knowledge chunking, and the learning of heuristics that together will facilitate problem solving processes.

Fourthly, the knowledge about what is in the environment that allows an agent to satisfy its internal needs is not built into it from the beginning and must be learned rapidly through a mechanism of unsupervised effective causal learning.

Also, the primary goals of the system provide an "internal ground" that defines the ultimate motivations of the system (this will be discussed further in the next section) and the information from the external environment through perception and actions provides an "external ground" to define knowledge of the external environment at the most fundamental level (this will be discussed further in Chap. 4).

With this understanding of the roles of the various functional blocks of processes in a noological system, we can construct a basic architecture of a noological system as shown in Fig. 1.7 consisting of the various components discussed above[7]:

A point to note is that the actions of a noological system can be directed toward the outside environment or toward its internal states. Moving toward food to satisfy its need for energy is an external action. There are also internal actions – e.g., the conceptualization process mentioned above is an internal action. There is a built-in need for a typical noological system to achieve maximum generalizations over the rules that it learns while interacting with the external environment in order to create

---

[6] If the system's actions consume energy, which they typically do, then non-purposeful actions are generated only when there is excessive energy remaining.

[7] Contrast this with the general views of an agent and its internal processing components in traditional AI – e.g., Chapter 2 of Russell and Norvig (2010).

1.3 A Basic Noological Processing Architecture

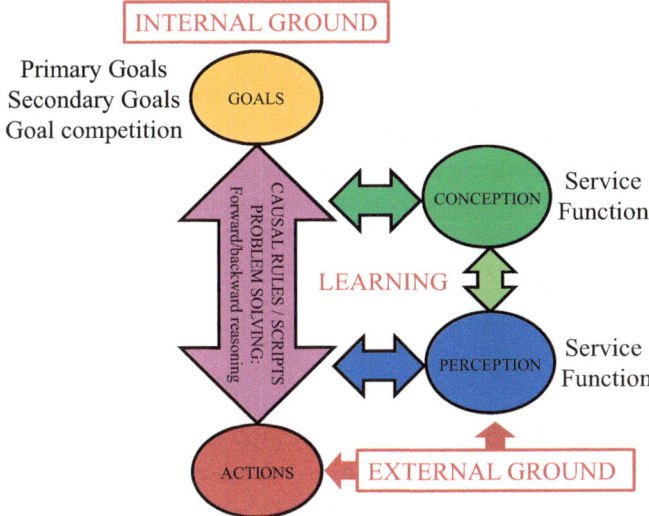

Fig. 1.7 An architecture for a noological system

internal knowledge representations that can assist in problem solving in as efficient a manner as possible. The example given in the previous section in which the agent learns a general rule Touch(Agent, Hexagon) → Increase(Agent, Energy) (location does not matter) is an example of an act of generalization directed at an internal representation – the earlier more specific rule, Location(Agent, L) and Touch(Agent, Hexagon) → Increase(Agent, Energy), has been transformed to the more general rule. Just like the external actions, the internal actions taken to change internal states is also subject to learning – if the results are wrong or unsatisfactory, a better process of generalization is learned (e.g., a system may over-generalize or under-generalize on specific knowledge). In a complex intelligent system like a human being, often a massive amount of internal actions may be taking place without any visible external actions – when the human being is involved in "thinking" processes. Cogitation is a built-in need in animals including human beings.

Based on the foregoing discussion, the basic operations underlying a noological system are summarized below:

- A noological system must consist of some explicitly represented motivational specifications that encode its needs/primary goals. These goals compete for priority to be attended to.
- A noological system's primary processing backbone is the problem solving processes that generate a sequence of actions to satisfy the needs arising from the motivational specifications.
- A noological system can emit external actions to effect changes in the environment or internal actions to effect changes in its internal states. These may lead to immediate or subsequent satisfaction of its needs.

- A noological system's perceptual apparatuses collect potentially useful information to improve the efficiency of the problem solving processes.
- A noological system's conceptual processes, using information from the perceptual apparatuses, create useful generalizations that further improve the efficiency of the problem solving processes.
- A noological system's perceptual and conceptual processes use unsupervised rapid effective causal learning to learn the necessary causal rules, scripts, and heuristics that can be used in problem solving to rapidly derive a sequence of actions to satisfy the system's internal needs.

Thus, the primary focus of a noological system is the problem solving processes that address the requirements of its internal motivations and the other processes such as the perceptual and conceptual processes provide a service function to the central problem solving backbone.

Other processes depicted in Fig. 1.2 such as the Attention Processes further provide a service function of conserving processing resources by directing the agent to collect information only from a subset of all possible sources of information. The Episodic Memory Processes serve the conceptualization processes by buffering a longer period of available information from perception as well as the internal states for the conceptualization processes to operate on. These mechanisms will be elucidated in further detail in Chap. 2 and the rest of the book.

An interesting observation can be made at this point with regards to the relative emphases on the various aspects of noological processes (e.g., those shown in Fig. 1.2) investigated by the various sub-disciplines of cognitive science. Earlier in the beginning of this chapter we observed that a high emphasis is usually placed on the issues on motivation in the field of psychology but no emphasis at all is placed on these issues in AI. On the other hand, the issues on problem solving is usually scantily treated in psychology or neuroscience but very extensively treated in AI. The architecture shown in Fig. 1.7 and the principles of noological systems that we articulate in this section thus unify the views of the different sub-disciplines of cognitive science and provide a framework for situating the different aspects of noological processing that captures a deep and proper understanding of their relative roles in the overall functioning of a noological system.

An interesting parallel can be drawn between the noological architecture of Fig. 1.7 and the vertebrate cerebral cortical architecture of Fig. 1.8a proposed by Fuster (2008). Figure 1.7 is repeated in Fig. 1.8b for comparison with Fig. 1.8a. Figure 1.8a shows two separate steams of cortical processing. On the left is the "motor" stream in which neural signals "originate" at the very apex of the prefrontal cortex, going through a few levels of processing, and finally arriving at the bottom levels of motor cortices, the bottommost level of which – the primary motor cortex – drives the various actions emitted by the animal involved. The word "originate" is in inverted commas because there are neural signals that come in from the right side – the "sensory" stream of processing. Neural signals in the cerebral cortex move in both directions due to the reciprocal connections between the different cortical areas. It is possible to distinguish a "forward" direction represented by the thicker

1.3 A Basic Noological Processing Architecture

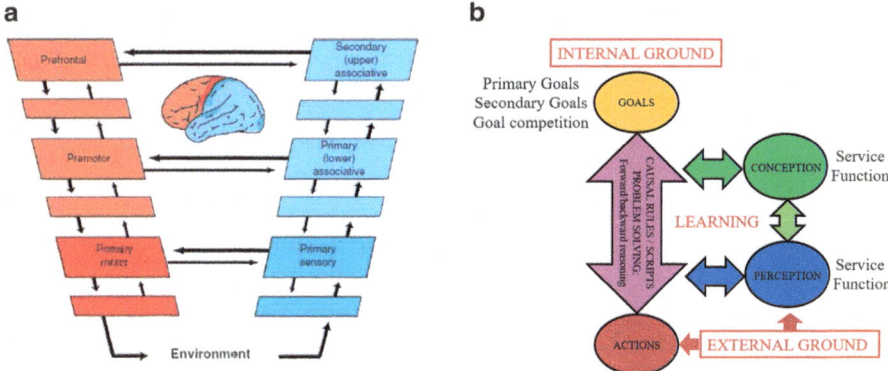

Fig. 1.8 (**a**) Vertebrate cerebral cortical architecture (Reprinted from The Prefrontal Cortex, Fourth Edition, Joaquin M. Fuster, Overview of Prefrontal Functions: The Temporal Organization of Action, Page 360, Figure 8.3, 2008, with permission from Elsevier.) (**b**) Architecture of a noological system (Fig. 1.7)

arrows in Fig. 1.8a and a "backward" direction represented by the thinner arrows. So, in the motor stream (left side) of the system, signals move from the top to the bottom of the figure and in the sensory stream (right side) they move from the bottom to the top of the hierarchy.

In the apical areas of the prefrontal cortex on the motor stream side can be found areas such as the orbital frontal cortex which is generally recognized to be involved in the representation and processing of the motivational aspects of the system. This corresponds to our GOALS block – the motivational processing area – in Fig. 1.8b. In the sensory stream on the right, the lower levels (in parallel with our PERCEPTION block in Fig. 1.8b) are involved in lower level perception and the higher levels (in parallel with our CONCEPTION block in Fig. 1.8b) are involved in "higher level" processing which represents generalized information – e.g., in the higher levels of visual processing, the neuronal receptive fields cover the entire visual field and they no longer respond to more localized kind of information like in the lower perceptual levels. The primary motor level corresponds of course to our ACTIONS block. Figure 1.8a does not specify where the problem solving processes take place nor where the learned causal rules/scripts/heuristics are located to assist with problem solving. Possibly, these are embedded in the various blocks of processing shown in Fig. 1.8a. There are also major subcortical areas such as the basal ganglia and the cerebellum that are not shown in Fig. 1.8a that may play a role in these processes. (The basal ganglia has been postulated to be involved in reinforcement learning (Houk et al. 1995), hence it has a role to play in search and problem solving.) In Fig. 1.8b, though, these processes are specified explicitly in the vertical purple arrow that links the GOALS to the ACTIONS blocks.

1.4 The Bio-noo Boundary and Internal Grounding

In the previous section we identified two levels of internal goals arising from motivations. One level consists of primary goals that are "born" with the agent if it is a biological agent or "built-in" by the builder if it is an artificial agent. The other level consists of secondary goals that are derived from the primary goals. For example, seeking money is a secondary goal as part of the purpose could be to exchange it for food, which is in turn then used to achieve the primary goal of energy increase. There could also be many levels of secondary goals, forming some kind of hierarchy.

The reason why something like seeking money is considered a secondary goal is that it is something that is contingent on the environment and not "inborn." In our world, after thousands of years of human development, money emerged as a medium of economic exchange that can be used to satisfy many kinds of internal needs from the acquisition of food to shelter. It is an invention of human beings. On the other hand, primary goals such as seeking food (energy) and shelter are born with the biological agents or in-built with the artificial agents (artificial agents also need "shelter" to prevent rusting and other kinds of damage, man-made or otherwise). However, for biological agents, there is a period, often very extensive, of learning through evolution before the primary goals become inborn with the agent – i.e., through natural selection, a kind of learning, organisms that do not have the inborn goal of seeking for food and shelter are quickly eliminated. This learning took place over an ensemble of biological agents through evolution. There is no possibility to learn that not increasing one's internal energy will lead to death at the level of individual biological agents because death erases the agent's existence. What does not exist cannot learn. There is of course the possibility of learning this through observing another agent's dying but this requires more complicated reasoning and knowledge acquisition processes, including knowing that the other agent's internal energy is plunging toward zero and that is the cause of its death.

For artificial agents, the builders are the ones that use their knowledge, also through learning, to decide the necessary built-in goals for the agents in order for them to function well. Artificial agents have an advantage over biological agents. Even if it runs out of energy and stops working, as long as it has a permanent memory system that can remember what happened in the period before its "death," there is a possibility that it can learn that allowing its energy to continue to decrease until death is not desirable. However, there is still the need for the human builders to define "death" as something undesirable for the artificial agent before the avoidance of it is meaningful to the agent. For natural agents, death leads to non-existence so the learning of it through evolution is implicit. The reason that the builders of the artificial agents want to penalize the non-functioning (death) consequence of the agent is that they are most likely to have built the agent to perform some functions continuously, and not programming in the idea of "death avoidance" in the agent defeats the purpose of building the agent to begin with.

1.4 The Bio-noo Boundary and Internal Grounding

We term the set of built-in primary goals and priorities the *bio-noo boundary*. ("Bio-noo" is pronounced as \'bī-ō-nō-ə\, in a similar manner as "bio-Noah.") They represent the ultimate purposes of an agent, biological or artificial, that come with the agent and need not be learned. The "learning" that results in the particular set of built-in goals takes place in evolution (for biological agents) or during the "design and construction" process (for artificial agents). If there are drastic changes in the environment that render this set of built-in goals meaningless or lead to negative consequences to the agent, adaptation would be difficult. For biological agents, they may cease to exist (unless evolution provides a way out). For artificial agents, a redesign by the builders is necessary. For example, if a kind of biological agents has the built-in mechanisms to look for energy in the form of biological "food" and the environment ceases to have biological food, even though perhaps "food" in the form of electromagnetic waves is available, it would lead to drastic consequences – total elimination of that kind of agents from the biosphere – unless a sub-population had earlier mutated and evolved the internal goal of seeking this kind of food and the corresponding physiological mechanisms to digest it. Whether it has a chance to mutate and adapt depends on whether the change of food source is gradual or rapid. Hence, there is a set of built-in needs/primary goals that defines the bio-noo boundary and the purposes of an agent's actions can be linked back ultimately to it.

Consider the case of lower biological life-forms. If one observes a sequence of actions emitted by them such as climbing a tree, one can characterize its goal as trying to reach the top of the tree. But why would it want to reach the top of the tree? One possibility is that there is food up there that it can consume. But why would it want to consume the food? Because food can lead to increase in its internal energy. And the explanation stops here. There is no need to go any further in search of deeper goals when one reaches this bio-noo boundary. Of course, the agent could also be climbing the tree in order to reach the tree top to absorb more sunlight. Sunlight could be a kind of food for some agents or it can satisfy other kinds of physiological needs such as the manufacture of vitamin D for the proper functioning of the biological system. Either way, the explanation stops here because one has reached the bio-noo boundary.

For higher life-forms such as human beings, this bio-noo boundary is a lot more complicated. We can use Maslow's hierarchy of needs (Maslow 1954) as a guideline of what this boundary may look like (Fig. 1.9a).

When a certain human being is trying to satisfy her need, whether it be at lower levels such as security or friendship, or at higher levels of self-esteem or self-actualization of the Maslow's hierarchy, and she carries out certain actions to satisfy them, explanations of the purpose of her actions do not go any further. For example, if someone, in her later years, takes on singing as a hobby with no possibility of economic gains, her explanation of the purpose is likely to be "self-actualization." There is no need and in fact it would be impossible to probe for a deeper motivation. This is the bio-noo boundary or the built-in goals.

There has been some recent controversy surrounding the Maslow hierarchy and some newer paradigms have been proposed to study basic needs and motivations (Reeve 2009) – e.g., there are questions on whether the hierarchy is a strict one and

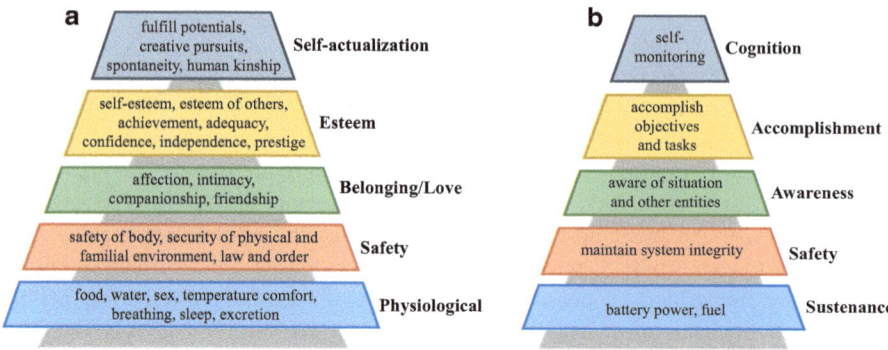

Fig. 1.9 (a) Maslow hierarchy of human needs (Maslow 1954). (b) Artificial autonomous system's needs (Quek 2006, 2008)

whether needs priorities change over time and under certain circumstances. However, the Maslow hierarchy still presents a very useful picture to discuss the issues involved in the bio-noo boundary. At the least, the various needs listed in the hierarchy are there to drive and motivate a noological system's behavior. The prioritization and adaption of these needs can be addressed separately.

Another example of built-in goals is Edmund Rolls' primary reinforcers such as shown in Table 1.1 (Rolls 2008). In contrast with Maslow's hierarchy which consists of desired goals to achieve, Roll's reinforcers consist of rewards and punishers. In the table, the items listed correspond mainly to the lower levels of Maslow's hierarchy and more categories of basic needs are listed.

Sometimes these built-in goals could be overridden by the agent involved in extreme circumstances. For example, a hunger strike or a suicide could lead to the overriding of the death avoidance goal. This could arise from other goals such as the goal to achieve certain social effects (such as drawing attention to certain social issues) through hunger strike or suicide. Tastes in fashion and art, for example, has a largely built-in component for each individual but change in some complex manner as a function of social and psychological forces. However, it is still useful to characterize an agent as starting with these basic goals and then consider the adaptation processes that may lead to the overriding of them and/or the change of their relative strengths separately. Thus, adaptation takes place at the bio-noo boundary.

If an artificial agent such as a UAV is programmed to have a primary goal of "seek and destroy the enemy," no matter how much "self-awareness" is built into its artificial "brain" one does not ask and it would not be able to answer the question: "Why do you want to destroy the enemy?" This is its bio-noo boundary. The answer to that may reside in the minds of the builders but not in the agent itself.

Figure 1.9b shows a hierarchy of needs used in an artificial autonomous system (Quek 2006, 2008). These include the levels of Sustenance (battery power, fuel, etc.), Safety (maintenance of system integrity), Awareness of situation and other autonomous entities, Accomplishment of objectives and tasks, and Cognition (self-

1.4 The Bio-noo Boundary and Internal Grounding

Table 1.1 Edmund Rolls' reinforcers

Taste
Rewards: Salt taste, sweet, umami
Punishers: Bitter, sour, tannic acid
Odor
Reward: Pheromones
Punisher: Putrefying odor
Somatosensory
Rewards: Touch, grooming, washing, suitable temperature
Punishers: Pain, too hot/cold
Visual
Rewards: Youthfulness, beauty, smile, blue sky, open space, flowers
Punishers; Snakes, threatening facial expression
Auditory
Rewards: Soothing vocalization, music
Punishers: Warning call, aggressive vocalization
Reproduction
Rewards: Courtship, sexual behavior, mate guarding, parental attachment
Punishers: crying of infant, departure of mate
Others
Rewards: Novel stimuli, sleep, altruism, reputation, control over actions, play, exercise, mind reading, cogitation, breathing
Punishers: unreciprocated altruism, betrayal

Based on Rolls (2008, Table 3.1, Page 114)

monitoring of internal system states). This has been applied successfully to the control of an autonomous robot in Quek (2006, 2008).

In contrast to a kind of noological/semantic grounding called external grounding that we will discuss later in this chapter and Chap. 4, which deals with the external aspects of the environment that has to do with the ontology of the physical world, we term this *internal grounding* – an internal layer of motivational specifications that present ultimate constraints on the agent's behaviors. This is labeled accordingly as INTERNAL GROUND above the GOALS block in Figs. 1.7 and 1.8b.

Figure 1.10 summarizes the idea of the bio-noo boundary using an affordance chain. A chain of affordances leads to the final satisfaction of an internal need.[8] At each stage of the affordance chain, each entity involved serves a certain secondary goal or need. Each of these secondary goals or needs in turn has a purpose that can be stated explicitly or represented explicitly in the agent's mind. Each stage also involves causal learning: the agent learns that cash can be used to buy food, food can be used to alleviate hunger, etc. At the internal ground level, the purpose of the primary goal does not need to be explicitly accounted for nor does it have to be

---

[8] Here we extend the narrower definition of "afford" as defined in Sect. 1.2 in connection with Fig. 1.6. The usage here is synonymous with "support the function of."

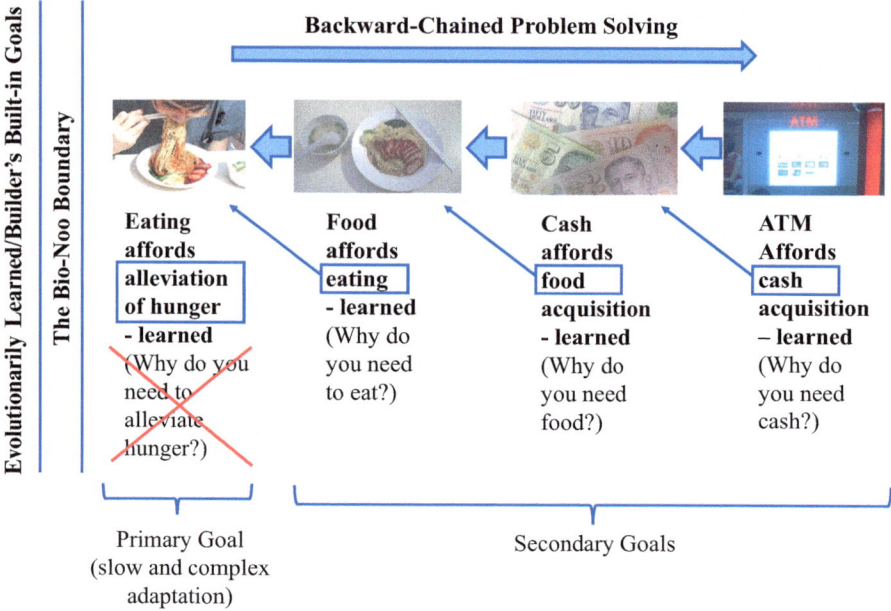

Fig. 1.10 Affordance chain illustrating the idea of bio-noo boundary. Each step consists of a causality that can be learned. E.g., one learns that ATM affords/causes case acquisition, eating affords/causes alleviation of hunger, etc. However, there is no need to provide an answer to the "why" at the boundary as it consists of strong built-in drives. In a backward-chained problem solving process, one starts with a primary or secondary goal, and then one uses the earlier learned steps of causality/affordance to chain together a solution

learned. The primary goal is subject to a slow and complex kind of adaptation, as mentioned above. Of course, in the case of hunger alleviation, a conscious being such as a human being typically knows that if hunger is not alleviated, it may lead to death or extreme suffering later, but the strong drive to alleviate hunger is not the result of any cogitation about possible future consequences.

The agent could learn the affordance at each step of the affordance chain separately and when a need to satisfy a goal arises, in this case the alleviation of hunger, the steps can be put together into a complete solution for the problem through a process of, say, backward-chained problem solving.

Figure 1.11 provides another example of an affordance chain that has a more abstract built-in goal – "jewelry enjoyment." Typically, when the internal needs of an agent are satisfied, the agent would arrive at the emotional state of "happiness."

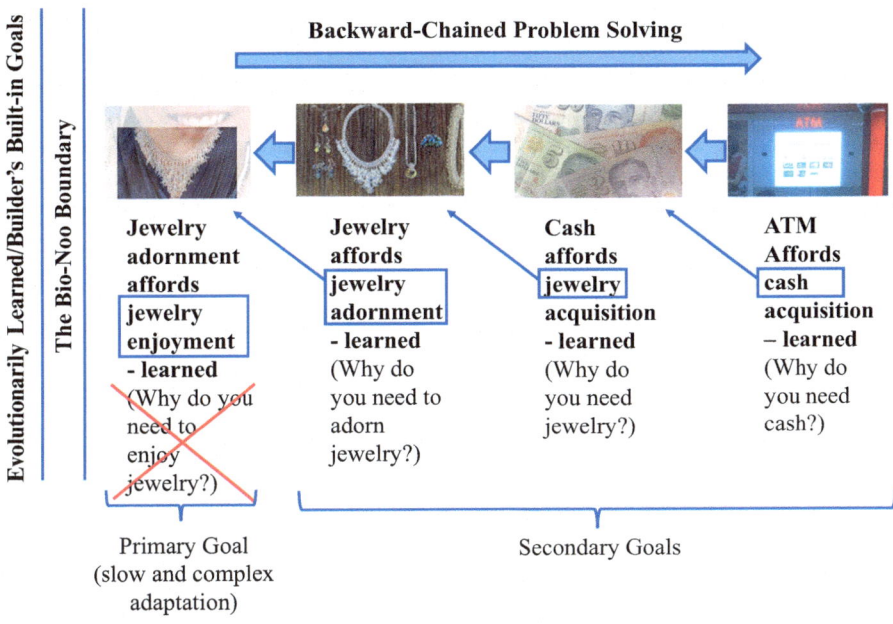

Fig. 1.11 Affordance chain with a built-in goal of "jewelry enjoyment"

1.5 Motivation/Goal Competition and Adaptation, and Affective Learning

As long as there exists more than one goal in an agent, whether they are at the same level or different levels of the Maslow hierarchy or an Autonomous Systems' Needs hierarchy (Fig. 1.9), or whether they are primary or secondary, the issue of goal competition arises. It is of course possible that two goals can be satisfied by the same sequence of actions but more often than not, the goals would lead to different sequences of actions. The result of goal competition dictates which sequence of actions would take priority and be executed first.

In simpler agents, the competition between goals might involve something like "to continue hunting for food" vs "to run away from an approaching enemy." In higher level agents such as human beings, the goal competition may become very complex. It may also lead to the suppression of some very basic and important goals such as the alleviation of hunger or avoidance of death, as discussed above. And, as also discussed above, the strengths of various motivations may change due to a variety of factors and this results in motivation adaptation.

Another related issue is *affective learning*. Certain emotional or affective states associated with driving the problem solving processes, such as the state of anxiousness of an agent, may change as a result of the outcome of the problem solving process. Consider the case of anxiousness. Suppose a situation is encountered which may lead to a negative consequence to the agent involved. The agent would first enter into a state of anxiousness. If the problem solving process leads to a successful identification of a solution and it could be executed in time, then the next time the same problem arises, the intensity of anxiousness will be reduced. For other emotional states, the process is similar. This is affective learning.

In this book, we will not provide a complete general treatment of motivation competition, motivation adaptation, or affective learning. However, in Chaps. 3 and 7, we will use a simple example from the StarCraft Game Environment that involves the competition between two goals to illustrate the idea of goal competition and affective learning and leave further investigations of this important topic to future expositions.

1.6 External and Internal Grounding: Breaking the Symbolic Circularity

A long-standing problem that haunts AI is the problem of meaning. Since computers are apt at processing symbols, AI thrives on symbol processing. In a typical conversation involving two human beings, strings of symbols are exchanged and usually the human beings involved "understand" what the other means from the symbols received. Sophisticated question-answering AI systems have been programmed to respond cleverly to questions posed to it in the form of natural language sentences consisting of strings of symbols, such as IBM's cognitive system Watson (Ferrucci et al. 2010). However, no matter how accurate or clever the answers given by Watson or similar question-answering machines are, the criticisms leveraged against them are that they do not really "understand" the contents involved in the sentences.

The crux of the problem is that it all depends on how these symbols are defined or represented within the supposedly "intelligent" machine. If the machine's innards consist of defining or representing symbols in terms of other symbols, such as how symbols are typically handled in dictionaries, a situation of circularity will arise. Let us consider a simple example using some definitions of words from the Merriam-Webster dictionary. For example, the Merriam-Webster defines the word Move as follows:

Move: To *go* from one place or position to another.

However, the word/concept *go* in this definition is in turn defined in the dictionary with the word *move*:

Go: To *move* on a course.

Therefore, Move is defined in terms of Go and Go is in turn defined in terms of Move. This would even puzzle a human being: so what does Move really *mean*? There is a bottom-line that cannot be avoided: For AI systems to be able to truly understand the meaning of the symbols it manipulates, some of the symbols must be

1.6 External and Internal Grounding: Breaking the Symbolic Circularity

represented by some sort of "ground-level" constructs that they refer to. Then, it will be alright for other symbols to be defined in terms of these symbols, because there will then be no circularity.

For example, the "real" idea of the concept Move is a change of physical location over time. Suppose we use spatiotemporal representations such as that in Fig. 1.12 to represent the concept Move. The representations in Fig. 1.12 represent 1-dimensional (1D) *spatial* movements over (1D) time. Each small square in the representation represents an elemental discrete spatial or temporal location. Figure 1.12a shows a *specific* kind of movement in which the Object moves one elemental spatial location over one elemental time frame and in the upward direction. Figure 1.12b shows a *general* movement over *any* number of spatial locations in *any* amount of time, and in either the upward or the downward direction (in 1D space, there are only two directions – upward or downward in this case). The horizontal and vertical Range Bars represent *any* amount of space and time intervals respectively, that correspond to *any* amount of spatial and temporal displacement respectively for the Object involved (see Chap. 4, Sect. 4.3.7.1 – discussion in connection with Figs. 4.14 and 4.15 that further elaborates on how this representation works). The "Up-Down Flip" symbol specifies that the pictorial representation can be flipped in the upward-downward direction, hence representing the upward and downward movements accordingly. The representation in Fig. 1.12b represents the general and grounded concept of Move.

Why do we claim that this is a ground level representation that truly captures the meaning of the concept Move? The reason is simple: Since Move is a spatiotemporal concept connoting a change in location of an object over time, just represent it accordingly. Of course, there has to be an attendant computational process that operates on the representations of Fig. 1.12 as well to fully utilize them in order for the intelligent system that has this concept to be able to claim to have a full understanding of it. For example, one question posed to the system could be: Since you know the meaning of Move, can you *move* an object? To do so, the system will retrieve the spatiotemporal representations of Fig. 1.12 and use it to operate on an object, and the consequence of which would be that the object would change position over time, thus showing that it "understands" by knowing how to "operate" with the representations. The circular definition above will not allow the system to achieve this. (If the process uses Fig. 1.12a, then the object changes its position by one elemental spatial location over one elemental time frame and in the upward direction. If Fig. 1.12b is used, then the object can change its position by any number of elemental locations over any amount of time and in any direction – upward or downward direction in the case of 1D movement.) Later, in Chap. 5, we will see more extensive examples of how a series of ground level representations allows a system to solve complex problems at the ground level, hence, in the *real* world.

Another question can also be asked: Since you know the meaning of Move, can you recognize an instance of the concept? The answer is, again, yes, as the spatiotemporal templates of Fig. 1.12 can be used to recognize that a movement has taken place for an object by matching these templates with events that take

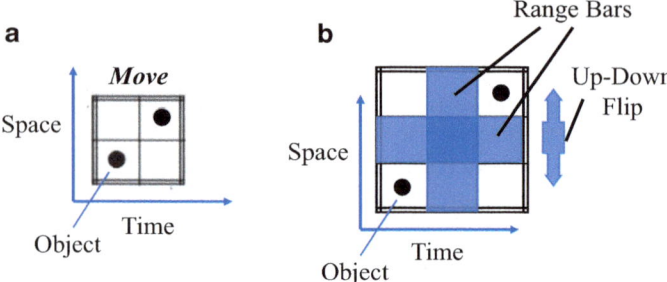

Fig. 1.12 The ground level (symbolic) representation of the concept Move. Each *square* represents an elemental discrete spatial or temporal location. (**a**) A *specific* Move in which the Object moves over an elemental spatial location over an elemental time frame and in the upward direction. (**b**) A *general* Move in which the Object moves over *any* number of spatial locations over *any* amount of time and in the up or down direction

place in the environment, either for a specific Move-ment such as in Fig. 1.12a, or a general Move-ment such as in Fig. 1.12b. Again, the circular definition above will not allow the system to achieve this.

An interesting question arises here: Are the spatiotemporal representations shown in Fig. 1.12 not also a kind of symbol? A *symbol* is "something that stands for or suggests something else" (Merriam-Webster). The spatiotemporal "pictures" of Fig. 1.12 stand for the "real world event" of "move." Therefore, the distinction is actually between *ground-level* symbolic representation and *high-level* (or non-ground level) symbolic representation. We submit that Fig. 1.12 depicts ground-level symbolic representations of the concept Move. In fact, the spatiotemporal representations of Move in Fig. 1.12 have equivalent predicate logic representations, which are made up of symbols of predicates and arguments. We will discuss this further in Chap. 4.

With Move having operationally represented, the other concepts such as Go above that is defined in terms of Move symbolically can then acquire the operational definition and the system hence "understands" what Go really means as well.

In the field of linguistics, many attempts in the past have been devoted to the study of meaning or semantics. Among the different paradigms, cognitive linguistics (Evans and Green 2006; Geeraerts 2006; Langacker 2008, 2009) stands out as having succeeded in providing a framework for grounded meaning representations. Chapter 4 explores the issue of semantic grounding more fully.

Note that the concept of Move such as represented in Fig. 1.12 can be applied to external events – e.g., the movement of a physical object in the real world – or internal events – e.g., the moving up and down of hunger level (this is further discussed in Sect. 4.3.8). Therefore, Move can be applied to internally grounded concepts such as hunger or externally grounded concepts such as a physical object and hence it can participate in both internally or externally grounded processes.

1.7 Perception, Conception, and Problem Solving

In the foregoing discussion we have emphasized that problem solving represents the processing backbone of a noological system (Fig. 1.7), and that perception and conception perform service functions to the system. Moreover, the primary purpose of problem solving is to address the basic needs and motivations of the noological system which represent the internal ground of the system. We have discussed this using a simple example in Sect. 1.2 (Figs. 1.5 and 1.6). In this section we use a more complex computational model of a *functional definition* for visual recognition to tie together these various issues discussed in the foregoing sections.

Ho (1987) raised the issue that AI had not adequately addressed a fundamental issue of visual recognition: despite the fact that some objects that purportedly belong to the same category look very different from each other in terms of their visual appearances, they are nevertheless classified under the same category.

Figures 1.13a–e show a number of chairs that do not all have the same set of necessary and sufficient visual features that are nevertheless all called "chairs." There may be a possibility of using the concept of "family resemblance" to explain the grouping of these objects into the same category (Rosch and Mervis 1975) – a first object shares some common features with a second object, and the second object shares yet another set of common features with a third object, but the third object may in turn share very few or no common features with the first object. The objects involved are thus grouped based on some kinds of disjunctive conditions – for example, a chair could have four legs *or* one leg, and the seat could be square *or* circular in shape, etc. However, without a set of necessary and sufficient conditions to group these objects together, it would be difficult to recognize novel instances. Suppose now there is a new kind of chair with a triangular seat, does it still belong to the chair category? More interestingly, Ho (1987) pointed out that there is a kind of chair called Balans chair, shown in Fig. 1.13f, that has very little visual similarity to those in Fig. 1.13a–e – it doesn't even have a *back*, which all the chairs in Fig. 1.13a–e have, otherwise those same chairs would have been called *stools*.

Interestingly, the way the Balans chair supports a human body, as shown in Fig. 1.13g, allows it to function like a chair. Normally the back of a chair prevents a person's upper body from falling backward when his muscles are relaxed. That is what fundamentally distinguishes a chair from a stool. However, due to the way the thigh and the lower leg portions are supported in a Balans chair, the muscles of a person can be relaxed and yet the upper body does not fall backward. Therefore, the Balans chair *functions* more like a chair than a stool. Figure 1.13h shows how a normal chair supports a human body for comparison.

Ho (1987) devised a computational model to recognize all kinds of chair based on a *functional definition* – how the object supports a human body comfortably in a sitting position. To achieve that, the recognition system attempts to fit a model of a human body onto the object involved, and if it can support the human body in a sitting position (say, like that shown in Fig. 1.13h), it is then recognized as an instance of chair. There are a few other criteria for a chair to be a good chair, such as

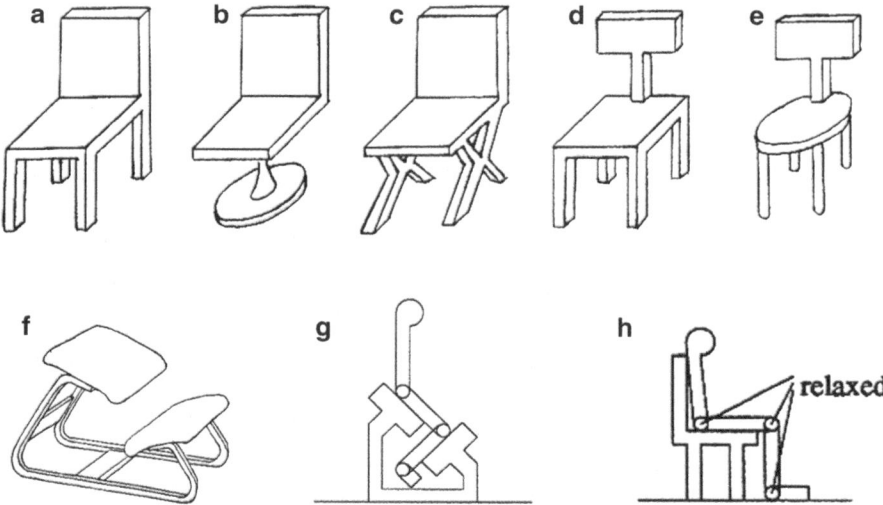

Fig. 1.13 (**a–e**) Various kinds of chairs. (**f**) A *Balans*® chair. (**g**) How a person sits in a *Balans*® chair. (**h**) How a person sits in a normal chair (From Ho 1987)

it allows not only just one position of human body support, but it also allows the human body to move around a bit and still be supported, etc. A fuzzy measure is used to characterize "how good a chair an object is."[9]

The process of functional reasoning is shown in Fig. 1.14. Firstly, the image of the object to be recognized is loaded into a Physical Configuration Array. There is a built-in Physical Reasoner that encodes the laws of physics – how objects interact in various physical situations of mutual contact and under the law of gravity as well. Then the Functional Reasoner, encoding the above functional definition of a chair, invokes the Physical Reasoner with an internal model of a human body (a three-sectioned jointed model as shown in the figure) and see how the human body interacts with the purported chair object and whether or how well the object satisfies the functional definition of a chair.

The complete structure of a concept is shown in Fig. 1.15 (Ho 1987) in which there are two portions. One is the *Functional Definition* portion and the other is called *Symptomatic Perceptual Conditions*. These correspond to what psychologists call *core* and *identification procedure* of a concept respectively (Miller and Johnson-Laird 1976; Nelson 1974). The core of a concept consists of the basic definition of the concept which would include a set of necessary and sufficient conditions and the identification procedure specifies how an instance can be identified, such as through perceptual characteristics. The possibly disjunctive set of symptomatic perceptual conditions for the concept is hence learned under the

---

[9] There are other constraints such as the "economy constraint" that states that a good chair should not have more parts than necessary to achieve the stated function – this means a chair with an extremely tall back is not a very good chair. These are discussed in Ho (1987).

1.7 Perception, Conception, and Problem Solving

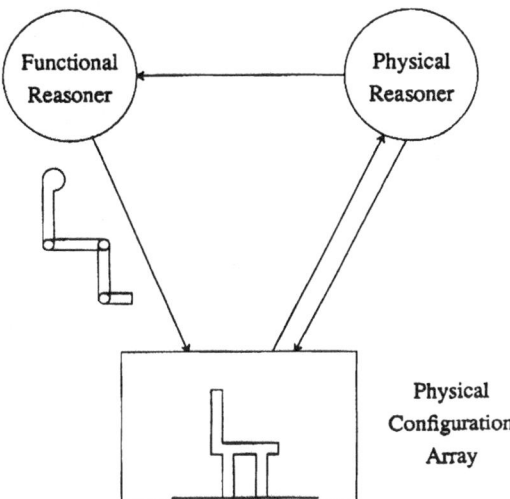

Fig. 1.14 The processing structure for functional reasoning (From Ho 1987)

supervision of the core definition. This kind of two-part structure is especially applicable to concepts of artifacts because artifacts are created to serve certain functions (Fig. 1.6 showed a simple case of functional vs visual attributes definition for a natural object – food.)

Therefore, consider the core functional concept of a *chair* is to basically support a human comfortably in a sitting position plus other constraints as discussed above. There are many ways a chair can be designed and many ways the human body can be supported in those chairs (in ways as different as that of Fig. 1.13g, h), and yet the human is able to relax her muscles and feel comfortable in it. Then, the identification procedure or the symptomatic perceptual conditions would be the conditions associated with various, possibly disjunctive, perceptual characteristics (from Fig. 1.13a–f, etc.) that allow an instance of the chair to be identified such that the functional or the core definition of *chair* can be satisfied.

This discussion ties in with our characterization of a noological system as a system that primarily attempts to satisfy its internal needs and the other internal activities of it including perception exist to support that primary function. So, for the case of chair, the purpose of a noological system in attempting to recognize an instance of it is so that the system can make use of it to satisfy one of the system's internal needs – to sit down and relax and maybe to work at a table to satisfy yet other needs. The visual recognition of a chair and the subsequent actions to make use of the chair for sitting is to serve that deeper purpose. Therefore, the conundrum of why some concepts such as *chair* seem not to have necessary and sufficient conditions if defined in terms of perceptual features can be resolved if the issue of recognition or categorization through perception is not considered alone but in connection with the entire functioning of the noological system involved. The discussion in this section involves a more complex physical object but the issues raised and engaged are similar to those in connection with Fig. 1.6 which involved a simpler object.

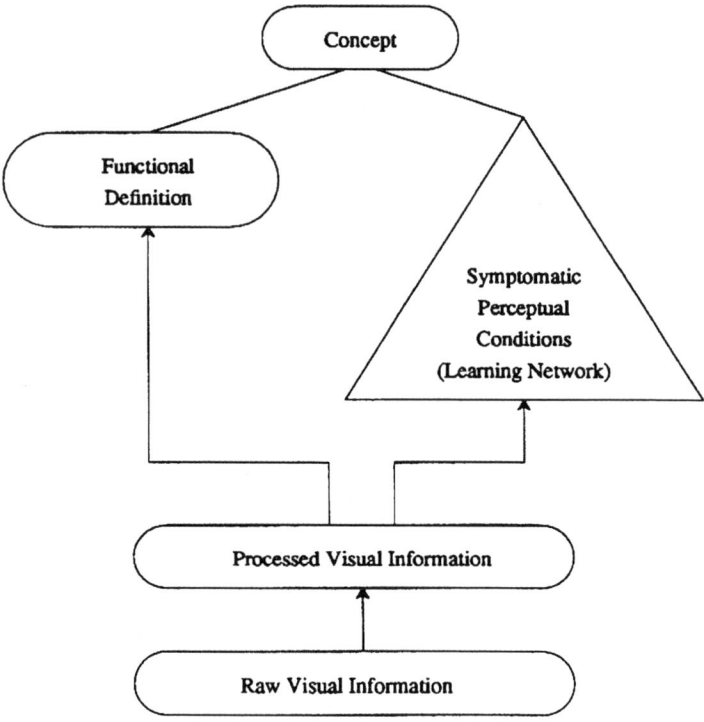

Fig. 1.15 Functional definition vs symptomatic perceptual condition of a concept. The learning network learns the possibly visually disjunctive instances of the concept, supervised by the functional definition (From Ho 1987)

In Fig. 1.16 we use a similar diagram as that in Fig. 1.6 to illustrate how a concept or artifact such as a chair is embedded in the various noological processes. The fact that a chair has certain visual form and physical characteristics is a consequence of the complex interactions between these processes. Figure 1.16 is minimally different from Fig. 1.6. Mainly, "Food" is replaced by "Chair," and the Primary Goal is a Resting Need. (There are different kinds of resting need – some may lead to the seeking of chairs while some to the seeking of beds, etc.) "Anxiousness" is in quotes because the level is usually lower than that for more pressing kinds of needs such as energy. Rules of physical interactions are assumed to have been learned earlier. One major item we have added in Fig. 1.16 is ACQUIRE/BUILD/INVENT CHAIR if chair cannot be found in the immediate environment or does not exist anywhere. This is the same process that led to humanity inventing and building millions of artifacts to satisfy all kinds of needs, and that also explains why objects of similar categories could have vastly different visual forms and physical characteristics.

1.8 Addressing a Thin and Deep Micro-environment for Noological Processing   33

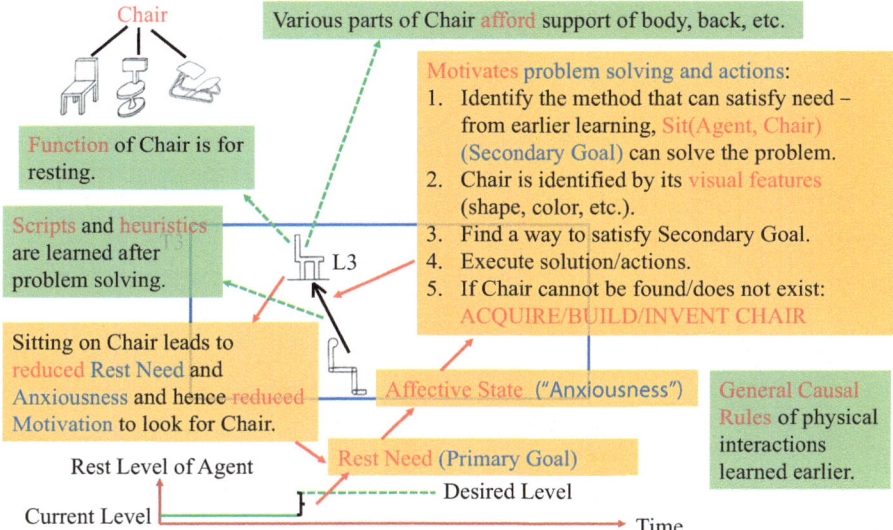

Fig. 1.16 Similar to Fig. 1.6 except that "Food" is replaced by "Chair" and the Primary Goal is a Resting Need. If Chair cannot be found, it can be acquired from somewhere else or built based on known design, or if such an artifact or design does not exist, it can be invented

1.8 Addressing a Thin and Deep Micro-environment for Noological Processing

We believe in order to fully understand a noological system, it is necessary to study it in "depth" in the sense of addressing all the noological processing aspects as laid out in Fig. 1.2. This is because all the aspects are intertwined and the functioning of each aspect is only meaningful when considered in the context of the entire operation of the noological system as illustrated in our earlier discussion such as that in connection with Fig. 1.6. However, for each of the processing aspects in Fig. 1.2, there are many issues in the "width" direction. Consider the perceptual level – for vision alone, there are issues related to object recognition as well as perception of depth, texture, motion, etc. (Wolfe et al. 2009). To study the entire depth and width of the various aspects of Fig. 1.2 would be a formidable task. Therefore, we propose to first address a *thin and deep* slice through the entire space of noological processing aspects as shown in Fig. 1.17.

However, our approach and focus here is on addressing critical general principles covering all the aspects of noological processing. At times we use a reduced and simplified version of the real environment to elucidate certain principles, but we consider the issues without sacrificing the generality of the principles involved. And then, hopefully many of the principles uncovered are applicable as more issues are brought in for consideration in the width direction of the noological processing aspects space. For example, one major emotional state that is addressed in

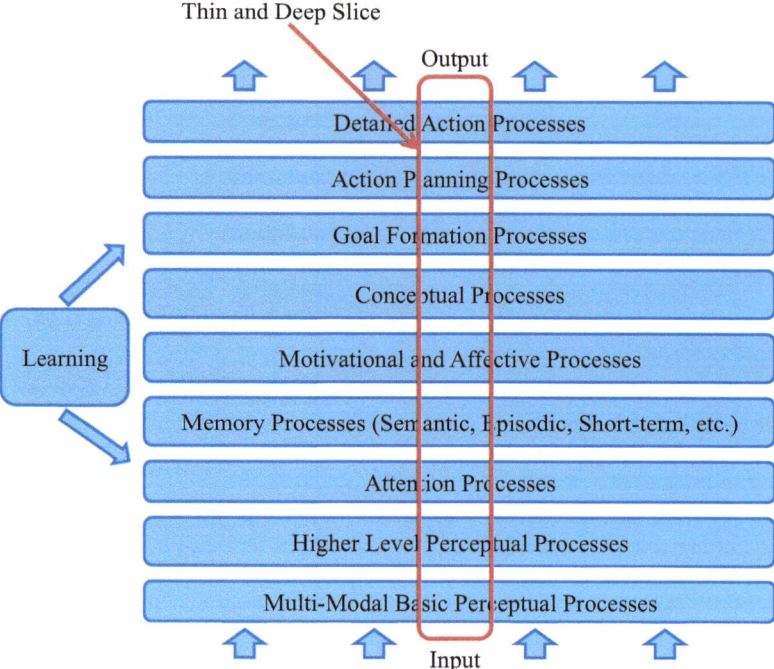

Fig. 1.17 A thin and deep slice of the noological processing aspects space (©2013 IEEE. Reprinted, with permission, from A Grand Challenge for Computational Intelligence: A Micro-Environment Benchmark for Adaptive Autonomous Intelligent Agents, Proceedings of the IEEE Symposium Series on Computational Intelligence – Intelligent Agents, Page 45, Figure 1)

considerable detail in this book is the state of *anxiousness*. This is addressed in a number of chapters. The function of anxiousness is formulated in the context of problem solving which concerns the engagement of various noological processing mechanisms including learning. Other emotions are not addressed in detail in this book but it is our belief that the fundamental principles established here can be extended to handle them. In the concluding chapter of Chap. 10 we discuss some issues on scaling up to more complex environments. Therefore, after addressing the issues of rapid effective causal learning in Chap. 2, which is the core learning mechanism for a noological system, we lay out a general framework that covers all the critical aspects of noological processing in Chap. 3.

At times, our approach to study certain issues smacks of the micro-world approach in traditional AI, in which a simplified micro-world is used to study various issues and then it is hoped that they could be scaled up to handle issues in the real world. This method did not meet with great success in traditional AI. Very often, methodologies and principles developed for the micro-world fall apart and are not applicable when the micro-world is scaled up to the real world (e.g., Winograd 1973). A favorite method of problem solving, the A* method (Hart et al. 1968), faces the issue of combinatorial explosion when the micro-world is

1.8 Addressing a Thin and Deep Micro-environment for Noological Processing

scaled up to the real world as the number of parameters involved become unmanageable.

Our approach and the micro-environment used differ from the earlier approaches in a few aspects. Firstly, as mentioned in the previous paragraph, our method studies the entire depth of processing from perception to action while addressing the very important issues on the internal environment – the primary motivations and goals. In the first place, while traditional AI is concerned only with the "world" out there – hence the "micro-world" approach – at the outset, the "environment" that we address involves not only the external environment but also the internal environment. The external environment consists of the events and processes that take place in the outside world while the internal environment consists of the internal goals and priorities of the agent, as articulated above. The agent optimizes its behavior between both the internal and external constraints – internally, there are build-in goals to satisfy, which directs its problem solving efforts, and externally, there are causalities about the world it has to learn in order to discover the right sequence of actions to take to concoct a solution. This is adapting to the environment to serve the internal needs. The earlier efforts of micro-world do not characterize the agents in such an integrated manner.

Secondly, to handle scalability, we address the issues of combinatorial explosion at the outset. Actually, combinatorial explosion does not necessary take place only when one scales things up to the real world. Even a simple micro-environment can give rise to combinatorial explosion. We will see this shortly in the next chapter when we address the simple and basic problem of spatial movement to goal. As opposed to many of the earlier micro-world approaches, our approach is to address the hairy issues at the outset. Therefore, the principles we uncover are scalable.

Traditionally one way to handle combinatorial explosion is to use heuristics (Russell and Norvig 2010). However, one will find out quickly that if the set of heuristics available to the agent is built-in and hence fixed in number, one will still run into the issue of complexity as one encounters more complex rules and situations in the environment and the heuristics are not applicable. Therefore, domain specific heuristics should be something that is learned, and as more heuristics are learned, they can continue to help reduce the complexities of search and problem solving. These are not issues that have been addressed in traditional AI but will be addressed here in the outset. A simple example of heuristics learning will be considered in Chap. 2 and the same learning method will be used again later in more complex situations in Chap. 6 (Sect. 6.4) and Chap. 7 (Sects. 7.4.3.5, 7.4.3.6, and 7.4.3.8).

The other method to reduce computational complexity is through knowledge chunking. Though there have been efforts in AI that deal with chunking (e.g., Alterman 1988; Carbonell 1983; Carbonell et al. 1989; Erol et al. 1996; Fikes et al. 1972; Hammond 1989; Laird et al. 1986, 1987), the domain specific rules that are used in the chunking processes of these earlier efforts are built-in and not learned. The issue is how to keep learning and chunking. The learning of chunked rules in the form of scripts using rapid effective causal learning will be addressed in Chap. 3, Sect. 3.5, as well as in other subsequent chapters (e.g., Chap. 7).

Even though we had said that the issues of motivation and goal are of primary importance, there is one issue to be addressed first before the issue of motivation can be addressed. We mentioned that an agent learns about entities in the external environment (e.g., food) that can have some causal impact on its internal environment (e.g., increase in energy). We mentioned that a rapid learning of this causal relationship requires a rapid effective causal learning process that is unsupervised. Therefore, this is the main issue toward which we devote our effort to explain in the next chapter. In any case, a central mechanism that permeates all the levels of noological processing is learning (Fig. 1.2). Therefore, the issue of learning has to be addressed first.

1.9 Summary of the Basic Noological Principles

In summary, the following are the principles we consider fundamental for a theoretical framework for characterizing noological systems (also stated at the beginning of this chapter). All these principles are addressed in computational terms in this book and we indicate below the places in the book where the issues involved are discussed. These principles contrast strongly with what have been addressed and emphasized in traditional AI as well as the cognitive sciences and will greatly enhance these disciplines[10]:

- A noological system is characterized as primarily consisting of a processing backbone that executes problem solving to achieve a set of built-in primary goals which must be explicitly defined and represented. The primary goals or needs constitute the bio-noo boundary. (This chapter.)
- Motivational and affective processes lie at the core of noological processing and must be adequately computationally characterized. (Sects. 1.4 and 3.3, Chaps. 7, 8, and 9.)
- Rapid effective causal learning provides the core learning mechanism for various critical noological processes. (Chap. 2).
- The perceptual and conceptual processes perform a service function to the problem solving processes – they generalize and organize knowledge learned (using causal learning) from the noologial system's observation of and interaction with the environment to assist with problem solving. (This chapter and Chap. 6.)
- Learning of scripts (consisting of start state, action steps, outcome/goal, and counterfactual information) from direct experience with the environment enables knowledge chunking and rapid problem solving. This is part of the perceptual/conceptual processes. Scripts are noologically efficacious

---

[10] For a discussion of how these principles contrast with what have been typically addressed and emphasized in AI and the cognitive sciences, see the discussion in connection with these principles stated at the beginning of this chapter.

fundamental units of intelligence that can be composed to create further noologically efficacious units of intelligence that improve problem solving efficiency, in the same vein that atoms are composed into molecules that can perform more complex functions. (Sect. 2.6.1, Chaps. 6, 7, and 8.)
- Learning of heuristics further accelerates problem solving. Similarly, this derives from the perceptual/conceptual processes. (Sects. 2.6.1, 6.4, and Chap. 7, specifically Sects. 7.4.3.5, 7.4.3.6, and 7.4.3.8.)
- All knowledge and concepts represented within the noological system must be semantically grounded – this lies at the heart of providing the mechanisms for a machine to "really understand" the meaning of the knowledge and concepts that it employs in various thinking and reasoning tasks. There exists a set of ground level atomic concepts that function as fundamental units for the characterization of arbitrarily complex activities in reality. (This chapter and Chap. 4 in general, and specifically Sects. 4.5, 5.4, and 7.5.)

Problem

Provide more examples of affordance chains similar to that of Fig. 1.10 that involve other needs in the Maslow hierarchy of Fig. 1.9a.

References

Aaron, S. (2014). *The cognitive and affect project*. http://www.cs.bham.ac.uk/research/projects/cogaff/
Alberts, B., Johnson, A., Lewis, J., Morgan, D., Raff, M., Roberts, K., & Walter, P. (2014). *Molecular biology of the cell* (6th ed.). New York: Garland Science.
Albrecht-Buehler, G. (1985). Is the cytoplasm intelligent too? *Cell and Muscle Motility, 6*, 1–21.
Albrecht-Buehler, G. (2013). *Cell intelligence*. http://www.basic.northwestern.edu/g-buehler/FRAME.HTM
Alterman, R. (1988). Adaptive planning. *Cognitive Science, 12*, 393–422.
Braun, D. A., Mehring, C., & Wolpert, D. M. (2010). Structure learning in action. *Behavioral Brain Research, 206*(2), 157–165.
Cambria, E., & Hussain, A. (2015). *Sentic computing: A common-sense-based framework for concept-level sentiment analysis*. Cham: Springer.
Cambria, E., Hussain, A., Havasi, C., & Eckl, C. (2010). *Sentic computing: Exploration of common sense for the development of emotion-sensitive systems* (LNCS, Vol. 5967, pp. 148–156). Cham: Springer.
Carbonell, J. G. (1983). Derivational analogy and its role in problem solving. In *Proceedings of AAAI-1983* (pp. 64–69).
Carbonell, J. G., Knoblock, C. A., & Minton, S. (1989). *PRODIGY: An integrated architecture for planning and learning* (Technical Report CMU-CS-89-189). Pittsburgh: Computer Science Department, Carnegie-Mellon University.

Chambers, N., & Jurafsky, D. (2008). Unsupervised learning of narrative event chains. In *Proceedings of the annual meeting of the association for computational linguistics: Human language technologies*, Columbus, Ohio (pp. 789–797). Madison: Omni Press.

Erol, K., Hendler, J., & Nau, D. S. (1996). Complexity results for HTN planning. *Artificial Intelligence, 18*(1), 69–93.

Evans, V., & Green, M. (2006). *Cognitive linguistics: An introduction*. Mahwah: Lawrence Erlbaum Associates.

Fages, F. (2014). Cells as machines: Towards deciphering biochemical programs in the cell. In *Proceedings of the 11th International Conference on Distributed Computing*, Bhubaneswar, India (pp. 50–67). Switzerland: Springer.

Ferrucci, D., Brown, E., Chu-Carroll, J., Fan, J., Gondek, D., Kalyanpur, A. A., Lally, A., Murdock, J. W., Nyberg, E., Prager, J., Schlaefer, N., & Welty, C. (2010). Building Watson: An overview of the DeepQA Project. *AI Magazine, 31*(3), 59–79.

Fikes, R. E., Hart, P. E., & Nilsson, N. J. (1972). Learning and executing generalize robot plans. *Artificial Intelligence, 3*, 251–288.

Ford, B. J. (2009). On intelligence in cells: The case for whole cell biology. *Interdisciplinary Science Reviews, 34*(4), 350–365.

Fuster, J. M. (2008). *The prefrontal cortex* (4th ed.). Amsterdam: Academic.

Gazzaniga, M. S., Ivry, R. B., & Mangun, G. R. (2013). *Cognitive neuroscience: The biology of the mind* (4th ed.). New York: W. W. Norton & Company.

Geeraerts, D. (2006). *Cognitive linguistics*. Berlin: Mouton de Gruyter.

Gibson, J. J. (1979). *The ecological approach to visual perception*. Boston: Houghton Mifflin.

Gleitman, H., Gross, J., & Reisberg, D. (2010). *Psychology* (8th ed.). New York: W. W. Norton & Company.

Hameroff, S. R. (1987). *Ultimate computing: Biomolecular consciousness and nanotechnology*. Amsterdam: Elsevier Science Publishers B.V.

Hammond, K. (1989). *Case-based planning: Viewing planning as a memory task*. San Mateo: Addison-Wesley.

Hart, P. E., Nilsson, N. J., & Raphael, B. (1968). A formal basis for the heuristic determination of minimum cost paths. *IEEE Transactions on Systems Science and Cybernetics SSC4, 4*(2), 100–107.

Ho, S.-B. (1987). *Representing and using functional definitions for visual recognition*. Ph.D. thesis, University of Wisconsin-Madison.

Ho, S.-B. (2014). On effective causal learning. In *Proceedings of the 7th International Conference on Artificial General Intelligence*, Quebec City, Canada (pp. 43–52). Berlin: Springer.

Houk, J. C., Davis, J. L., & Beiser, D. G. (1995). *Models of information processing in the Basal Ganglia*. Cambridge, MA: MIT Press.

Laird, J., Rosenbloom, P. S., & Newell, A. (1986). Chunking in soar: The anatomy of a general learning mechanism. *Machine Learning, 1*, 11–46.

Laird, J., Rosenbloom, P. S., & Newell, A. (1987). SOAR: An architecture for general intelligence. *Artificial Intelligence, 33*(1), 1–64.

Langacker, R. W. (2008). *Cognitive grammar: A basic introduction*. Oxford: Oxford University Press.

Langacker, R. W. (2009). *Investigations in cognitive grammar*. Berlin: Mouton de Gruyter.

Manshadi, M., Swanson, R., & Gordon, A. S. (2008). Learning a probabilistic model of event sequences from internet weblog stories. In *Proceedings of the 21st FLAIRS conference*, Coconut Grove, Florida (pp. 159–164). Menlo Park: AAAI Press.

Maslow, A. H. (1954). *Motivation and personality*. New York: Harper & Row.

Miller, G. A., & Johnson-Laird, P. N. (1976). *Language and perception*. Cambridge, MA: Harvard University Press.

Nelson, K. (1974). Concepts, word, and sentence: Primacy of categorization and its functional basis. In P. N. Johnson-Laird & P. C. Wason (Eds.), *Thinking*. Cambridge: Cambridge University Press.

References

Norman, D. A. (1988). *The psychology of everyday things*. New York: Basic Books.

Pan, S. J. & Yang, Q. (2010). A survey on transfer learning. *IEEE Transactions on Knowledge and Data Engineering, 22*(10), 1345–1359.

Quek, B. K. (2006). Attaining operational survivability in an autonomous unmanned ground surveillance vehicle. In *Proceedings of the 32nd Annual Conference of the IEEE Industrial Electronics Society*, (pp. 3969–3974). Paris: IEEE Press.

Quek, B. K. (2008). *A survivability framework for autonomous systems*. Ph.D. thesis, National University of Singapore.

Reeve, J. (2009). *Understanding motivation and emotion*. Hoboken: Wiley.

Regneri, M., Koller, A., & Pinkal, M. (2010). Learning script knowledge with web experiments. In *Proceedings of the 48th annual meeting of the association for computational linguistics*, Uppsala, Sweden (pp. 979–988). Stroudsburg: Association for Computational Linguistics.

Rolls, E. (2008). *Memory, attention, and decision-making*. Oxford: Oxford University Press.

Rosch, E., & Mervis, C. B. (1975). Family resemblances: Studies in the internal structure of categories. *Cognitive Psychology, 7*, 573–605.

Russell, S., & Norvig, P. (2010). *Artificial intelligence: A modern approach*. Upper Saddle River: Prentice Hall.

Schank, R., & Abelson, R. (1977). *Scripts, plans, goals and understanding*. Hillsdale: Lawrence Erlbaum Associates.

Sutton, R. S., & Barto, A. G. (1998). *Reinforcement learning: An introduction*. Cambridge, MA: MIT Press.

Tu, K., Meng, M., Lee, M. W., Choe, T. E., & Zhu, S.-C. (2014). Joint video and text parsing for understanding events and answering queries. *IEEE MultiMedia, 21*(2), 42–70.

Winograd, T. (1973). A procedural model of language understanding. In R. C. Schank & K. M. Colby (Eds.), *Computer models of thought and language*. San Francisco: W. H. Freeman and Company.

Wolfe, J. M., Kluender, K. R., Levi, D. M., Bartoshuk, L. M., Herz, R. S., Klatzky, R. L., Lederman, S. J., & Merfeld, D. M. (2009). *Sensation & perception* (2nd ed.). Sunderland: Sinauer Associates.

Chapter 2
Rapid Unsupervised Effective Causal Learning

Abstract This chapter introduces a novel learning paradigm that underpins the rapid learning ability of noological systems – effective causal learning. The learning process is rapid, requiring only a handful of training instances. The causal rules learned are instrumental in problem solving, which is the primary processing backbone of a noological system. Causal rules are characterized as consisting of a diachronic component and a synchronic component which distinguishes our formulation of causal rules from that of other research. A classic problem, the spatial movement to goal problem, is used to illustrate the power of causal learning in vastly reducing the problem solving search space involved, and this is contrasted with the traditional AI A* algorithm which requires a huge search space. As a result, the method is scalable to real world situations. Script, a knowledge structure that consists of start state, action steps, outcome/goal, and counterfactual information, is proposed to be the fundamental noologically efficacious unit for intelligent behavior. The discussions culminate in a general forward search framework for noological systems that is applied to various scenarios in the rest of the book.

Keywords Causality • Effective causality • Causal learning • Diachronic causal condition • Synchronic causal condition • Desperation and generalization • Spatial movement to goal problem • Heuristic • Heuristic generalization • Learning of heuristic • Script • Counterfactual information • Forward search framework

Currently in AI, a number of learning paradigms have been used for a variety of tasks. Reinforcement learning (Sutton and Barto 1998) has been used for learning a correct sequence of actions to obtain a certain reward. Supervised and unsupervised learning have been used for data classification such as image classification in computer vision. Bayesian reasoning/learning has been used for cause recovery in a *closed* domain (Pearl 2009). In Bayesian learning, the reasoning/learning process proceeds as follows. Firstly, the probabilities of causal relationships between some variables (called "likelihoods") are known, which are typically obtained through statistical data, with hand-selection of relevant parameters. Then, together with some a priori probabilities of the variables involved, cause recovery then involves identifying which known causal relationship is more likely.

However, *rapid* learning of causality in an *open* domain has not been studied in detail. For example, learning that when the rain comes, things may get wet, or when one gets hit by a rock, one usually bleeds and feels pain, etc., are open domain causality learning and learning these rapidly (within one to two instances of observation) is critical for survival.

Let us consider an example to contrast closed domain cause recovery vs open domain cause recovery. Suppose one observes that rain can cause wetness with a certain probability, so can a flood. If one observes wetness in one's house, the Bayesian process allows one to find out whether it was due to rain or flood. To do that, one needs to have observed/learned the probabilities of rain and flood causing wetness (the "likelihood") and the "a priori" probabilities of rain and flood themselves. However, this begs the question of how one learns the various likelihoods to begin with, especially if they were to be learned rapidly for the purpose of survival – when a noological system is facing an immediate problem to be solved, there is usually no time to collect a lot of data.

Scientists have observed an interesting phenomenon in the field of urban ecology. The situation involves animals that normally live in the wild and sometimes they find their way into the city. To survive, they learn many things very quickly, chief among which is the ability to cross roads safely. Observations revealed that animals such as coyotes are able to learn the traffic patterns and how stoplights work (Greenspan 2013). Obviously, they need to learn causality without a large amount of data and they also need to do so without the numerous positive and negative reinforcement signals typically required in reinforcement learning (Sutton and Barto 1998) because typically one negative reinforcement signal means death for these animals and further learning is obviously impossible.

In any case, in the context of problem solving such as that depicted in Fig. 1.6 which is the main aim of a noological system, the above issues with Bayesian causal inference is moot as the normal inference that can be derived from the Bayesian method does not assist with problem solving directly (this will be explained in Sect. 2.1). In this chapter we will first review the Bayesian method using a relevant example to place in context the purpose of the topic of this chapter.

Following the review of the Bayesian method and the development of rapid *effective* causal learning, we will show that a kind of search called *causal learning augmented search* is more akin to a typical search process carried out by noological systems such as human beings. Human beings do not carry out very extensive blind search or even the relatively extensive traditional A* heuristic search (Hart et al. 1968). Typically, a small amount of search is carried out and carried out quickly, then some intelligent learning process kicks in and obviates any need for further lengthy search. We show how this is possible with causal learning augmented search. Also, in this chapter, we will show that heuristics can be learned in the process of causal learning augmented search, unlike in traditional AI in which heuristics are typically built-in knowledge. We use these mechanisms to motivate a general noological processing framework that we will gradually develop more fully in the rest of the book.

2.1 Bayesian Causal Inference and Rapid Causal Learning

Figure 2.1 illustrates a typical Bayesian causal inference scenario. Suppose there are two actions that can be performed on a block – a push (Pu) action and a twist (Tw) action. Suppose also that each action can result in two possible movements of the block – a translational (Tr) movement or a spinning (Sp) movement. (For simplicity assume that the resultant movement is exclusively one or the other type and not a combination of both.)

In Fig. 2.1, there are likelihoods such as the probabilities of a push action giving rise to a translational movement P(Tr/Pu) or a spinning movement P(Sp/Pu) and the probabilities of a twist action giving rise to a translational movement P(Tr/Tw) and a spinning movement P(Sp/Tw). Given a priori probabilities of the push and the twist actions, P(Pu) and P(Tw), respectively, one can calculate the probabilities of, say, given that a translational movement has been observed, how likely a push action that was effected earlier (i.e., *push* was the cause) using the Bayes rule – P(Pu/Tr) = P(Tr/Pu) * P(Pu)/P(Tr) – or, say, given that a spinning movement has been observed, how likely it was a twist action that was effected (i.e., *twist* was the cause) – P(Tw/Sp) = P(Sp/Tw) * P(Tw)/P(Sp).

This is causal inference in a closed domain – the likelihoods P(Tr/Pu), P(Tr/Tw), P(Sp/Pu), P(Sp/Tw) have all been determined earlier (in a "learning" process, and as mentioned above, typically by looking at correlations between hand-picked parameters), and the causal inference is to infer which of the two possible actions is more likely to have been performed given certain observation. This begs the question of where all the likelihoods come from to begin with and how they can be learned rapidly.

On the other hand, in an open domain in which we begin with no knowledge of what causes what, identifying that there is indeed a causal connection between say, a push action, and say, either the translational or spinning consequence is the

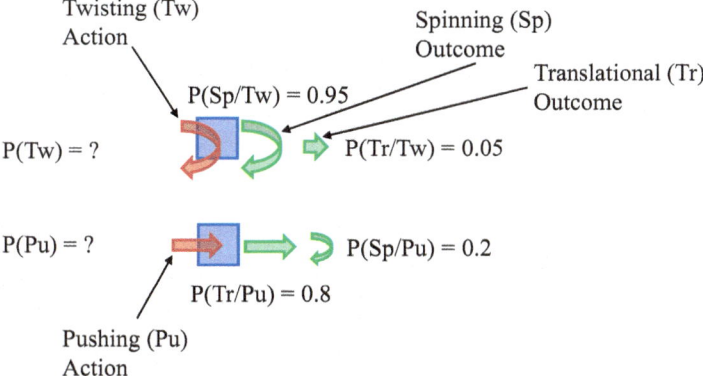

Fig. 2.1 A typical Bayesian causal inference scenario involving pushing (Pu) or twisting (Tw) a block and observing whether it spins (Sp) or moves translationally (Tr)

primary concern. How do we know that the cause is not, say, the flash of light from the headlamp of a car that happens to shine into the room where and when the translational or spinning consequence is observed to take place?

Moreover, in the context of problem solving, suppose one has already known P(Tr/Pu), P(Sp/Tw), etc. (through first establishing the causal connection between Pu and Tw on the one hand and Tr and Sp on the other, and then collecting some probabilities), the concern is, if one wants to create a translational movement (Tr), which action is a better one to take? Suppose P(Tr/Pu) is larger or much larger than P(Tr/Tw), then it is clear that one would choose Push as the action to effect the translation – one has a higher chance of success. The other concern of a noological system is to project into the causal future. Suppose one observes a twist action, what is more likely to happen in the future? In the example of Fig. 2.1, the answer would be a spinning movement, according to the probabilities given in Fig. 2.1: P(Sp/Tw) > P(Tr/Tw).

Yet another concern of a noological system is understanding the intention of others. Given that one system observes that another system is taking a certain action, say, push, and assuming that the other system has the same knowledge of the various likelihoods of Fig. 2.1, what can the first system say about the intention of the other? The answer would be the other intends the block to move translationally, since P(Tr/Pu) > P(Sp/Pu).

Therefore, if one had already known P(Tr/Pu), P(Sp/Tw), etc., Bayesian inference has no role to play in problem solving, anticipation of future events, or inferring intention. And a rapid method of cause discovery is needed to establish the connection between Pu or Tw, the causes, and the possible effects, Tr or Sp.

However, Bayesian inference is useful in other tasks, and both problem solving and it require the likelihoods involved. The crucial question is, how are the likelihoods obtained? Is the usual method of obtaining these through statistical parameter correlations based on hand-picked parameters adequate? Is there a more rapid method that does not require hand-picked parameters, as the process of "intelligent" hand-picking itself should be a consequence of intelligent actions to start with? This is an issue to which we will attend in the next section.

2.2 On Effective Causality

It has often been said that correlation does not imply causality (e.g., Moore et al. 2009; Agresti and Franklin 2007, etc.). The situation is depicted in Fig. 2.2. In Fig. 2.2, suppose there is a factor, A, that is observed to be correlated with another factor, B. One has to be careful about concluding that A causes B because there could yet be a third factor, C, that "actually causes" A and B (assuming first that somehow this particular causality can indeed be established). This gives rise to a correlation between A and B but this correlation is not due to A causing B directly.

One good example would be smoking (A) and lung cancer (B) (Ho 2014). Statistical data have shown that there is a correlation between smoking and lung

2.2 On Effective Causality

Fig. 2.2 Correlation and causality. Factor A is correlated with factor B. Factor C is correlated with both factors A and B

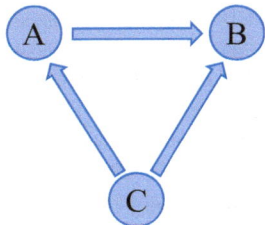

cancer. However, many a tobacco company has argued against interpreting the correlation as causality (Milberger et al. 2006). Perhaps there exists a third factor (C), which is a genetic predisposition of both smoking and lung cancer. This factor causes both the A and B factors, thus creating a correlation between A and B. Thus, the tobacco companies argue, the fault of lung cancer (B) does not lie in them who supply the cigarettes, leading to factor A (smoking), since A does not cause B.

The issue of causality is a hairy one. Let us ignore the more complex correlations between factors that are widely separated in time and that are probabilistic in nature (even if it is true that there is a strong correlation between smoking and cancer, for sure not everyone who smokes would ultimately develop cancer), and consider a more proximal daily example. Suppose on a sunny afternoon you are not thinking of achieving anything in particular and a thought emerges in your mind (brain) that corresponds to the following content, "Hey, what if I push this object in front of me? Would it move?" You then emit an action of pushing the object and it moves. Because the thought originates in your head – at least you think you are sure about that – and the pushing action is clearly a consequence of your thought and the movement of the object is seemingly strongly correlated with your action, you have no doubt that "Push(Object) causes Move(Object)." The issue of causality here seems so clear-cut that there is no need to raise the question "correlation may not imply causality" of Fig. 2.2. However, can we be so sure about that?

Suppose we postulate the following, that there is a God of action watching over all of us. This God creates the thought in you that is "Hey, what if I push this object in front of me? Would it move?" and this subsequently leads to your pushing action. At the same time, this God emits a force to move the object (after waiting for your hand to contact the object to deliberately mislead you that your action has an efficacy). Would the God not be the factor C in Fig. 2.2? How could I ever prove that it is not, other than through the criterion of parsimony: perhaps arguing along the line that one could not imagine God to be so free to do this in every instance of our life to help us emit actions and create their effects? Therefore, in principle, one can go on and on to search for the "true" underlying causality of all things in nature and not be certain that they are indeed "true" causes. However, from the point of view of whether one can *effectively* use the correlation between the pushing and movement of an object to solve certain problems at hand, does one care if God is really involved?

Therefore, we adopt the stance that what really matters is *effective causality* (Ho 2014). Consider the case of non-probabilistic situations. If every time I emit an action to push an object or observe that when others do likewise, the object moves, I can say that *effectively* the action causes the movement. It is effective in the sense that I can rely on this rule to achieve successful problem solving every time – i.e., if I want something to move, I push it, or get someone to push it. Even if the situation is probabilistic and for some reason it is only the case that most of the time I get the movement when I push the object (much like the case in Fig. 2.1 regarding translational or spinning movement), as long as there are no other actions that can give rise to a higher probability of movement, I would still choose to use pushing to achieve movement. The causal rule, probabilistic or not, *effectively* helps me to achieve my goal (of moving an object).

We will use an example to contrast *effective* causal recovery on the one hand and *statistical* cause recovery on the other (Ho 2014). Figure 2.3a shows a situation in which a person uses a remote control to control a toy car. Suppose when he pushes the control in a certain manner (an event, which connotes a change of state of something), he observes that the wheels move in a certain manner (another event), and no other events happen in the environment. (The actions of toy car remote control sticks typically move the car forward or backward and the front wheels left or right.) And suppose he tries again, and the same thing happens again. He then concludes that a manner of pushing the control stick (e.g., to the left) *effectively* causes the moving of the wheels in a particular manner (e.g., turning to the left). The confidence in the causal relationship that makes it effective – useful for subsequent applications – is enhanced by further repetitions of the cause-effect. However, not many instances of this observation is needed to more or less establish this causality.

There is no prior knowledge necessary for this cause discovery. Even if this remote control and the toy car are given to a young child or placed in a jungle and discovered by some primitive people who have never seen a car (toy or otherwise) or a remote control, or had knowledge of electromagnetic radiation (which is the underlying mechanism that transmits the signal to be translated into turning of the car's wheels), he would still be able to work out the causality between the pushing of the stick and the turning of the wheels (and for that matter, the forward and backward movements of the car when the stick is pushed in certain manners accordingly).

And what about a hidden cause in all these? As discussed above, it is certainly possible that some hidden causal agent, a "God," could be behind the emergence of the thought of pushing the stick and the movement of the wheels of the car. Typically, we would ignore this possibility as far as possible unless some other discovery warrants our search for this hidden cause.

There is the possibility that some confounding factors are present when the situation of Fig. 2.3a takes place. There could be wind, for example, at the same time as the pushing of the remote control stick which might be interpreted as the cause of the movement of the car's wheels. However, when the person pushes the stick again, it is unlikely that these confounding factors happen again in precisely the same way. A noological system should have the built-in mechanisms to filter these spurious causal factors away.

2.2 On Effective Causality

Fig. 2.3 Rapid effective causal learning. (**a**) A person controlling a toy car and is able to observe the effect(s) of his action proximally. (**b**) A person controlling a toy car placed some distance away and can only infer the effect(s) of his action based on the smoke emitted by the car (Ho 2014)

Now consider the situation in Fig. 2.3b in which the toy car has been moved to a distance and the only way to observe its movement is through some smoke emitted from its top that rises upward. Because of the lag in the movement of the smoke relative to the movement of the car and this lag is not always constant given the same movement (had it been a rigid stick that sticks upward from the car's top, there will be no lag), it would be difficult for the person to infer the causality very quickly. More observations of the smoke's behavior would have to take place before some correlation/causality between the remote control stick's movements and the movements of the car and its attendant parts could be established. This is when statistical cause recovery comes into play.

Returning to the smoking and lung cancer example, suppose one has the instrument that can reduce a human observer to the size of the smoke particles and one follows the particles into the lungs of people. Suppose one then observes the particles entering the cells, interacting with the DNA, ripping the bonds and resulting in the reconfiguration of the molecules, and this is followed by a sequence of molecular activities that finally leads to the cell becoming cancerous (with every step distinctly observed by the human observer), then one can establish the effective causality between smoking and lung cancer in one series of observations, bypassing statistics.

Therefore, provided that the senses can be trusted, we can identify effective causality in a situation of *proximal* observation and rapidly as well. This is carried out in an unsupervised manner. When the objects and activities related to the effects are distal, such as in the cases shown in Fig. 2.3b or in the smoking/lung cancer situation, and observations are made through some intervening activities that bear only probabilistic relations to the activities of concern, then effective causality cannot be established and one has to defer to statistical techniques.

2.3 Opportunistic Situation and Rapid Cause Recovery

In the situation of Fig. 2.3a, basically what is being used to establish effective causality is a temporal association. The temporal association is possible because of a relatively tranquil environment save for a small amount of noise (e.g., the example of wind mentioned above). We posit that in a very noisy environment, cause recovery would be impossible without prior knowledge that could remove spurious causes. Therefore, effective cause recovery or cause-effect identification relies on an "opportunistic situation." After the establishment of some fundamental causalities based on a tranquil environment, subsequent cause recovery in a nosier environment becomes possible – the earlier knowledge can assist in ruling out certain spurious temporal correlations as possible candidates of causality in the later situations.

In the situation of Fig. 2.3a, the temporal interval between the pushing of the remote control stick and the movements of the car/wheels is small. The effect follows the cause almost immediately. Moreover, the temporal intervals from instance to instance (each instance being a push-movement event) are more or less consistently the same as far as a typical human is able to discern. This temporal closeness is a strong hint of causality, and the confidence in the causal relationship that makes it effective – useful for subsequent applications – is further enhanced by repetitions with a consistent interval.

Other than the consistency of temporal intervals in repetition, the event types themselves involved are often sufficient to establish causality. Assuming that in the first instance, there is a relatively short interval (at most seconds, not minutes or longer) between the pushing of the remote control stick and the movement of the car's wheels. A noological system with some simple in-built causality heuristics can first assume that there is a causal relationship between these two events. Suppose now the noological system pushes the stick in the same manner again and suppose this time the movement of the wheels is delayed by a few seconds. Also suppose in subsequent repetitions the delays of the wheels' movement are not the same (and occasionally perhaps the wheels do not even respond at all), an effective causal relationship can still be established between these two events provided that the second event happens *most* of the time *after* the first event. One can make good *practical* use of the situation – as far as the system knows, pushing the stick in a certain manner is still the best way to make the wheels move in a certain manner as a consequence.

One can assign a degree of confidence of a certain causal relationship based on the consistency of the repetition interval and the probability of whether the second event is being repeated. A simple formula could be: *Confidence in causal rule = Consistency of repetition of causal relationship* minus *Average variability of the interval between cause and effect*. Figure 2.4 shows the changes of the Confidence measure in a particular cause-effect rule as more cause-effect pairs are observed, and there are inconsistencies in the repeatability of the cause-effect relationship and variability in the intervals between the cause and effect. Fire and Zhu (2015) also

2.3 Opportunistic Situation and Rapid Cause Recovery

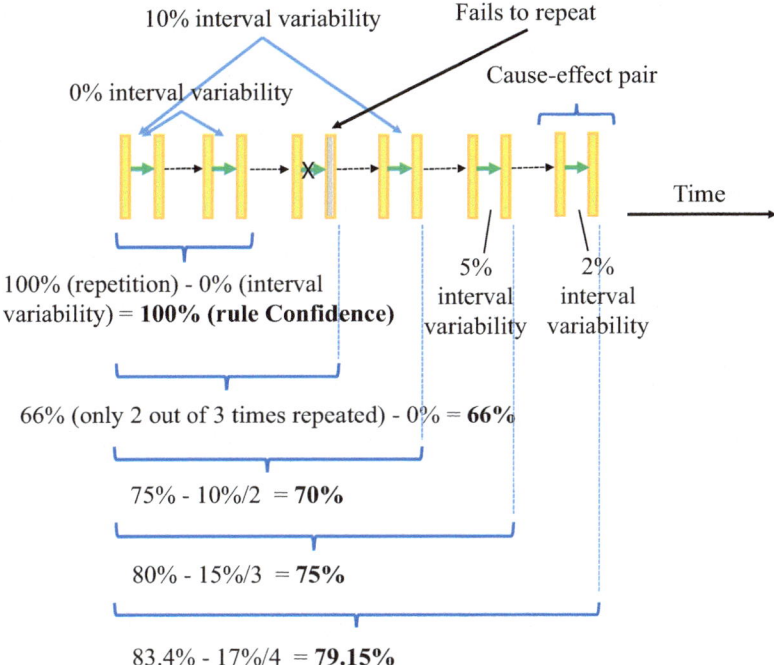

Fig. 2.4 Probabilistic situation and rule confidence. Each pair of cause-effect is represented as two *yellow vertical bars* (each *bar* representing an event's onset) connected by a *green arrow*. After the cause-effect pair has repeated twice, it fails to repeat the third time, and the failed effect repetition is indicated as a *vertical gray bar*. The variability in interval is measured with respect to the interval of the first cause-effect pair. The rule Confidence reduces and then increases as more observation is made

considers the learning of causal relationship from video in the presence of noise in the video data.

A more complicated situation can arise in a situation similar to that of Fig. 2.3. Let's call the pushing of the remote control stick a Type 1 event and the movement of the car's wheels a Type 2 event. Suppose there are two remote controls A and B (hence two Type 1 events when the sticks are pushed), each of which controls one of two cars C and D (hence potentially two Type 2 events when, say, the wheels on each of the cars move). There may be confusion if two Type 1 events take place first that are then followed by two Type 2 events. Even after a few repetitions, it is hard to establish whether it is remote control A that causes the movement of car C and remote control B that causes the movement of car D or vice versa, or whether perhaps only one of the two remote controls is causing the movements of *both* cars C and D, and the other one is a dummy. The situation is illustrated in Fig. 2.5a. To clarify the confusion, an opportunistic situation has to be created by the agent (the noological system) involved. It has to activate only *one* of the two Type 1 events – i.e., pushing the stick of one of the remote controls, say, A – first, and wait for one or both of the two Type 2 events (i.e., movement of either car C or D, or both) before

a A(Type 1) → B(Type 1) → C(Type 2) → D(Type 2)

b A(Type 1) → C(Type 2) → B(Type 1) → D(Type 2)

Fig. 2.5 (a) A confusion arises as to the connection between the potential causes A and B on the one hand and the effects C and D on the other. (b) Agent carries out "scientific experiments" by activating A and then observes whether it is correlated with C or D

activating the second Type 1 event to clarify which Type 1 event causes which Type 2 event and whether in fact one of the Type 1 events is non-efficacious. The situation is illustrated in Fig. 2.5b. This lies at the foundation of the "controlled" experimental situations scientists often enact to establish various causes-and-effects.

In the above situation, because one of the two types of events involved in the causal relationship is created by the agent (i.e., A(Type 1) and B(Type 1) are actions emitted by the agent), the agent is not confused by the directionality of causality – i.e., suppose two consecutive Type 1–Type 2 event pairs are activated (i.e., Type 1 – Type 2 – Type 1 – Type 2), it cannot be the case that the second Type 1 event is caused by the first Type 2 event because the agent *knows* that it did not activate the second Type 1 event as a result of observing the first Type 2 event. However, there may be situations in which both Type 1 and Type 2 events are external observations made by the agent and the observations take place in the middle of a long chain of alternating Type 1 and Type 2 events. The agent may be confused by the directionality of causality unless there is a much longer interval between the last Type 2 event and a current Type 1 event observed compared to the interval between the current Type 1 event and the next Type 2 event. The agent then assumes it is the Type 1 event that is the cause.

To use a more often encountered real-world event to illustrate this situation, the example of lightning causing thunder is used as shown in Fig. 2.6. In Fig. 2.6, the horizontal axis is time and an event is marked with a yellow vertical bar when it begins (say, when a light or sound appears) and another yellow vertical bar when it ends (say, when a light or sound disappears). On the right side of the figure there are two pairs of observed consecutive lightning-thunder events and on the left side of the figure it is shown that the last event that happened a long time ago was a thunder event (time interval $T1 \gg T2$). Therefore, it is assumed that it is the lightning that causes the thunder and not the other way around. We may never know the "true" correct causal direction – after all, it can still be the last thunder a long time ago causing the current "first" lightning – but this effective causality can be used to correctly anticipate thunder on the observation of lightning. Further observations on the correlation between the distance of the lightning event from the observer and the temporal delay in the ensuing thunder event may also help to establish the directionality. Today, our confidence in the correct directionality is enhanced by a deeper knowledge of nature – the knowledge on electricity, etc.

2.4 Diachronic and Synchronic Causal Factors in Rapid Causal Learning

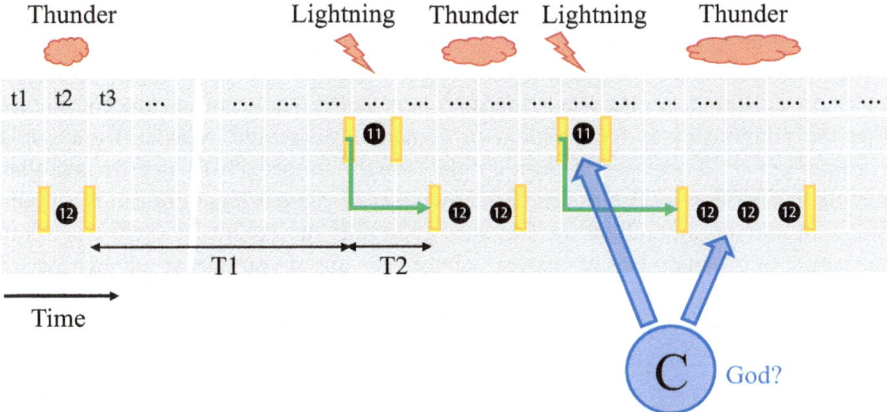

Fig. 2.6 The establishment of the causal relationship between lightning and thunder. Time interval T1 >> T2. *Yellow bars* indicate the coming into existence and the disappearing of a sound or light

Again, in Fig. 2.6 we show a third possible agent – a "God" (as a factor "C" as in Fig. 2.2) – that could be the "true" underlying cause of both lightning and thunder. However, as discussed above, in terms of *effective* causality, it is not necessary to bring this factor in. Other factors may also enter into the consideration of the causal connection between lightning and thunder. If confounding factors are many and incessant – e.g., rain, appearance of dark clouds, etc. – an agent that arrives on this world the first time may be confused initially about what causes what. However, in the case of lightning and thunder, an agent probably does not need a very long time of observation to be able to filter away wind and other potential participating factors in the causal relationship.

2.4 Diachronic and Synchronic Causal Factors in Rapid Causal Learning

Having established how the manner of the basic temporal relationship between two events, the nature of the events (agent-emitted action or observed event), and opportunistic situations can feature in the recovery of a basic causal relationship, we now turn to a discussion on diachronic and synchronic causal factors (Ho 2014).

To motivate the discussion, the example of gravity is used. Since time immemorial, human beings know that when a held object is released in mid-air (an action event), the consequence of it falling (an observed event) always follows. However, consider that an agent discovers this for itself for the first time in a certain context (e.g., a room or a certain location). How does it know that other specific factors – such as being in the room or at a certain location – are not the necessary conditions for the causal relationship? After all, context specific effects must have been observed before – e.g., rain does not wet you if you are in a room. Also, it is prudent not to over-generalize first.

However, the moment that the same release-fall relationship is observed for the object in a different context – e.g., outdoor or at a different location – the confidence that the relationship is independent of context is increased. We term this *rapid causal recovery/learning* – requiring only a small number of instances to establish that the law of gravity is seemingly independent of context/location. But this could still lead to over-generalization. Humanity certainly has been very confident about the nature of gravity – i.e., no matter "where" you are, if you release a held object it will fall – over a long period of time until they discovered a vastly different "where" in which the same gravitational effect does not hold. Firstly, there is the free fall situation in which gravity disappears. There are also locations far away from Earth in outer space that gravity is vastly reduced or becomes non-existent (e.g., a location in between celestial objects at which the gravitational forces cancel out). When this new situation is discovered, the generalized-away context needs to be brought back in some form. Now, the basic rule of gravity becomes contingent upon context and location, even though there is a large number of contexts in which it works – i.e., all of the locations on or near earth's surface, at least, and also in most locations in the cosmos.

We term the earlier discussed pre-condition such as an action or an earlier event that is causally linked with a later event the *diachronic* causal condition (this is a connection across *time*). The contextual factor – not an event, which connotes a change of state of something, but simply a continually present factor – that *enables* the diachronic causal relationship is termed the *synchronic* causal condition. (There is also something called the "synchronic effect" that will be discussed below).

We term the action of bringing back the earlier discarded synchronic conditions, such as the location/context in the case of gravity above, the *retroactive restoration* of synchronic conditions. In the case of gravity, the earlier causal rule that says "no matter where you are, if you release a held object it will fall to the ground" still served humanity very well over a vast, long period of time before the modern understanding came into play. The context is retroactively restored to some extent – "if you are not very far away from earth or in an orbit around earth, the gravity effect takes place." This is the spirit behind *effective* causality. This term is partly inspired by the writings of the physicist Lee Smolin (Smolin 2013) in which he concluded that "all physical theories are *effective* and approximate that apply to *truncations* of nature that cover only a limited domain of phenomenon."

In Fig. 2.7 we provide a similar example to that of the gravity example above. Figure 2.7a shows a force being applied to an object in a room and the object moves. An agent that observes this the first time would assume that the context – the room – is part of the precondition/context for the force to be efficacious. So, the application of the force is the diachronic causal condition, the movement of the object is the diachronic effect, and the room is the synchronic causal condition. A causal rule is formed accordingly. Later, suppose the force application takes place outdoor as shown in Fig. 2.7b and the object likewise moves, the synchronic causal condition of the room is removed and a general rule is obtained: the diachronic effect which is

2.4 Diachronic and Synchronic Causal Factors in Rapid Causal Learning

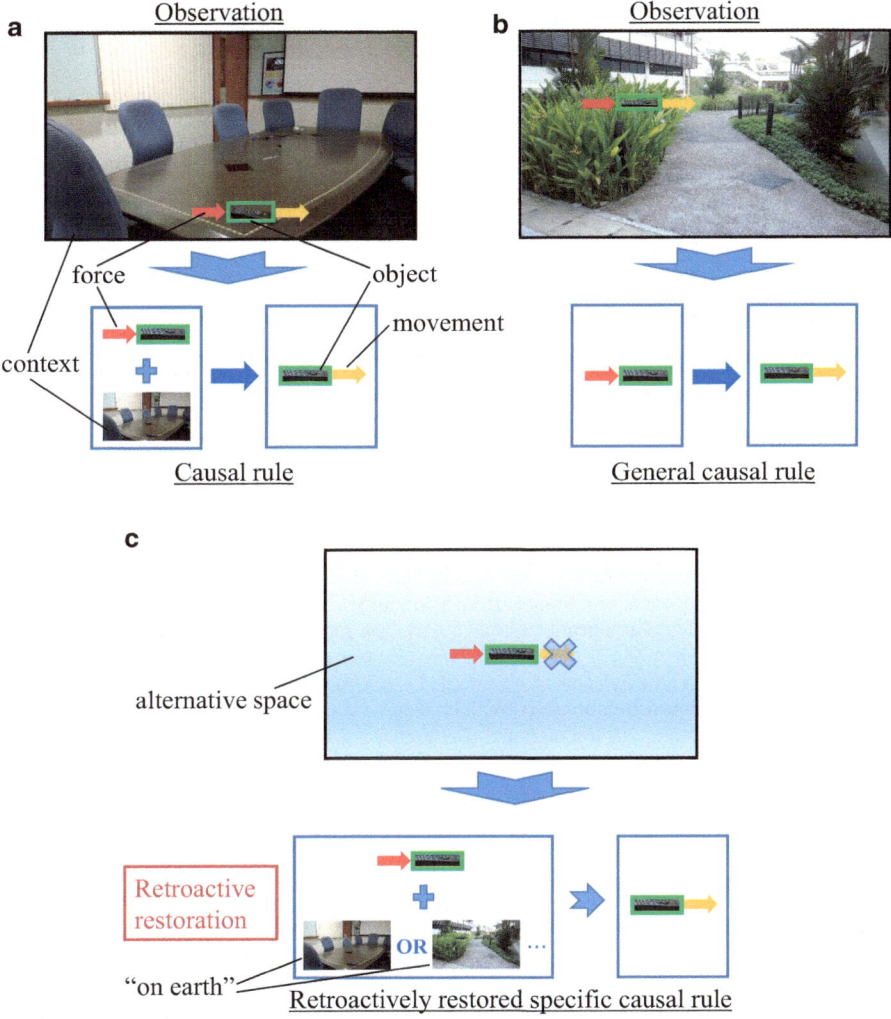

Fig. 2.7 Another example illustrating retroactive restoration of synchronic conditions. (**a**) Initially an object is being moved by a force in a room (a context) and a causal rule is formed with the room being the synchronic causal condition. (**b**) When the force also has the same efficacy in a different environment – outdoor – a more general causal rule is formed with the earlier context discarded. (**c**) If the force loses its efficacy in an "alternative space" (a different context), the disjunctive sum of all the earlier synchronic conditions have to be restored and the causal rule becomes more specific

the movement of an object would *always* follow the presence of a diachronic causal condition which is a force applied to it in *any* context – outdoor, indoor, and wherever. If many more instances like that of Fig. 2.7a, b are experienced, then the confidence in the general rule is increased. Suppose yet some time later, the situation of Fig. 2.7c is observed in some "alternative space" (yet another context)

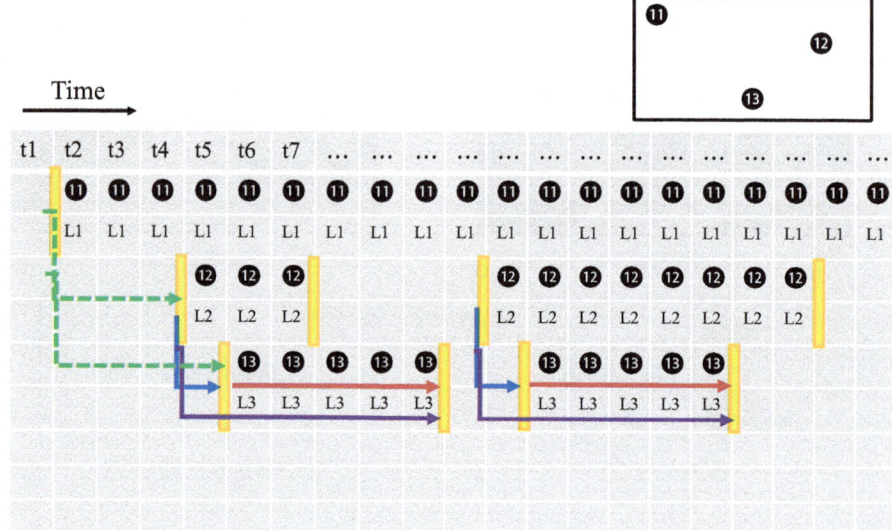

Fig. 2.8 Diachronic and synchronic causal conditions. *Black circles* are objects – $O(11)$, $O(12)$, and $O(13)$, and they appear at or disappear from locations L1, L2, and L3 respectively. *Yellow bars* indicate the appearance or disappearance of an object. The appearance of object $O(13)$ followed by the appearance of object $O(12)$ form a *diachronic* causal relationship ($Appear(O(12), L2) \rightarrow Appear(O(13), L3)$). The existence of object $O(11)$, $Exist(O(11), L1)$, is a *synchronic* enabling condition for the diachronic relationship (With kind permission from Springer Science + Business Media: Proceedings of the 7th International Conference on Artificial General Intelligence, 2014, Page 46, Seng-Beng Ho, Fig. 1.)

in which the force application is *not* followed by a movement of the object involved, then retroactive restoration will add a synchronic causal condition to the force movement rule: *"on earth"* or *"in space as we know it"* (i.e., the disjunctive sum totality of all the contexts that have been experienced before are brought back), a force application to an object would be followed by the movement of the object.[1]

We use Fig. 2.8 to further illustrate the basic idea of diachronic and synchronic causal conditions. In Fig. 2.8, three simple events are shown to take place at three different locations: The appearances and disappearances of three simple objects, $O(11)$, $O(12)$, and $O(13)$, at different locations L1, L2, and L3 respectively are shown in the top right figure as well as in the spatiotemporal representation below. In the spatiotemporal representation, an event – an appearance or a disappearance of an object – is shown with a yellow vertical bar, as in Fig. 2.6. An event is basically a

---

[1] If earlier the places at which the efficacious force was experienced were in Place 1, Place 2, ... to Place N, then the disjunctive sum totality of them would be Place 1 *or* Place 2, ... *or* Place N, and if these are places on earth or "in the space as we know it," then the synchronic precondition is "on earth" or "in space as we know it."

2.4 Diachronic and Synchronic Causal Factors in Rapid Causal Learning

change of state, and in this case it is the change of the state of existence of these objects.

Figure 2.8 shows that object $O(11)$ appears at location L1 between time frames t1 and t2, represented as *Appear*($O(11)$, L1). And then, between time frames t4 and t5, object $O(12)$ appears at location L2, and between t5 and t6 object $O(13)$ appears at location L3. Now, an agent observing this could tentatively establish a possible causal relationship between *Appear*($O(11)$, L1) and *Appear*($O(12)$, L2) (represented as a *causal rule Appear*($O(11)$, L1) → *Appear*($O(12)$, L2)) and another possible causal relationship *Appear*($O(12)$, L2) → *Appear*($O(13)$, L3).

Suppose now further along, objects $O(12)$ and $O(13)$ disappear ($O(12)$ disappears first followed by $O(13)$) and then appear again with the same temporal order and the same temporal delay as in the earlier situation – i.e., object $O(12)$ appears first followed by $O(13)$ after one time frame. Suppose also that object $O(11)$ does not disappear and continues to exist in the time interval of interest. In this time interval of interest, it is shown that objects $O(12)$ and $O(13)$ later disappears, but in the reverse order as their earlier disappearances (object $O(13)$ disappears first).

Even though in the time interval of interest in Fig. 2.8, the appearance of object $O(12)$ the second time is not followed from the appearance of object $O(11)$ as in the first time (at t5, i.e.), using the probabilistic formula of Fig. 2.4 we can still accord a certain likelihood for the causal rule *Appear*($O(11)$, L1) → *Appear*($O(12)$, L2). This likelihood will reduce to an insignificant level should further observations reveal that *Appear*($O(11)$, L1) and *Appear*($O(12)$, L2) have no more temporal relationship like what was observed in the time frames t1–t5. The process is similar for the relationship between *Appear*($O(11)$, L1) and *Appear*($O(13)$, L3). (Both of these relationships or the lack of them are shown in dotted green arrows in Fig. 2.8).

Let us now focus our attention on the causal rule *Appear*($O(12)$, L2) → *Appear* ($O(13)$, L3). This relationship is observed twice and the attendant causal rule has a high degree of confidence as both times the two objects involved appear with the same temporal delay (100 % rule Confidence, based on the calculation of Fig. 2.4). *Appear*($O(12)$, L2) and *Appear*($O(13)$, L3) are then said to be in a *diachronic* causal relationship. (These are indicated in blue arrows in Fig. 2.8). *Appear*($O(12)$, L2) is the diachronic causal condition and *Appear*($O(13)$, L3) is the diachronic *effect*. There are also other causal relationships as indicated by red and purple arrows in Fig. 2.8.

Now, as mentioned above, when the diachronic relationship *Appear*($O(12)$, L2) → *Appear*($O(13)$, L3) takes place, there is one *synchronic* factor that is "always" present – the existence of object $O(11)$ at location L1 (at least within the interval of time shown in Fig. 2.8). Therefore, the current synchronic and diachronic preconditions for the causal relationship is *Exist*($O(11)$, L1) Λ *Appear*($O(12)$, L2) → *Appear*($O(13)$, L3). (A fuller representation includes the labeling of the type of causal condition – *Exist*($O(11)$, L1) [SYN] Λ *Appear*($O(12)$, L2) [DIA] → *Appear* ($O(13)$, L3) – but this is omitted in most places in the following discussion.)

Now, consider the situation in Fig. 2.9, in which *Exist*($O(11)$, L1) is not always the case when *Appear*($O(12)$, L2) → *Appear*($O(13)$, L3) takes place. In a second instance of *Appear*($O(12)$, L2) → *Appear*($O(13)$, L3), instead another object O

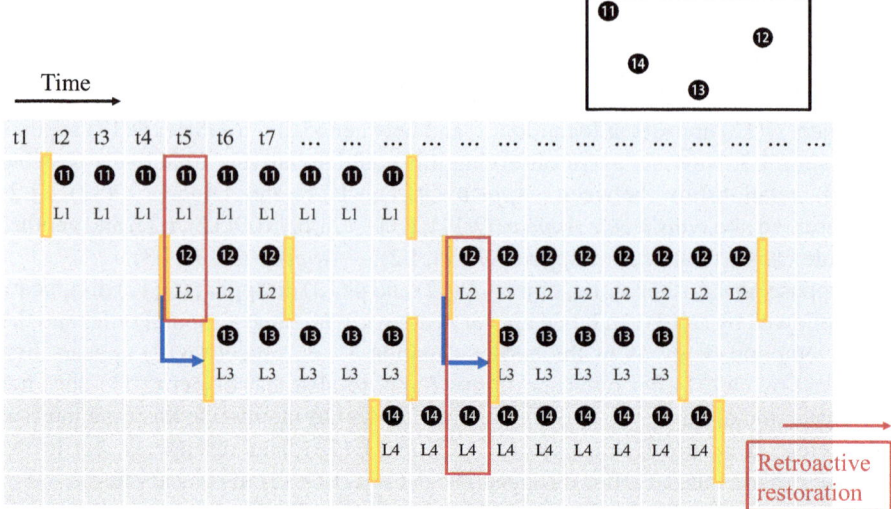

Fig. 2.9 Retroactive restoration of disjunctive synchronic causes. The *dark red vertical boxes* show the synchronic preconditions involved (With kind permission from Springer Science + Business Media: Proceedings of the 7th International Conference on Artificial General Intelligence, 2014, Page 48, Seng-Beng Ho, Fig. 2)

(14) is present at location L4 – *Exist*(O(14), L4). In our framework of rapid effective causal learning, we generalize over the factors that are not always present and both the conditions – *Exist*(O(11), L1) and *Exist*(O(14), L4) – are removed as potential synchronic preconditions for *Appear*(O(12), L2) → *Appear*(O(13), L3).

However, strictly speaking, at this point in time we cannot be sure that the two potential synchronic preconditions *Exist*(O(11), L1) and *Exist*(O(14), L4) are not *disjunctively* necessary – i.e., it could have been *Exist*(O(11), L1) or *Exist*(O(14), L4) before *Appear*(O(12), L2) → *Appear*(O(13), L3) can take place. If at a future time outside the time interval shown in Fig. 2.9, when neither *Exist*(O(11), L1) or *Exist*(O(14), L4) is true and the diachronic cause *Appear*(O(12), L2) no longer causes *Appear*(O(13), L3), then we would need to activate a *retroactive restoration* of *Exist*(O(11,) L1) *or Exist*(O(14), L4) as *disjunctive* synchronic preconditions for *Appear*(O(12), L2) → *Appear*(O(13), L3). That is: (*Exist*(O(11), L2) V *Exist*(O(14), L4)) [SYN] Λ *Appear*(O(12), L2) [DIA] → *Appear*(O(13), L3).

This is reminiscent of the examples of gravity or force discussed above – when we discovered that the law of gravity in its original form did not work in certain conditions – in a free fall orbit or certain locations in outer space – we restore the earlier discarded synchronic preconditions in a disjunctive arrangement. Suppose we had experienced gravity in its original form (i.e., when an Object is released, it falls with a certain acceleration, AccelX – *Release*(Object) → *Fall*(Object, AccelX) – in all the places, Places 1 – N, on Earth, then the restored disjunctive synchronic condition is (Place 1 V Place 2 V ... Place N) and the corresponding causal rule is: (Place 1 V Place 2 V ... Place N) [SYN] Λ *Release*(Object) [DIA] → *Fall*(Object,

2.4 Diachronic and Synchronic Causal Factors in Rapid Causal Learning

AccelX). The disjunctive sum totality of Places 1- N on Earth is non-other than Earth itself. Therefore, the causal rule can be rewritten: *On-Earth*(Object) [SYN] $\wedge$ *Release*(Object) [DIA] $\rightarrow$ *Fall*(Object, AccelX).

Suppose A, B, C, D, E, ... are synchronic causes, the above method can recover arbitrary combinations of these causes such as "(A V B) $\wedge$ (C V D V E)..." Also, if a series of disjunctive synchronic causes are parameter values that are close in values, (e.g., a location value LX = 1.1 or 1.2 or 1.3, ...), they can be combined into a range (e.g., "1.1< LX <2.0"). This is similar to the situation discussed above in connection with gravity in which the object being at various places on Earth are combined into *On-Earth*(Object).

In the above example of events and causal rules in Fig. 2.8, we have shown that the same objects had always appeared at the same location. The locations of objects themselves are also a kind of synchronic precondition. Suppose the objects can appear at different locations. Now, suppose we observe that both *Appear*($O(12)$, L2) and *Appear*($O(12)$, L2') lead to *Appear*($O(13)$, L3), then the location of object $O(12)$ when it appears can be discarded as a synchronic causal condition – i.e., *Appear*($O(12)$, *) $\rightarrow$ *Appear*($O(13)$, L3) – the "*" representing a "don't care" condition. I.e., no matter where the object $O(12)$ appears, it always leads to *Appear*($O(13)$, L3). Similarly, suppose object $O(13)$, when caused by object $O(12)$ to appear, can appear anywhere, then we relax the *synchronic effect* – *Appear* ($O(12)$, *) $\rightarrow$ *Appear*($O(13)$, *) – which basically says that the appearance of object $O(12)$ anywhere can lead to the appearance of object $O(13)$ anywhere.

A general algorithm that attempts to establish causalities between events could begin with an observed event and look backward in time to look for a potential cause (See Appendix B.) In this framework, it is interesting to note that suppose the scenario depicted in Fig. 2.8 is a micro-universe with time beginning at t1, then *Appear*($O(11)$, L1) would not have any preceding identifiable cause. It could have spontaneously emerged as a "first-cause" of all subsequent activities and events in the micro-universe. This is the customary concept of "God" in many cultures. Or, it might be postulated that the cause of *Appear*($O(11)$, L1) is outside the space-time continuum of Fig. 2.8, and that would be the first cause, the concept of God, outside the known space-time continuum.

There is a situation that is a dual of the retroactive restoration situation depicted in Fig. 2.9. Figure 2.10 depicts a situation in which there was an earlier observation (in the t2–t4 time frame) that when object $O(12)$ appeared at location L2 alone, it did not cause another event. Subsequently, when object $O(11)$ was present at L1, object $O(13)$ appeared shortly after object $O(12)$'s appearance. Now, when another *Appear*($O(12)$, L2) $\rightarrow$ *Appear*($O(13)$, L3) event takes place later without the presence of object $O(11)$ but with the presence of object $O(14)$, much like in the situation of Fig. 2.9, *Exist*($O(11)$, L1) or *Exist*($O(14)$, L4) are *proactively installed* as the disjunctive causal condition of *Appear*($O(12)$, L2) $\rightarrow$ *Appear*($O(13)$, L3), instead of being removed as in the initial situation in Fig. 2.9. This is like a *counterfactual* causal observation: "Had object $O(11)$ or object $O(14)$ not being present, *Appear*($O(12)$, L2) would not have caused *Appear*($O(13)$, L3)."

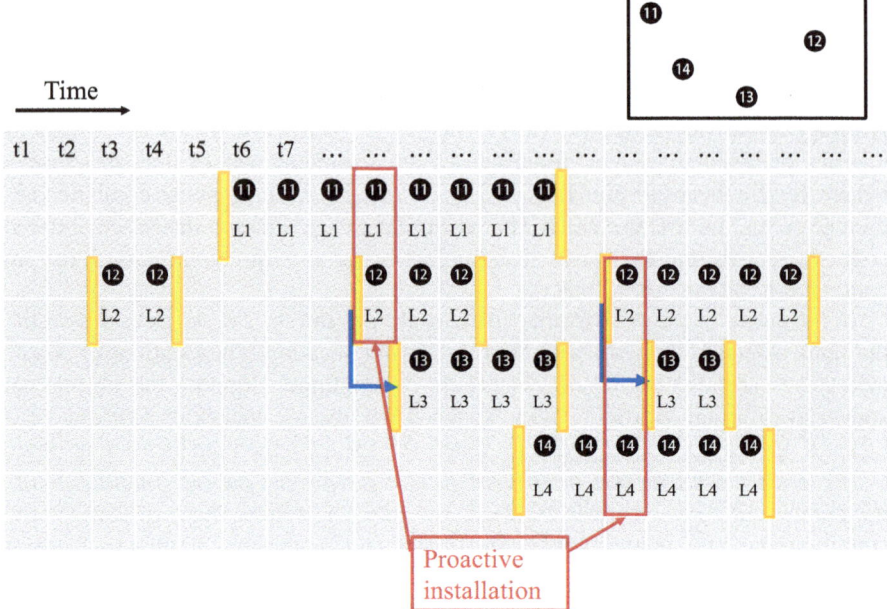

Fig. 2.10 Proactive installation of disjunctive synchronic causes – the disjunctive synchronic condition (*Exist*(11, L2) V *Exist*(14, L4)) is kept as there is an earlier situation in which the appearance of object $O(12)$ itself did not cause object $O(13)$ to appear

In the situations we considered above in Figs. 2.8, 2.9, and 2.10, there were always only one diachronic cause/causal condition. There could also be situations in which more than one diachronic cause participates in a diachronic causal relationship. Figure 2.11 shows a situation in which the appearance of object $O(12)$ at location L2 in the presence of object $O(11)$ at location L1 did not cause the appearance of object $O(13)$, but when another event, the appearance of $O(14)$ at location L4 occurred at the same time as the appearance of object $O(12)$ at L2, then object $O(13)$ subsequently appears. The rule is: *Exist*($O(11)$, L1) [SYN] ∧ (*Appear*($O(12)$, L2) ∧ *Appear*($O(14)$, L4)) [DIA] → *Appear*($O(13)$, L3). The diachronic condition consisting of the two appearances is a *conjunctive* diachronic condition.

An often encountered situation in the real world that corresponds to the situation in Fig. 2.11 is shown in Fig. 2.12. In the top right corner of the figure, it is shown that there are two forces acting on an object $O(11)$ simultaneously and the object moves by an elemental step. Had there been only one of the two forces present, the object would not have moved. The causal-temporal description of the situation is shown below. The momentary appearance of a force is shown as an appearance and disappearance event over one time frame. The force has two parameters, a magnitude, F1, and a location, L1. The object $O(11)$ starts at location L1. Earlier, when only one force appeared, object $O(11)$ did not move. When two forces appear later, object $O(11)$ moves from L1 to L2. (Assuming in this case there is friction and the object does not have the momentum to continue the movement farther. The case in

2.5 Desperation, Generalization, and Rule Application

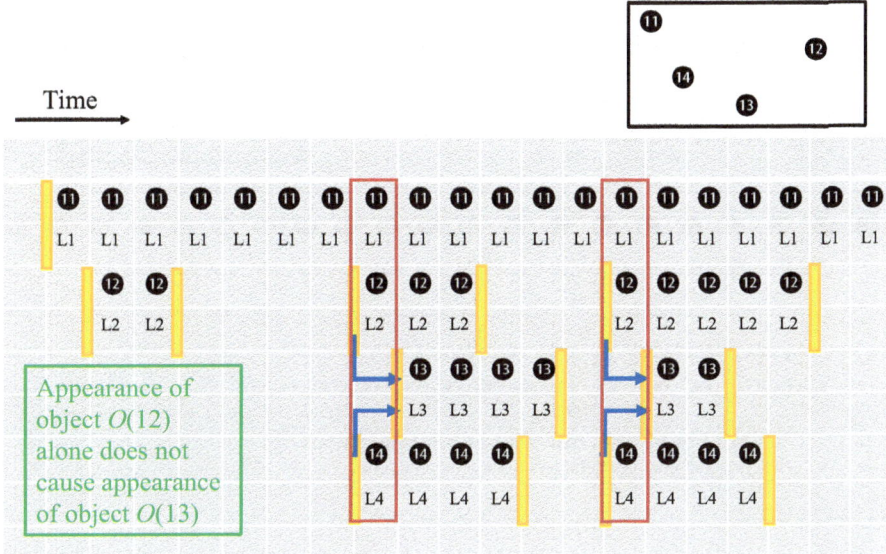

Fig. 2.11 Conjunctive diachronic condition

which the object has a momentum and it continues to move for a long distance after the application of a force is considered in Chap. 4.) These two forces participate in a *conjunctive* diachronic causal condition.

There is a corresponding *disjunctive* diachronic condition. If the same event can be triggered by one *or* another event, then both the preceding events participate in a disjunctive diachronic condition for the causal rule.

2.5 Desperation, Generalization, and Rule Application

In the discussions above in connection with Fig. 2.9, we have used what is termed "dual-instance" generalization – i.e., we generalize over the factors that are not always present and both the conditions – $Exist(O(11), L1)$ and $Exist(O(14), L4)$ – are removed as potential synchronic preconditions for $Appear(O(12), L2) \rightarrow Appear(O(13), L3)$ after comparing situations in two instances. (Based on the formula of Fig. 2.4, the generalized rule also has a Confidence of 100%). Whether a noological system needs to effect a generalization to derive a general form of a rule depends on how *desperate* it is to apply the rule. Consider the situation in which a noological system is dying of hunger and there is an action that can allow it to acquire food (i.e., the action *causes* food to be acquired.) Now, suppose the system has only observed one instance of the action → food rule. If it is desperate it will carry out "single-instance" generalization – i.e., assume that the action can indeed lead to food and proceed to try the action. Because there had been only one

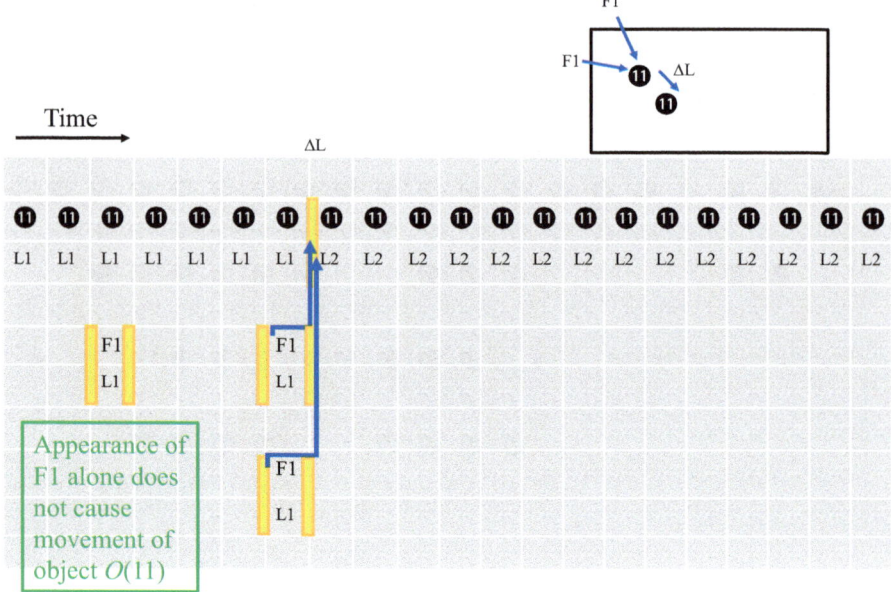

Fig. 2.12 Forces participating in a conjunctive diachronic condition

instance of observation, the confidence in the rule is small, but because of the state of desperation it is in, it goes ahead and effects the generalization anyway – i.e., assumes the rule is true and applies it.

Figure 2.13 shows three situations – high, medium and low desperation situations – in which generalization is effected. As mentioned above, in a high desperation situation, single instance generalization is effected. In a medium desperation situation, generalization can be based on two or a few instances (dual+ instances) of observation. If the system is in a low desperation situation, then it can afford to wait for more observations.

The graph on the top right side of Fig. 2.13 shows that the state of desperation is a function of time and situation. The graph on the bottom right side of the same figure shows how a specific kind of desperation – Desperation(Hunger) – is derived from the state of Hunger of the system.

Another emotion, *eagerness*, is related to desperation. If one is eager to achieve some effects, one is also prone to carry out generalization based on fewer instances.

As noted in Chap. 1, the process of generalization cannot be divorced from affective processes such as those that might lead to emotions such as desperation and eagerness. The discussion in this section demonstrate just such a connection between desperation and generalization.

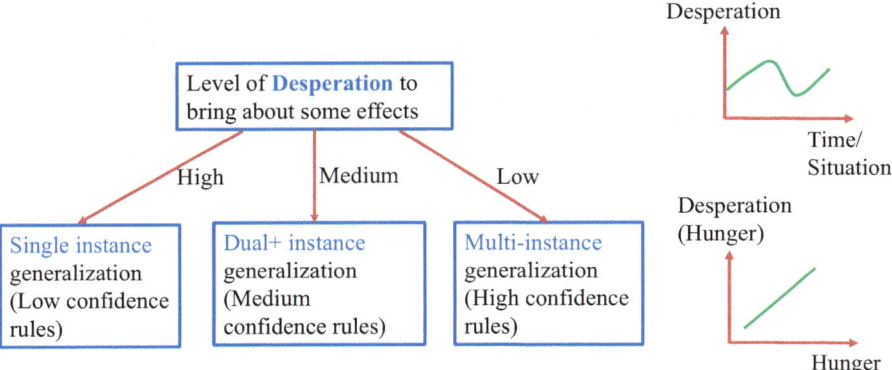

Fig. 2.13 Desperation and generalization – different levels of desperation determine how many instances of a cause-effect situation need to be observed before generalization is effected. The two graphs on the *right side* of the figure show the change of desperation with respect to time and situation and the derivation of a specific type of desperation – hunger desperation – from the Hunger signal of a noological system

2.6 Application: Causal Learning Augmented Problem Solving Search Process with Learning of Heuristics

In this book we will be applying the basic rapid unsupervised effective causal learning method discussed above to a few problems often addressed in AI to demonstrate its viability. In this section we apply the method to a so-called "spatial-movement–to-goal" (SMG) problem and we relegate the application of this method to other problems in subsequent chapters (e.g., Chaps. 6, 7, and 9). The SMG problem is typically addressed in AI using the A* heuristic search algorithm (Hart et al. 1968). Here, we are assuming that there are no obstacles involved in the SMG problem and the application of this method to a typical variant of the basic spatial-movement-to-goal problem – the "spatial-movement-to-goal-with-obstacle" (SMGO) problem – is relegated to Chap. 6. The A* algorithm does not make use of causal information to improve the search and problem solving efficiency. In this section we show how the application of the rapid effective causal learning method can greatly improve the efficiency of the forward search and problem solving processes.

2.6.1 The Spatial-Movement-to-Goal (SMG) Problem

Figure 2.14a illustrates the SMG problem. Basically, an agent, starting from a certain START location in space, is supposed to find a way to move to a GOAL location. As mentioned above, in this section we assume there are no obstacles between the START and GOAL locations, and the case with obstacles – the SMGO

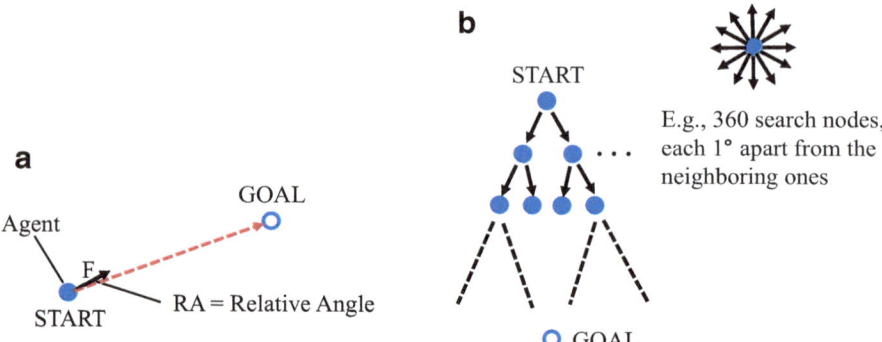

Fig. 2.14 (**a**) The spatial-movement-to-goal (*SMG*) problem. (**b**) A search process for solving the problem

problem – will be treated in a later (Chap. 6). To solve this problem, one way is to rely on a built-in a heuristic rule that says that the best solution is to move in a straight line continuously from the START to the GOAL position. However, that would be too contrived. In fact, this begs the question of how this heuristic rule could be learned or emerge from a problem solving process in the first place. Below, we show that in fact a heuristic like that can be learned from a search process enhanced with effective causal learning, which is a general learning framework as discussed above. And, subsequently, an agent can then always apply the learned heuristic, obviating any further need to carry out search.

Figure 2.14b shows how a typical search process can be used to solve the problem. In the worst case, a *blind* search process can proceed as follows. First, all possible directions of movement of the Agent from the START location is considered. In our framework, the fact that a force, either internally generated within the Agent or externally applied to the Agent, can be used to move the Agent across space is learned from experience through the causal learning process (such as that illustrated in Fig. 2.12), whereas in a traditional AI search problem this piece of knowledge is built-in. Suppose the system considers only 360 possible directions – 1° apart – as shown in Fig. 2.14b. Suppose also that the system considers an elemental move (i.e., an elemental spatial displacement) each time. This would take the Agent to 360 elementally displaced locations from the START location, only two of which are shown in the search tree in Fig. 2.14b for clarity. Then, from each of these 360 locations, each of which is called a "node" in the search process, a further consideration of 360 possible locations "expanded" out from this current node is carried out. Out of these 360 possible elemental moves, one of them would bring the Agent back to the original node from which the current node was expanded. The search process can ignore this and in a subsequent expansion considers only the other 359 nodes. This process continues until one of the moves reaches the GOAL location. If the record to all the elemental moves is kept in the search process, when the expansion finally reaches the GOAL location,

2.6 Application: Causal Learning Augmented Problem Solving Search Process with...

the path making up of all the elemental moves from the START to the GOAL locations would be the solution to the problem, and this would turn out to be more or less a straight line from the START to the GOAL. (It would be "more or less" a straight line and not an "exact" straight line due to the discrete number of angles considered – 360 discrete angles – for the elemental movement expansion from each mode.)

The blind search suffers from a major problem – the search space is incredibly big. Therefore, typically an A* search algorithm (Hart et al. 1968) is used that would vastly reduce the search space. This algorithm uses a basic search framework augmented by a "heuristic" measure at any given point in the search. The heuristic is typically a general domain independent heuristic such as "pick a next move that is closest to the GOAL." (In the SMG problem, this "closeness" translates into a distance measure from the current location of consideration to the GOAL. In other problems, there might be other kinds of measures for closeness to the GOAL.) The A* search algorithm assumes that there is a way, for each domain encountered, to generate such a heuristic measure.

Our *causal learning augmented search algorithm* will take advantage of this heuristic measure feature of A* search but it is able to further reduce the entire search space drastically through the general causal learning mechanism described above in Sect. 2.4. Let us first consider the typical amount of computation needed in an A* search for this SMG problem that is built upon the basic search framework depicted in Fig. 2.14b. In A*, firstly, similar to that discussed above, at each node 360 nodes are expanded (strictly speaking 359 nodes as explained above, but the difference is immaterial), each of which corresponds to an elemental movement in one of the directions. Then, having expanded the node, a selection of a "best" node out of the 360 expanded nodes for further expansion is carried out. Since the purpose is finally to reach the GOAL, a reasonable heuristic for a current node would be the distance from the location corresponding to the current node to the GOAL – h(node). This selected best node – the one with the smallest heuristic value – is then expanded and the same process is repeated. A* typically also includes the distance from the START node to the current node as part of the heuristic measure as well – g(node). The heuristic measure used is then f(node) = g(node) + h(node). Because not every node at every level of expansion is selected for further expansion, which would be the case for a totally blind search as discussed above, and only the heuristically selected node is selected for further expansion, this A* algorithm reduces the search space drastically compared to that of the blind search. This is a good baseline against which to compare any further improvement to the search process. Even though the A* search represents a great reduction in the search space compared to the blind search, it still has to *expand* the heuristically selected node into 360 nodes at each level of the search. This could still represent a lot of computational effort, especially if the goal is far from the starting location.

Now, consider the process of the causal learning augmented search depicted in Fig. 2.15. The top part of the figure shows a physical scenario which is a process of an Agent searching for a path to go from a START position to a GOAL position. In line with our general causal learning framework, through causal learning through

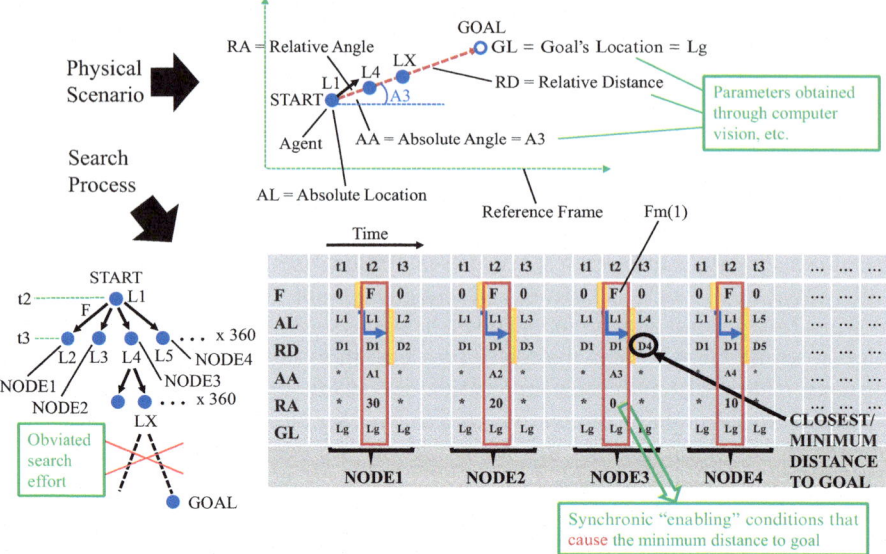

Fig. 2.15 Causal learning augmented search process with learning of heuristics applied to the SMG problem. See text for explanation. "*" represents undefined value (With kind permission from Springer Science + Business Media: Proceedings of the 7th International Conference on Artificial General Intelligence, "On Effective Causal Learning," 2014, Page 49, Seng-Beng Ho, Fig. 3)

experience (e.g., such as that shown in Fig. 2.12), it is learned that a "force" – "F" – can physically push the Agent involved toward an elemental next location in the process of finding a good path to the GOAL. This could be the situation of an agent actuating a force within itself to propel itself or an agent applying a force externally to push another agent. Throughout the discussion below, we will use the case of the Agent propelling itself to represent both situations.

Suppose the sensory system (computer vision, natural vision, proprioceptive sensors, etc.) is able to supply the various parameters associated with the Agent and the force. These would be parameters such as the absolute location of the Agent, AL, the relative distance of the Agent to the GOAL, RD, the absolute angle (with respect to the entire frame of reference), AA, of the applied force, and the relative angle, RA, of the applied force with respect to a straight line joining the START and GOAL positions. These supplied parameters from the sensory system(s) would also include the START and GOAL locations, L1 and GL, respectively.

The bottom left corner of the figure has a search tree that represents the basic search framework. The picture depicts four nodes expanded at the first level – NODE1 – NODE4 – and these respectively correspond to the locations of L2, L3, L4, L5 to which the object is moved by F from the START location, L1. Similar to the A* process described above in Fig. 2.14, we also assume that there are 360 possible directions to move the Agent to and they are separated by 1°. There

2.6 Application: Causal Learning Augmented Problem Solving Search Process with...

is also a temporal aspect to the process – at t1 the object stays stationary at L1, at t2 F is applied, and at t3 the object is moved to one of the 360 possible locations.

On the bottom right of the figure is a tabular representation of the search process much in the spirit of the temporal representation of Figs. 2.6 and 2.8, etc. showing the causalities involved. In this representation, each node and the parameters associated with it occupy a small segment of the temporal space. The physical process associated with each node is a force, F, applied between t1 and t2, which is a change of state or an event labeled with a short yellow vertical bar. The parameters associated with the force are the AA and RA which are defined only at t2 when the force is in effect. The starting parameters AL and RD of the Agent are indicated in the t1 column and after the application of the force, AL and RD change to other values in t3. These changes are also indicated with a short yellow vertical bar. In the parlance of our effective causal learning framework described in Sects. 2.2, 2.3, and 2.4, it is established that F *causes* the particular movement or the changes in the associated parameters AL and RD of the Agent and is a *diachronic* cause. The *synchronic* "enabling" causal conditions are the values of all the parameters observed in the environment at t2 which include AL, RD, AA, RA, and GL.

Thus, at NODE1, say, the force was applied with its associated synchronic parameter values F(L1, D1, A1, 30, Lg) and this causes a movement in which the Agent moves from (L1, D1) to (L2, D2) – *Move*(Agent, L1, D1, L2, D2). ("30" is the value of the relative angle RA, which is 30°, that corresponds to NODE1.) Applying the nearest distance to goal heuristic, NODE3 is chosen for further expansion in which the Agent is at (L4, D4) which is the closest location to the GOAL among all the locations (L2, D2), (L3, D3), (L5, D5), ... corresponding to the other nodes NODE1, NODE2, NODE 4, ... respectively. Physically, this node corresponds to the Agent at location L4, which lies on a straight line connecting START to GOAL as shown in the physical scenario. This is a movement in the absolute direction, AA = A3, say. The associated force for NODE3 with its synchronic parameters is F(L1, D1, A3, 0, Lg). We use Fm(N) to represent the force that *causes* the resultant location to be selected by the minimum distance heuristic at level N of the node expansion and hence Fm(1) = F(L1, D1, A3, 0, Lg).

Next, consider that this NODE3 has been selected for expansion and the process creates a next-level 360 nodes for evaluation by the minimum distance heuristic. Suppose now it is the node at this expanded level that corresponds to the Agent at location LX shown in the physical scenario that satisfies the heuristic of minimum distance from the GOAL. This location LX is arrived at by applying F in the direction of A3 from location L4. (This direction actually lies on the straight line connecting the START and the GOAL locations and that is why it is the one selected by the heuristics but we want the system to discover this for itself.) The full synchronic parametric specification of the force for this movement to LX is Fm (2) = F(L4, D4, A3, 0, Lg). Now, comparing this force and the one above, Fm(1) = F(L1, D1, A3, 0, Lg), that also satisfies the heuristic but at a different level of node expansion and absolute location, one can see that some of the synchronic parameters are different – L1 vs L4 and D1 vs D4 – while others stay the same – A3, 0, and

Lg. Both these forces *cause* the selection of the resultant node by the minimum distance heuristic. Using the rapid effective causal learning process described above (Sects. 2.2, 2.3, and 2.4), through dual-instance generalization (at a medium level of desperation, say), we derive a more general force rule that causes the selection of the resultant node by the minimum distance heuristic, namely Fm(*) = F(*, *, A3, 0, Lg). The "*" indicates that the corresponding parameters, namely N (the level of search expansion), AL (absolute location) and RD (relative distance to the GOAL from the Agent current location), are *not relevant* as synchronic enabling conditions for the force that causes the resultant node to be selected by the heuristic. The "*" in Fm(*) implies that no matter at what *level N* of search expansion, the generalization holds.

Therefore, at this point, after two levels of expansion, the general causal rule learned is that no matter at what level of expansion, the force needed to satisfy the minimum distance heuristic is F(*, *, A3, 0, Lg), i.e., Fm(*) = F(*, *, A3, 0, Lg). Hence, for the next level onward, there is no further node expansion needed. One just needs to keep applying F(*, *, A3, 0, Lg) – i.e., whatever location (represented by the parameter AL and RD that are now labeled "*") is reached at whatever level of search expansion, one just keeps applying the force in the A3 direction – namely the direction pointing toward the current GOAL at Lg – with a relative angle RA = 0 that is subtended between the direction of the force and the straight line connecting START and GOAL. The last enabling synchronic condition, Lg, is always satisfied as the GOAL is stationary. This indirectly implies that the A3 direction is the correct one provided the GOAL does not move in the process of the search, otherwise the desired AA will keep changing as the Agent continues its movement in the direction where the GOAL moves to. Hence there is just one node that needs to be expanded/considered at all levels of movement all the way to the GOAL. This obviates the need for continued expansion of 360 nodes at every level of node expansion all the way to the GOAL, which is needed in A* (shown with a big red cross in the search tree in the figure), and drastically reduces the search space compared to A* search – only two levels of 360 node expansions each are needed in total.

Now, consider that the entire process is repeated at a different START (L11) and GOAL (Lg1) location as shown in Fig. 2.16 with a different absolute angle, A31, for the straight line joining START and GOAL, and a different starting relative distance between START and GOAL. Let us define this as a different *situation*. Now, in this situation, Fm(1) = F(L11, D11, A31, 0, Lg1) and Fm(2) = F(L41, D41, A31, 0, Lg1), assuming that the first minimum distance heuristic selected node corresponds to location L41, and the corresponding relative distance to GOAL is D41. This results in a generalization to Fm(*) = F(*, *, A31, 0, Lg1) and it obviates any expansion beyond the second level just like the situation above in Fig. 2.15.

A further generalization can be made after comparing the above two situations of Figs. 2.15 and 2.16 (let us call them Situation(1) and Situation(2)). Fm(*)[Situation (1)] = F(*, *, A3, 0, Lg) and Fm(*)[Situation(2)] = F(*, *, A31, 0, Lg1) implies that Fm(*)[Situation(*)] = F(*, *, *, 0, *). What this means is, in general in *any* situation

2.6 Application: Causal Learning Augmented Problem Solving Search Process with...

Fig. 2.16 A different START and GOAL location (a different *situation*) for the SMG problem

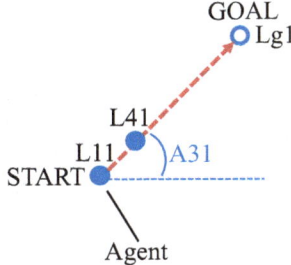

with *any* START and *any* GOAL locations, the solution is to move straight in the direction of the GOAL.

In a sense, $F(*, *, *, 0, *)$ is a very general *domain specific heuristic* (in the domain of spatial movement) that is derived from the general causal learning augmented search process. After these two situations and a total of two times of 360 node expansions, this heuristic extracted *obviates all further search in* any *SMG problem without obstacles*. This heuristic is thus learnable through the causal learning process to expedite all future problem solving processes as regards spatial movement to goal and does not have to be built-in in a contrived manner. As for the basic A*, not only much more search is required in each *situation*, the entire A* search has to be repeated in a different situation when the START and GOAL locations are changed. There is no transfer of any kind of information or learning from one situation to the other. (We can think of the learning of heuristics as a kind of transfer learning.)

Thus, in this section, we have not only demonstrated that it is possible to drastically reduce the amount of search through rapid effective causal learning, which results in a kind of search process that is more similar to a typical human intelligent search process, it is also possible to learn the heuristic(s) involved that will further reduce the need for search (and in fact in this case of the SMG problem, obviate entirely the need for search) for future similar situations.

In the above discussion of using either the conventional A* search or our causal learning augmented search for the SMG problem, we have assumed that the system has already encoded the knowledge of using a force to cause an elemental movement and the search process uses this elemental movement rule to try various possibilities to arrive at the solution. As mentioned above, the learning of the elemental movement rule, in fact, can also be achieved using causal learning as discussed in Sect. 2.4 in connection with Fig. 2.12.

2.6.2 Encoding Problem Solution in a Script: Chunking

In the solution for the case of the SMG problem as discussed above or for that matter, for the case of any general and typical problem, there is usually a series of actions to be performed. An efficient strategy for a noological system would be to encode this solution so that no repeated future effort in looking for the solution is needed should the same problem situation be encountered again. Moreover, this sequence of actions forms a knowledge "chunk" that can participate as part of the solution to more complex problems that require even longer sequences of actions to solve. That way the solutions to ever increasingly complex problems can be discovered faster and more efficiently. This "incremental" chunking method will be described in more detail in Chap. 3. In this section, we focus on describing the structure of a "script" that encodes this chunked piece of knowledge.

There are two ways a script can be learned/created. One is through a problem solving process of an agent as discussed above. The other is through observation of other activities that may be performed by other agents in their problem solving processes using basically the same effective causal learning process as discussed above in Sects. 2.2, 2.3, and 2.4, except that it may need to be applied to an extended period of time that results in a longer sequence of activities. This will be described in detail in Chap. 7.

There is some similarity between our idea of scripts and that proposed by Schank and Abelson in their work in the 1970s (Schank and Abelson 1977). The main difference in the idea of scripts here compared to that of Schank and Abelson is that the scripts here are learned from an agent's problem solving processes or from observations of activities taking place in the environment which could in turn be executed solutions of other agents' problem solving processes. In Schank and Abelson (1977), no detailed computational process was described for the learning of scripts. In addition, the focus of Schank and Abelson's application of scripts is on question-answering, and for us, it is on problem solving. As a result, there are other elements in our version of the script, such as counterfactual information, which will be discussed in Sect. 2.6.3 and which are not present in Schank and Abelson's script.

The basic structure of a script here consists of four main portions – SCENARIO, START, ACTIONS, and OUTCOME – as shown in Fig. 2.17a, which is based on the movement scenario of the SMG problem.

For comparison, Fig. 2.17b shows a part of a script structure devised by Schank and Abelson (Schank and Abelson 1977) and in this particular instance, it is a Restaurant Script – encoding knowledge about going to a restaurant. We use blue rectangles to indicate the portions of their script that correspond to the various parts of our script in Fig. 2.17a. The Restaurant Script consists of an Entry conditions portion, which corresponds to our START condition, that specifies the start state of the main role – the person S visiting the restaurant – and in this case it specifies that the person is hungry and has money. Though not specified in Fig. 2.17b, the START state could conceivably also include the state of the restaurant – such that it is located within a reachable location and that it is open. The Results portion of the script corresponds to our OUTCOME portion. In this case, the person S ends up having less money, not hungry, and so on. The Scene 1 portion and the rest of the script in Fig. 2.17b that are not shown correspond to our ACTIONS portion which consists of a causal chain of actions and events specifying, in this case, the activities

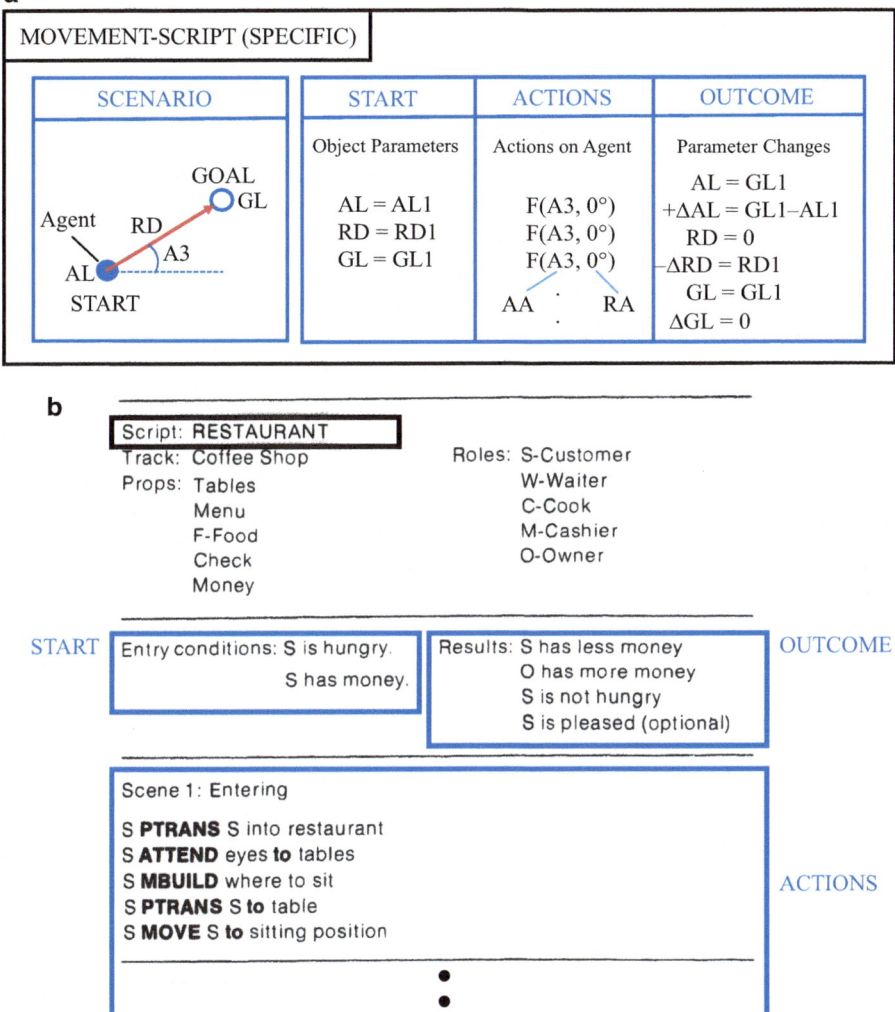

Fig. 2.17 (**a**) A specific SMG Script – MOVEMENT-SCRIPT (SPECIFIC). All parameters AL, AA, RD, RA, and GL have the same definitions as those in Fig. 2.15. (**b**) Part of Schank's Restaurant Script (Republished with permission of Taylor and Francis Group LLC Books, from Scripts Plans Goals and Understanding, Roger Schank and Robert Abelson, 1977; permission conveyed through Copyright Clearance Center, Inc.) The blue rectangles and words show correspondences with our version of script. See Fig. 10.1 in Chap. 10 for a complete version of the Restaurant Script

that take place in a restaurant. The header portion of the script corresponds approximately to our SCENARIO portion.

In our process, a script is first created from a problem solving process by encoding the specific values of the parameters involved. For example, when the problem situation of a SMG problem is presented to/encountered by an Agent the first time, the Agent has a specific START location, say AL1, and a specific relative distance to the GOAL, say, RD1. The GOAL also has a specific location, say GL1, as defined by the problem. Together these constitute the START state (Fig. 2.17a). At the end of the movement process, the Agent's final location is the same as that of the GOAL location, GL1. In the OUTCOME portion we also encode the *changes* or *absence of changes* in the parameters involved. The ACTION portion of the script encodes the specific actions taken to solve the problem – i.e., go from the START to the GOAL location. Figure 2.17a shows the *specific* script encoded in the above script creation process. Each elemental action is a force, F(AA, RA), that takes two arguments, AA = Absolute Angle and RA = Relative Angle, just like in Fig. 2.15. (There are other synchronic causal parameters, such as the specific value of GL that have been included with F in the process described in Fig. 2.15, that are now omitted for clarity.)

Note that at any given time when an agent is exploring and interacting with the environment, there could be many other entities and their associated parameters that can potentially participate in any script that the agent is attempting to encode. However, in the current example, the reason why only the Agent and its GOAL's associated parameters are encoded in the script is that in a problem solving process, there is always a focus on certain aspects of the environment. For the case of the current SMG problem, the Agent begins with a *need* to be satisfied. This need is for it to change its current location to a GOAL location and this can be achieved by movement. It *knows* that for itself, as far as movement is concerned, the relevant parameters are the AL, RD, AA, RA, etc. mentioned above, as it has earlier learned the causal connections between the force applied to itself and the changes in these parameters. The GOAL's parameters are also relevant as the reaching of the GOAL is a need to be satisfied in this problem solving process. The fact that any problem solving process is a purposeful act, together with some earlier knowledge learned (say, in the form of heuristics to be discussed further in Sect. 6.4 of Chap. 6), allow the script with the relevant entities and their associated parameters that it contains to be carved out of the environment. If more than enough entities/parameters are included initially, they will be weeded out in subsequent learning processes. If insufficient entities/parameters are included, subsequent learning processes will discover this and include them. In Chap. 6, notably Sect. 6.4, there will be more discussion on the learning of knowledge and heuristics that allows the system to discover relevant or irrelevant causal entities and parameters in the environment that may influence a problem solving process and hence the encoding of scripts.

In Fig. 2.17a, it is shown that the absolute location, AL, of the agent has changed (increased) from AL1 to GL1 and the change is indicated as $+\Delta AL = GL1 - AL1$. The relative distance from the agent to the GOAL, RD, has decreased by an amount $-\Delta RD = RD1$ and the final value is $RD = 0$. GL remains unchanged. The actions in the ACTIONS portion consist of a sequence of force applications in the absolute direction of A3, say, and a relative direction of $0°$ (zero degree) with respect to the direction connecting the START and GOAL locations (as in Fig. 2.15). The script is named MOVEMENT-SCRIPT and it is a SPECIFIC script because it captures

2.6 Application: Causal Learning Augmented Problem Solving Search Process with...

movement starting from a specific location – AL1 – and ending in a specific location – GL1.

The SCENARIO portion of the script basically describes the scenario involved – that there is an Agent that is currently at a START location and that it intends to move to a GOAL location. This portions also contains the specification of the parameters involved. The representation of this portion could be in analogical/pictorial or other forms.

Now, suppose another similar problem situation of an SMG problem is presented to/encountered by the agent with different START and GOAL locations, and, based on the causal learning augmented search process discussed above, the agent finds the same solution – i.e., at each elemental step of movement, always elect to move directly toward the GOAL. Based on a moderate level of desperation and dual instance generalization as depicted in Fig. 2.13, a general movement script can be created as shown in Fig. 2.18.

The "*'s" in front of the values of AL1, RD1, and GL1 indicate that these can take on *any* value. The forces in the ACTIONS portion are now F(*, 0°), indicating that they can be pointing at *any* absolute angle but they must be pointing toward the GOAL (RA = 0°). This general script in Fig. 2.18 basically encodes the result of the problem solving process as discussed in Sect. 2.6.1 with the general learned heuristic of always moving in the direction of the GOAL if the minimum distance requirement is to be satisfied.

Figure 2.19 shows the basic process by which a script is constructed. In the case of a known problem solving process initiated by the system, the START and GOAL states and the SCENARIO (e.g., a spatial movement problem) are known (these are at a meta-level to the problem solving process itself). These are then used to construct the SCENARIO section of the script. In the case of Fig. 2.17, the SCENARIO description may contain an analogical representation of the problem itself as this is a problem of a spatial nature, as with most of the problems that we will encounter in this book. A vector representation can still be used to store the relationships between different entities in the scenario for the sake of storage efficiency but later when the need arises to match a newly encountered SCENARIO to the scripts stored in the Script Base (assuming this is a place where the earlier learned scripts are stored), an analogical/spatial matching may be carried out for the SCENARIO portion. After the problem solving process has completed, the ACTIONS and OUTCOME portions of the script are then constructed accordingly.

As indicated in the SCRIPT CONSTRUCTION process in Fig. 2.19, a script can be constructed either as a consequence of a problem solving process or through observation – i.e., observing the activities in the environment such as in this case of spatial movement of an agent, or observing the spatial movements or other activities of other agents including interactions between agents, etc. and constructing the script accordingly.

In the SCRIPT QUERY process (Fig. 2.20), presumably there is a problem at hand and the SCENARIO and START of the problem are matched to the SCENARIO, START and OUTCOME portions of the scripts in the Script Base (where earlier learned scripts are stored). The matched script will then be retrieved and the

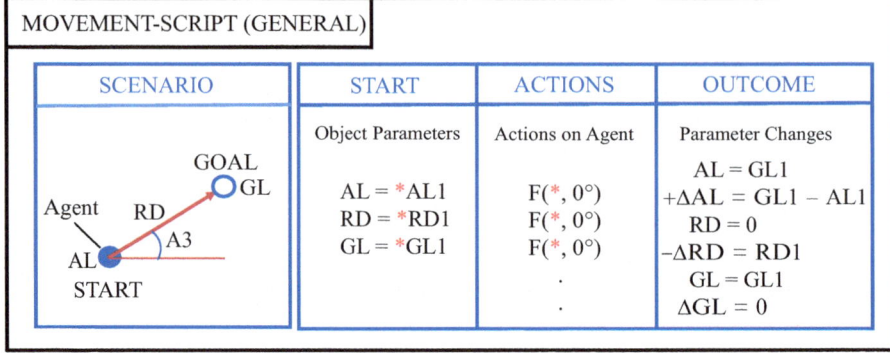

Fig. 2.18 A general MOVEMENT-SCRIPT. "*" indicates that the corresponding parameter can be of any value

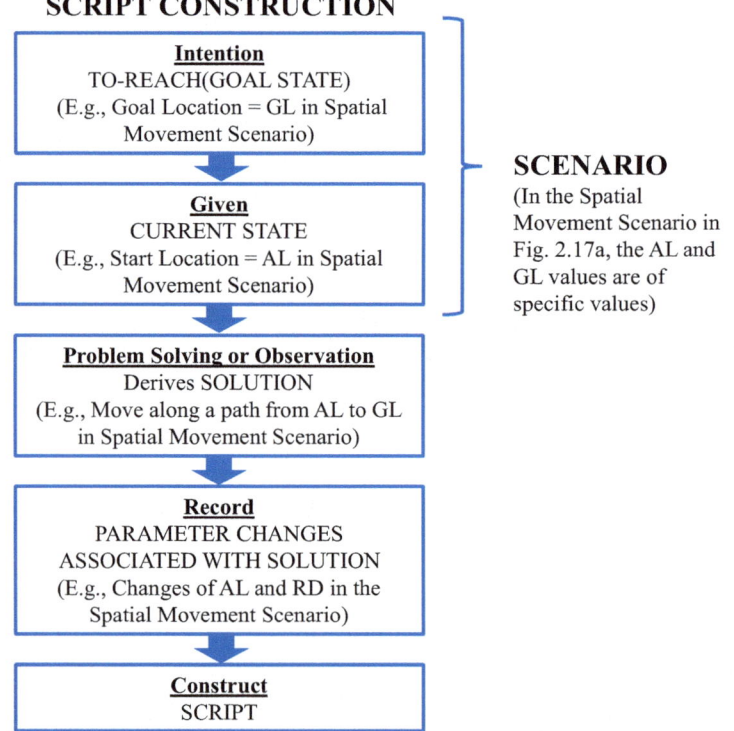

Fig. 2.19 The script construction process

solution retrieved accordingly as shown in Fig. 2.20. In the case of the MOVEMENT-SCRIPT (SPECIFIC) of Fig. 2.17, it can only be used to solve a spatial movement problem that matches its AL = AL1 and GL = GL1 values exactly, because it is a specific script that has not been generalized yet. For the

2.6 Application: Causal Learning Augmented Problem Solving Search Process with...

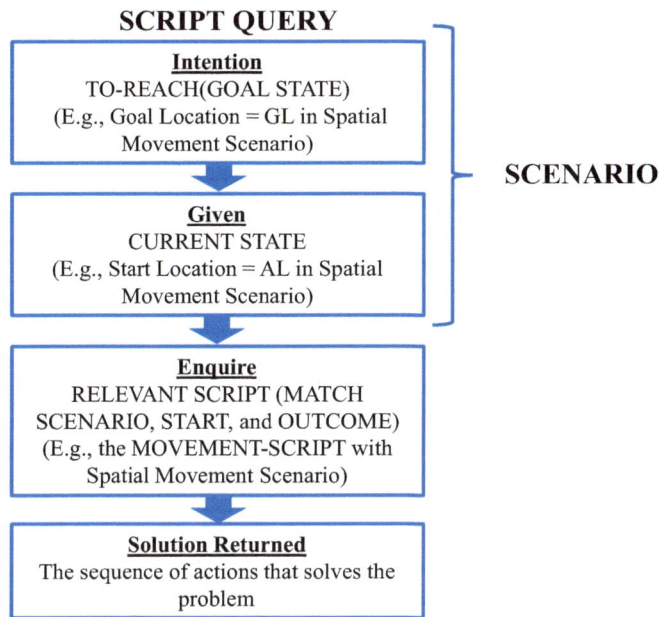

Fig. 2.20 The script query process

MOVEMENT-SCRIPT (GENERAL) of Fig. 2.18, it can be used to move from any AL (*AL1) to any GL (*GL1). In both cases, the solution to the problem can be applied immediately without any further lengthy problem solving processes.

Another way to apply the script would be to work "backward" in a process of backward reasoning as shown in Fig. 2.21. For example, if a certain change in parameter is desired, such as the change of absolute location of an agent, a PROBLEM SPECIFICATION is set to be $+\Delta AL$, and the OUTCOME portions of all the scripts in the Script Base are queried to see if any script results in the change of absolute location and if so, it is retrieved (The MOVEMENT-SCRIPT contains just such as OUTCOME of $+\Delta AL$). Then, the current START state is checked to see if it matches the script's START portion. If so, the ACTIONS portion of the script is immediately retrieved as the solution. If not, the difference between the current START state and the retrieved script's START portion is set as a new PROBLEM SPECIFICATION to begin another process of searching for a script in the Script Base that can solve the problem (a typical backward chained problem solving). There could be more than one script that provides a potential solution at any time (i.e., with an OUTCOME portion that matches the PROBLEM SPECIFICATION) and each would spawn off a branch of backward chaining. This backward chaining process would continue until the problem is solved – i.e., the current START state matches the last found script's START portion.

In the consideration of the SCENARIO portion of the script above, we have focused on the entities and parameters that are of concerned to the problem itself –

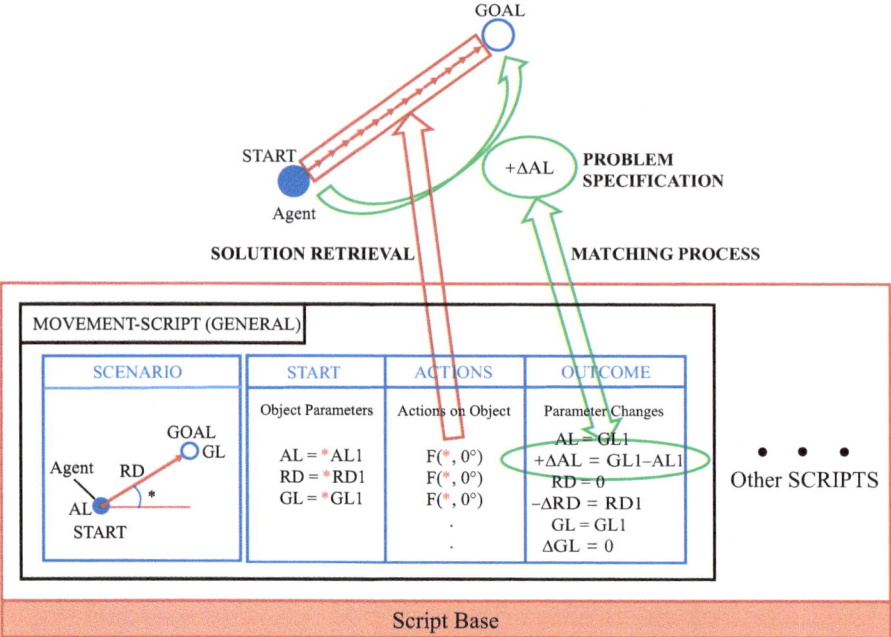

Fig. 2.21 Script matching and solution retrieval in a backward chaining process

e.g., the parameters associated with the Agent's START and GOAL states for the spatial movement problem. In general, any problem of interest may be embedded within the context of other objects and activities that may have consequence on the problem solution. In Chap. 6, for example, we will discuss a situation in which action at a distance created by electrical charges can influence the movement of an object (Sect. 6.4), as well as a situation in which a "wall" present between the START and GOAL locations becomes an impediment to the execution of the simple straight-line solution above (Sect. 6.5). These will complicate the SCENARIO matching process.

Consider the simple situation in Fig. 2.22a in which the SMG problem is embedded in a simple environment represented by a rectangle with no other objects or activities. Contrast that with the more complicated situation in Fig. 2.22b in which there is an ObjectX present which does not block the path from START to GOAL. In both situations, because the ObjectX does not interfere with the solution execution, the Agent will encode a script that will ignore any other object in the environment and will not include these other objects in the SCENARIO portion of the script. Later, in Chap. 6, when we consider the spatial movement to goal with obstacle (SMGO) problem, the Agent will learn that another object being present in certain locations will interfere with the solution found here – going in a straight line toward the GOAL. Hence, the current script encoding will become an overgeneralization that will be corrected later.

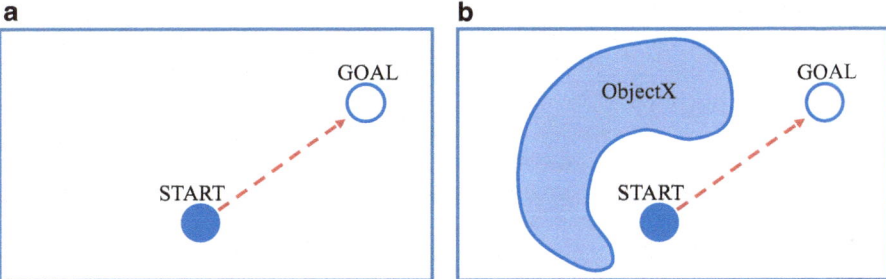

Fig. 2.22 (**a**) A scenario in which there are no other objects in the context of the SMG problem. (**b**) A scenario in which there are other objects in the context of the SMG problem that do not interfere with the SMG solution

One way to characterize scripts is that they are a kind of "fundamental units of intelligence" of a noological system. They encode a fundamental piece of knowledge structure that has noological efficacy – it specifies a start state, a series of actions and the outcome. This knowledge construct provides a fundamental chunk of information to address the fundamental concern and purpose of a noological system – "if I am in this state, and I desire a certain outcome, what do I do?" Scripts are also compositional – larger chunks can be built up from smaller chunks (as we shall describe in subsequent discussions) – and when larger chunks become available, they provide further noological efficacy by further improving the subsequent problem solving efficiency of the noological system. This is much like building complex molecules from simpler atoms (fundamental units of matter) and deriving more functionalities from the complex units.

2.6.3 Elemental Actions and Consequences: Counterfactual Script

The core general principle behind causal learning is the discovery and establishment of the causal relationships between various parameters observed in the environment, external and internal, through observations made on their temporal correlation, subject to certain diachronic and synchronic conditions, as we have illustrated so far with a few examples in the foregoing sections. However, in the node expansion process depicted in Fig. 2.15 in the process of solving the SMG problem, in considering which direction of movement gives rise to the minimum distance toward the GOAL, we have discarded the information on the other directions of movement as it was not relevant to the task at hand. In the most general situation, however, a noological system should take note of all of its actions and their attendant consequences (as far as its memory capacity permits) – i.e., what do their actions *cause* – so that the information may be of use later.

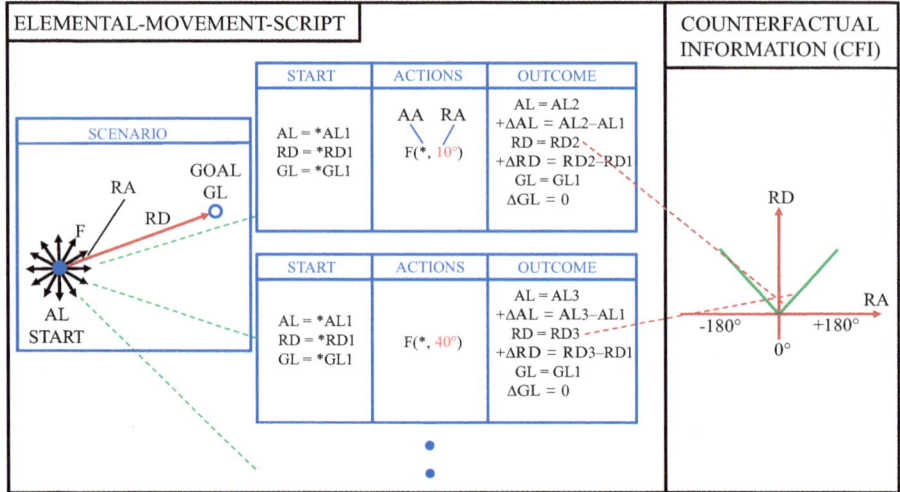

Fig. 2.23 ELEMENTAL-MOVEMENT-SCRIPT with COUNTERFACTUAL INFORMATION (*CFI*). All parameters AL, AA, RD, RA, and GL have the same definitions as in Fig. 2.14. RA is measured relative to the direction of the GOAL and in a clockwise direction from the line joining START and GOAL

Hence, in the process of causal learning augmented problem solving as described above in Sect. 2.6.1 for the SMG problem, a kind of *counterfactual* information can be encoded to help reduce future problem solving effort in other related situations. In the process of node expansion in which a force is applied to move the agent elementally in each one of the 360 directions to determine which direction satisfies the minimum distance heuristic, it is found that the direction toward the GOAL gives rise to the minimum distance. There are also other directions that do not give rise to the minimum distance. In fact, there is a direction that gives rise to a *maximum* distance from the GOAL, which is the direction 180° away from the direction of the GOAL. This information is captured in an ELEMENTAL-MOVEMENT-SCRIPT with COUNTERFACTUAL INFORMATION (CFI) – i.e., "had the agent moved in this direction, the distance to the GOAL would be of such a value, etc." This is shown in Fig. 2.23.

In Fig. 2.23, in the ACTIONS portion of the script, there is only one action involved – an elemental application of the force, F. This is in contrast with the MOVEMENT-SCRIPT of Fig. 2.17 in which the ACTIONS portion encodes a series of actions. The first argument of the force, F, is the absolute angle AA and the second argument is the relative angle, RA (i.e., F(AA, RA)). (These angles have the same definitions as in Fig. 2.15.) Therefore, F(*, 10°) means that the absolute angle is a "don't care" and the relative angle, RA, is 10° which corresponds to a movement 10° off to the "right" of the line joining the START location and the GOAL location. The right portion of the script labeled COUNTERFACTUAL INFORMATION (CFI) consists of a graph showing how the resultant relative distance to the GOAL, RD, changes with RA. When RA is 0, RD reaches a

2.6 Application: Causal Learning Augmented Problem Solving Search Process with...

Fig. 2.24 Counterfactual Script construction process. Contrast this with Fig. 2.19 – the differences are highlighted in *dark red*

COUNTERFACTUAL SCRIPT CONSTRUCTION

Intention
TO-REACH(GOAL STATE)
(E.g., Goal Location = GL in Spatial Movement Scenario)

↓

Given
CURRENT STATE
(E.g., Start Location = AL in Spatial Movement Scenario)

↓

Try or Observe
ACTIONS(1, 2, 3, ... N)
(E.g., Move an elemental distance from AL in Spatial Movement Scenario)

↓

Record
PARAMETER CONSEQUENCE ASSOCIATED WITH ACTIONS
(E.g., Closeness of distance to goal after movement in the Spatial Movement Scenario)

↓

Construct
SCRIPT and SCRIPT'S COUNTERFACTUAL GRAPH

minimum value and when RA is $+180°$ or $-180°$, which is a direction directly *away* from the GOAL, RD reaches a maximum value. If the GOAL is something desired and a minimum time is desired to reach it, one would elect to move directly toward the GOAL. If instead it is something to be avoided, one would select to move away from it and if a fastest possible direction is desired, then $+180°$ or $-180°$ would be selected. Other directions between these two extreme ends give rise to intermediate consequential relative distances from the GOAL.

The process that creates the counterfactual script of Fig. 2.23 is shown in Fig. 2.24. The difference between the construction process of the counterfactual script in Fig. 2.24 and that of a basic script depicted in Fig. 2.19 is that here there are multiple instances considered and their effects are recorded. The parts highlighted in dark red in Fig. 2.24 are the differences between Figs. 2.19 and 2.24.

In Fig. 2.25 it is shown how the script is used when a spatial movement situation is encountered and how some information about the consequence of various actions that is needed can be enquired from the script. There are two kinds of queries. For Value Query, the agent could obtain the resultant parameter value when a certain action is taken based on the counterfactual graph – e.g., in the case of spatial

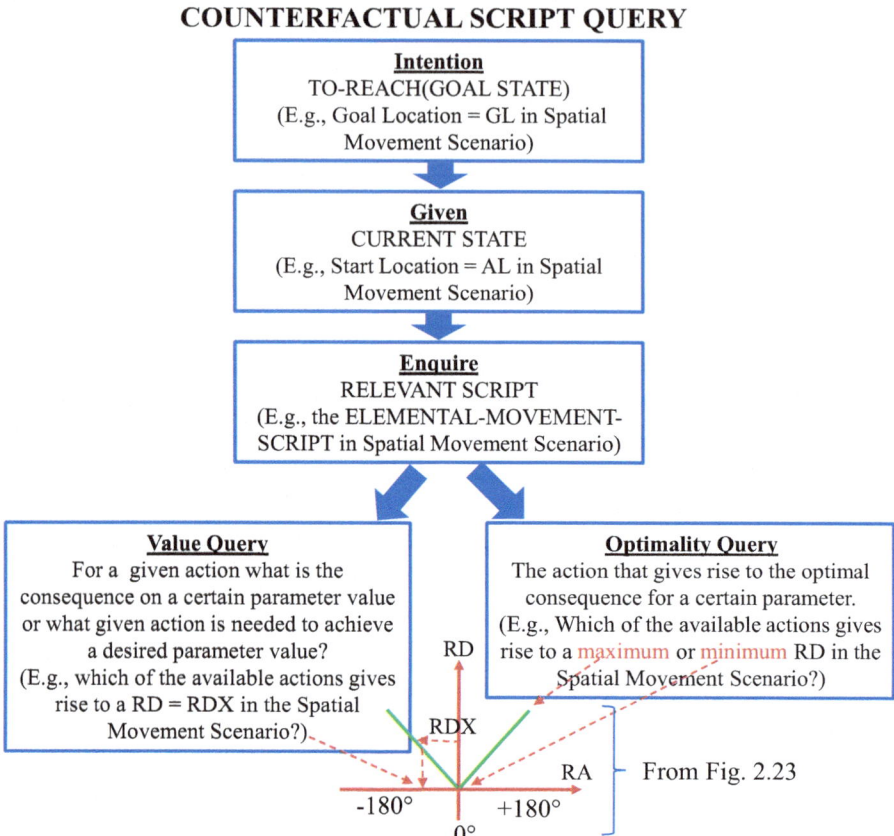

Fig. 2.25 Counterfactual Script query process

movement, the resultant RD (relative distance) value when a movement of an agent in a certain direction is effected. Another use of Value Query is the reverse of the above – the agent could query for the action needed to obtain a certain desired parameter value – e.g., the direction of movement that is needed to achieve a certain value of RD, say, RDX. For Optimality Query, the agent could query for the action that results in the maximum or minimum of the parameter of interest – say, which direction of movement would give rise to the maximum or minimum RD in the spatial movement scenario. In Chaps. 6 and 7 more examples of the use of the counterfactual information will be described.

The counterfactual information in Counterfactual Scripts will be useful in the rapid learning of optimal movement paths in the SMG with obstacles problem to be discussed in Chap. 6.

Counterfactual information on action consequences need not, of course, be physical-spatial in nature. Suppose someone is being pinched, typically she would let out a yelp. If tickled, a laugh. And if she is yelled at, she gets angry.

2.6 Application: Causal Learning Augmented Problem Solving Search Process with... 79

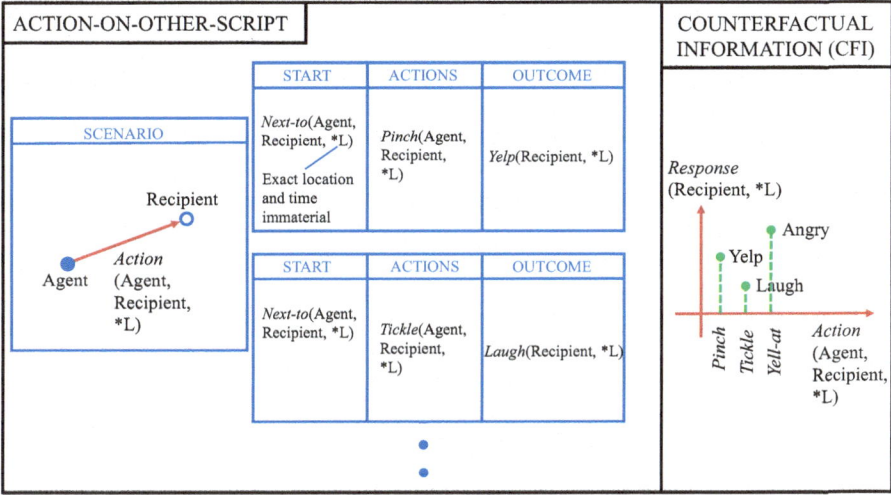

Fig. 2.26 Counterfactual Script for non-spatial consequences. *L shows that the actions and outcome are location independent

These can also be likewise encoded in a Counterfactual Script as shown in Fig. 2.26 (*L in the script indicates that the location is unimportant for the actions and outcomes involved). Some important details are being left out in the script of Fig. 2.26, such as there are intensities of the various responses involved and that they are typically proportional to the intensity of the corresponding actions. There would also be finer distinctions between different kinds of responses as a result of the corresponding actions, such as the kind of Anger arising from being Yelled-at may be different from the kind arising from being Slapped (which may contain a component of being humiliated), etc.

2.6.4 Maximum Distance Heuristic for Avoidance of Anti-goal

More often than not, problem solving issues explored in AI involve processes that try to achieve a certain goal. However, in the complete functioning of a noological system, there are also "anti-goal" to be avoided by the system. This is to avoid negative consequences, such as poison, land mines, undesirable agents such as "bad people," etc. Therefore, for the case of the SMG problem, there is an opposite, antithetical "spatial movement away from anti-goal" (SMAG) problem.

Typically, in the process of solving an SMG problem, a minimum distance heuristic is employed as discussed above. A general piece of knowledge on "action priority" that should be built into a noological system is that in a problem solving process, one strategy is to strive to move (spatially or otherwise) in a way that is closest to the goal at any given time. This is a piece of *domain independent* heuristic. Translated to the case of spatial movement, it would become the minimum distance heuristic. In general, this "distance" need not be spatial distance but could any measure of distance between the current state and the goal state.

We suggest that likewise, a corresponding built-in maximum distance heuristic be used to avoid an anti-goal. This information can be obtained from the same script as in Fig. 2.23 – when the goal is an anti-goal, the desired course of action at every step of a move would be the direction directly away from the goal – when $RA = 180°$.

It is possible for the maximum distance heuristic not to be built-in and instead learned from the experience with the physical world. Imagine a natural noological system, say an animal, being pursued by a predator. Consider the situation in which the prey's and the predator's speeds are kept constant but the predator's speed is higher. The prey would learn very quickly that any movement in a direction other than that corresponding to an RA of 180° would result in the predator closing on it.

2.6.5 Issues Related to the SMG Problem and the Causal Learning Augmented Search Paradigm

In fact, even though we have begun this discussion by considering what happens if A* is used for the SMG problem, it is used mainly for illustration and comparison purposes. This is because it is practically impossible to use A* for the problem involved as even with the minimum distance heuristic, A* quickly becomes unmanageable if the distance between START and GOAL becomes something sizeable in a real-world robot movement problem with numerous discrete steps from START to GOAL. A* has been used in problems such as game playing relatively more effectively as there is a much smaller number of discrete steps from START to GOAL, albeit an extremely large search space is still often encountered. The causal learning augmented version of A* as described above can manage a real world situation for the SMG problem as only two levels of search are needed and the rest is basically straight-line movement all the way to the GOAL. Because, as illustrated above, this straight-line movement heuristic has to be learned only once, often in practice in a real world robotic problem, this heuristic is built in. Our intention though in this section is to use this as a simple example to explain the causal learning augmented search paradigm. In subsequent chapters this paradigm will be scaled up to more complex problems (Chaps. 6 and 7) in which it will be illustrated that domain specific heuristics can indeed be learned along the way to achieve a vastly reduced search process augmented by causal learning.

At the outset it may appear to be the case that the process of discovering the domain specific heuristic above is a tautology (the same applies to the heuristic used by the A* search algorithm when applied to the same SMG problem). Since at any given instance of the node expansion process, a (built-in) minimum distance

2.6 Application: Causal Learning Augmented Problem Solving Search Process with...

heuristic is used, it is expected that the outcome of the entire problem solving process would result in a heuristic that requires that the agent moves in a direction that always points at the GOAL. However, we should treat the provision of the minimum distance heuristic and the causal determination and generalization of the best action to take at every step as two separate processes. The minimum distance heuristic is *domain independent* and it is an idea of "making progress toward the goal." In this case, the problem is a spatial one and the general idea of "making progress toward the goal" is translated into a distance measure to the goal. It may be translated into something else in a different context. We are assuming that in the context of the SMG problem, some sensory system is able to provide the information on the agent's distance to the goal at every step of the process. This information provider has no idea of the action taken or the context in which it is taken to arrive at the minimum relative distance or any other relative distance (i.e., if the action taken was not along a straight line from START to GOAL, the relative distance obtained would not be the minimum relative distance). It simply reports on the relative distance detected. This distance report module does not encode any generalization, such as that discovered by the causal learning process, that the minimum distance is achieved irrespective of location in the search space by pointing the force in a certain direction.

It will be instructive to consider a slightly different example of problem solving – how can a Person-A take a sequence of actions to generate the most amount of smile from another person, Person-B's, face? Similar to the case of the example above, firstly we need a visual detector which can provide the problem solver (Person-A) a measure of the amount of smile Person-B is generating. Since the GOAL is to achieve the largest amount of smile, at each search step the problem solving process tries to optimize its action locally – i.e., picks the one which is nearest to the GOAL – namely the action that generates the most amount of smile in Person-B. (The action could very well be Person-A's own smiling action, or certain actions of her that amuses Person-B, e.g., a funny face or action.) Therefore, just like in the case of the SMG problem, the brain system of Person-A must first provide, through the visual sense, the measure of the amount of perceived smile of Person-B (like the measure of the relative distance). Then the brain system must make use of a built in "domain independent" heuristic – at each step do something that results in a state closer to the goal – that has been translated into the particular domain – the "most smile" heuristic. In this case, it is easier to see that there is no tautology – Person-A has to learn what causes the satisfaction of the "most smile" heuristic – which could be a smile on her part or some other facial expressions or actions.

Now, there is an issue that relates to the motivational aspects of a noological system that is typically not considered in traditional AI research. This is the "meta-level" issue of why in the first place does an agent want to find a shortest distance to a goal or why does it want to maximize the amount of smile in someone else's face in the first place. This will be addressed in Chap. 3.

Another issue concerns the parameters used in causal learning, such as AL, AA, RD, RA, and GL in Fig. 2.15. The causal learning augmented search paradigm of course requires the availability of some very basic parameters and their associated values connected to the agent and the environment. These parameters include those that are derived from the external environment such as its location and distance to other objects (e.g., AL, AA, RD, RA, etc. as defined in Fig. 2.15) as well as those derived from the internal environment such as the force that it exerts/applies (e.g., F, in the above) or other needs-related signals – (e.g., hunger, pain, etc.). We assume and require that these parameters be supplied to the agent through the (internal as well as external) sensory systems, without which the causal learning process with its attendant drastic reduction in search effort cannot be achieved. We assume and require that this be built-in information that does not require learning. This is a totally reasonable assumption and requirement – it has already been illustrated using an extremely simple example in Chap. 1 (Fig. 1.3) that sensory information lies at the root of all problem solving efficiency and the built-in internal signals provide drives and purposes to an agent. This is fundamental information a noological system exploits, as enshrined in one of the principles laid out in Chap. 1. As the complexity of problems encountered by an agent increases, an agent will exploit more of this information. In nature, as the behavioral complexities of agents/organisms increase, their sensory systems also supply more and more detailed sensory information to support its problem solving processes – e.g., the visual and other sensory cortices of higher animals such as humans and primates occupy a relatively larger area of the surface of the cerebral cortex compared to that of the lower animals (Nolte 2009).

Based on the architecture of a noological system depicted in Figs. 1.7, 1.8, Fig. 2.27 provides an overview of the sources of the various built-in internal and external sensory parameters. In the foregoing discussion, we have shown how some of these parameters are causally related and the associated causal rules can be learned and used for problem solving – e.g., applying a force, F, in a certain direction and it changes the relative distance to a goal, RD, in a certain manner, or touching a piece of food and it decreases hunger. In the next chapter, we will discuss other causalities between other parameters – e.g., actions that change certain parameters such as the absolute location, AA, that may lead to changes in the internal Pain and Hunger states/parameters. This would lead to the discussion on issues on goal satisfaction and the motivation of noological system that we will explore more deeply in the next chapter.

The built-in internal and external parameters also represent a *ground* level characterization of the internal and external environment on which higher level characterizations could be built. More of this will be discussed in Chap. 4 on semantic and knowledge grounding.

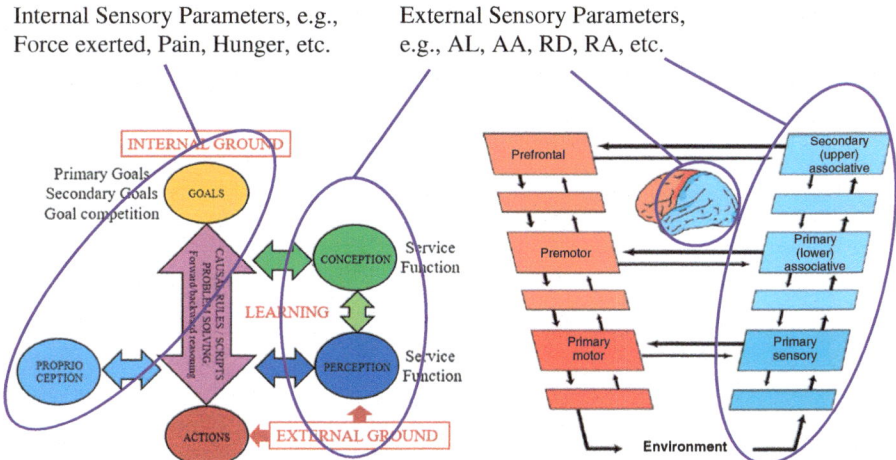

Fig. 2.27 Built-in internal and external sensory parameters made available in a noological system as a foundation for the causal learning process (Based on Fig. 1.8. *Right* illustration reprinted from The Prefrontal Cortex, Fourth Edition, Joaquin M. Fuster, Overview of Prefrontal Functions: The Temporal Organization of Action, Page 360, Figure 8.3, 2008, with permission from Elsevier)

2.7 Toward a General Noological Forward Search Framework

In the foregoing, as we discussed the issues of causal learning which is the main topic of this chapter, as well as its application to a problem solving scenario in the form of causal learning augmented forward search (Fig. 2.15), we touched upon a number of other related topics such as the learning of heuristic rules, the learning of scripts and knowledge chunking, and the built-in internal and external parameters that support causal learning. Each of these topics represents a processing component in the learning and problem solving framework that we have outlined above. As we have emphasized in Chap. 1, problem solving with respect to some built-in goals is the backbone of noological processing, and learning, of course, is necessary for a noological system to be adaptive to the environment. It will be instructive to put all of the above-discussed processing components in a total noological processing framework to illustrate their inter-relationships. In the subsequent chapters we will then have more in-depth discussions on each of these topics and other related topics that are relevant to a complete characterization of a noological system.

Exploring the world in the forward direction is important as it allows us to discover causality: If an action is taken now, what is the consequence next? But forward search is also difficult as there are many possible routes to the goal. Therefore, a number of devices are necessary to help to constrain the search.

Figure 2.28 shows a general and scalable forward search noological framework consisting of processing components that parallel the various processing steps discussed above. The process begins with a Problem Statement (Goal). In the

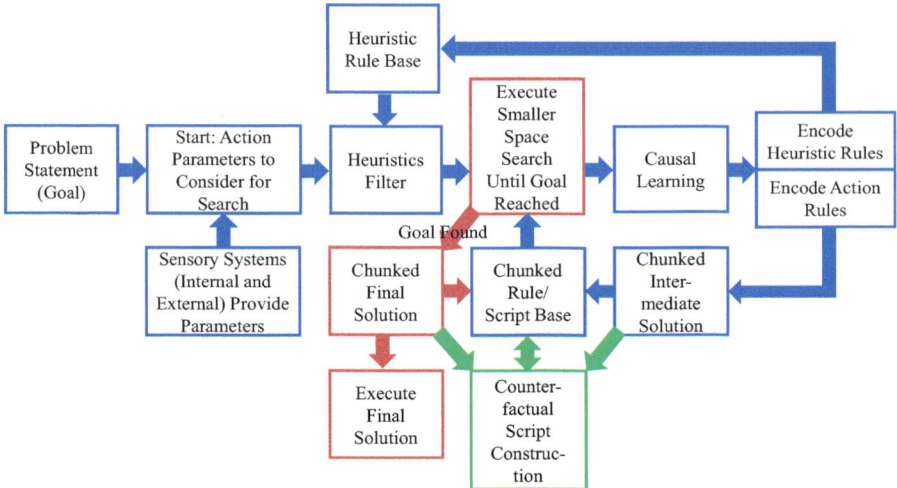

Fig. 2.28 A general and scalable noological forward search framework

case of the SMG problem (Fig. 2.15), it would be to move to a GOAL location from a START location. The second step of the process – Start: Action Parameters to Consider for Search – would correspond to the step in a typical forward search process in which a set of action parameters are selected to be varied to see if a path can be found that changes the current state of the agent and the environment toward the goal state. In the case of the SMG problem, this would correspond to considering the change of the direction of the force on the Agent to attempt to move the Agent toward the GOAL, as well as other parameters associated with the application of the force such as the absolute location and the absolute angle relative to a certain reference frame, and the relative distance to the GOAL of the agent at the time of the force application, as was discussed in Sect. 2.6.1. As discussed in Sect. 2.6.5, these built-in parameters (locations, angles, pain, hunger) come from the internal and external sensory systems.

The next step, the Heuristics Filter, applies whatever heuristics that have been discovered earlier, either from an earlier time in the current search process or from earlier problem solving processes, to narrow the search space. Therefore, the $Fm(*) = F(*, *, A3, 0, Lg)$ heuristic in connection with Fig. 2.15 in Sect. 2.6.1 would be a heuristic that was learned in the first two levels of expansion of the SMG problem solving process that provides the information to drastically reduce the subsequent search space of the search process. This is the next step – Execute Smaller Space Search Until Goal Reached – which is the core of the problem solving process. The $Fm(*)[Situation(*)] = F(*, *, *, 0, *)$ heuristic learned in connection with Fig. 2.16 in Sect. 2.6.1 would be a heuristic that was learned in earlier, separate episodes of problem solving processes that provides the information to drastically reduce the search space in any further instances of the SMG problem.

2.7 Toward a General Noological Forward Search Framework

In the process of executing the search, Causal Learning is applied, as shown in the next step in the figure and also as we have discussed in Sect. 2.6.1 for the SMG problem. In this process, two kinds of rules are learned and encoded. One kind is the Heuristic Rules, as described above and in Sect. 2.6.1 and this is stored in the Heuristic Rule Base. Another kind is the Action Rules, such as the Elemental Force-Movement Rule as mentioned in Sect. 2.6.3 and at the end of Sect. 2.6.1. More than one Action Rule could be chunked together into a longer chunked "script" in the process of problem solving and this could reduce the search space involved. This is like the MOVEMENT-SCRIPT of Fig. 2.18 discussed in the current chapter but more on this will be discussed in the next chapter. Sometimes, a Chunked Intermediate Solution could arise: Suppose in the problem solving process, the system has not reached the final GOAL but has reached an intermediate stage with a sequence of actions. These actions could be encoded as a chunked script that may facilitate the subsequent search process (we will see examples of this in the next chapter). These chunked scripts are stored in a Chunked Rule/Script Base.

The difference between Heuristic Rules and Action Rules is that Heuristic Rules encode "meta-level" information. E.g., in the SMG example in Sect. 2.6.1, the Heuristic Rule learned is: "the best direction to move to satisfy the minimum distance heuristic." Action Rules are rules that capture the causality between the activities at the level of the problem itself – e.g., in the SMG example of Sect. 2.6.1, how an applied force moves the agent involved.

Finally, when the GOAL has been found/reached in the problem solving process, the entire Final Solution would be encoded as a script – a chunked piece of knowledge – such as the MOVEMENT-SCRIPT of Fig. 2.17 and then the Final Solution is executed. The Chunked Final Solution is also stored in the Chunked Rules/Script Base.

Also indicated in Fig. 2.28 is a green-outlined box that is labeled "Counterfactual Script Construction." This is implementing the process discussed in connection with Fig. 2.24 and also in a number of other examples to be discussed in later chapters (e.g., Sect. 3.1 of Chap. 3, Sect. 6.5.2.2.2.1 of Chap. 6, and in various places in Chap. 7). In this process certain measures associated with the consequences (OUTCOME) of different actions or action sequences for similar scenarios are recorded to provide useful information for problem solving processes as discussed in Sect. 2.6.3 and also in the other examples in later chapters. Basically each time a new script is created, it is compared with existing scripts in the Chunked Rule/Script Base to see if a similar script with the same SCENARIO has been encountered before with different ACTIONS and OUTCOME (it is like "Had this action been taken instead in this scenario, what would the outcome/consequence have been?"). If so, counterfactual information is added to create a counterfactual script. However, in subsequent diagrams, this green box is sometimes omitted because the Counterfactual Script Construction process is part of the processes that proceed from either the Chunked Intermediate Solution or the Chunked Final Solution modules to the Chunked Rule/Script Base module (the blue and red arrows to the right and left of this module respectively).

The scalability of the forward search process is dependent on two critical mechanisms – one is the learning of Heuristic Rules and the other is the chunking of learned Action Rules into longer sequences of actions in the form of scripts. Both these mechanisms could drastically cut down the search space of any problem solving situation.

In a typical human problem solving process, even though we sometimes engage a forward search process, it is usually nothing like an extensive or exhaustive search process as in a typical traditional AI problem solving situation using search (Hart et al. 1968). We usually carry out a small amount of forward search augmented by earlier learned heuristics and solution chunks. This noological process is reflected and captured in the general and scalable noological forward search framework of Fig. 2.28. We will discuss knowledge chunking in more detail in Sect. 3.5 of Chap. 3.

Problem

The crawling robot problem is defined as follows. In Fig. 2.29a a robot, resting on the ground, is shown to have a body and two hinged arms extending from its "front" (its right "face"). A first arm, which is the arm that is hinged to the front of the body, makes an angle α with the vertical face. The second arm is hinged to the first arm as shown and it makes an angle β with a vertical line. The second arm's distant tip is labeled "Tip." The desired movement for the robot is to keep moving to the right. Both arms have to be moved in a correct sequence to allow the desired movement to take place – i.e., given the states of the arms as shown in Fig. 2.29a, if the second arm were to "swing outward," resulting in a reduction of β, then the robot will move to the left, which is undesired. The correct solution is first to lift the first arm – reducing the angle α – as shown in Fig. 2.29b. Then, the second arm will not be touching the ground (Tip is above the ground), and now it can swing outward (reducing β) until it swings pass the vertical line as shown in Fig. 2.29c. Now the first arm can be lowered until the second arm's Tip touches the ground as shown in Fig. 2.29d. After this, if the second arm swings inward toward the body of the robot (reducing the current β), the robot will move to the right. If this correct sequence of actions is repeatedly applied, the robot will keep moving to the right. The problem is to learn/discover this correct sequence of actions.

Reinforcement learning (Sutton and Barto 1998) has been successfully applied to this problem (Kranf Site: www.applied-mathematics.net/qlearning/qlearning.html) – each time the robot emanates a sequence of actions that leads to the desired direction of movement, a positive reinforcement is given, and when it makes an error (i.e., the sequence of actions results in the robot moving to the left), a negative reinforcement is given. However, reinforcement learning requires many learning episodes. Apply rapid causal learning to the problem to obtain a faster solution (See Ho (2014) for hints.)

Fig. 2.29 The crawling robot problem

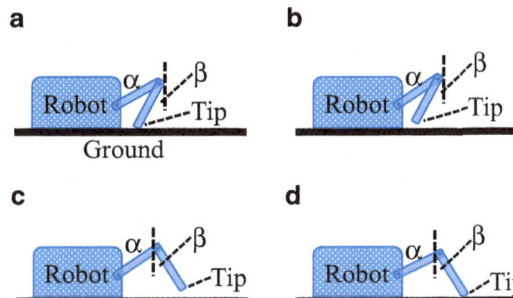

References

Agresti, A., & Franklin, C. (2007). *Statistics: The art and science of learning from data* (3rd ed.). Boston: Pearson Education, Inc.
Fire, A., & Zhu, S.-C. (2015). Learning perceptual causality from video. *ACM Transactions on Intelligent Systems and Technology, 7*(2), 23. doi:10.1145/2809782.
Greenspan, J. (2013). Coyotes in the crosswalks? Fuggedaboutit! *Scientific American, 309*(4), 17. New York: Scientific American.
Hart, P. E., Nilsson, N. J., & Raphael, B. (1968). A formal basis for the heuristic determination of minimum cost paths. *IEEE Transactions on Systems Science and Cybernetics SSC4, 4*(2), 100–107.
Ho, S.-B. (2014). On effective causal learning. In *Proceedings of the 7th international conference on artificial general intelligence,* Quebec City (pp. 43–52). Berlin: Springer.
Milberger, S., Davis, R. M., Douglas, C. E., Beasley, J. K., Burns, D., Houston, T., & Shopland, D. (2006). Tobacco manufacturers' defence against plaintiffs' claims of cancer causation: Throwing mud at the wall and hoping some of it will stick. *Tobacco Control, 15*(4), iv17–iv26. doi:10.1136/tc.2006.016956.
Moore, D. S., McCabe, G. P., & Craig, B. A. (2009). *Introduction to the practice of statistics* (6th ed.). New York: W. H. Freeman.
Nolte, J. (2009). *The human brain: An introduction to its functional anatomy* (6th ed.). Philadelphia: Mosby Elsevier.
Pearl, J. (2009). *Causality: Models, reasoning, and inference* (2nd ed.). Cambridge: Cambridge University Press.
Schank, R., & Abelson, R. (1977). *Scripts, plans, goals and understanding*. Hillsdale: Lawrence Erlbaum Associates.
Smolin, L. (2013). *Time reborn: From the crisis of physics to the future of the universe*. Boston: Houghton Mifflin Harcourt.
Sutton, R. S., & Barto, A. G. (1998). *Reinforcement learning: An introduction*. Cambridge, MA: MIT Press.

Chapter 3
A General Noological Framework

Abstract A number of issues are discussed in this chapter that build on previous discussions toward the characterization of a general noological framework. Spatial movement is used as an example to discuss the connection between the needs and motivations of a noological system and its attendant behaviors. It is shown how the alternative consequences of a problem solving process can be encoded in the counterfactual portion of a script – a knowledge chunk – to accelerate future problem solving. The idea of knowledge chunking is discussed in detail using a computer simulation of a micro-world. A noological system usually has more than one need that competes for attention. Anxiousness arises in the process of attempting to satisfy needs and the process of affective competition to resolve the attentional priority is illustrated using an example from the StarCraft game environment. Neuroscience research is reviewed to illustrate the current understanding in general brain architecture and it is shown how this can map nicely onto our general noological framework constructed from computational considerations.

Keywords Spatial movement • Need satisfaction • Affective computing • Affective competition • Affective learning • Script • Knowledge chunking • Incremental knowledge chunking • Brain architecture • Noological processing framework

At the end of the previous chapter, we have used the causal learning framework discussed in the early part of the chapter to motivate a noological forward search framework. In this chapter, we will discuss in more depth some of the basic issues involved in connection with the forward search framework, such as how needs affect choices of optimal actions, motivation/need competition, incremental knowledge chunking, and need-satisfaction learning (learning how to satisfy one's needs, given the existence of certain causalities in the environment). We will also review some neuroscience literature to compare and contrast our noological framework with the current state of the art neuroscience framework for structuring the understanding of the functioning of different parts of the brain.

3.1 Spatial Movement, Effort Optimization, and Need Satisfaction

Within our noological paradigm, we emphasize the idea as enshrined in one of the principles articulated in Chap. 1 that the needs and motivations of a noological system lie at the core of its behavior and have to be defined computationally and considered upfront. Even though we would provide more complex examples later in this book to illustrate the function of need and motivation, it is instructive to use the simple example of movement to throw some fundamental light on the issue.

In the previous chapter we employed an often invoked general, *domain-independent* heuristic – the minimum distance to goal heuristic – to illustrate the principles of causal learning augmented problem solving search and how it can discover the shortest path to the goal from a start position (the SMG problem) with drastic reduction of search space compared to traditional methods such as A* (Hart et al. 1968). The process also results in the learning of a general *domain specific* heuristic for spatial movement that can obviate all future searches in similar situations. The same principle of using the minimum distance heuristic was also used in the usual A* search algorithm in AI. In Sect. 2.6.5 of Chap. 2 we also used the example of taking actions to maximize another agent's smile (using the "maximum smile" heuristic) to contrast with the processes involved in the SMG problem. The question then arises: Why does an agent want to find the minimum path solution to start with? Or, for that matter, the maximum smile solution?

A deeper mechanism is at work here. In Sects. 1.3 and 1.4, Chap. 1 we mentioned that there is an internal ground level – the ultimate needs/motivations – in a noological system that is the fundamental determinant of its behaviors. The purposes of its various behaviors can only be understood when traced back to this internal ground level which we term the bio-noo boundary. In the case of minimizing the distance to goal in an SMG problem, the purpose would be to minimize the effort or energy consumed. Otherwise, any other non-straight-line path can also be used to reach the goal. For a biological organism, the purpose of minimizing the effort or energy consumed in any task is to delay the need to look for food or to conserve the energy level for contingencies (e.g., a sudden attack from an enemy). This has already been learned in evolution and has been built into the organism at the bio-noo boundary. Of course, in a situation of ample food or energy reserve, or when there are other higher priority needs in action, an organism's behavior would sometimes seem to be non-optimal with respect to certain measures or "non-effort conserving" – much like in the case of aimless wandering behavior mentioned in Sect. 1.2, Chap. 1 or hunger strike mentioned in Sect. 1.4, Chap. 1. In these situations, there may be other built-in motivations to override the fundamental motivations. For example, while walking in a nature park, one does not always walk in a straight line from point to point. This may seem to be "aimless wandering" but there could be other purposes at work – one often wanders off to smell the scent of the woods, to enjoy a view of the river, etc. These are other built-in needs that can compete with the basic energy minimization need and overcome the minimum path length priority. But if these other priorities are not at work and only the basic effort one is active, then the minimum path length behavior is the priority. (If one is a lumber jack and cutting wood in a forest, one would definitely choose the shortest distance path from point to point to carry the cut lumber.) Sometimes, built-in effort conservation priority is manifested or characterized as "laziness."

A deeper question concerning the priority to conserve energy as much as possible is whether this tendency can be learned. In Chap. 1 we mentioned that the priority to look for food or energy in a biological system has to be built-in, so that when the food level has lowered to a certain critical level the system would take the necessary actions, otherwise if the organism is dead, there is no further learning possible and therefore learning must have taken place at the evolutionary level and built-in at the bio-noo boundary. (And as mentioned above, this priority is not absolute, especially in the case of higher form organisms such as human beings in which hunger strike or hunger strike leading to death is possible.) And we have also noted that for an artificial system, if we want it to continue to function normally and do not want the interruption of "death from lacking in energy" we would also have put in place a built-in high priority for it to look for energy when the level is low.[1] Therefore for both biological and artificial agents, the recharging priority should be built-in. However, the energy conservation priority is tied to the recharging priority. It is possible for an agent to learn that if it exhausts itself in one task unnecessarily (such as going along a longer path to a desired destination unnecessarily), then it will have to carry out recharging more often, and that would affect the satisfaction of other internal needs, leading to an overall suboptimal situation. It will then adjust its behavior accordingly.

Each biological agent such as a human being probably has some components of "built-in energy conservation" and "learning to conserve energy." As discussed above, for the travel from start to goal situation, typically there exists a built-in priority to conserve energy unless it is overridden by other priorities. There could be other task situations in which a person may know initially that there are ways to conserve energy but does not pursue them. Later, some associated negative consequences may ensue, and then she would learn from this and change her behavior accordingly.

---

[1] As discussed in Chap. 1, in principle learning to give priority for "recharging" for an artificial system is possible since memory of previous experiences can be carried "across death" for an artificial system. If the artificial agent can remember that it did not give priority to its energy level maintenance and it led to its death in its "previous life" – and the establishment of this causality can be made through our causal learning method – it can adjust its priority accordingly in the "new life." But since the recharging priority is not difficult to build-in it is probably better for the creator of the artificial being to have it built-in rather than have the agent go through the trouble of learning it.

3.1.1 MOVEMENT-SCRIPT with Counterfactual Information

Figure 3.1 shows a knowledge structure, a MOVEMENT-SCRIPT, that contains a COUNTERFACTUAL INFORMATION (CFI) portion that encapsulates the concept of a movement event in which an entity (inanimate object or animate Agent) moves from one location to another (much like the counterfactual script of Fig. 2.23 of Chap. 2). One the leftmost part of the figure is a pictorial representation of the various paths that allow the movement from one specific location to another specific location to be executed. In the middle of the figure are basic MOVEMENT-SCRIPTS such as the one in Fig. 2.18 with a detailed specification of the START state, ACTIONS sequence and OUTCOME of the movement event. These are *alternative* MOVEMENT-SCRIPTS to capture the alternative movements from START to GOAL. In addition to the parameters in the MOVEMENT-SCRIPT of Fig. 2.18 for the START and OUTCOME portions, an energy parameter, E, is included that represents the energy level of the Agent involved (this is indicated in red). The action sequence specifications contain different action sequences to encode the various possible paths. On the rightmost side of the script is a graph showing "counterfactual information" in which the alternative paths and their consequences in terms of energy consumption are captured. The graph shows that there is an optimal path – namely the straight line path from the starting location to the ending location – which corresponds to the minimum amount of energy/effort consumed. (One way to interpret the CFI portion would be to read each point on the graph as corresponding to the description – "had this path been taken, the energy consumed would have been...," etc.)

Much like the MOVEMENT-SCRIPT of Fig. 2.18, the MOVEMENT-SCRIPT of Fig. 3.1 can be constructed through a problem solving process or learned through causal learning from an agent's experience with the environment including consideration of its internal states' changes (e.g., the changes in the energy, E). Figure 3.1 contains the knowledge structure that fully specifies the concept of movements and their consequences, and from it one can see that the reason to select the minimum distance path is to achieve an optimal consumption of energy. This MOVEMENT-SCRIPT or conceptual specification is one of the simplest examples we will encounter in this book that embodies the noological principle of including the motivations and purposes as part of the consideration whenever we address the issues of concepts. Minimum distance could in itself be a mathematical concept. However, noological systems carry out actions for a purpose. The ground level reason why a noological system prefers a minimum distance when moving from point to point has to do with its internal needs and priorities. A noological system's actions are noological in nature, not mathematical. Therefore, there is always a "why" associated with a noological system's behavior and this has to be adequately addressed before a proper understanding of noological systems can be achieved.

Similar to Fig. 2.24, Fig. 3.2 shows how the MOVEMENT-SCRIPT of Fig. 3.1 containing counterfactual information is constructed from either a problem solving

3.1 Spatial Movement, Effort Optimization, and Need Satisfaction   93

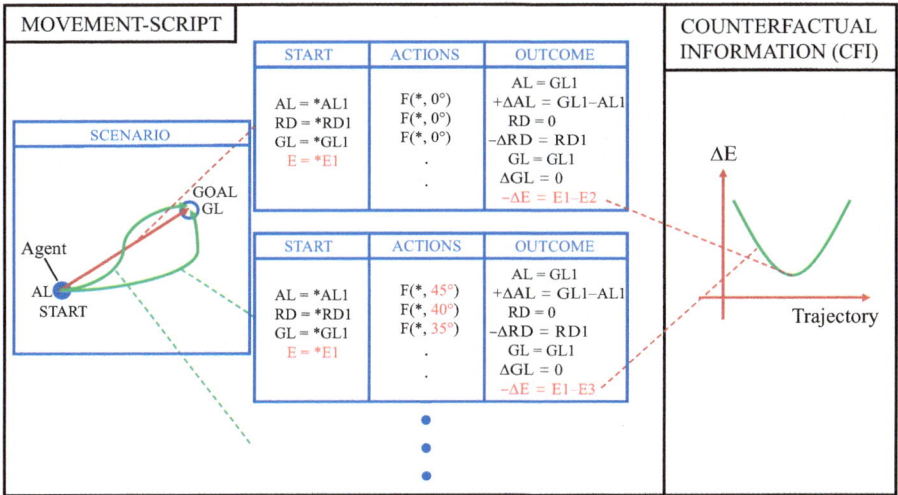

Fig. 3.1 MOVEMENT-SCRIPT with energy (**E**, in *red*) and COUNTERFACTUAL INFORMATION (CFI)

process or from observation. The differences between Figs. 2.24 and 3.2 are highlighted in dark red. The script can also be queried as shown in Fig. 3.3 in a similar manner as in Fig. 2.25. It can be seen that the processes shown in Figs. 2.24 and 2.25 on the one hand and Figs. 3.2 and 3.3 on the other are very similar, and hence they are general and would be applicable to other situations (the main difference is, whereas for Fig. 2.24 the focus is on "actions," in Figs. 3.2 and 3.3 the focus is on "solutions.")

In the script of Fig. 3.1, other than the information on the change in effort or energy for each of the paths going from a START location to a GOAL location, there can also be other parameters of interests. For example, Fig. 3.4 shows, in the CFI portion, the graph of another parameter, ΔAD, that is the total absolute distance traveled from the START to the GOAL location, that varies with the different paths/trajectories. (ΔAD is derived from ΔAX and ΔAY, the change in the absolute X and Y locations of the Agent or object undergoing the movement.) In this case, it happens to have the same shape as the ΔE vs trajectory graph. This graph conveys the information that suppose the Agent needs to find the shortest distance to the goal, it should use the script that contains elemental movements that are all in a direction pointing at the GOAL location (the topmost script in Fig. 3.4). In general, the CFI portion of the script can contain many graphs capturing the counterfactual information of various parameters' relationships to the alternative movement paths/trajectories. These can be generated automatically in anticipation of their potential uses (method outlined below) or from other processes that require some relevant information for their purposes.

In addition to the energy and distance traveled counterfactual information stored with a script as shown in Fig. 3.4, there could be other "instance statistics" stored as

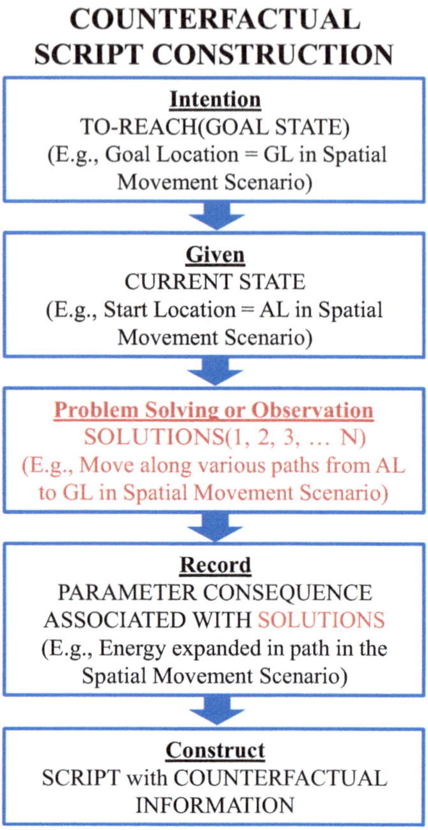

Fig. 3.2 Counterfactual script construction process for scripts containing solutions to problems. Contrast this with Fig. 2.24 – the differences are highlighted in *dark red*

COUNTERFACTUAL SCRIPT CONSTRUCTION

Intention
TO-REACH(GOAL STATE)
(E.g., Goal Location = GL in Spatial Movement Scenario)

Given
CURRENT STATE
(E.g., Start Location = AL in Spatial Movement Scenario)

Problem Solving or Observation
SOLUTIONS(1, 2, 3, … N)
(E.g., Move along various paths from AL to GL in Spatial Movement Scenario)

Record
PARAMETER CONSEQUENCE ASSOCIATED WITH SOLUTIONS
(E.g., Energy expanded in path in the Spatial Movement Scenario)

Construct
SCRIPT with COUNTERFACTUAL INFORMATION

well. This information could include any pattern or context observed in the usage of the script. For example, the time and location at which the script is usually triggered, the other scripts that are usually triggered together with this script in some earlier problem solving processes, the frequency the script is used in the course of a certain typical time interval (such as in the course of a day), etc.

Below we outline a method by which the counterfactual information is constructed. Basically, parameter changes (or lack of changes) across instances (e.g., different trajectories from start to goal states) are kept track off, whether it be something related to "internal needs," such as energy ΔE, or something that is a physical parameter, such as ΔAD – the total distance traveled. In a simple situation such as that of Fig. 3.1, there are not many parameters of interest and the system can simply list all possible parameters that are available (supplied by the visual and other sensory systems) and that either change or do not change across the instances. When the situation scales up to something more complex like the real world situation (such as in the case of the Restaurant Script to be described below shortly), extra knowledge will be needed to narrow down the relevant parameters to be considered, otherwise there will be an explosion in the number of parameters being listed in the CFI portion. This will be part of a "building knowledge from ground

3.1 Spatial Movement, Effort Optimization, and Need Satisfaction

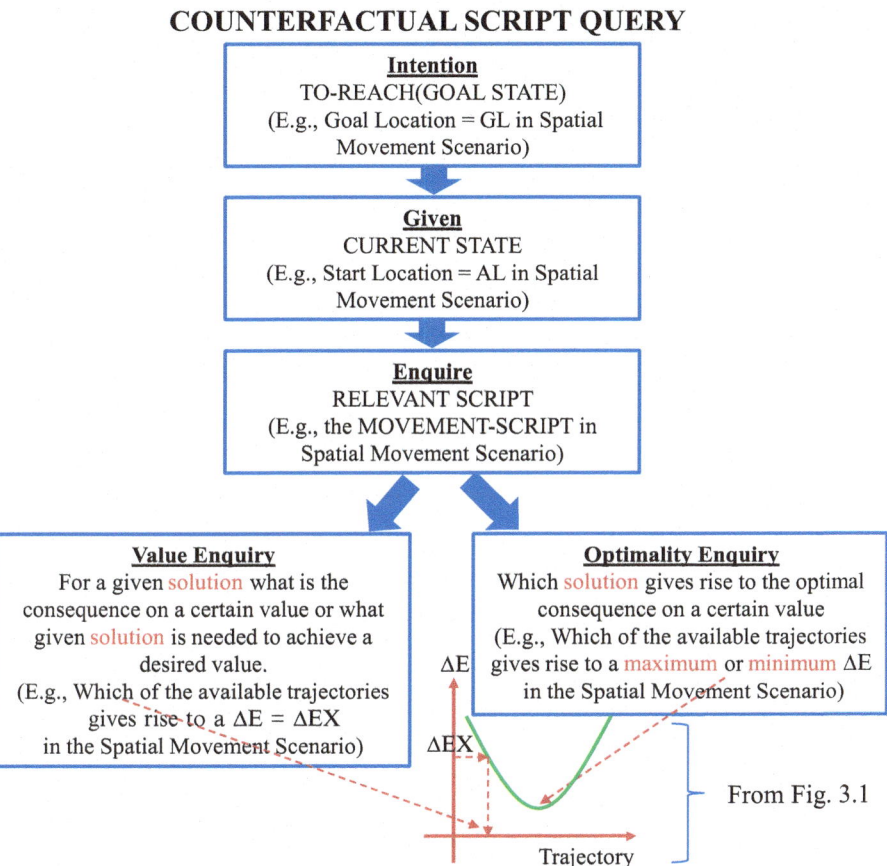

Fig. 3.3 Counterfactual script query process for scripts containing solutions to problems. Contrast this with Fig. 2.25 – the differences are highlighted in *dark red*

up" process that includes learning of meta-level heuristics that we will discuss later in this book. Suffice it to note that for the current purpose, we assume that this information is available for problem solving and other noological tasks.

As mentioned in Chap. 2, in traditional AI, the concept of script has been investigated by Schank and Abelson (1977) and as was also pointed out in Chap. 2, one major difference between our script and the earlier script is that we investigate the process that allows scripts to be learned, either through a problem solving process or through observation/experience. Another extension to the script idea compared to that of Schank and Abelson (1977) is that we add the CFI portion to the script structure as shown in Figs. 3.1 and 3.3. In Fig. 3.5, therefore, we show how a CFI portion can be added to Schank and Abelson's script.

In Fig. 3.5 it is shown that a number of counterfactual information graphs have been added to the traditional RESTAURANT-SCRIPT. These include the change of amount of money on the part of both the customer and owner, and the change of

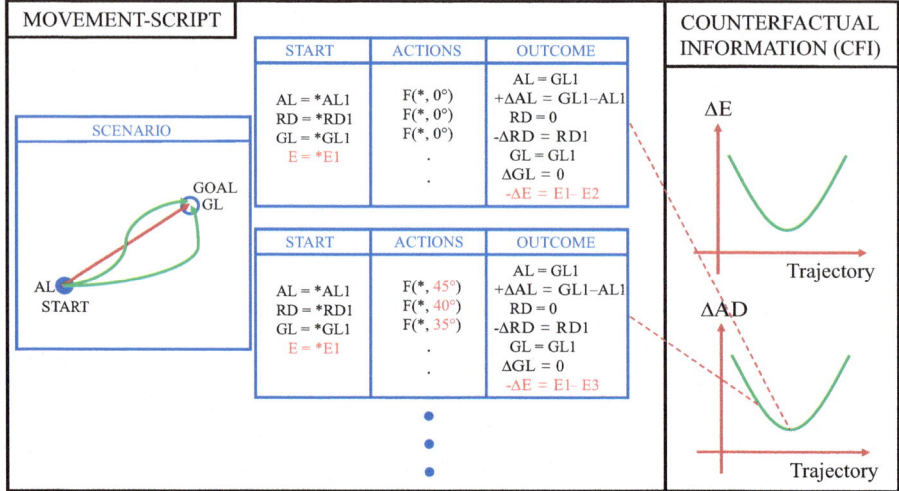

Fig. 3.4 MOVEMENT-SCRIPT with distance traveled counterfactual information, ΔAD

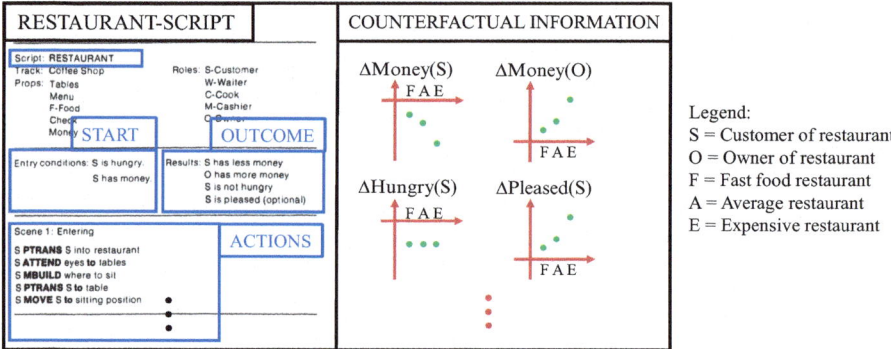

Fig. 3.5 Schank and Abelson's (1977) RESTAURANT-SCRIPT enhanced with counterfactual information (RESTAURANT Script portion republished with permission of Taylor and Francis Group LLC Books, from Scripts Plans Goals and Understanding, Roger Schank and Robert Abelson, 1977; permission conveyed through Copyright Clearance Center, Inc.)

the level of hunger and the level of being pleased (by the service or the quality of the food) on the part of the customer.

As will be seen later in the book, notably in Chap. 7, this CFI portion of the script which provides more information than just the OUTCOME portion of the script contains very important information to allow the problem solving process to select the right/desired script in a backward chaining process.

As mentioned above, we would like to emphasize that the example given in Fig. 3.5 is just to illustrate how the basic idea of counterfactual information is applicable in a general situation. While the automatic extraction of the counterfactual information such as the change of energy and distance traveled with respect to the various trajectories in Figs. 3.1 and 3.4 is relatively simple as there are not many variables involved in those situations, and the system can simply record the changes over a few "built-in" designated variable of relevance, the restaurant situation, on the other hand, involves many variables and the selection of the relevant ones to create the CFI portion of the script may require higher level reasoning. Our current paradigm dictates that higher level knowledge is built on grounded level knowledge. The issues of ground level semantics are discussed in the next chapter.

Another issue concerns the competition of needs alluded to in Sect. 1.5, Chap. 1. One good example would be none other than the restaurant-going situation as encoded by the RESTAURANT-SCRIPT of Fig. 3.5. Suppose one needs to make a decision on which or what kind of restaurant to have a meal at. By examining the CFI portion, one can see that the three kinds of restaurants – the fast food, average, and expensive restaurants – all provide the same degree of satisfaction of hunger. However, the expensive restaurant provides a more "pleased" outcome (desired) whereas it also ends up with the most money spent (undesired). Hence a need competition arises between the need for being pleased with the food consumed versus the need to conserve as much money as possible. We will return to address this issue in more detail in Sect. 3.3.

3.2 Causal Connection Between Movement and Energy Depletion

In Fig. 3.1 we have extracted the connection between energy depletion on the part of an agent and its movement through a certain trajectory/path. This is done basically through recording the change (difference between final and initial values) of the energy of the system across different instances of trajectory/path execution. However, in a general situation, it is important to the system also to be able to understand and encode the knowledge that it is the movement itself that *causes* the depletion in the energy and not because of other parameters, such as the agent *singing* all the way from the beginning to the end of the trajectory. (In the case of energy, of course, almost every activity executed by the agent including singing will cause a depletion of energy of some degree, but there are activities that deplete energy more extensively than others). There are two ways that the connection between energy depletion and movement can be established. One way is, since energy depletion as a result of movement is very fundamental to the "survival" of a system, whether it be a natural or an artificial system, we can simply build-in the knowledge of this connection. The other is, through our causal learning process as outlined in Chap. 2, the connection between movement and energy depletion can be established as shown in Fig. 3.6.

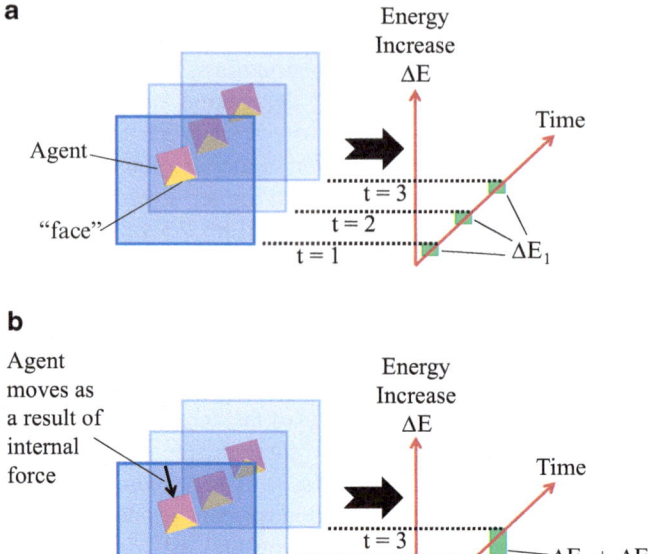

Fig. 3.6 (a) Agent is stationary but there is energy depletion due to internal processes. (b) Agent is being moved by an internal force that consumes extra energy. Agent is depicted to have a "face" that defines a direction of movement

Firstly, let us consider a situation in which the Agent does not move and yet there is an energy depletion due to some other internal processes that consume the energy as shown in Fig. 3.6a.[2] In the figure, it is shown that for every elemental time change from, say, $t = 1$ to $t = 2$, there is a decrease in energy by an amount ΔE_1 and that amount is registered at the end of the time frame – at $t = 3$ (the first ΔE_1 shown at $t = 1$ is due to the process between $t = -1$ and $t = 0$). For natural *living* agents, internal biological processes are always ongoing and consuming energy. For artificial agent, even when there is no movement, the internal processor could still be operating (perhaps it is still "thinking" and solving problems). This situation could be encoded as *Stationary*(Agent, $t = 1$ to 2) $\wedge$ *Thinking/Living* (Agent, $t = 1$ to 2)) $\rightarrow$ *Energy-Change*(Agent, ΔE_1, $t = 2$ to 3). The "action" that causes the Agent's change of energy is the thinking/living process. However, this causal rule could only be learned using the causal learning process described in Chap. 2 provided there was a situation earlier in which there was no thinking processes and there was no energy depletion, and when the thinking processes were switched on, the is an energy depletion that followed.

---

[2] The Agent in Fig. 3.6a is shown with a "face" indicating a "forward" direction. This feature will become more useful in subsequent discussions in this chapter.

Note that the "thinking" or "ongoing activities of an agent's internal processor" is a meta-level process that has to be observed and characterized at a meta-level in order for the system to be able to identify its causal connection to the energy depletion process. If that is not done, then a simpler characterization of the process would be *Stationary*(Agent, t = 1 to 2) → *Energy-Change*(Agent, ΔE_1, t = 2 to 3).

Next, we consider the situation in which the Agent moves. In Fig. 3.6b the movement of the Agent is shown to take place between time slices t = 1 and t = 2 due to an internally generated force that consumes energy. Over the elemental time change from t = 1 to t = 2, the internal energy of the agent decreases by an amount $\Delta E_1 + \Delta E_2$ and this is registered at t = 3. Hence an event – a change in position of the Agent – between t = 1 and t = 2 is followed by another event – a further change in the energy level – between t = 2 and t = 3. A causal connection can be established and causal rule encoded based on the effective causal learning method described in Chap. 2 such as *Move*(Agent, t = 1 to 2) → *Energy-Change*(Agent, ΔE_2, t = 2 to 3). (ΔE_2 is shown to be larger than ΔE_1.) This illustrates the learning of the causal connection between movement and energy depletion.

3.3 Need and Anxiousness Competition

As mentioned in Sect. 3.1, typically a noological system has multiple internal needs that compete for priority of being satisfied. The example given earlier was based on the RESTAURANT-SCRIPT (e.g., to please oneself by going to a more expensive restaurant or to save money by going to a cheaper one?). In this section we use a scenario from the StarCraft battle game environment (StarCraft II 2015) to illustrate the idea of need competition. In Chap. 7, there is more detailed discussion on learning and problem solving using the StarCraft environment.

In Fig. 3.7 we present some re-represented screenshots from the StarCraft game environment. Figure 3.7a shows three agents, a Self agent, an Enemy agent, and a Medic agent, that are placed at some distance away from each other in the environment. There are many kinds of agents in the StarCraft environment and there are also many kinds of activity that can take place. Typically, in a game, the human player would control the agents on the "Self" side and the system will react by controlling the agents on the "Enemy" side and the battle will ensue with many shooting and tactical movement events, etc. that will result in one side winning over the other side. The system also keeps track of many parameters associated with each agent such as its location, orientation, state of attack, and energy level.

In our current discussion, however, we will focus on a small number of agents and a small number of relevant parameters associated with the agents. One parameter of particular interest is what is called Health Point (HP) which reflects the state of health of the agent involved and if HP reaches zero (0), the agent "dies." Thus, maintaining a high/minimum level of HP is a priority for the agent.

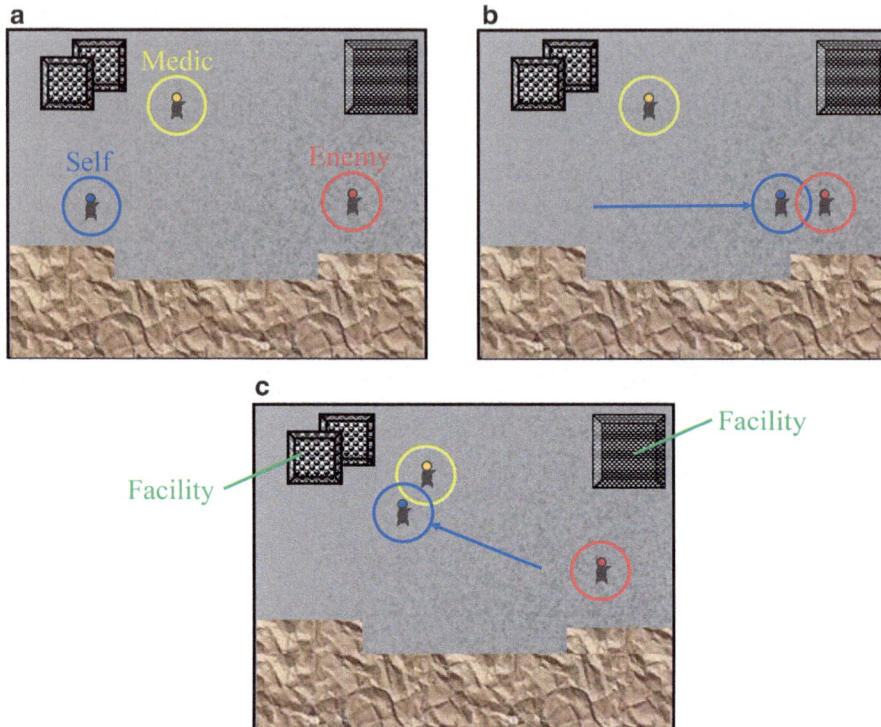

Fig. 3.7 (**a**) Three kinds of StarCraft agents: Self, Enemy, and Medic agents (Due to copyright reason, we did not capture the exact screenshots and illustrate them here. Instead, the same visual contents were redrawn and re-represented here. To have an idea of what the original visual output looks like, please follow some of the hyperlinks provided in Chap. 7) (**b**) Self agent moves to Enemy agent and both are then engaged in a shooting event. (**c**) Self agent moves to Medic agent to recharge its Health Point (HP). There are some objects in the StarCraft environment that are the "facilities" that we will ignore in the current discussion

In Fig. 3.7b it is shown that the Self agent has moved close to the Enemy agent and they are engaged in a shooting event in which they shoot at each other and if this is continued all the way it will lead to one of them being "killed," unless they disengage from each other before that happens. In Fig. 3.7c it is shown that the Self agent has disengaged itself from the shooting with the Enemy agent and approached the Medic agent to charge up its HP. (The Medic agent here only charges the HP of the Self agent, not that of the Enemy agent.)

The causal rules that are associated with these two situations – Self agent shooting with Enemy agent and Self agent's HP being charged by the Medic agent – both triggered by certain causal preconditions, are learnable and this has been described in Ho and Liausvia (2014) and will also be discussed in detail in Chap. 7. For this section, we are assuming that these rules are already learned and the focus will be on need/affective competition.

3.3 Need and Anxiousness Competition

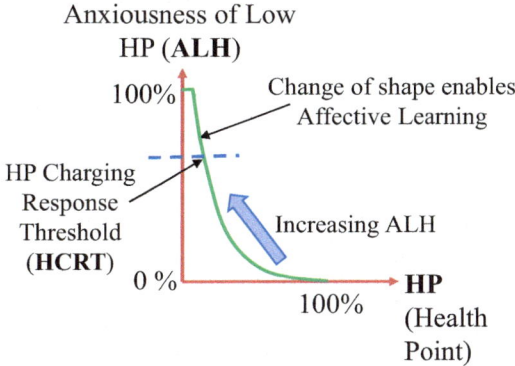

Fig. 3.8 Anxiousness of Low HP (ALH) vs Health Point (HP) graph

Figure 3.8 introduces a concept called "Anxiousness of Low HP"[3] (ALH) that captures the "emotional state" of the agent with regards to its state of HP level. Generally, the relationship is an inverse one – the lower the HP level, the higher the level of ALH. The relationship is also typically a non-linear one. For example, there is a range of HP reduction from the maximum level of 100 % in which ALH does not increase very much, but as HP nears 0, ALH would shoot up rapidly.

As the HP of an agent decreases, there is also a threshold of ALH above which the agent will respond to the situation and typically that means it would seek ways to recharge its HP. We term this HP Charging Response Threshold (HCRT).

The shape of the ALH vs HP curve is changeable with experience. For example, the agent may learn that it was not anxious enough at a given HP level and that led to some disastrous consequences (such as losing a battle or being killed), it would then "up-adjust" the curve. The HCRT is also learnable with experience – if response is taken, say, at too high an anxiousness level and it leads to undesirable consequences, the threshold would be up-adjusted.

There is also the situation that the agent may be "over anxiousness" – i.e., assigning too high an ALH level for a given HP level or setting two low an HCRT. This leads to a different kind of undesirable consequence – over attending to the task of recharging the HP and foregoing other priorities, which may in turn lead to other kinds of disastrous consequences (e.g., abandoning the assigned task of engaging the enemy and keep going to recharge the HP level, leading to the inability to win the battle.) In this case, the agent needs to "down-adjust" the curve or the HCRT.

In principle, suppose the purpose of the ALH-HP curve and the HCRT is for deciding when to recharge the agent's HP, then we could either change the HCRT as a result of learning while keep the ALH-HP curve constant or vice versa. The reason why we advocate having two adjustable aspects – the HCRT and the

---

[3] *Fear* and *anxiousness* are used synonymously here and the rest of the book. There are subtle differences which we will not particularly address in this book. E.g., typically it is "fear of the snake causes him to run" and "I am anxious that I cannot finish my homework by tonight." In these examples, it seems to be a matter of degree.

ALH-HP curve's shape is that it gives the agent flexibility in using these parameters to control its behavior. The use the agent puts the ALH-HP curve to could be multi-level. We show one response threshold in Fig. 3.8 above which the agent would look to recharge its HP. There could be other ALH levels at which the agent may carry out other activities – e.g., there may be a threshold above which the agent "begins to wonder" if it should disengage from whatever it is doing and start planning for a recharging plan, before carrying out any actual actions for recharging. Therefore, we could have different kinds of RTs (respond thresholds) set up on the same ALH-HP curve. And then, we could change one of the respond thresholds, say HCRT, while keeping the other thresholds constant. But in some situations we would adjust the entire ALH-HP curve upward or downward (changing the anxiousness levels across the board) and that would have effects on all the responses simultaneously.

Now, consider the situation in Fig. 3.7b when the two agents have been engaging in shooting at each other, and both sides have been experiencing the reduction of their HP values. Suppose the Self agent's behavior is controlled by the ALH-HP curve of Fig. 3.8 and at some point its HP level has lowered to the extent that its ALH level reaches the HCRT. At this point in time, the Self agent would disengage itself from the shooting engagement and look for a way to recharge its HP, because it is "anxious" enough about its state of HP level. Suppose it has earlier learned the knowledge that there is a kind of agent called the Medic agent that can help it to recharge its HP and suppose it also knows how to use a script to execute a movement to the Medic agent to allow it to begin the recharging process.[4] It would then disengage from the shooting event with the Enemy agent and move to the Medic agent to begin the recharging process as shown in Fig. 3.7c. Suppose there are no other needs to worry about at this point, ideally the Self agent would then keep recharging its HP and keep reducing its ALH until the HP is full, so that it is in a better position to engage in other future battles.

However, suppose another need arises to compete with this need to recharge the HP. Typically, the Enemy agent(s) will not just sit there waiting for the Self agent to charge up its HP. As the Self agent is charging its HP, the Enemy agent would typically begin to move toward the Self agent to attempt to engage it in a battle (and what other better time to do so than when the Self Agent is "vulnerable" – having a low HP and engaging in a re-charging activity?) Suppose the Self agent has earlier encoded an Anxiousness of Being Attacked (ABA) vs Enemy Distance (ED) relationship as shown on the right side of Fig. 3.9. The earlier graph of ALH vs HP from Fig. 3.8 is shown together in the same figure on the left side of Fig. 3.9. On the ALH vs HP graph, it is shown that the ALH is reducing (from some higher level) as a result of the HP charging and on the ABA vs ED graph, it is shown that the ABA is increasing (from some lower level, corresponding to when the Enemy agent is far away) as the Enemy agent is approaching.

---

[4] As mentioned earlier, the learning of this knowledge through causal learning has been described in Ho and Liausvia (2014) and will also be discussed in Chap. 7.

3.3 Need and Anxiousness Competition

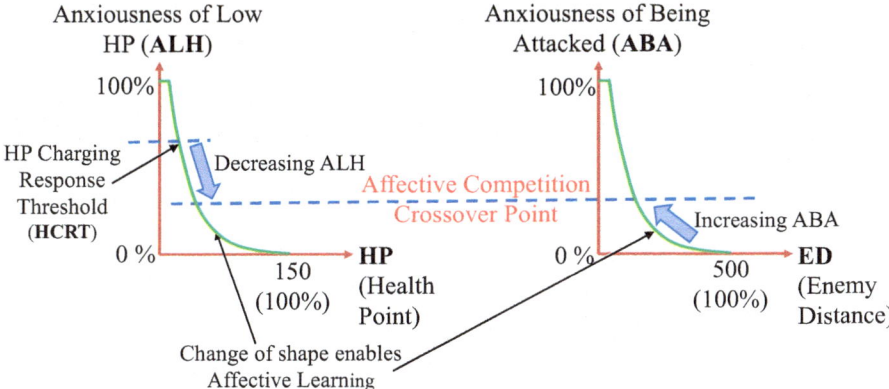

Fig. 3.9 Affective competition between Anxiousness of Low HP (ALH) and Anxiousness of Being Attacked (ABA)

When the ABA level has increased to a point which is higher than the ALH level, which means now the Self Agent is more anxious about being attacked than about being "hungry," a change of its behavioral priority ensues. Now, it should and would disengage from the "feeding" (HP charging) behavior and face or move toward the Enemy Agent to engage it in battle, in order not to be caught in a situation in which it is likely to be defeated. We term this "Affective Competition Crossover Point." This can take place not just between ALH and ABA as illustrated but also between many other forms of affect – e.g., when a noological system "likes/loves" one thing/object more than another in some situation, the thing/object involved would win in the affective competition over the other thing/object and changes the systems' behavior in its favor.

This, in a nutshell, is the process of affective competition shown through a StarCraft example. This video shows affective competition at work in a computer simulation programmed for the StarCraft environment – https://www.youtube.com/watch?v=lo0woEt7L78 (or see video on https://noologyblog.wordpress.com: StarCraft Battle: Affective Competition) – with the Self agent controlled by our program and the Enemy agent controlled by the StarCraft built-in procedures. In Chap. 7, Sect. 7.4.2 there is more detailed discussion on this video and further discussion on affective competition.

3.3.1 Affective Learning

As discussed above, the change of the level of HCRT or the shapes of the curves of the ALH vs HP and ABA vs ED graphs allow affective learning. Affective learning is a complex issue and in this book we will not develop a complete treatment of it. We relegate this to future investigations.

However, a binary search process can provide a quick and dirty method to learn the correct HCRT level or anxiousness level given a certain value of a parameter that causes the anxiousness, such as HP or ED.

Let us use the level of HCRT as an example. Firstly, it is beneficial to have as high an HCRT as possible because otherwise the agent involved would keep disengaging from a current activity to go and charge its HP ("find food"). On the other hand, waiting for HP to become too low (high anxiousness) to begin looking for a way to charge its HP (i.e., having a high HCRT) exposes the agent to the danger of dying and also being killed by an Enemy agent.

Using the scenario of Fig. 3.7, suppose on one occasion, based on a certain level of HCRT, the Self agent disengages itself in the middle of a shooting event with an Enemy agent to charge its HP. Suppose this does not lead to any bad consequence – i.e., the Self agent is able to subsequently successfully defend itself against the Enemy agent or even kill it. This could mean that the initial level of HCRT was too low – i.e., the Self agent could have been less anxious. It can then first make an attempt to increase its HCRT by a value that is half of the difference between the current value (a number between 0 and 100 – see Fig. 3.8) and the maximum value – 100. If this attempt does not lead to any disastrous consequences in the future, such as being killed by an Enemy agent, it can keep this level of the HCRT or it can increase it further in the same manner. However, if this leads to it being killed by an Enemy agent, it can reduce this HCRT by half the amount of the difference between the current value and the earlier, initial value. (Artificial agents could retain the memory of an "earlier life" and use the learned experience in the next one.)

However, each time the agent changes the HCRT level, there may not be an immediate clear-cut consequence that it can observe. The consequence also tends to be probabilistic. Therefore, affective learning is complex and we will not develop a complete treatment in this book other than to mention this binary search method which may provide a basis for a more complete treatment.

3.4 Issues of Changing Goal and Anti-goal

The previous section discussed a situation in which a goal can change, i.e., when the agent involved has assessed a situation and decided that another goal has a higher priority due to the need to address a high priority need. There is also a situation in which the basic goal does not change but the agent still has to adjust its sub-goals as a result of external environmental changes in order to address the basic goal. One example is when the agent needs to alleviate its hunger by consuming food. However, the first piece of food found may be inedible due to some reason, e.g., it is rotten. The agent then has to find another piece of food. Another example would be one of spatial movement in which the target of capture by an agent changes its location. The basic goal of capturing the target remains but the agent has to re-plan its movement to achieve the goal. This is shown in Fig. 3.10a.

3.4 Issues of Changing Goal and Anti-goal

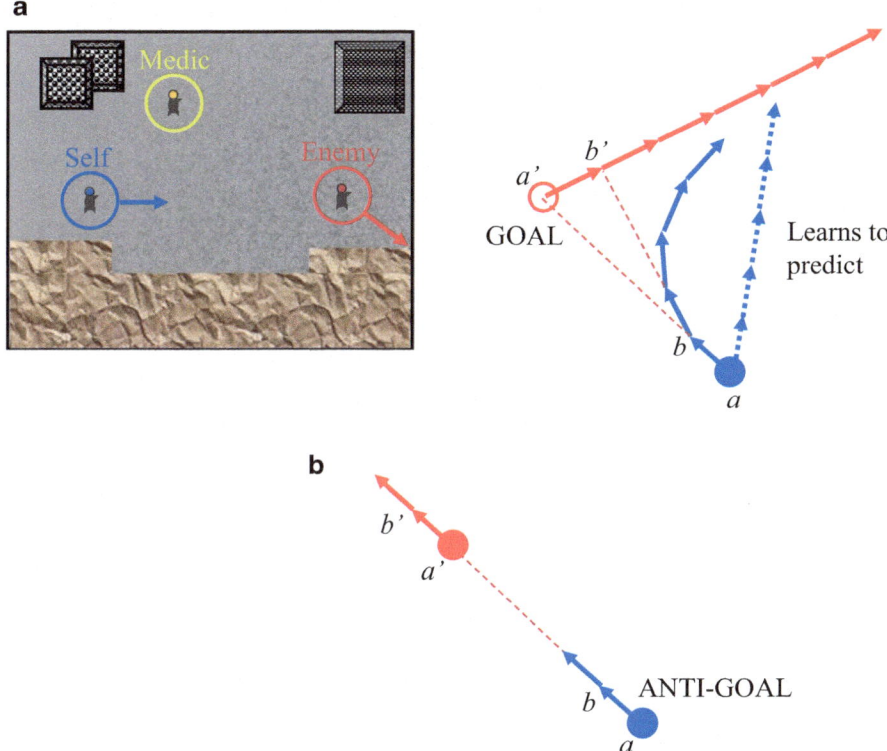

Fig. 3.10 (a) Situation in which the goal keeps changing – the GOAL of the *blue* agent – the *red* agent – keeps moving and hence the *blue* agent's GOAL keeps changing. (b) An anti-goal to be avoided – the *red* agent avoids engaging the *blue* agent

In Fig. 3.10a we use a similar StarCraft example as that in Fig. 3.7. Whereas in Fig. 3.7 we show that the Enemy agent is stationary, here the Enemy agent moves as soon as the Self agent moves toward it. On the right side of Fig. 3.10a we show how, the pursuing agent (e.g., the Self agent, in blue) first plots a straight-line path toward the GOAL (e.g., the Enemy agent, in red), using, say, the MOVEMENT-SCRIPT of Fig. 2.18. Suppose the blue agent starts at location a and the red agent is initially at location a'. Now, when executing the MOVEMENT-SCRIPT's ACTIONS portion, after the blue agent has executed the first elemental step to location b, the red agent moves to location b'. From location b, the blue agent uses the new GOAL location of b' to compute another straight-line path toward this new GOAL. However, on executing one elemental step, the GOAL moves again, and so on. The resulting path of the blue agent is a curved path slowly converging on the red agent as shown in the figure.

Figure 3.11 shows a general algorithm for problem solving given the situation of changing goal. This algorithm can also be used as a top-level control algorithm for the case of changing goal as a result of affective competition as discussed in the previous section.

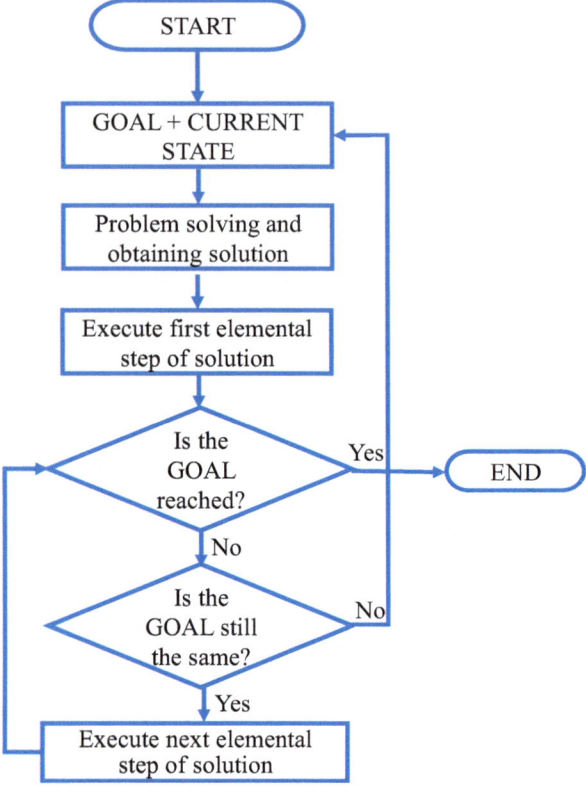

Fig. 3.11 General, top-level control algorithm for problem solving with changing goal

Figure 3.10a also shows that, suppose the blue agent has observed the red agent behaving in this manner – i.e., moving away in a certain direction as soon as the blue agent begins to move toward it – and has then encoded the behavior of the red agent in a form of a script, the blue agent can then use the script to predict the red agent's future path and then plot to intercept the red agent at a specific point by going in a straight-line as shown. This way, the blue agent does not have to use the goal changing algorithm of Fig. 3.11 that generates a curved path to intercept the red agent, which is less efficient in terms of computation and execution. In Chap. 9 we will further explore the use of scripts to anticipate other agents' behaviors and pre-compute and generate anticipatory responses to deal with them.

Figure 3.10b illustrates the concept of an anti-goal as discussed in Sect. 2.6.4 using the current scenario. In the battle situation of Fig. 3.10a, suppose the red agent wants to avoid engaging the blue agent (perhaps because its HP is low and it does not want to risk being killed), one reason for it to keep going in the straight-line path shown and risk being intercepted by the blue agent would be perhaps because there is some shelter or reinforcement farther along the straight-line path or there is a plan to ambush the blue agent by luring it to some place. Otherwise, the red agent would move in a direction that creates a maximum distance between it and the blue agent as shown in Fig. 3.10b. (See Sect. 2.6.4 on maximum distance heuristic.) The blue agent may give chase, of course, and the red agent would continue to move in that direction. In this situation, the blue agent is something to be avoided, rather than something to head toward. The blue agent, whether stationary or pursing the red agent, would be an "anti-goal" for the red agent.

3.5 Basic Idea of Incremental Knowledge Chunking

In Sects. 2.6 and 2.7 of Chap. 2 we discussed the idea of knowledge chunking and the acceleration of problem solving processes through the use of chunked knowledge in the form of scripts. In this section, we discuss and illustrate this idea in further detail by the use of a simple micro-environment with a spatial object manipulation scenario (Ho and Liausvia 2013b). In line with the general approach in this book, the issues concerned would include both the *learning* of the knowledge chunks as well as their application to problem solving situations.

In Fig. 3.12a we define a simple micro-environment consisting of an Agent and some Walls. The Agent consists of a "face" that defines its forward direction of movement. In Fig. 3.12b we depict a goal for the Agent, which is to build a Shelter consisting of the Walls in Fig. 3.12a. In Fig. 3.12c we show that the Agent needs to learn how to manipulate the Wall by interacting with it. Through a causal learning process, it learns to move the Wall forward, turn the Wall to face another direction, etc. Currently we define the following physical interactions: When the Agent is touching the Wall, it can *push* the Wall forward or *turn* the Wall to the left or right. For every *elemental* application of a force, there will be an elemental translational displacement or elemental rotational displacement. There is also an *attach* action – when the Agent is attached to the Wall, it can move it forward and turn it left or right as usual, but in addition to these, it can also *pull* the Wall.

Figure 3.12d illustrates how a simple causal rules could be extracted and learned from the actions and consequences that take place in the micro-environment through a causal learning process similar to that discussed in Chap. 2 (or similar to the learning of the causal rule between movement and energy depletion in Fig. 3.6b). At time frame $t = 1$, the Agent propels itself forward while in contact with the Wall (i.e., exerting a "force" on the Wall) and at time frame $t = 2$ both the Agent and the Wall are displaced elementally in the direction of the force. A causal rule "Push-Wall" is then encoded in the form of an IF-THEN rule as shown. This rule encodes the scenario and the entities (the Agent and the Wall) involved in a raw pictorial form.

Figure 3.13 depicts the subsection (the area in dashed outline) of the general forward search framework of Fig. 2.28 that is involved in the current discussion. We have mentioned causal rule learning and encoding in the micro-environment of Fig. 3.12a. In the subsequent discussion we will address the issues of the chunking of intermediate solutions and the application of chunked knowledge to problem solving (the bottom few blocks in the subsection of Fig. 3.13). In this section, we will not be discussing heuristic rule learning.

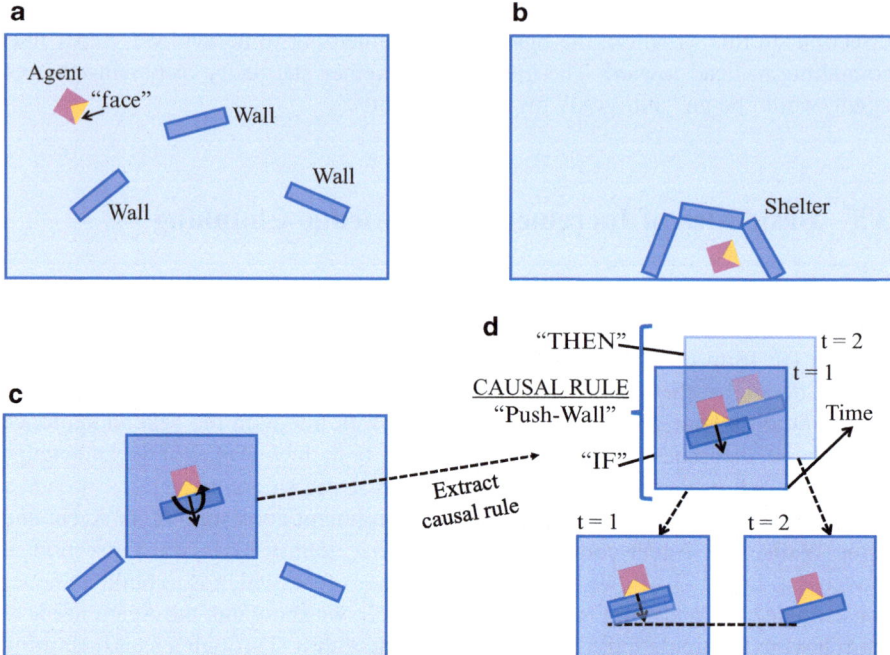

Fig. 3.12 (**a**) A simple micro-environment consisting of an Agent and some Walls. A "face" is identified on the Agent. (**b**) The goal for the Agent – to build a Shelter. (**c**) Actions that can be performed by the Agent on the Wall – push the Wall, turn the Wall, etc. These and the consequences are extracted, learned, and encoded through causal learning. (**d**) The "Push-Wall" causal rule learned and encoded through causal learning (©2013 IEEE. Reprinted, with permission, from Ho, S.-B. and Liausvia, F., "Incremental Rule Chunking for Problem Solving," Proceedings of the 1st BRICS Countries Conference on Computational Intelligence, Pages 324–325, Figs. 1, 2, and 5)

In Fig. 3.14, we illustrate the basic incremental rule chunking process. In Fig. 3.13a, we begin with a relatively simple problem of moving the Wall from a START position to a GOAL position. Suppose we use a relatively straightforward blind search process in which all the available elemental actions (such as pushing, turning left, turning right, etc.) are tried (shown in the top right corner of Fig. 3.14a), and then after a number of steps the GOAL is reached.[5] We then record the entire sequence of steps (in this case, a continued application of the "push forward" actions) as a chunked set of causal rules consisting of the elemental steps. These combined actions are basically a script.

---

[5] We use a blind search process for discussion here to show that even with a blind search process, incremental chunking can drastically reduce the search space involved. If heuristics and other information are added to the search process as discussed in Chap. 2, the search space naturally can be further reduced.

3.5 Basic Idea of Incremental Knowledge Chunking

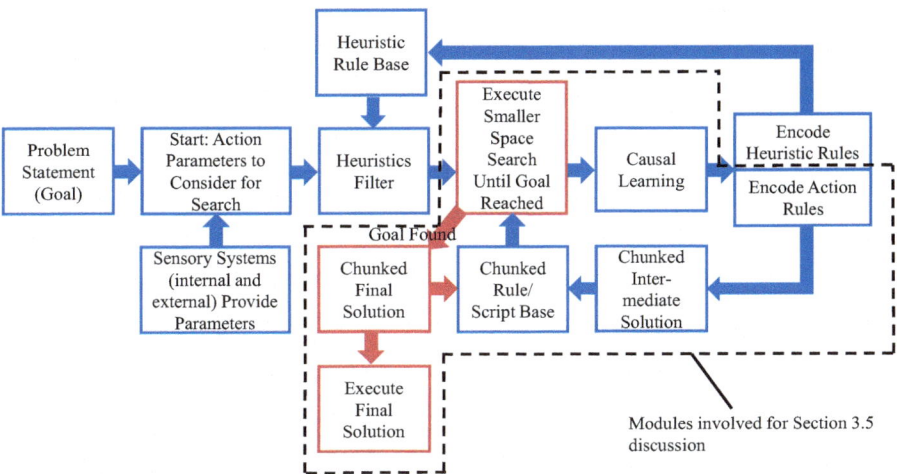

Fig. 3.13 The subsection (the *dashed* outline) of the general noological forward search framework of Fig. 2.28 that is involved in the discussion in the current section

Now, suppose the Agent is presented with a slightly more complicated problem as shown in Fig. 3.14b in which the GOAL is to move the Wall slightly further away and also have it ending up in a different orientation relative to the initial orientation. Using a similar search process but now using the earlier chunked causal rule in the process of search (as shown in the search tree in the top right corner of Fig. 3.14b), the Agent is able to find the GOAL much faster as the earlier learned chunked causal rule can bring the Agent to an "Intermediate State" that is much closer to the GOAL. For this round of search process, the sequence of steps between the Intermediate State and the GOAL still has to be found in a relatively lengthy blind search process, but the earlier chunked causal rule has already helped to cut down the search space. After the successful discovery of the solution to this problem, the steps between the Intermediate State and the GOAL will be stored as another chunked set of causal rules (another script) to be available for use in subsequent problem solving processes.

Therefore, if at any time the Agent is confronted with a relatively "hard" problem that requires many steps of actions to solve, the search space will be formidable (e.g., the problem of going from the state of Fig. 3.12a to the state of Fig. 3.12b). However, if each time the Agent encounters instead an incrementally difficult problem and the earlier learned chunked rules/script/solution can assist in the solution to the problem, the search space will be manageable. It is posited that this is how human beings learn to solve problems from infanthood to adulthood.

Let us now consider the problem of Fig. 3.12a, b. Suppose earlier there was a series of learning episodes with various incrementally difficult scenarios/problems of START and GOAL states and in all these problems the solutions were stored as chunked rules or scripts in a Chunked Rule Base or Script Base (Fig. 3.13). Now, if the Agent is presented with the scenario of Fig. 3.12a, b – a relatively complicated

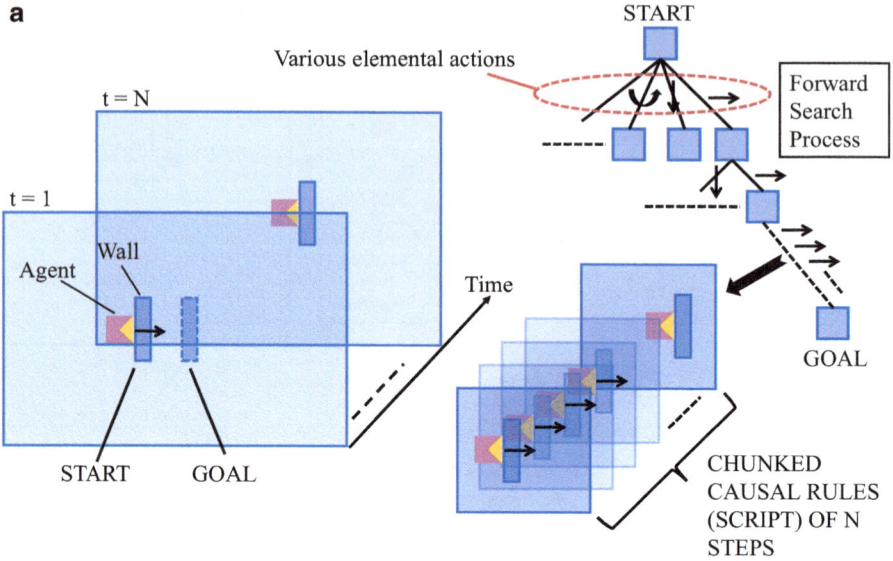

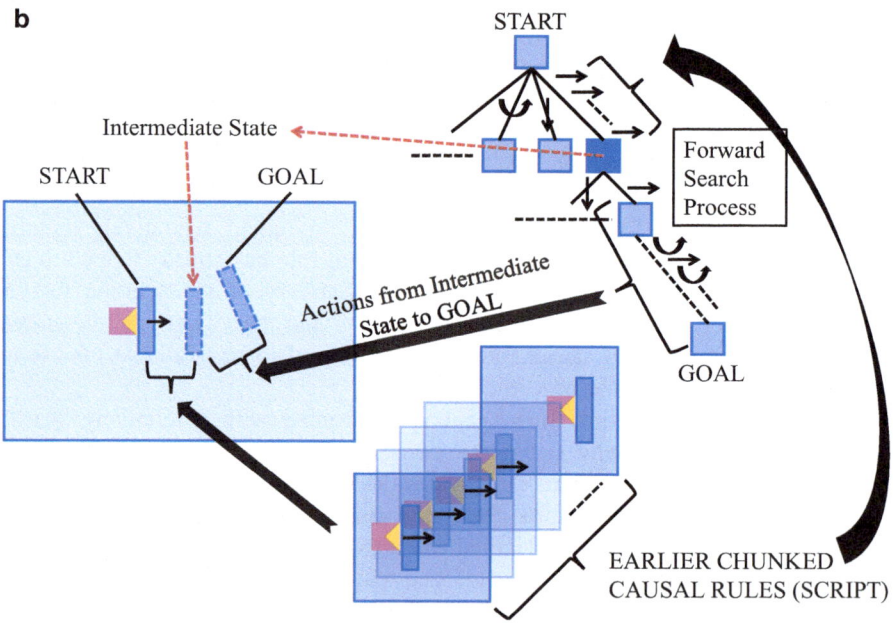

Fig. 3.14 (a) A simple problem of moving the Wall from a START to a GOAL position. (b) A slightly more difficult problem in which the GOAL is to move the Wall slightly farther away and also have it ending up in a different orientation relative to the initial orientation (©2013 IEEE. Reprinted, with permission, from Ho, S.-B. and Liausvia, F., "Incremental Rule Chunking for Problem Solving," Proceedings of the 1st BRICS Countries Conference on Computational Intelligence, Page 325, Fig. 6)

3.5 Basic Idea of Incremental Knowledge Chunking 111

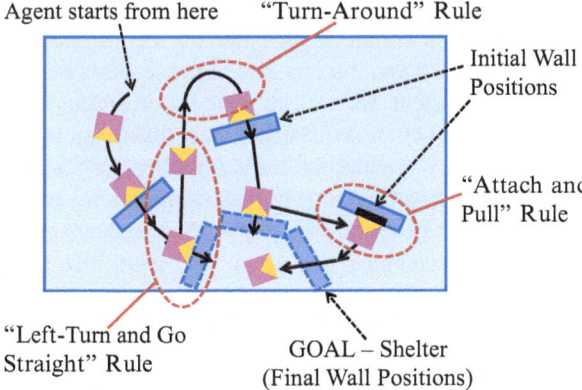

Fig. 3.15 Application of chunked rules in a more complex problem of building a Shelter by pushing the Walls in various manners from their initial positions to the final positions constituting the Shelter (©2013 IEEE. Reprinted, with permission, from Ho, S.-B. and Liausvia, F., "Incremental Rule Chunking for Problem Solving," Proceedings of the 1st BRICS Countries Conference on Computational Intelligence, Page 326, Fig. 7)

problem of building a Shelter by pushing the Walls in a certain manner from some arbitrary initial positions – the Agent would then be able to arrive at a solution relatively easily. Figure 3.15 shows just such a possible solution.

In Fig. 3.15 it is shown that there are chunked rules such as "Left-Turn and Go Straight," "Turn-Around," and "Attach and Pull" that were learned earlier and that are applied to the current Shelter building problem, reducing an otherwise formidable large search space to a more manageable proportion in the problem solving process.

3.5.1 Computer Simulations of Incremental Chunking

In this section we use a number of computer simulations carried out in a microenvironment similar to that depicted in Fig. 3.12 to illustrate the above idea of incremental chunking (Ho and Liausvia 2013b), culminating in a relatively complex computer problem solving process similar to that in Fig. 3.15 – Shelter building. We begin with the very fundamental learning of the basic physical causal rules depicted in Fig. 3.12c. In this "Initial Infant Learning Phase," the Agent, much like an infant, learns about the basic causal physical rules of its environment. There are actions that the Agent can perform on itself and actions that the Agent can perform on other objects. The Agent learns to activate certain actions on itself to push itself forward or backward or to turn left and right. A YouTube video illustrates this process: http://youtu.be/xPYPloHh67c. (or see video on https://noologyblog.wordpress.com: Initial Infant Learning for Agent Itself). Altogether eight elemental actions are used: push (itself) forward (and move by 1 pixel), push backward, push to the left, push to the right, turn right (by 1°), turn left, attach, and detach. When the Agent is in contact with other objects, it learns causal physical rules such as pushing an object and the object will move, and twisting an object and the object will turn. The Agent also learns that by attaching itself to an object (a Wall), it can *pull* the object. A YouTube video illustrates this process: http://youtu.be/tYvVvfLjUGA (or see video on https://noologyblog.wordpress.com: Initial Infant Learning for Agent and Wall). In these simulations, there is an underlying "Physics Simulator" that generates the physical consequences of actions. If the physics of the environment changes, the Agent will learn different physical rules and apply them accordingly to various problem solving scenarios. For ease of retrieval, all these learned elemental causal rules consisting of one step of elemental action and movement are stored in the Chunked Rule Base, along with chunked rules consisting of longer sequences of actions and movements, the learning of which will be discussed below.

In a Second Infant Learning Phase, the Agent learns to solve relatively simple problems incrementally through knowledge chunking. The problem used in the simulations is basically the spatial movement problem – from a START state consisting of a location/orientation specification,[6] find a path to a GOAL location/orientation – for the Agent itself or for an object manipulated by the Agent. The A* algorithm (Hart et al. 1968) is used for the problem solving process. The cost of each state reached is computed based on the sum of the number of elemental steps needed to reach that state and a closeness measure of that state to the goal state. The closeness measure is based on a weighted sum of the translational and rotational difference between the current state and the GOAL state. With this cost measure, chunked rules in the Chunked Rule Base that consist of the smallest number of elemental steps and that can bring the search process closer to the GOAL would be considered first.

For the scenario in which the Agent moves itself in various ways (consisting possibly of a sequence of translational and rotational movements) to reach a final GOAL state of a particular location and orientation, we trained the Agent to reach intermediate goals starting from three elemental translational or rotational steps away (e.g., move 1-pixel forward, rotate 1° clockwise, and move 1-pixel right, each of which is considered "1-step" of action) and in 100 random combinations of these steps, followed by intermediate goals that are 6, 12, 24, and 48 steps away from the initial starting state and in 100, 100, 50, and 50 random combinations respectively.

Figure 3.16 shows the traces of the movements/solution paths of the Agent. For the sake of clarity, only a subset of the traces is shown. Basically, having learned to tackle the simpler problem, say, of three steps of actions from START to an intermediate goal, the Agent then uses the learned chunked knowledge (stored in the Chunked Rules Base of Fig. 3.13) to solve problems of the next level of difficulty – those requiring six steps of actions – and so on in the same manner as

---

[6] A "state" is defined as a location plus an orientation of the Agent, the Wall, or any object involved.

3.5 Basic Idea of Incremental Knowledge Chunking

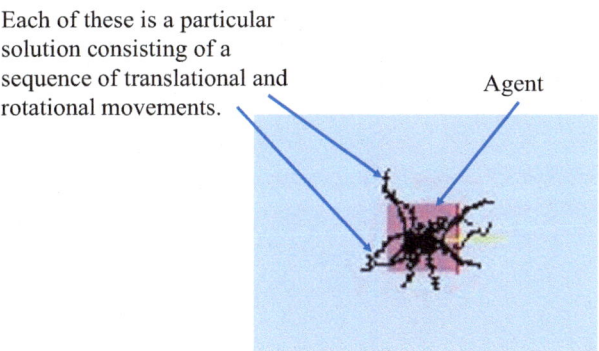

Each of these is a particular solution consisting of a sequence of translational and rotational movements.

Agent

Fig. 3.16 Incremental chunking in the Second Infant Learning Phase (computer simulation). The Agent moves 3, 6, 12, 24, and 48 steps away from the starting location/orientation, each step is either an elemental translational or an elemental (1°) rotational displacement. The chunked rules learned earlier (e.g., 3-step chunked rules) are used for the problem solving for the subsequent, more "complex" (in terms of number of steps) problems (e.g., problems with 6, 12, 24, etc. steps). The videos are shown on YouTube: http://youtu.be/ttJnTq9Puso and http://youtu.be/1pgPURS8pPc (or see https://noologyblog.wordpress.com) (©2013 IEEE. Reprinted, with permission, from Ho, S.-B. and Liausvia, F., "Incremental Rule Chunking for Problem Solving," Proceedings of the 1st BRICS Countries Conference on Computational Intelligence, Page 326, Fig. 8)

described in Fig. 3.14.[7] This YouTube video shows the learning of chunked rules for 3, 6, and 12 steps: http://youtu.be/ttJnTq9Puso (or see https://noologyblog.wordpress.com: Second Infant Learning Phase for Agent – Chunking 3–12 Steps Rules), and this video shows the same process for 24 and 48 steps: http://youtu.be/1pgPURS8pPc (or see https://noologyblog.wordpress.com: Second Infant Learning Phase for Agent – Chunking 24–48 Steps Rules). Training for 3–48 steps problems was also carried out in a similar manner for the entire assembly of the Agent attached to the Wall and this is shown in this video: http://youtu.be/aOYycWDir4M (or see https://noologyblog.wordpress.com: Second Infant Learning Phase for Agent and Wall – Chunking 3–48 Steps Rules).

After the Initial and Second Infant Learning processes, we put the system to the test in an "Adult Problem Solving and Learning Phase" in which much harder problems were given.[8] Figure 3.17 shows an Agent Moving problem involving the Agent itself moving from a START location/orientation to a GOAL location/

---

[7] In Fig. 3.16, it does not appear that the Agent, represented by the square, has moved very much. This is because the size of the Agent is 50 pixels by 50 pixels, and therefore even for the "longest" displacement – 48 elemental steps of actions away – from the original location, it does not result in a large amount of displacement from the starting location relative to the size of the Agent.

[8] Our criterion for considering what constitutes "Infant Learning" or "Adult Learning" is somewhat arbitrary here. Basically we trained the Agent in the "Infant" phase up till 48 steps of elemental movements and in this first "Adult" problem solving and learning scenario of Fig. 3.17, the problem presented consists of a separation of START and GOAL locations of about 300 pixels of elemental translational displacements.

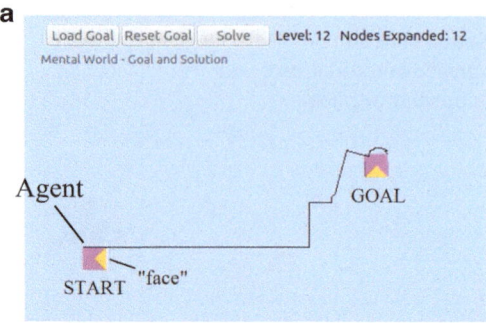

Trace of solution steps

Fig. 3.17 Agent Moving problem (computer simulation). (**a**) Agent is to move from a START location/orientation (a "face" defining its orientation) to a GOAL location/orientation. The movement of the Agent is traced by showing the movement of its "left-back" corner. (**b**) The trace of the solution steps. Numbers not in parenthesis are elemental steps. Numbers in parenthesis are chunked rules. The video is shown on YouTube: http://youtu.be/pkzoNnViHG4 (or see https://noologyblog.wordpress.com: Agent Moving) (©2013 IEEE. Reprinted, with permission, from Ho, S.-B. and Liausvia, F., "Incremental Rule Chunking for Problem Solving," Proceedings of the 1st BRICS Countries Conference on Computational Intelligence, Page 327, Fig. 9)

orientation – the Agent is required to not only move to a relatively distant location (about 300 pixels away) but it has to change its orientation (note that there is a "face" on the Agent defining its orientation). With this relatively "difficult" problem, if there is no earlier learned chunked knowledge to assist in the problem solving process, the solution search space would be combinatorially too large to be handled.

Figure 3.17a shows the final simulation solution. The solution is represented by the jagged line connecting the START and the GOAL states and this line is derived from tracing the movement of one of the "corners" of the Agent – in this case the "left-back" corner. Figure 3.17b shows the trace of the detailed solution steps. In the trace, the numbers not in parentheses are elemental steps, each number representing

3.5 Basic Idea of Incremental Knowledge Chunking 115

a different kind of elemental movement (e.g., forward, backward, left, right, etc.), and the numbers in parentheses are the numbers assigned to the chunk rules learned earlier and used here in the solution (rule numbers were assigned sequentially as new chunked rules were learned and added to the Chunked Rule Base of Fig. 3.13). It can be seen that quite a number of chunked rules were used in the solution. It can be seen in the solution trace that in total 12 chunked rules were used. Without these chunked rules, the search space would have been combinatorially large. Some of the chunked rules consist of about 50 steps of elemental movements, each elemental movement advancing the Agent toward the GOAL by about 1 pixel on the average (the rotational movements do not advance the Agent translationally). The video of the process can be seen here: http://youtu.be/pkzoNnViHG4 (or see https://noologyblog.wordpress.com: Agent Moving).

After the solution is found to the above problem, other than executing it as a solution, it is also stored as a chunked piece of knowledge in the Chunked Rule Base as shown in Fig. 3.13.

Next, we put the system to the test with a "Shelter Building Problem" similar to that of Fig. 3.12b as shown in Fig. 3.18. Figure 3.18a shows the START and GOAL states of the problem. The START state is represented by the yellow Walls scattered around in various locations and orientations in an environment. The GOAL state is represented by the orange Walls in a closed configuration. The Agent, as before, is the purple colored square with a "face." The Agent is to move the Walls from the START state to the GOAL state, represented by the orange Walls – forming an enclosed "Shelter." Similarly, using the A* algorithm (Hart et al. 1968) and the available chunked knowledge in the Chunked Rule Base, the Agent was able to solve the problem within a reasonable amount of time and this is shown in the sequence of steps in Figs. 3.18a–h as well as in the video on YouTube: http://www.youtube.com/watch?v=W0YVSOu1xbo (or see https://noologyblog.wordpress.com: Shelter Building).

Figure 3.18a, b show the Agent moving the first Wall into place, and Fig. 3.18c, d show the Agent moving the second Wall into place, and so on. Between Fig. 3.18a, b, it can be seen that the Agent pushes the first Wall from its starting position to the final position by some translational as well as rotational movements. Between Fig. 3.18b, c, the Agent moves from the first Wall's final position to the second Wall's starting position. Between Fig. 3.18c, d, the Agent attaches itself to the second Wall and "pulls" it toward the final position. Between Fig. 3.18d, e the Agent moves from the second Wall's final position to the third Wall's starting position. Between Fig. 3.18e, f, the Agent attaches itself to the third Wall and turns it around and pushes it into its final position. Between Fig. 3.18g, h, the Agent pulls and moves the fourth Wall into position and then detaches itself from it. A shelter around the Agent is thus formed.

Unlike the kind of precise and minimal distance movements generated by, say, a typical robotic problem solving method using inverse kinematics (Russell and Norvig 2010), what can be seen in these movements used by the Agent to move the Wall is something akin to the trial-and-error movements used by humans/animals to move objects into designated positions in the real world situation. This

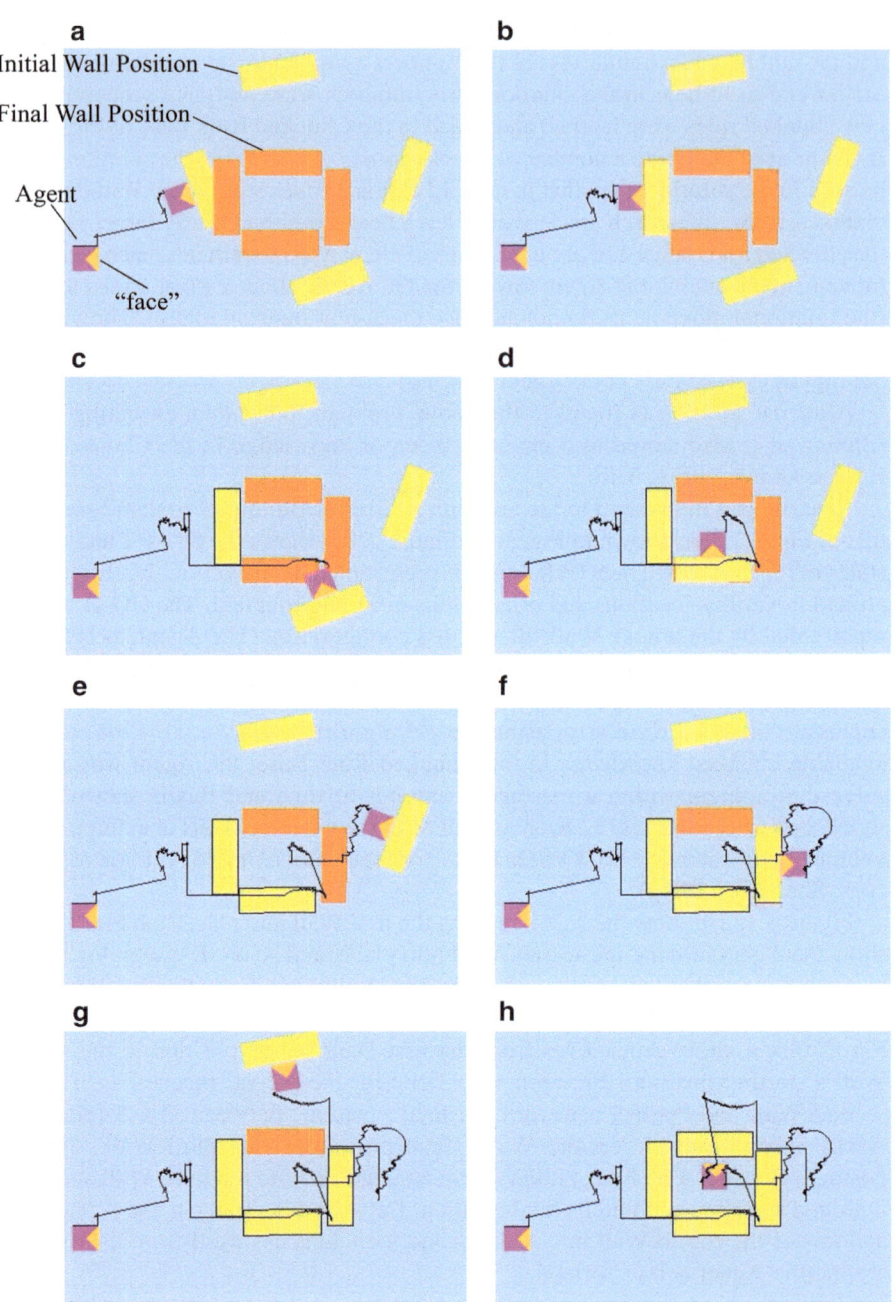

Fig. 3.18 Shelter Building problem (computer simulation). The *yellow Walls* represent the starting positions of the Walls and the *orange Walls* represent the GOAL positions – the configuration that makes up the Shelter. The *black lines* trace the movements of the "left-back" corner of the Agent. (**a**) Agent moves to and touches the first Wall. (**b**) Agent moves/pushes the first Wall into position. (**c**) Agent moves to the second Wall. (**d**) Agent attaches itself to the second Wall, and "pulls" and moves it into position. (**e**) Agent moves to the third Wall. (**f**) Agent attaches

3.5 Basic Idea of Incremental Knowledge Chunking

process is more clearly seen in the video http://www.youtube.com/watch?v=W0YVSOu1xbo on YouTube (or see https://noologyblog.wordpress.com: Shelter Building). The jagged black lines in the video tracing the movements of the Agent and the Wall (represented by the movement of the "left-back" corner of the Agent) are shown in the figures of Fig. 3.18.

For the problem of Fig. 3.18, the solution obtained consists of 226 consecutive chunked or elemental rules. Even though parts of the solution path are jagged and do not represent shortest distance paths, the search process is tractable because chunked rules were used.

Therefore, what we are trading off here is that using our method of causal learning and incremental chunking, the problem solution can be obtained very quickly, even though they may not be "optimal" in a "shortest distance," mathematical sense. Also, our system encodes the entire process of learning the basic physics to the incrementally chunked pieces of knowledge and their applications to problem solving, whereas typically in traditional robotic problem solving, the physical rules are built-in by human beings and not learned by the system, and no knowledge chunking is used. As before, as soon as this problem is solved, its solution becomes a chunked piece of knowledge to be used in future, more difficult problems. The method of causal learning is also general and can be applied to other domains. This basic framework of causal learning and incremental chunking is scalable to more complex environments and problem solving situations.

In Table 3.1, we tabulate the effects of changing the amount of Second Infant Phase learning on the amount of effort needed to solve the two Adult phase problems – the Agent Moving and the Shelter Building problems. The amount of Second Infant phase learning is defined as the longest chunked rules being learned. The table shows three learning regimes: (1) chunked rules of 3, 6, 12, 24, and 48 steps; (2) chunked rules of 3, 6, 12, 24 steps; and (3) chunked rules of 3, 6, 12 steps. When there was less Second Infant Phase learning, there were more chunked or elemental rules needed for the solution, and hence the time spent for the search process increased accordingly, as expected.

We also varied another parameter of the scenarios of Figs. 3.17 and 3.18. Recall that in the above simulation, the elemental translational movements that result from an action comprise movements of just 1 pixel. Now, we kept everything else constant but introduce a "large force" – an action that results in 3-pixel translational movements either in the forward, the backward, the left, or the right direction, for

---

Fig. 3.18 (continued) itself to and turns the third Wall around and moves it into position. (**g**) Agent moves to the fourth Wall. (**h**) Agent pulls and moves the fourth Wall into position and then detaches itself from it. Now the Agent has successfully built a Shelter around itself. The video is shown on YouTube: http://www.youtube.com/watch?v=W0YVSOu1xbo (or see https://noologyblog.wordpress.com: Shelter Building). The video shows that the solution obtained is human/animal-like and noologically realistic (©2013 IEEE. Reprinted, with permission, from Ho, S.-B. and Liausvia, F., "Incremental Rule Chunking for Problem Solving," Proceedings of the 1st BRICS Countries Conference on Computational Intelligence, Page 327, Fig. 10)

Table 3.1 The relationship between the amount of Second Infant Phase learning and the number of steps needed for the solutions of "Adult" problems: the Agent Moving and Shelter Building problems. When there is less Second Infant Phase learning, there are more steps needed for the solution

Problem	Number of steps to intermediate goals in Second Infant Phase learning		
	3,6,12,24,48	3,6,12,24	3,6,12
Agent moving	12[a]	17	42
Shelter building	226	252	331

[a]Number of rules (chunked or elemental) needed to solve the problems (©2013 IEEE. Reprinted, with permission, from Ho, S.-B. and Liausvia, F., "Incremental Rule Chunking for Problem Solving," Proceedings of the 1st BRICS Countries Conference on Computational Intelligence, Page 327, TABLE I)

the Agent or the Wall when moved by the Agent. We modified the Physics Simulator to implement this new physics.

We then tested the system with the same two problems – the Agent Moving problem of Fig. 3.17 and the Shelter Building problem of Fig. 3.18. The solutions obtained were similar except that now the Agent is able to incorporate this "faster" movement into many of the elemental rules as well as the chunked rules and the resulting solutions allow the GOAL states to be reached much faster. These can be seen in the following two videos: http://youtu.be/pAZ4l4TzP3Q and http://youtu.be/0XQZuAJvR8A (or see https://noologyblog.wordpress.com: the Large Forward Translational Force videos), corresponding to the scenarios of Figs. 3.17 and 3.18 respectively. Other similar "new physics" can be added – such as much larger translational forces or rotational forces that translate and rotate the Agent or the Wall by many more pixels or many more degrees respectively each time they are applied.

Thus this section successfully demonstrates that incremental knowledge chunking can greatly facilitate problem solving search process. The process also generates human/animal-like and noologically realistic solutions, as expected because natural noological systems also exploit this paradigm of learning and problem solving.

The incremental knowledge chunking process demonstrated in this section is for problems involving spatial movement. Incremental knowledge chunking, of course, is general and can be applied to all kinds of knowledge processed by a noological system. In Chap. 7 we demonstrate knowledge chunking for a different domain of knowledge.

Also, as mentioned in Chap. 1, Sect. 1.8, there have been earlier efforts in AI that deals with chunking (e.g., Alterman 1988; Carbonell 1983; Carbonell et al. 1989; Erol et al. 1996; Fikes et al. 1972; Hammond 1989; Laird et al. 1986, 1987). However, in these efforts, the domain specific rules that are used in the chunking processes are built-in and not learned. The issue is how to keep learning and chunking as we have demonstrated here.

3.6 Motivational Learning and Problem Solving

In the previous section, the learning and problem solving processes take place in a physical domain of spatial movement. In this section, we use a similar learning and problem solving paradigm to illustrate how the learning and problem solving can take place at a motivational level (Ho and Liausvia 2013a).

In Fig. 3.19, we show a process of causal discovery for an Agent in which a certain item with a certain shape in the environment can serve to increase its internal energy (effectively functioning as food) much like the situation discussed in Sect. 1.2 of Chap. 1. Figure 3.19 shows that at certain time $t=1$, the Agent is 1 pixel away from the Food (the hexagonal shape), and at time $t=2$, the Agent touches the Food. Internal to the Agent, there is an energy level that it keeps track of and at an elemental time interval after $t=2$, at $t=3$, its energy level goes up to a relatively high level as shown in the graph on the right side of the figure. Through a causal learning process like that described in Chap. 2, the Agent then encodes a causal rule as shown in the figure. This causal rule consists of a L.H.S. (left hand side) – the "IF" side – which depicts the external event and a R.H.S. (right hand side) – the "THEN" side – which depicts an internal event. We term this an EX-IN CAUSAL RULE (an EXternal event causing an INternal event). This rule is also stored in the system's Chunked Rule Base (Fig. 3.13).

Therefore, unlike in the case of physical processes depicted in Figs. 3.12 and 3.14 in which the effects of causes take place in the "outside" world (the physical world), in the motivational domain, the effects (such as "energy gain," "pain," "sadness," "anger," "depression," and "anxiousness") of various causes (such as "food," "pain-causing-event," "sadness-causing event," "anger-causing event," and "impending doom") are changes to some internal states of the Agent.

In Fig. 3.19 we show a kind of "food" that increases the internal energy of the Agent by a certain amount. There could be other kinds of food that increase the internal energy of the Agent by a different amount (more or less than that in Fig. 3.19). In Fig. 3.19 we also show that the increase of internal energy is one elemental time step after the external event and it is a step-wise increase. There could also be internal processes reacting to different kinds of food that result in a slower onset of energy increase or a gradual increase over some longer time interval rather than the step-wise increase as depicted in Fig. 3.19. The temporal representational scheme for the changes of internal states such as that depicted in Fig. 3.19 can represent all kinds of complex change profiles for the changes in internal states (see Sect. 4.3.8, Chap. 4, for some discussion on this.)

Figure 3.20 depicts a process of motivational problem solving that involves a level of motivational problem solving per se and another level of physical problem solving to support the motivational level problem solving in a backward chaining process.

The scenario begins with an Agent at a certain location in an environment consisting of a few pieces of Food placed in a few random locations as shown in the left side of Fig. 3.20. The Agent has a Current Energy Level, and the goal is to

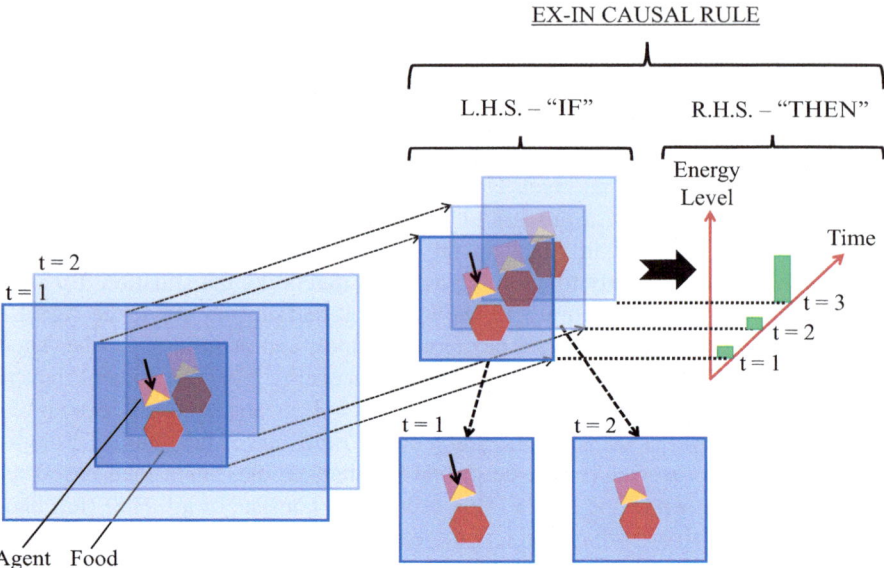

Fig. 3.19 An agent touches a piece of food (at t = 2) and its internal energy level is increased after that (at t = 3). This is encoded in an EX-IN CAUSAL RULE (With kind permission from Springer Science+Business Media: Proceedings of the 6th International Conference on Artificial General Intelligence, "Knowledge Representation, Learning, and Problem Solving," 2013, page 66, Seng-Beng Ho and Fiona Liausvia, Fig. 5)

increase that ultimately to a higher level (the ENERGY GOAL) as shown in the Energy vs Time diagram at the bottom right of Fig. 3.20. Because there is a gap between the Current Energy Level and the ENERGY GOAL, the system looks for causal rules in the Chunk Rule Base (Fig. 3.13) that has a R.H.S. that consists of energy increases. It finds the EX-IN CAUSAL RULE as learned earlier and as depicted in Fig. 3.19. It then begins a backward chaining problem solving process. The next step is to satisfy the EX-IN CAUSAL RULE's L.H.S. The EX-IN CAUSAL RULE's L.H.S. is a certain configuration of the environment that must be achieved and this consists of the Agent being in contact with a piece of Food (any one of those pieces of Food labeled 1, 2, and 3 in Fig. 3.20). Now, since the Agent's starting location is some distance away for any one of these pieces of Food, the system looks for a rule to change the location of the Agent to one that satisfies the L.H.S. of the EX-IN CAUSAL RULE. It finds the MOVEMEN-SCRIPT (Fig. 2.18) in the Chunked Rule Base that was learned earlier and applies that in a Physical Problem Solving process to satisfy the L.H.S. of the EX-IN CAUSAL RULE of Fig. 3.19 – which is to move the Agent to touch the Food.

In Fig. 3.20 it is shown that after consuming the first piece of Food, the Agent still has not satisfied the ENERGY GOAL. This is because of the "Energy Depletion Due to Movement Rule" of Fig. 3.6b which dictates that in executing the physical movement to the Food from the starting location, some energy is being

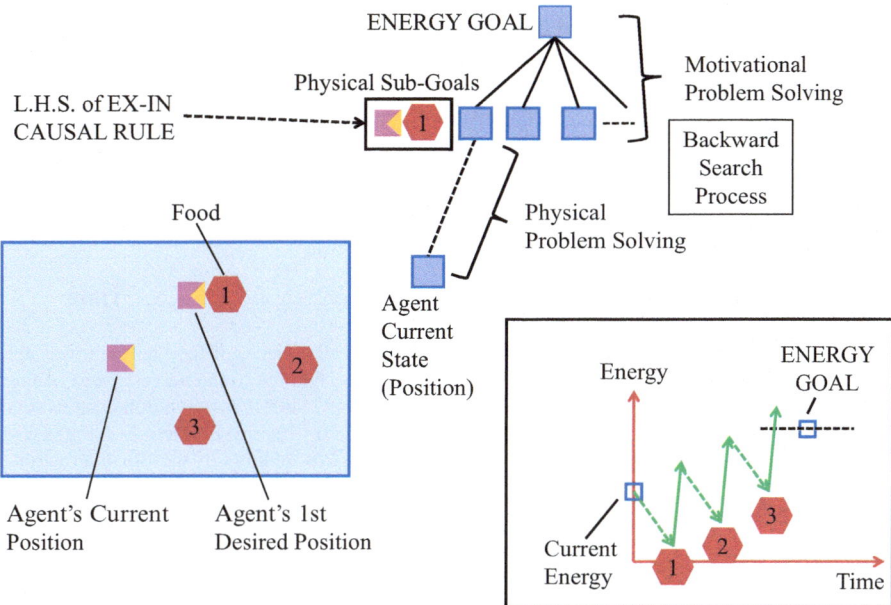

Fig. 3.20 Motivational problem solving process that involves two-level processes – a motivational level problem solving process per se and a physical level problem solving. The physical level problem solving involves invoking a MOVEMENT-SCRIPT (Fig. 2.18) to move the Agent to touch the Food based on the requirement of the EX-IN CAUSAL RULE (Fig. 3.19) invoked by the motivational level problem solving. There are 3 pieces of Food labeled 1, 2, and 3. Agent needs to eat all 3 pieces of Food to reach the ENERGY GOAL (With kind permission from Springer Science+Business Media: Proceedings of the 6th International Conference on Artificial General Intelligence, "Knowledge Representation, Learning, and Problem Solving," 2013, page 67, Seng-Beng Ho and Fiona Liausvia, Fig. 6)

depleted. Since the energy increase is more than the energy depleted, this is still considered a partially successful problem solving process. The process then continues to look for a second piece of Food to see if the ENERGY GOAL can be reached, and so on. Suppose at some point in the process of going from one piece of Food to the next the energy level dips below 0, it is considered a search dead end and the process is backtracked to the previous Food selection and a different piece of Food is selected to be tried next. The system carries out this trial and error forward simulation process until the ENERGY GOAL is reached. Figure 3.20 shows that after consuming the three pieces of Food in a certain order the ENERGY GOAL is exceeded.

We have encoded these processes in a computer simulation and the result of the simulation is shown in Fig. 3.21 – the simulation can be seen on YouTube: http://www.youtube.com/watch?v=exoVU0dX4RQ (or see https://noologyblog.wordpress.com: Energy Problem Solving). Four pieces of Food were used in the simulation and the Agent needed to consume all four of them to reach the ENERGY GOAL. It reached out to the nearest piece of food first. We have built in an EX-IN

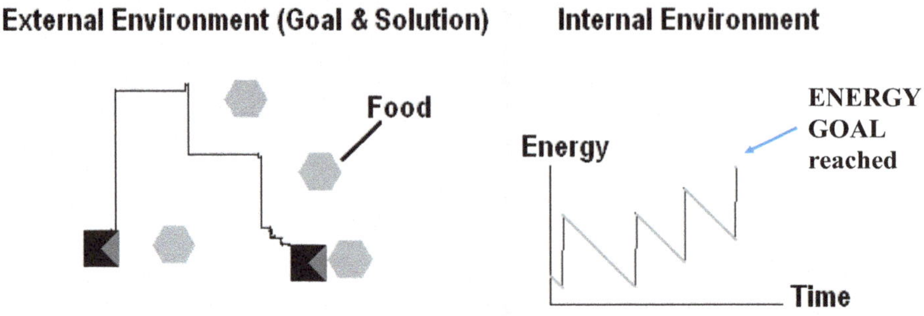

External Environment (Goal & Solution)   **Internal Environment**

Fig. 3.21 Computer simulation: the Agent had to consume all 4 pieces of food before it was able to increase its energy to exceed the ENERGY GOAL. It reached out to the nearest piece of food first. The simulation can be seen on YouTube: http://www.youtube.com/watch?v=exoVU0dX4RQ (or see https://noologyblog.wordpress.com: Energy Problem Solving) (With kind permission from Springer Science+Business Media: Proceedings of the 6th International Conference on Artificial General Intelligence, "Knowledge Representation, Learning, and Problem Solving," 2013, page 68, Seng-Beng Ho and Fiona Liausvia, Fig. 7)

CAUSE-EFFECT simulator (much like the Physics Simulator that simulates the cause and effect of physical processes in the previous section) for the purpose of this simulation. The Agent then learns the EX-IN CAUSAL RULE of Fig. 3.19 in its process of exploration. More kinds of food with different effects on the Agent's energy level can be added to the system.

Figure 3.22 shows a different, more real-world like scenario similar to the above food acquisition scenario, except that now the need to be satisfied by the Agent is to purchase various Items in a Shopping Mall. Each purchase of an Item satisfies a certain amount of "Purchasing Need." (In reality, different kinds of item purchased may satisfy different kinds of purchasing need but we simplify the scenario here.) There is a double-track planning situation here in which each purchase of an Item causes the Agent's Money Level to be decreased to the level that it is not enough for the next purchase (assuming the Agent had earlier learned the cost of the next item). Therefore, a visit to the ATM is needed after each purchase. Money functions like the Energy in the food acquisition scenario and the physical energy expanded in the movements in the Shopping Mall is assumed to be negligible. There is a PURCHASING GOAL to be achieved or PURCHASING NEED to be satisfied much like the ENERGY GOAL of Figs. 3.20 and 3.21. This GOAL is achieved after purchasing the three Items shown.

3.7 A General Noological Processing Framework

Having discussed in further details the various issues on needs/affective competition, incremental knowledge chunking, motivational learning and problem solving through backward search, etc. we now place everything in a more complete general

3.7 A General Noological Processing Framework

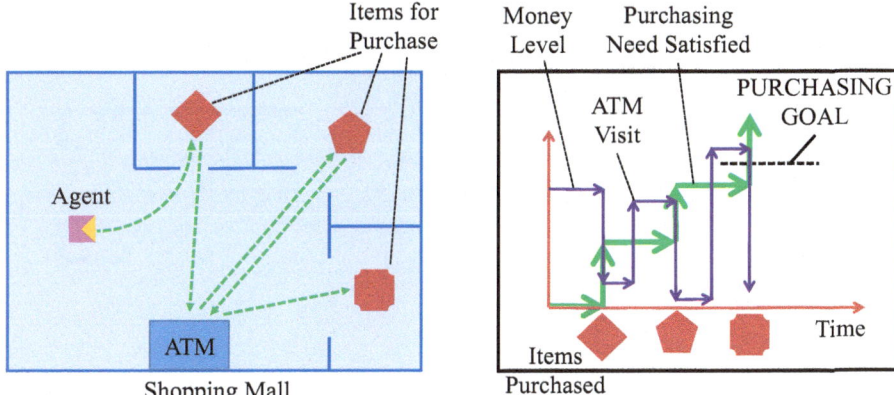

Fig. 3.22 A "real-world" scenario in which there is a need for an Agent to purchase certain Items in a Shopping Mall in order to achieve a PURCHASING GOAL or satisfy a PURCHASING NEED, similar to the food acquisition scenario in Figs. 3.20 and 3.21. Money is needed to purchase the Items and visits to the ATM is necessary to replenish the Money Level of the Agent. Physical energy expanded in the movements in the Shopping Mall is assumed to be negligible

noological processing framework as shown in Fig. 3.23. This is a further refinement of the general noological architecture of Fig. 1.7 in Chap. 1.

Figure 3.23 echoes the general architecture of the noological system as illustrated in Fig. 1.7 of Chap. 1, in which problem solving is the main processing backbone of the system. And then, there is a higher or highest level at which the problem to be attended to at any given instance depends on the outcome of needs/motivations competition. This is dictated by something akin to a Maslow hierarchy (Maslow 1954) – needs are organized into a hierarchical structure and the lower, more basal needs typically take precedence to be satisfied and within each level, various needs also compete to be addressed first. After a top priority need has been identified, the corresponding problem solving process is then engaged to solve the problem. The problem can be solved in a forward search or a backward search direction. As discussed, the forward search process and plan generation (Figs. 2.28 or 3.13, here embedded within a larger block of processes) is assisted/accelerated by heuristics and script-type knowledge learned earlier and stored in the Heuristic Rule Base and Chunked Rule Base respectively.

The Chunked Rule/Script Base is built-up through an incremental knowledge chunking process discussed in Sect. 3.5. The backward search process makes use of the Script Base for problem solving and plan generation. The (internal and external) perceptual/sensory processes provide the information of the real-world (including internal environment) in the form of various parameters encoding the states of the entities involved. The conceptual process helps to create general causal rules from experience through rapid and appropriate generalization (Sects. 2.4 and 2.5 of Chap. 2). Both perceptual and conceptual processes are thus participating in a form of *service* function to the noological system's backbone problem solving processes.

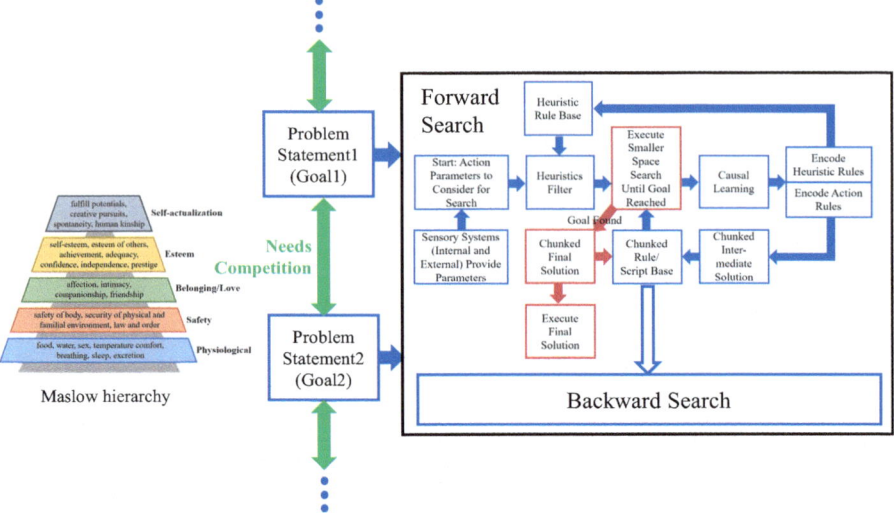

Fig. 3.23 A general noological processing framework (See text for explanation)

This framework is scalable due to heuristic rule learning and incremental knowledge chunking. Another important aspect of the system, namely semantic grounding, will be discussed in the next chapter. This provides robustness to the system in terms of its ability to "understand" the concepts involved and hence process concepts not only at the symbolic level but also at the functional and operational levels.

3.8 Neuroscience Review[9]

The main thrust of this book is to develop a framework embodying general principles for noological systems. The main purpose of this section is to survey the neuroscience literature to see if the neuroscientists have uncovered any interesting general principles of operations of the brain that may be compared and contrasted with the principles we develop here. As it turns out, there are indeed general structures of brain function embodying certain general principles of learning mechanisms that cut across vastly disparate domains of brain operations that some neuroscientists believe exist in the brain. The consensus in the neuroscience community in this regard is not unanimous and the general principles of operations

---

[9] About 50 % of the paragraphs in this section are derived, with minor modifications, from the following source: ©2013 IEEE. Reprinted, with permission, from Ho, S.-B., "A Grand Challenge for Computational Intelligence: A Micro-Environment Benchmark for Adaptive Autonomous Intelligent Agents," Proceedings of the IEEE Symposium Series on Computational Intelligence – Intelligent Agents, Pages 49–52, Section V.

as currently characterized in neuroscience do not seem to map 100 % onto our architecture here such as that articulated in Fig. 3.23. The neuroscience architecture is computationally at a coarser level and ours is more computationally detailed, and this is perhaps why it is difficult to map some of the mechanisms in Fig. 3.23 onto known neurobiological functions currently. We have also stated in the very beginning of this book, in Chap. 1, that part of our motivation for investigating detailed computational noological mechanisms is to guide neuroscience, as well as the other cognitive sciences, to further characterize natural neural systems in concrete computational details. Nevertheless, it is interesting to see that there are indeed general architecture and principles of operations that are deemed to exist in the brain and we believe it would be instructive to review this literature and attempt to map some of the features of the general brain architecture onto some of the mechanisms in our foregoing discussions.

3.8.1 The Basal-Ganglionic-Cortical Loops

Traditionally, neuroscientists focus their attention on the cerebral cortex when studying the nervous systems of higher animals, especially with regards to attempting to understand higher cognitive functions. There are probably two main reasons for it. One is that it is relatively accessible compared to the subcortical structures such as the basal ganglia and the thalamus. Many cognitive phenomena have also been observed to be correlated with activities on the cortical surfaces – for example, the processes associated with the various modalities of perception, language, decision making, planning, etc. all have cerebral cortical level correlates (Gazzaniga et al. 2013). Traditionally, major subcortical structures such as the basal ganglia and the cerebellum have been associated with motor functions (Allen and Tsukahara 1974; Bhatia and Marsden 1994; Brooks and Thach 1981; DeLong and Georgopoulos 1981; Glickstein 1997). However, in the past 20 years or so, an increasing body of evidence has accumulated that pointed to the important and indispensable roles of the basal ganglia and the cerebellum in higher level cognitive processes (Alexander et al. 1986; Allen and Courchesne 2003; Andreasen and Pierson 2008; Houk 1997; Ito 2008; Kim et al. 1994; Middleton and Strick 2000; Owen et al. 1998; Strick et al. 2009). What is particularly interesting about this new understanding is that the structures and processes discovered seem to embody some general principles – that no matter what modality of sensory processing is involved (e.g., whether it is visual, auditory, or somatosensory) or what type of cognitive processing is involved (e.g., whether it is sensory or motor in nature, whether it is lower level motor execution or higher level planning in nature, or even whether it is more "affective" or more "cognitive" in nature) similar structures and processes seem to underlie these functions (Alexander et al. 1986; Middleton and Strick 2000; Seger 2008; Seger et al. 2010). This idea is illustrated in Fig. 3.24.

Figure 3.24 depicts four parallel corticostriatal loops that course through the various parts of a group of prominent subcortical structures, the basal ganglia. The

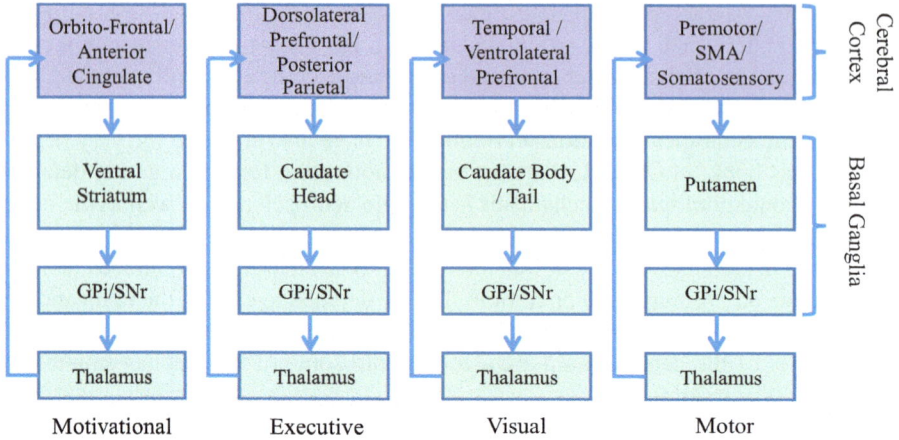

Fig. 3.24 General structures across various aspects of cognitive processing – four primary corticostriatal loops in the brain engaging in motivational, executive, visual, and motor functions. GPi: Globus pallidus, internal segment. SNr: Substantia nigra pars reticulata (Reprinted from Neuroscience and Behavioral Reviews, Vol. 32, Carol A. Seger, "How do the basal ganglia contribute to categorization? Their roles in generalization, response selection, and learning via feedback," Pages 265–278, Figure 2, 2008, with permission from Elsevier)

basal ganglia lie at the center of the brain and has extensive connections with the cerebral cortex. Substructures of the basal ganglia include the caudate nucleus, the putamen, the ventral striatum, the globus pallidus, etc. Signals originating from the cerebral cortex project onto the basal ganglia and after coursing through the various substructures within the basal ganglia region, return via the thalamus to the cerebral cortex. In Fig. 3.24, each of the four loops have been identified to subserve a different domain of cognitive processing – from that of the motivational to the executive, the sensory, and the motor – but all of them have similar loop structures. The top boxes of the loops show the corresponding areas on the cerebral cortex from which the projections to the basal ganglia originate. Each of these cortical areas – Orbito-Frontal/Anterior Cingulate, Dorsolateral-Prefrontal/Posterior Parietal, Temporal/Ventrolateral Prefrontal, and Premotor/SMA/Somatosensory – have been observed to participate in affective/motivational, executive/planning, sensory, and motor functions respectively. The first stop of the projections from the cerebral cortex in the basal ganglia is the striatum area, and different loops engage different parts of the striatum – the ventral striatum, the head of the caudate nucleus, the body and tail of the caudate nucleus, and the putamen respectively for the four loops illustrated in Fig. 3.24. The next stops are the globus pallidus internal segment (GPi) and substantia nigra pars reticulata areas (SNr). The output of the basal ganglia from the GPi then projects through the thalamus and back to the same cortical areas from which the corresponding projections originate. There are some more complex connectivities and side loops within the basal ganglia that have been omitted in Fig. 3.24 but the bulk of the recurrent structure is captured succinctly in Fig. 3.24. There are also connections between the cerebral cortical areas (the top

3.8 Neuroscience Review

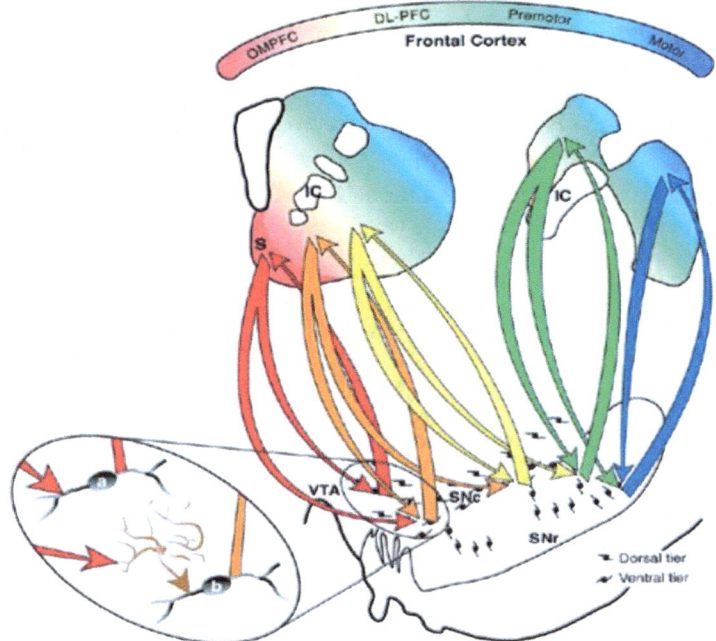

Fig. 3.25 Spiral connections between the various basal ganglia loops (Republished with permission of Society of Neuroscience, from "Striatonigrastriatal Pathways in Primates Form an Ascending Spiral from the Shell to the Dorsolateral Striatum," The Journal of Neuroscience: The Official Journal of the Society for Neuroscience, Suzanne N. Haber, Julie L. Fudge, Nikolaus R. McFarland, Vol. 20, No. 6, 1981; permission conveyed through Copyright Clearance Center, Inc.)

boxes in Fig. 3.24) that are not shown. Other research has also identified many more loops than those shown in Fig. 3.24 (Alexander et al.1986; Dum and Strick 2009) but Fig. 3.24 suffices for the purpose of our discussion.

An important point to emphasize is that the parallel loops depicted in Fig. 3.24 do not function independently nor are they anatomically separate. Firstly, there is a massive amount of connections between the cortical areas. Secondly, research has shown that the loops are connected through structures deep within the basal ganglia in a form of a spiral – "upstream-loop," e.g., the affective and motivational loop, courses through the lower portions of the basal ganglia (the substantia nigra area) and then back to the upper level of the "downstream-loop," such as the executive/planning loop, and so on as depicted in Fig. 3.25 (Haber et al. 2000). It has been shown that in a typical cognitive process such as a visual categorization task, all the loops are engaged to address the task (Seger 2008; Seger et al. 2010).

Our paradigm for a noological system as articulated in the previous chapters as well as the Shield-and-Shelter (SAS) micro-environment to be described in Chap. 8 map very well onto the general neuronal structures depicted in Figs. 3.24 and 3.25. Firstly, the general noological paradigm articulated in Chap. 1 as well as in the SAS micro-environment of Chap. 8 recognize that affective and motivational factors are the driving force behind an organism's actions to improve its chances of survival (e.g., avoidance of undesirable value signals such as *pain*) and they drive the activation of its cognitive processes (e.g., problem solving processes) to discover a solution. Signals such as *pain* and their related higher level characterizations such as *anxiousness, stress, relief*, and *comfort* (Chap. 8) provide the motivation for the organism to carry out actions on itself and on the world (e.g., seeking *shields* and building *shelters*) to effect certain desired affective goals. Conceptualization processes ensue (such as forming the fully embodied, situated, and grounded concepts *shield* and *shelter*) to aid future behavior. It is no coincidence that we have placed Affective Processes strategically at the center among all the levels in Fig. 1.2 in Chap. 1 because of their central importance. Mapping this onto Figs. 3.24 and 3.25 we observe that the affective/motivational loop being at the most "upstream" position, is in a position to send signals to and direct the more "downstream" loops to effect certain cognitive and motor processes.

The various processes outlined in Fig. 1.2 show planning, motor, sensory, and memory processes all acting to satisfy the organism's internal affective goals. (Similarly, for the processes associated with the SAS micro-environment in Chap. 8, they also act to satisfy the agent's internal affective goal.) These can be mapped onto the various basal ganglia loops depicted in Figs. 3.24 and 3.25 – the affective/motivational, the executive/planning, the sensory, and the motor loops. The precise nature of this mapping has not been identified but Figs. 3.24 and 3.25 dictate that the mechanisms involved must be general – they are applicable to different types of cognitive processes from the affective to the motor.

What then do the basal ganglia do, if they lie at the prominent center of the brain, are connected to all regions of the cerebral cortex, and seem to function in a general manner? A current consensus is that it is involved in reinforcement learning (Houk et al. 1995). Research in the computational domain had established the mathematical foundation of reinforcement learning (Sutton and Barto 1998) and efforts have been ongoing to attempt to map the basic reinforcement mechanisms to the various physiological and anatomical data of the basal ganglia (e.g., Joel et al. 2002). Reinforcement learning is a general mechanism that can be applied to any domain of cognitive processing. It is conceivable that the relatively similar anatomical structures (and attendant similar physiological processes) across the different subregions of the basal ganglia (i.e., those that are engaged in the various loops) subserve a similar function and implement reinforcement learning. At the heart of reinforcement learning is action selection – the selection of the "best" action with respect to some total future reward. This is applied to the signals from the cerebral cortex in a general way across all domains of cognitive and affective processing.

3.8.2 The Cerebellar-Cortical Loops

As mentioned earlier, there is also increasing evidence that another major subcortical structure – the cerebellum – which is traditionally thought to be involved only in motor function is also involved in cognitive functions ranging from affective processes to language processes, general thinking processes, and attentional processes (Allen and Courchesne 2003; Andreasen and Pierson 2008; Houk 1997; Kim et al. 1994; Strick et al. 2009). The cerebellum is likewise connected to the cerebral cortex in loops much like the basal ganglia loops depicted in Figs. 3.24 and 3.25 – i.e., signals originating from a certain area of the cerebral cortex would first course through a brainstem structure called the pons and input to the cerebellum. The signals then course through certain regions of the cerebellum and output through some cerebellar deep nuclei such as the dentate nuclei back to the same regions of the cerebral cortex from which they originate. (Different cerebellar deep nuclei are involved in different "loops.") In phylogeny, when the prefrontal cortex expanded, the deep cerebellar output nuclei that are normally connected to the prefrontal regions – the dentate nuclei - also expanded accordingly. This signifies the importance of the cerebellum in higher functions such as decision making, planning, and various affective and motivational processes, because it has been established that the prefrontal cortex is involved in these processes (Fuster 2008).

3.8.3 The Houk Distributed Processing Modules (DPMs)

Combining both the basal ganglia and the cerebellar loops, Houk (2005) proposed that the brain is actually made up of many similar distributed processing modules (DPMs), each of which consists of a piece of the cerebral cortex connected to a piece of the basal ganglia and a piece of the cerebellum as illustrated in Fig. 3.26. Traditionally, the cerebellum has been understood to be performing the function of motor refinements – generating fine motor commands to direct the limbs to effect specific accurate actions. To achieve this, the cerebellum is being trained by error signals in a supervised learning process to learn to model the skeletal-muscular subsystem accurately so that it can send signals to produce actions in a feedforward and fast manner. When applied to cognitive processes, the equivalent mechanisms would be to produce the accurate outcomes of some thought processes – hence thought processes can be conceived as "internal actions" which has to be controlled and refined like external actions.

Houk (2005) further suggests that the basal ganglia's role is to select the ballpark action from many possibilities present in the cerebral cortex, and the cerebellum then refines the action. Accordingly, this is how the basal ganglia and the cerebellum coordinate their functions in the DPMs. Houk (2005) proposes that the entire brain consists of about 100 of these DPMs, each of which is connected to about

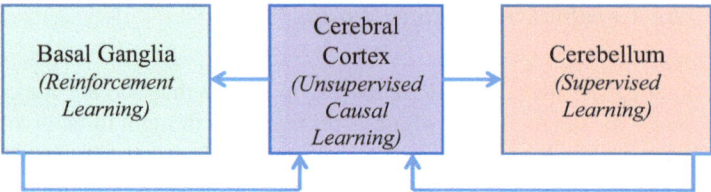

Fig. 3.26 A distributed processing module (DPM) (Houk 2005) and the learning mechanisms with its 3 subparts – reinforcement learning, unsupervised causal learning, and supervised learning – that are the main learning mechanisms in the basal ganglia, the cerebral cortex and the cerebellum respectively (©2013 IEEE. Reprinted, with permission, from Ho, S.-B., "A Grand Challenge for Computational Intelligence: A Micro-Environment Benchmark for Adaptive Autonomous Intelligent Agents," Proceedings of the IEEE Symposium Series on Computational Intelligence – Intelligent Agents, Page 51, Figure 16)

7 other modules, and together they form the bulk of the mechanisms for noological processing in the brain.

Doya (2009) also looks at the three subsystems of the brain – the basal ganglia, the cerebellum, and the cerebral cortex – from a learning mechanism point of view. He proposes that the basal ganglia are involved in reinforcement learning, the cerebellum is involved in supervised learning and the cerebral cortex is involved in unsupervised learning. Research on cortical plasticity had revealed that the various areas on the cerebral cortex, despite the fact that they seem to have definite functions in ordinary circumstances, is able to be reprogrammed to function in an entirely different domain and manner under other circumstances. For example, if the auditory input to what is normally functioning as the auditory cortex is being severed and instead visual input is connected to the auditory cortical regions, the auditory cortex can acquire the abilities to process visual information (Roe et al. 1992; Sur et al. 1988). Likewise, in a blind person whose visual cortex no longer receives any visual input, his visual cortex can be taken over by the somatosensory modality to process somatosensory signals (Sadato et al. 1996). This evidence strongly indicates that the cerebral cortex functions in an unsupervised manner and learns to classify and process whatever information is being received by it.

Among the three major subsystems – the basal ganglia, the cerebellum, and the cerebral cortex – the cerebellum has been found to have a very uniform anatomy across its entire bulk. Therefore, it does not matter which region of the cortex it is connected to, it seems to function in exactly the same manner (in performing supervised learning, as mentioned above). The uniformity of the basal ganglia has not been established with certainty but it would appear that it is quite uniform as well (and they perform reinforcement learning across all the various aspects of noological processing). The cerebral cortex has variations in its cytoarchitecture in different regions of the cortex and neuroanatomists have identified many "cortical areas" (e.g., the Brodmann areas) based on these cytoarchitectonic variations even though there is a lot of similarity in the neuronal elements and the microcircuits across all these areas. One can identify a canonical cortical circuit that is a general and basic circuitry that subserves some general functions (Douglas and Martin 1989, 2007; Mountcastle 1982). Mountcastle (1982) observes that the cerebral cortical circuits in various parts of the brain are basically similar, despite the fact that they serve functions as disparate as vision, audition, touch, language, thinking, problem solving, and motor output. As mentioned above, the generality, plasticity, and reprogrammability of the various areas of the cortex can be seen in various experiments (Roe et al. 1992; Sur et al. 1988; Sadato et al. 1996). Therefore, the cerebral cortex basically performs unsupervised learning (as indicated in Fig. 3.26) which picks up regularities in its input signals without a "supervisor."

Recently, a number of researchers have proposed that the cerebral cortex's general operation is that of "hierarchical Bayesian inference" (e.g., Friston 2010; George 2008; George and Hawkins 2009; Kersten et al. 2004; Lee and Mumford 2003) and regularities in the environment can be learned by the cerebral cortex and then subsequently used for inference in the sensory levels as well as other levels of the cerebral cortical hierarchy that involve language processing, problem solving and motor output. This provides more finesse to the basic idea of the cerebral cortex performing unsupervised learning. The purpose of the Bayesian inference is to uncover *causes* in the environment (Friston 2010; George 2008; George and Hawkins 2009; Kersten et al. 2004; Lee and Mumford 2003). Thus, the cortex should be characterized as performing "unsupervised causal learning" as indicated in Fig. 3.26. Passingham and Wise (2012) also highlighted that the granular prefrontal cortex (a more advanced kind of cortex found in higher animals such as primates and humans) facilitates "single event" learning based on *causal analysis* and this is a vast improvement over the evolutionarily older kind of learning – reinforcement learning – that requires many repetitions. This is the kind of learning that we have specified in Chap. 2 as well as for the SAS micro-environment in Chaps. 8 and 9.

3.8.4 The Computational Power of Recurrent Neural Networks

Doya (2003) highlighted one fundamental commonality among the three major brain structures – the basal ganglia, the cerebellum, and the cerebral cortex – which is that they are all structured in a recurrent fashion in their gross anatomy (e.g., the recurrent loop structure for the basal ganglia shown in Fig. 3.24). This is illustrated in Fig. 3.27. In the field of neural network research, it has been established that recurrent neural networks consisting of neuronal elements with sigmoidal activation functions are Turing computable (Kilian and Siegelmann 1996; Siegelmann and Sontag 1991). Moreover, if the neural network weights are allowed to take on real values, i.e., if the recurrent neural network is analog, then it is super-Turing computable (Siegelmann 2003; Siegelmann and Sontag 1994) which means it can compute a class of functions that is a superset of recursive functions.

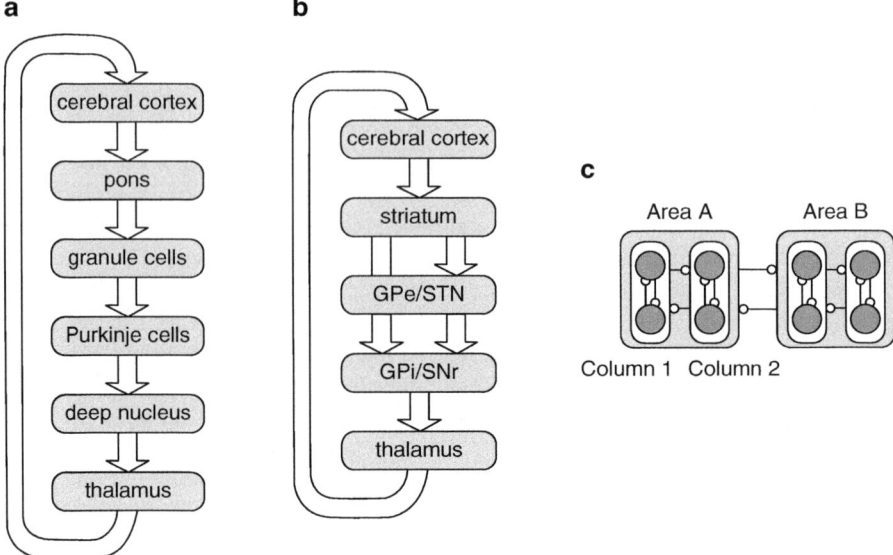

Fig. 3.27 Recurrent network structures in the cerebellum (**a**), the basal ganglia (**b**) and the cerebral cortex (**c**) (Arbib, Michael A., ed., *The Handbook of Brain Theory and Neural Networks*, *second edition*, figure 2, page 959, ©2002 Massachusetts Institute of Technology, by permission of the MIT Press)

Therefore, a recurrent neural network can potentially implement general and powerful functions and since computation lies at the core of the processes in a noological system, this implies that a connectionist intelligent system must have a recurrent network at its core to have general computational capabilities. From the point of view of generating behavior, a recurrent network can emit a sequence of correct and controlled micro-actions, guided by the requirements of the various goals and plans of the noological system involved. Thus, it should be central to the noological system's functions.

Mapping the computational insight from neural network research – i.e., there must be a *core* recurrent neural network guiding overall behavior – to the observation that the three major substructures in the brain – the basal ganglia, the cerebellum, and the cerebral cortex – are all structured in a recurrent manner, which substructure is the most likely candidate for the core recurrent neural network? We postulate that the basal ganglia system is the core recurrent neural network as it implements reinforcement learning that controls and directs actions to achieve the most optimized behavior for the organism with respect to some total reward and goal functions.

Neuroscience data support the view that the basal ganglia lie at the core of noological processing. Figure 3.28 shows the locomotor network of a lower vertebrate – the lamprey – in which the basal ganglia sit atop of and is intimately connected to the lower locomotor neural circuits to direct and control their

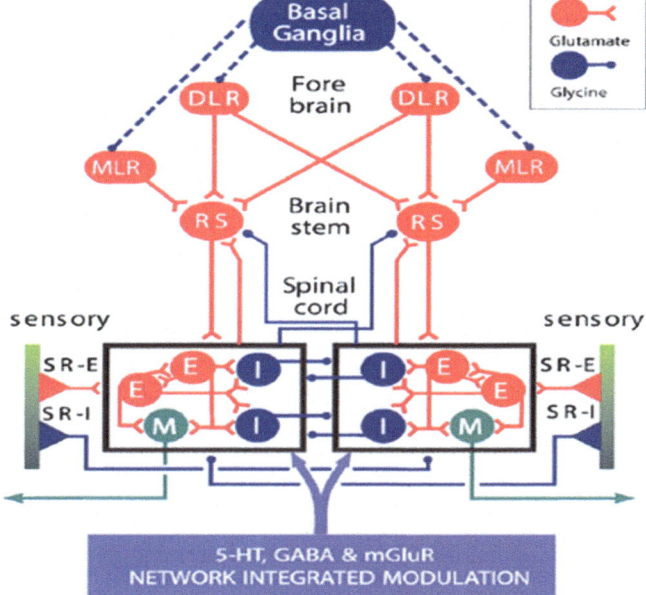

Fig. 3.28 The lamprey's locomotor neural network with the basal ganglia sitting at the *top* (Reprinted from Brain Research Reviews, Vol. 57, Sten Grillner, Peter Wallen, Kazuya Saitoh, Alexander Kozlov, Brita Robertson, "Neural bases of goal-directed locomotion in vertebrates-An overview," Pages 1–12, Fig. 6, 2008, with permission from Elsevier)

operations that ultimately translate to the lamprey's goal-directed actions and behavior (Grillner et al. 2008). Whether an animal is "higher" or "lower" in the evolutionary hierarchy, goal-directed behavior lies at the core of its intelligent functioning (Fig. 1.7). Higher form animals have more complex repertoires of sensory and motor behavior and this is subserved by their larger and more elaborate cerebral cortices. It is instructive to consider the neural structures of lower form animals because their cerebral cortices are relatively less prominent (thus limiting their sensory and motor repertoires) and hence the "core" of the neural structures that drives goal-directed behavior is less obstructed by the cerebral cortex. Putting this in the context of Fig. 1.17 of Chap. 1, lower form animals have a smaller *width* of processing but still must have enough depth of processing – including the *core* recurrent processing – to generate the necessary intelligent behavior to ensure survival. All vertebrates, no matter whether they are lower or higher in the evolutionary hierarchy, possess the three basic neural structures – the basal ganglia, the cerebellum, and the cerebral cortex – necessary for their intelligent functioning.

Therefore, in constructing artificial systems in the spirit of natural noological systems as revealed by the above neuroscience insights, we should target to design and build general modules such as that of the basal ganglia, the cerebellum, and the

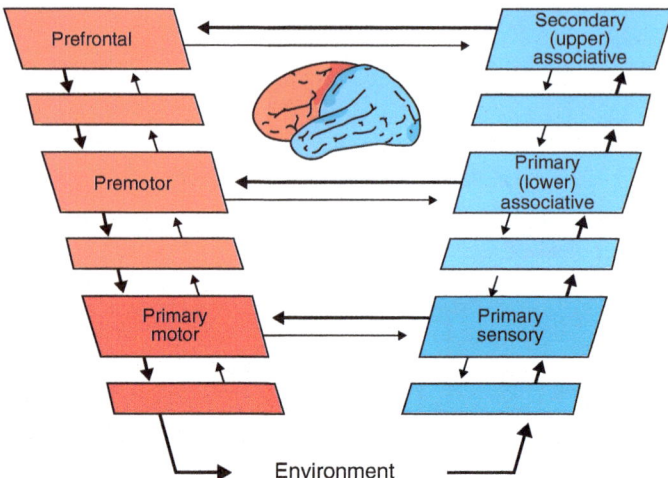

Fig. 3.29 Hierarchical, reciprocal, parallel, and serial organizations of the areas in the cerebral cortex (Reprinted from The Prefrontal Cortex, Fourth Edition, Joaquin M. Fuster, Overview of Prefrontal Functions: The Temporal Organization of Action, Page 360, Figure 8.3, 2008, with permission from Elsevier)

cerebral cortex and wire them together in some manner to perform the required noological functions as depicted in Fig. 1.2. As mentioned, detailed computational models already exist for these modules separately. For the basal ganglia, the reinforcement learning computational mechanisms have been quite well understood (Sutton and Barto 1998). For the cerebellum, the supervised learning mechanism has been extensively investigated by the neural network community (e.g., Bishop 2006). And for the cerebral cortex, a relatively sophisticated computational model of the cerebral cortical circuit is the hierarchical temporal memory (George 2008; George and Hawkins 2009). In Doya (2009), some models are proposed in which computational modules incorporating the three types of learning are integrated into a goal-directed general learning system. A general computational module that implements a general function such as that of the DPM (Fig. 3.26) that is scalable could use these known computational mechanisms as a starting point.

3.8.5 Overall General Brain Architecture

We are now ready to integrate the foregoing ideas on cortical and subcortical structures into an overall brain architecture. Firstly, we consider the current understanding of how the various cortical areas are organized on the cerebral cortex. Fuster (2008) summarized this most succinctly in a picture shown in Fig. 3.29 (same as Fig. 1.8a of Chap. 1). Firstly, the environmental input enters through the Primary sensory areas. From there it goes on to the Secondary sensory areas, etc. in

Fig. 3.30 Phylogenetic changes of the cerebral cortex – overall, cortical areas increased in size and specifically, the frontal areas increased more in size (Reprinted from The Prefrontal Cortex, Fourth Edition, Joaquin M. Fuster, Anatomy of the Prefrontal Cortex, Page 12, Figure 2.2, 2008, with permission from Elsevier)

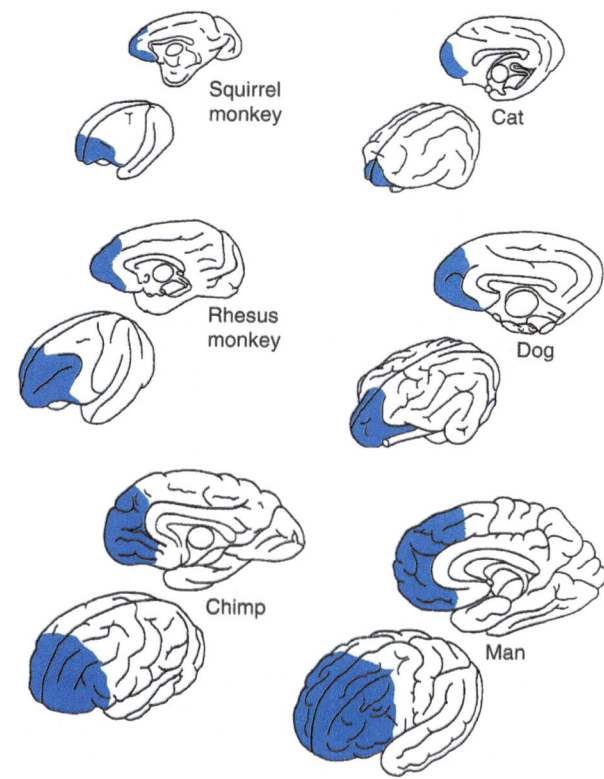

a hierarchy with both bottom up and top down connections. The various levels of the sensory areas also connect to the frontal cortex (the horizontal connections in the figure). The sensory cortices at the various levels connect to the frontal cortices at the corresponding levels reciprocally. The various areas, sensory and frontal, thus form a parallel-serial kind of hierarchy with reciprocal connections everywhere.

Phylogenetically, it has been observed that the higher form animals not only have more cortical areas, there are also relatively more frontal cortical areas as shown in Fig. 3.30. However, the connections from the frontal cortical areas to the basal ganglia and the cerebellum remain the same – i.e., from the point of view of the cerebral cortex-basal ganglia-cerebellum DPMs as shown in Fig. 3.26, there are just more of these modules and they form the corresponding frontal areas DPMs. As organisms evolve, their actions become more complex, their goals driving their actions more remote in space and time relative to the sensory triggers, and the reasons or motives for attaining these goals also become more indirect and more based on prior experience than on immediate instinctual needs. These are related to the increased frontal areas. What is interesting is that these increases in complex behavior are brought about by more of the same kind of cortical modules (i.e., the DPMs), reflecting some general mechanisms at work.

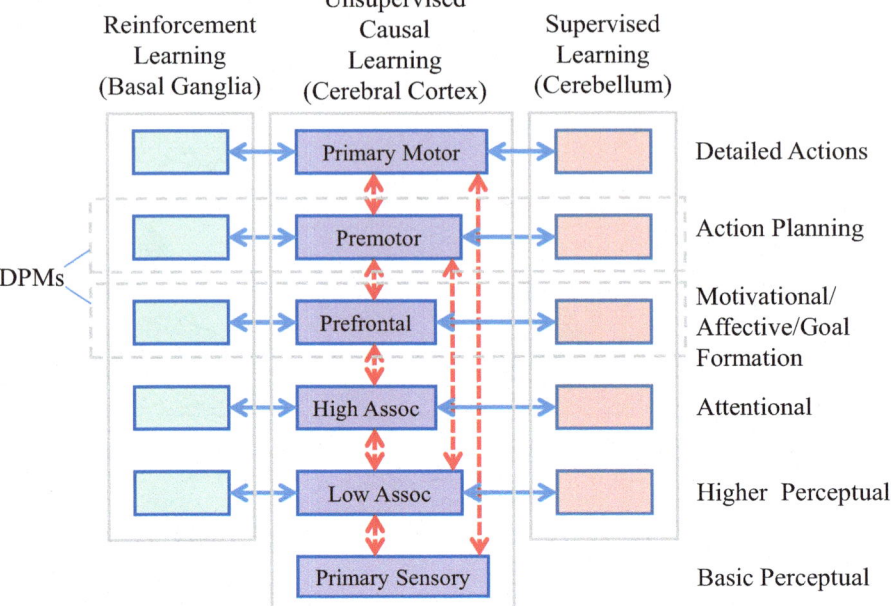

Fig. 3.31 A neuroscience-inspired general noological architecture that incorporates, in a hierarchical structure, general learning modules (DPMs) that are applicable to a wide range of noological processes and that is hence scalable. The cerebral cortical areas in the "unsupervised causal learning" section are typical cortical areas performing various functions from perception to action (©2013 IEEE. Reprinted, with permission, from Ho, S.-B., "A Grand Challenge for Computational Intelligence: A Micro-Environment Benchmark for Adaptive Autonomous Intelligent Agents," Proceedings of the IEEE Symposium Series on Computational Intelligence – Intelligent Agents, Page 52, Figure 17)

The overall general brain/noological architecture is thus shown in Fig. 3.31 (Ho 2013), combining the hierarchical-reciprocal-parallel-serial organization of the various cortical areas (Fig. 3.29) with the subcortical modules made up of the basal ganglia and the cerebellum (The Houk modules – Fig. 3.26). Organisms exhibiting different levels of complex behaviors have more or fewer of these modules but they are organized in a similar manner.

In Fig. 3.31 it is shown that the Houk module is not present for the primary sensory areas as anatomical evidence shows that the primary sensory areas generally do not connect to the basal ganglia and the cerebellum (Nolte 2009).

3.9 Mapping Computational Model to Neuroscience Model

In Sect. 3.6 we presented a computational model and simulation of motivational learning and problem solving that involve a two-level problem solving process – the motivational level per se and a physical level that is engaged to serve the

3.9 Mapping Computational Model to Neuroscience Model

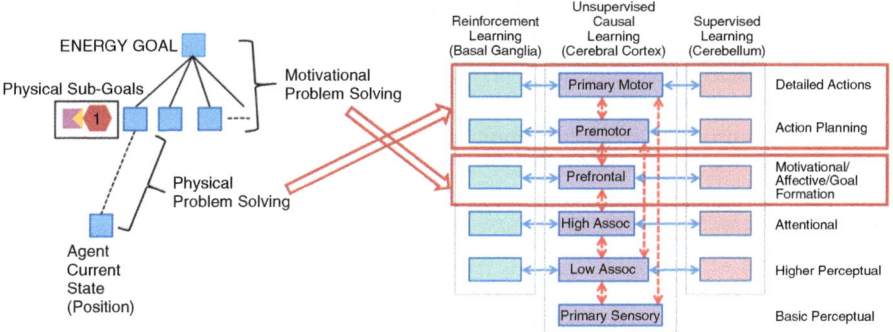

Fig. 3.32 The mapping of the two-level motivational problem solving processes of Fig. 3.19 (*left*) to the neuroscience-inspired general noological architecture of Fig. 3.30 (*right*). The motivational problem solving level per se maps onto the prefrontal cortical level, and the physical problem solving level maps onto the premotor and primary motor areas (*Left* side of figure: With kind permission from Springer Science+Business Media: Proceedings of the 6th International Conference on Artificial General Intelligence, "Knowledge Representation, Learning, and Problem Solving," 2013, page 67, Seng-Beng Ho and Fiona Liausvia, Fig. 6; *right* side of figure: ©2013 IEEE. Reprinted, with permission, from Ho, S.-B., "A Grand Challenge for Computational Intelligence: A Micro-Environment Benchmark for Adaptive Autonomous Intelligent Agents," Proceedings of the IEEE Symposium Series on Computational Intelligence – Intelligent Agents, Page 52, Figure 17)

requirement of the motivational level. This two-level process can be mapped nicely onto the neuroscience-inspired general noological architecture of Fig. 3.31 as shown in Fig. 3.32.

In Fig. 3.32 it can be seen that the *motivational* learning and problem solving level (left side of the figure) is mapped onto the Houk module at the level of the prefrontal cortex (right side of the figure). Specifically, the *orbitofrontal* cortex is involved in motivational processes (Zald and Rauch 2006). Presumably, both learning and problem solving processes in connection with motivational processes are present at this level of the brain. As for the computational *physical* learning and problem solving level (left side of the figure), it is mapped onto the Houk modules at the premotor and primary motor cortices (right side of the figure).

Within each Houk module, we propose that perhaps the causal learning mechanisms of the computational system as illustrated in, say, Fig. 3.23 are mapped onto the cerebral cortex's unsupervised causal learning mechanisms. (As discussed in Sect. 3.8, one interpretation of the mechanisms of the cerebral cortex is unsupervised causal learning.). The Chunked Rule Base of Fig. 3.23 is mapped onto the cerebellum's supervised learning mechanisms as the acquisition of the rules or chunked rules involves supervised learning (association of certain desired output – the goal – to a certain method to achieve the goal, corresponding to the R.H.S. and the L.H.S. of a causal rule respectively such as that illustrated in Fig. 3.19). The search and problem solving processes of Fig. 3.23 map onto the basal ganglia's reinforcement learning mechanisms as the function of reinforcement learning is to carry out search to generate an optimal sequence of actions to achieve a certain reward or goal. This is consistent with our earlier observation that problem solving processes are the backbone of noological processes (Sect. 1.3, Chap. 1) and the basal ganglia with a recurrent structure, that presumably implement reinforcement learning, lie at the core of intelligent processes (Sect. 3.8.4 and Fig. 3.28).

Figure 3.32, and for that matter, Fig. 3.31 are of course a simplification of the full complexity present in the brain and in the computational processes needed to instantiate a full noological system. However, this initial simplified model could serve as a starting point for a more refined model, both from the computational and the neuroscience perspectives, in future investigations.

3.10 Further Notes on Neuroscience

We have mentioned in Sect. 3.8.3 that the general operations of the cerebral cortex have been recognized by a number of neuroscientists to be carrying out "cause recovery" of the phenomena observed in the environment. We observed that this tallies well with the causal learning processes that we discussed in Chap. 2 which we think lies at the heart of learning about causality in the environment for the purpose of problem solving for the survival of a noological system. Another aspect of causal learning is uncovering the causes of perceived sensory information. For example, for the visual domain, the images on the retinas are *caused* by certain objects and lighting conditions out there in the real world, and the role of the sensory cortices is to uncover these causes, thus providing useful information for the noological system. In our characterization, the sensory cortices and hence perceptual processes are providing a *service function* to the problem solving processes – Fig. 1.7 of Chap. 1 and associated discussion. Section 3.8.3 also highlighted the fact that the various areas of the cerebral cortex are similar in cytoarchitecture and hence various kinds of cause recovery for different sensory or other non-sensory domains make use of more or less the same processing mechanisms, and that these cortical areas are reprogrammable to handle information of a different nature.

However, what is lacking in the discussion in Sect. 3.8.3 with regards to the cortical functions is that there was no distinction made between the sensory and the motor parts of the cortex. As is known in neuroanatomy and as shown in Fig. 3.29, the cerebral cortex can be divided into roughly the sensory areas at the "back" (shown in blue on the brain picture, and the corresponding processing stages also shown in blue on the right side of the figure) and the motor areas in the "front" (shown in red, and the corresponding processing areas also shown in red on the left side of the figure), separated by a central sulcus. If the sensory side of the cortex is doing "cause recovery" based on sensory input from the environment, what is the motor side doing using the same underlying mechanisms?

3.10 Further Notes on Neuroscience

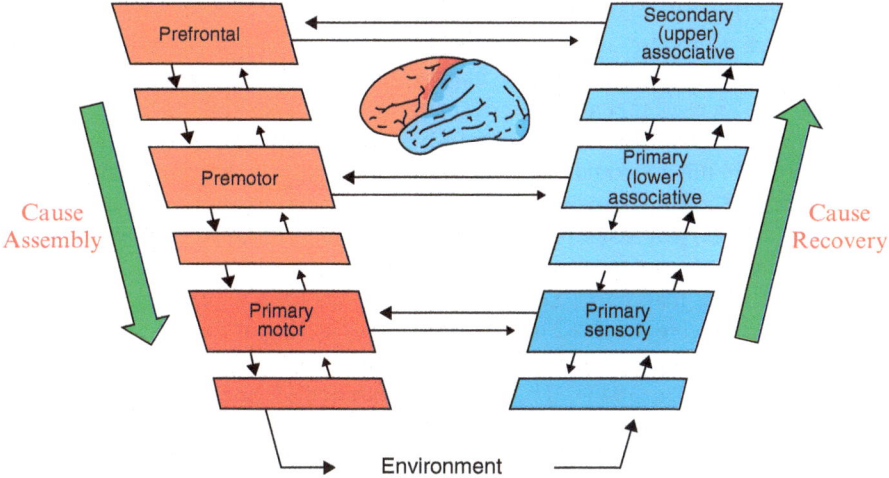

Fig. 3.33 Fuster's (2008) cerebral cortex hierarchy of Fig. 3.29 elaborated with a Cause Recovery (sensory) and a Cause Assembly (motor) sides (Reprinted from The Prefrontal Cortex, Fourth Edition, Fuster 2008, with permission from Elsevier)

As we know, the motor side of the cortex is basically involved in generating motor output to effect changes in the environment. The motor side, like the sensory side, also consists of a hierarchy of stages, with the "ideas to effect actions" generated in the highest levels and the final control signals of the motor output generated at the lowest levels to effect actions on the environment. We propose that the best way to characterize this process is "cause assembly" – assembling a series of "causes" that effect changes in the environment. This idea is illustrated in Fig. 3.33.

Hence the sensory side and the motor side have a nice symmetry or anti-symmetry to them. Note that since the various cerebral cortical areas have been observed to have similar cytoarchitecture, what this implies is that the same cortical mechanisms can be used not only for various different domains of sensory processing but also for motor processing as well. This is elegance and beauty of natural systems' mechanisms that should be a target for artificial systems to emulate.

3.10.1 The Locus of Memory Is Not the Synapse

In our review of neuroscience in this section and in Sect. 3.8, we have only looked at the brain at a gross architectural level and attempted to map the purported functions of these architectural subparts to noological processes. The functioning of these subparts presumably arise from the functioning of neurons. Since more than a century ago, neurons have been presumed to be the fundamental units of the nervous system's information processing. The connections between neurons are the synapses and memory has been assumed to be stored in the strengths of these connections, in the form of "synaptic efficacy" – the stronger the connection, the more likely a "presynaptic" signal would trigger a "postsynaptic" impulse (Cajal 1894).

However, recently there is a twist to this story. Armed with more advanced experimental methods, some experiments such as that conducted by Chen et al. (2014) have revealed that the synapses are not the locations at which memory is stored. The locus of memory goes deeper into the cell, possibly at the DNA level. The exact mechanisms have yet to be fully understood but if further experiments were to confirm this, it would bring about a revolutionary new understanding of the mechanisms of information processing in the nervous system. This will overturn more than a century of assumptions on the role of the most basic cellular unit of the nervous system, the neurons.

Emerged more than half a century ago, there has been an entire generation of researchers who were involved in mathematical modeling of neurons and neuronal networks in the form of artificial neurons and the associated networks (e.g., since McCulloch and Pitts 1943). The "synaptic strengths," the presumed locus of memory, are modeled as the "connection weights" between the artificial neurons. Successful learning algorithms and applications of these ideas have been devised and demonstrated (Deng and Yu 2014; LeCun et al. 2015; Rumelhart et al. 1986; Werbos 1974). The systems thus devised are often touted as "bio-inspired artificial/cognitive systems." However, if further experimental work shows that the locus of memory is not in the synapses but are instead buried deeper in the cellular machineries, a total revision of mathematical modeling of information processing in the brain would have to be made.

3.10.2 Function of the Prefrontal Cortex and the Entire Brain

As discussed in the foregoing discussions since Chap. 1, a fundamental noologically efficacious unit of processing is the script. Where might the locus of script processing in the human brain be? A number of neuroscientists have identified it to be in the prefrontal cortex (Barbey et al. 2008; Krueger and Grafman 2008; Wood and Grafman 2003; Wood et al. 2005). They believe the human prefrontal cortex stores Structured Event Complexes (SECs), and these SECs are representations composed of higher-order *goal-oriented sequence of events* that are involved in the planning and monitoring of complex behavior. From a computational perspective, the SECs are non-other than *scripts* in our noological processing framework. The contents of scripts include not only "high level" organizational constructs (which map onto the "higher level" cortices such as those in the prefrontal cortex) but also include grounded concepts, as can be seen in the various scripts in the foregoing discussions. These grounded concepts involve sensory level representations and processing. Therefore, the representational and computational processes associated with scripts involve not only the prefrontal cortex but the entire brain as well (Wood and Grafman 2003).

Often neuroscience experimental efforts and theorization begin with the sensory cortices and proceed "upward" toward the prefrontal cortex. This is partly because

Fig. 3.34 A robotic arm movement problem

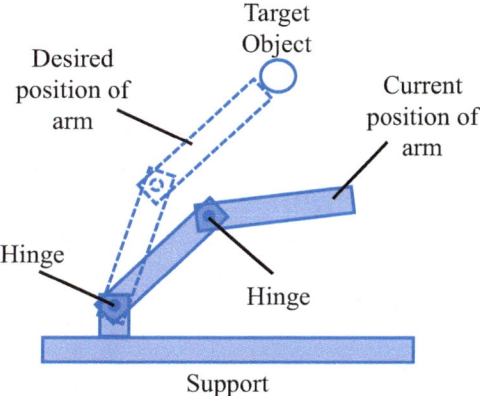

it is easier to conduct definitive experiments in the sensory cortices as the nature of information being processed there is more well defined. However, to understand the entire brain's function, it may serve to look at it from the point of view of the noologically efficacious fundamental units of processing – the scripts.

Problems

1. Devise a better affective learning process than that discussed in Sect. 3.3.1.
2. Figure 3.34 shows a typical robotics problem in which an arm consisting of two hinged "sub-arms" is used to reach for objects. The problem is to move the arm from the current position/configuration as shown to the final desired position/configuration so that its tip touches the Target Object. Without using the method of inverse kinematics (Russell and Norvig 2010) and using instead the method of incremental knowledge chunking as described in Sect. 3.5, show how a set of causal rules/chunked rules governing the movements of the arm/sub-arms can be learned through causal learning and used to solve the problem in the same manner as the Shelter Building problem of Fig. 3.18.

References

Alexander, G. E., Delong, M. R., & Strick, P. L. (1986). Parallel organization of functionally segregated circuits linking basal ganglia and cortex. *Annual Review of Neuroscience, 9*, 357–381.

Allen, G., & Courchesne, E. (2003). Differential effects of developmental cerebellar abnormality on cognitive and motor functions in the cerebellum: An fMRI study of autism. *American Journal of Psychiatry, 160*, 262–273.

Allen, G. I., & Tsukahara, N. (1974). Cerebrocerebellar communication systems. *Physiological Review, 54*, 957–1006.

Alterman, R. (1988). Adaptive planning. *Cognitive Science, 12*, 393–422.
Andreasen, N. C., & Pierson, R. (2008). The role of the cerebellum in schizophrenia. *Biological Psychiatry, 64*, 81–88.
Barbey, A. K., Krueger, F., & Grafman, J. (2008). Structured event complexes in the medial prefrontal cortex support counterfactual representations for future planning. *Philosophical Transactions of the Royal Society Series B, 364*, 1291–1300.
Bhatia, K. P., & Marsden, C. D. (1994). The behavioral and motor consequences of focal lesions of the basal ganglia in man. *Brain, 117*, 859–876.
Bishop, C. M. (2006). *Pattern recognition and machine learning*. New York: Springer Science.
Brooks, V. B., & Thach, W. T. (1981). Cerebellar control of posture and movement. In V. B. Brooks (Ed.), *Handbook of physiology, section 1, the nervous system, vol. 2, motor control, part II* (pp. 877–946). Bethesda: American Physiological Society.
Cajal, S. R. (1894). The Croonian Lecture: La fine structure des centres nerveux. *Proceedings of the Royal Society of London, Series B: Biological Sciences, 55*, 444–467.
Carbonell, J. G. (1983). Derivational analogy and its role in problem solving. In *Proceedings of AAAI-1983* (pp. 64–69).
Carbonell, J. G., Knoblock, C. A., & S. Minton, S. (1989). *PRODIGY: An integrated architecture for planning and learning* (Technical Report CMU-CS-89-189), Computer Science Department, Carnegie-Mellon University.
Chen, S., Cai, D., Pearce, K., & Sun, P. Y.-W. (2014). Reinstatement of long-term memory following erasure of its behavioral and synaptic expression in *Aplysia. eLife, 3*, e03896. doi:10.7554/eLife.03896.
DeLong, M. R., & Georgopoulos, A. P. (1981). Motor functions of the basal ganglia. In V. B. Brooks (Ed.), *Handbook of physiology, section 1, the nervous system, vol. 2, motor control, part II* (pp. 1017–1061). Bethesda: American Physiological Society.
Deng, L., & Yu, D. (2014). *Deep learning methods and applications*. Delft: Now Publishers.
Douglas, R. J., & Martin, K. A. C. (1989). A canonical microcircuit for neocortex. *Neural Computation, 1*, 480–488.
Douglas, R. J., & Martin, K. A. C. (2007). Mapping the matrix: The ways of the neocortex. *Neuron, 56*, 226–238.
Doya, K. (2003). Recurrent networks: Learning algorithms. In M. A. Arbib (Ed.), *The handbook of brain theory and neural networks*. Cambridge, MA: MIT Press.
Doya, K. (2009). What are the computations of the cerebellum the basal ganglia and the cerebral cortex? *Neural Networks, 12*, 961–974.
Dum, R. P., & Strick, P. L. (2009). Basal ganglia and cerebellar circuits with the cerebral cortex. In M. S. Gazzaniga (Ed.), *The cognitive neurosciences* (4th ed.). Cambridge, MA: MIT Press.
Erol, K., Hendler, J., & Nau, D. S. (1996). Complexity results for HTN planning. *Artificial Intelligence, 18*(1), 69–93.
Fikes, R. E., Hart, P. E., & Nilsson, N. J. (1972). Learning and executing generalize robot plans. *Artificial Intelligence, 3*, 251–288.
Friston, K. (2010). The free-energy principle: A unified brain theory? *Nature Reviews Neuroscience, 11*, 127–138.
Fuster, J. M. (2008). *The prefrontal cortex* (4th ed.). Amsterdam: Elsevier.
Gazzaniga, M. S., Ivry, R. B., & Mangun, G. R. (2013). *Cognitive neuroscience: The biology of the mind* (4th ed., p. 2002). New York: W. W. Norton.
George, D. (2008). *How the brain might work: A hierarchical and temporal model for learning and recognition*. Ph.D. thesis, Stanford University.
George, D., & Hawkins, J. (2009). Towards a mathematical theory of cortical micro-circuits. *PLoS Computational Biology, 5*(10), e100532.
Glickstein, M. (1997) Mossy-fibre sensory input to the cerebellum. In C. J. de Zeeuw, P. Strata, & J. Voogd (Eds.), *Progress in brain research* (Vol. 114, pp. 251–259). Amsterdam: Elsevier Science BV.

Grillner, S., Wallen, P., Saitoh, K., Kozlov, A., & Robertson, B. (2008). Neural basis of goal-directed locomotion in vertebrates – an overview. *Brain Research Review, 57*, 2–12.
Haber, S. N., Fudge, J. L., & McFarland, N. R. (2000). Striatonigrastriatal pathways in primates form an ascending spiral from the shell to the dorsolateral striatum. *The Journal of Neuroscience, 20*(6), 2369–2382.
Hammond, K. (1989). *Case-based planning: Viewing planning as a memory task*. Boston: Academic Press.
Hart, P. E., Nilsson, N. J., & Raphael, B. (1968). A formal basis for the heuristic determination of minimum cost paths. *IEEE Transactions on Systems Science and Cybernetics SSC4, 4*(2), 100–107.
Ho, S.-B. (2013). A grand challenge for computational intelligence – A micro-environment benchmark for adaptive autonomous agents. In *Proceedings of the IEEE symposium series on computational intelligence – Intelligent Agents*, Singapore (pp. 44–53). Piscataway: IEEE Press.
Ho, S.-B., & Liausvia, F. (2013a). Knowledge representation, learning, and problem solving for general intelligence. In *Proceedings of the 6th international conference on artificial general intelligence*, Beijing, China (pp. 60–69). Berlin: Springer.
Ho, S.-B., & Liausvia, F. (2013b). Incremental rule chunking for problem solving. In *Proceedings of the 1st BRICS Countries Conference on Computational Intelligence*. Ipojuca, Pernambuco, Brazil (pp. 323–328). Piscataway: IEEE Press.
Ho, S.-B., & Liausvia, F. (2014). Rapid learning and problem solving. In *Proceedings of the IEEE symposium on computational intelligence,* Orlando, Florida (pp. 110–117). Piscataway: IEEE Press.
Houk, J. C. (1997). On the role of the cerebellum and basal ganglia in cognitive signal processing. In C. J. de Zeeuw, P. Strata, & J. Voogd (Eds.), *Progress in brain research* (Vol. 114, pp. 543–552). Amsterdam: Elsevier Science BV.
Houk, J. C. (2005). Agents of the mind. *Biological Cybernetics, 92*, 427–437.
Houk, J. C., Davis, J. L., & Beiser, D. G. (1995). *Models of information processing in the basal ganglia*. Cambridge, MA: MIT Press.
Ito, M. (2008). Control of mental activities by internal models in the cerebellum. *Nature Reviews Neuroscience, 9*, 304–313.
Joel, D., Niv, Y., & Ruppin, E. (2002). Actor-critic models of the basal ganglia: New anatomical and computational perspectives. *Neural Networks, 15*, 535–547.
Kersten, D., Mamassian, P., & Yuille, A. (2004). Object perception as Bayesian inference. *Annual Review of Psychology, 55*, 271–304.
Kilian, J., & Siegelman, H. T. (1996). The dynamic universality of sigmoidal neural networks. *Information and Computation, 128*, 48–56.
Kim, S.-G., Ugurbil, K., & Strick, P. L. (1994). Activation of a cerebellar output nucleus during cognitive processing. *Science, 265*(5174), 949–951.
Krueger, F., & Grafman, J. (2008). The human prefrontal cortex stores structured event complexes. In T. F. Shipley & J. M. Zacks (Eds.), *Understanding events: From perception to action*. Oxford: Oxford University Press.
Laird, J., Rosenbloom, P. S., & Newell, A. (1986). Chunking in soar: The anatomy of a general learning mechanism. *Machine Learning, 1*, 11–46.
Laird, J., Rosenbloom, P. S., & Newell, A. (1987). SOAR: An architecture for general intelligence. *Artificial Intelligence, 33*(1), 1–64.
LeCun, Y., Bengio, Y., & Hinton, G. E. (2015). Deep learning. *Nature, 521*, 436–444.
Lee, T. S., & Mumford, D. (2003). Hierarchical Bayesian inference in the visual cortex. *Journal of the Optical Society of America, 20*, 1434–1448.
Maslow, A. H. (1954). *Motivation and personality*. New York: Harper & Row.
McCulloch, W., & Pitts, W. (1943). A logical calculus of ideas immanent in nervous activity. *Bulletin of Mathematical Biophysics, 5*(4), 115–133.
Middleton, F. A., & Strick, P. L. (2000). Basal ganglia and cerebellar loops: Motor and cognitive circuits. *Brain Research Reviews, 31*, 236–250.

Mountcastle, V. B. (1982). An organizing principle for cerebral function: The unit module and the distributed system. In G. M. Edelman & V. B. Mountcastle (Eds.), *The mindful brain*. Cambridge, MA: MIT Press.

Nolte, J. (2009). *The human brain: An introduction to its functional anatomy* (6th ed.). Philadelphia: Mosby Elsevier.

Owen, A. M., Doyon, J., Dagher, A., Sadikot, A., & Evans, A. C. (1998). Abnormal basal ganglia outflow in Parkinson's disease identified with PET: Implications for higher cortical functions. *Brain, 121*, 949–965.

Passingham, R. E., & Wise, S. P. (2012). *The neurobiology of the prefrontal cortex*. Oxford: Oxford University Press.

Roe, A. W., Pallas, S., Kwon, Y. H., & Sur, M. (1992). Visual projections routed to the auditory pathway in ferrets: Receptive fields of visual neurons in primary auditory cortex. *The Journal of Neuroscience, 12*(9), 3651–3664.

Rumelhart, D. E., McClelland, J. L., & and the PDP Research Group. (1986). *Parallel distributed processing: Exploration in the microstructure of cognition, vol. 1 & 2*. Cambridge MA: MIT Press.

Russell, S., & Norvig, P. (2010). *Artificial intelligence: A modern approach*. Upper Saddle River: Prentice Hall.

Sadato, N., Pascual-Leone, A., Grafman, J., Ibanez, V., Deiber, M.-P., Dold, G., & Hallett, M. (1996). Activation of the primary visual cortex by Braille reading in blind subjects. *Nature, 380*, 526–528.

Schank, R., & Abelson, R. (1977). *Scripts, plans, goals and understanding*. Hillsdale: Lawrence Erlbaum Associates.

Seger, C. A. (2008). How do the basal ganglia contribute to categorization? Their roles in generalization, response selection, and learning via feedback. *Neuroscience and Biobehavioral Reviews, 32*, 265–278.

Seger, C. A., Peterson, E. J., Cincotta, C. M., Lopez-Paniagua, D., & Anderson, C. W. (2010). Dissociating the contributions of independent corticostriatal systems to visual categorization learning through the use of reinforcement learning modeling and Granger causality modeling. *NeuroImage, 50*, 644–656.

Siegelmann, H. T. (2003). Neural and super-turing computing. *Minds and Machines, 13*, 103–114.

Siegelmann, H. T., & Sontag, E. D. (1991). Turing computability with neural nets. *Applied Mathematics Letters, 4*, 77–80.

Siegelmann, H. T., & Sontag, E. D. (1994). Analog computation via neural networks. *Theoretical Computer Science, 131*, 331–360.

StarCraft II (2015): http://us.battle.net/sc2/en/. Blazzard Entertainment.

Strick, P. L., Dum, R. P., & Fiez, J. A. (2009). Cerebellum and nonmotor function. *Annual Review of Neuroscience, 32*, 413–434.

Sur, M., Garraghty, P. E., & Roe, A. W. (1988). Experimentally induced visual projections into auditory thalamus and cortex. *Science, 242*(4882), 1437–1441.

Sutton, R. S., & Barto, A. G. (1998). *Reinforcement learning: An introduction*. Cambridge, MA: MIT Press.

Werbos, P. J. (1974). *Beyond regression: New tools for prediction and analysis in the behavioral sciences*. Ph.D. thesis, Harvard University.

Wood, J. N., & Grafman, J. (2003). Human prefrontal cortex: Processing and representational perspectives. *Nature Reviews Neuroscience, 4*, 139–147.

Wood, J. N., Tierney, M., Bidwell, L. A., & Grafman, J. (2005). Neural correlates of script event knowledge: A neuropsychological study following prefrontal injury. *Cortex, 41*(6), 796–804.

Zald, D. H., & Rauch, S. L. (2006). *The orbitofrontal cortex*. Oxford: Oxford University Press.

Chapter 4
Conceptual Grounding and Operational Representation

Abstract Conceptual grounding refers to the process of linking concepts represented in a noological system to its real-world referents. This imbues the system with the ability to "truly understand" the concepts that it manipulates and enables it to apply them for problem solving and other noological processes effectively. This chapter describes how "operational representations" can be used to represent a variety of concepts at the ground level. These include concepts of existence, movement, propulsion, materialization, reflection, obstruction, penetration, attachment, etc. Other more complex concepts can be built upon these grounded concepts, achieving the effect of grounding through them. The convergence of our method of conceptual grounding and that of cognitive linguistics is discussed.

Keywords Conceptual grounding • Semantic grounding • Meaning • Knowledge representation • Operational representation • Spatiotemporal representation • Atomic operator • Cognitive linguistics

In this chapter, a novel operational representational framework is proposed to encode knowledge at the epistemic ground level – the most fundamental level that characterizes activities and interactions in the real world (Ho 2012, 2013). The basic constructs of operational representation are spatiotemporal representations that encode spatiotemporal features of activities and interactions. Specifically, and most importantly, the temporal aspects of activities and interactions are laid out explicitly along a spatial dimension in a noological system's internal memory structure for the purpose of conceptual organization, reasoning, and problem solving. The spatiotemporal representations together with some operations defined in the form of procedures to operate on them constitute the complete operational representational mechanisms. This framework provides the constructs to represent physical and mental activities at a high spatiotemporal resolution essential for conceptual grounding. This chapter, together with the next chapter, illustrate how the characterization of various concepts through an explicit temporal representation is natural, straightforward, and effective. The focus in this chapter is on the representations themselves and in the next chapter, causal rules built from operational representations that encode the generalizations of physical behaviors of objects are developed and their applications to reasoning and problem solving are described. These are developed in detail for elemental and extended objects[1] behaving in certain manners over one dimensional (1D) space and 1D time to illustrate the basic principles involved. The considerations derived from 1D space and 1D time capture the basic principles involved that can be extended to 2D and 3D space plus 1D time scenarios in a straightforward manner.

4.1 Ground Level Knowledge Representation and Meaning

The physical world is the source of empirical knowledge. Intelligent agents such as human beings are endowed with a rich variety of senses – from the visual to the auditory, the somatosensory, the gustatory and the olfactory – to pick-up and learn about objects, events and processes in the physical world. Among the senses, the visual sense is the most developed in higher level natural intelligent systems. The visual world provides most of the empirical information we have of the world, and this information is dynamic in nature, consisting of animate and inanimate objects exhibiting various forms of activities and interacting with each other in myriads events and processes.

A lot of research has been carried out in the area of computer vision in which the information in the visual world is processed for the purposes of object recognition, scene description, and activity classification (e.g., Forsyth and Ponce 2011; Shapiro and Stockman 2001; Szeliski 2010, etc.) but these efforts have not emphasized characterizing the conceptual aspects of the visual world based on the activities and interactions that take place between various entities in the world. (Although, recently, some effort has emerged in this direction: www.visionmeetscognition.org.) Psychologists and linguists have understood for some time that perception is important for conceptualization (e.g., the work by Miller and Johnson-Laird 1976) but no effort has been forthcoming in these fields in terms of computational characterizations of the ideas involved. In recent years, cognitive linguistics has arisen as a new paradigm for the study of meanings through perceptual and cognitive characterizations (e.g., the work by Croft and Cruse 2009; Evans and Green 2006; Geeraerts 2006; Johnson 1987; Langacker 1987, 1991, 2000, 2002, 2008, 2009; Ungerer and Schmid 2006, etc.). This brings us closer to establishing a link between perception and conception but detailed computational characterizations are still currently lacking.

Linking concepts to percepts is a process of grounding. Let us consider what we can observe of the world around us using our biological senses before we invented instruments to probe deeper and see farther. Consider a tree such as that illustrate in

---

[1] Extended objects are objects consisting of more than one elemental part.

4.1 Ground Level Knowledge Representation and Meaning 147

Fig. 4.1. We can characterize the tree as consisting of three major parts – the Trunk, the Branches, and the Leaves. However, there are actually a lot more details of the tree that we can see than this high level description. There is a lot of texture on the trunk, branches, and leaves that we can see, there are details on how the branches bend and twist at various points, there are details on the shapes of the leaves, etc. We also perceive the dynamic nature of the tree – *movements* of the leaves, branches, and perhaps even the trunk when the wind blows at them. These static and dynamic characterizations need to be of as high a resolution as is perceptually or cognitively possible, at an "atomic" level, before good use can be made of them.

With this detailed perception or *understanding* of the tree and its various subparts, we can perhaps imagine using some combination of leaves and branches as a "broom" to sweep floor or as a "fan" for cooling purposes (taking advantage of the certain "softness" of the leave-branch complex that we perceive), and perhaps proceed to do so when the needs arise. There could also be decorative purposes that we can use the subparts of the tree for because of certain detailed visual configurations of them that are "aesthetically pleasing." Using the tree to solve problems at this level would not be possible from the mere high level word-symbolic description of the tree as consisting of Trunk, Branches, and Leaves. The full visual (static and dynamic) characterization of the tree through our normal perceptual system at a high resolution level is a kind of epistemic ground level characterization. It allows us to understand and operate with things in the world as best we can before using complex instruments to characterize these things at even finer levels. It allows a noological system to function intelligently and solve problems effectively at that level.

As we probe deeper into the natural world with instruments that reveal their workings at finer levels, we begin to gain another level of understanding of things in the world. For the tree, we find out about cells, chlorophyll, photosynthesis, etc. This is a different level of static and dynamic characterizations of the tree that involve smaller physical entities – the various molecules – and their attendant interactions. With this level of understanding, we carry out corresponding intelligent problem solving at that level, and we reach yet another level of epistemic ground. Whether it is this epistemic ground level or the more "macro" one described above, they both consist fundamentally of forces and objects that interact with each other that give rise to events and processes. And, at each level, spatial locations, shapes, intensities of forces, etc. can all change elementally at a fine resolution. Knowledge represented appropriately at these ground levels engenders certain kinds of problem solving to take place, such as the broom and fan example above, or some other problem solving processes involving the biochemistry of the tree.

As mentioned in Chap. 1, Sect. 1.6, grounding concept breaks the circularity of the "dictionary-type" definitions of concepts often used in traditional AI, such as in the form of "semantic networks" (e.g., Russell and Norvig 2010). Using similar examples from Sect. 1.6, the following is a dictionary definition taken from Merriam Webster for the concept Move:

<u>Move</u> To *go* from one *place* or position to another.

However, the word/concept *go* in this definition are in turn defined with Move in Merriam Webster:

<u>Go</u> To *move* on a course.

Similarly, Place is defined as:

<u>Place</u> A specific *area* or region of the world.

And Area in turn is defined as:

<u>Area</u> A part or section within a large *place*.

Thus, one wonders what these concepts really mean when they are circularly defined in terms of each other, and the question of whether an AI system using word-like symbols to define other word-like symbols really "understands" the contents involved. For example, recently there was a seemingly successful demonstration of machine intelligence – the IBM Watson System – in which a highly sophisticated question answering system was able to outperform human participants in a TV game called Jeopardy (Ferrucci et al. 2010). The system is able to provide natural language answers – emitted in the form of a string of word symbols – to questions posed in natural language and often with much better accuracy than the human participants. However, the emitted sequence of symbols in the natural language answers seem to have "meaning" only in the minds of the human listeners, be they the human participants in the game or the audience.

So what really is "meaning"? We do not plan to delve into lengthy and complicated philosophical discussion in this book. We wish to adopt a more balanced view and the position that perhaps "meaning" can be found at multiple levels. So suppose you have a simplified and high-level symbolic representation of the concept Tree as something consisting of a Trunk, some Branches, and some Leaves (e.g., Fig. 4.1), which are themselves just word symbols, we will accept that perhaps to some extent you have captured the meaning of the concept Tree. Some kinds of problem solving at this symbolic level can still proceed with this symbolic representation (e.g., IBM's Watson has successfully solved the problem of answering the questions involved to some extent). However, what we wish to elucidate here in this chapter is what we call "meaning at the epistemic ground level." As discussed above, the meaning of the subparts of the Tree or for that matter any concept can be further elucidated symbolically using words (e.g., a Branch is that "which *grows* from the Trunk" and that "*supports* Leaves," etc.). And ultimately, a level is reached where further high-level word-like symbolic elaboration will become circular, such as demonstrated above with the example of Move and Place. (E.g., continue to define *grow* and *support* in the Branch example.)

4.1 Ground Level Knowledge Representation and Meaning

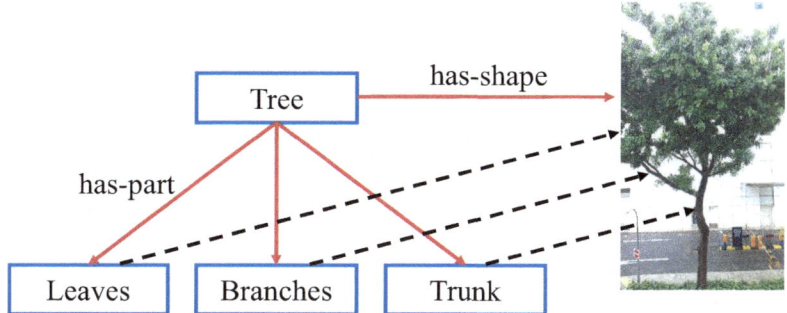

Fig. 4.1 High level concept of a tree connected to its detailed representation

Therefore, we posit that to represent, say, the concept Move in a grounded manner, devoid of further high-level word-like symbolic elaboration, it is better to just encode it directly in a spatiotemporal form as shown in Fig. 4.2.

In Fig. 4.2a, the concept is a *specific* Move and in Fig. 4.2b it is a *general* Move. The representations in Fig. 4.2 represent 1-dimensional (1D) *spatial* movements over (1D) time. Each small square in the representation represents an elemental discrete spatial or temporal location. Figure 4.2a shows a *specific* kind of movement in which the Object moves one elemental spatial location over one elemental time frame and in the upward direction. Figure 4.2b shows a *general* movement over *any* amount of spatial locations in *any* amount of time and in either the upward or the downward direction (in 1D space, there are only two directions – upward or downward in this case). The horizontal and vertical Range Bars represent *any* amount of space and time intervals respectively – that correspond to *any* amount of spatial and temporal displacement respectively for the Object involved (see Sect. 4.3.7.1 – discussion in connection with Figs. 4.14 and 4.15 that further elaborates on how this representation works). The "Up-Down Flip" symbol specifies that the pictorial representation can be flipped in the upward-downward direction, hence representing the upward and downward movements accordingly. The representation in Fig. 4.2b represents the general and grounded concept of Move.

Now, note that the spatiotemporal representation shown in Fig. 4.2 is still "symbolic" in the sense that it is "something that stands for or suggests something else" (Merriam-Webster) – i.e., it is a "picture" that stands for an "external event." Moreover, one can use a symbolic predicate logical representation to represent the spatiotemporal template of Fig. 4.2a as:

$$\forall Object, l, t \quad Spatial\text{-}Location(Object, l, t) \\ \land \ Spatial\text{-}Location(Object, l + \Delta, t + \Delta) \quad (4.1) \\ \rightarrow \ Move(Object)$$

where l and t are the spatial and the temporal locations respectively of the object involved at the left bottom square of the spatiotemporal representation of Fig. 4.2a.

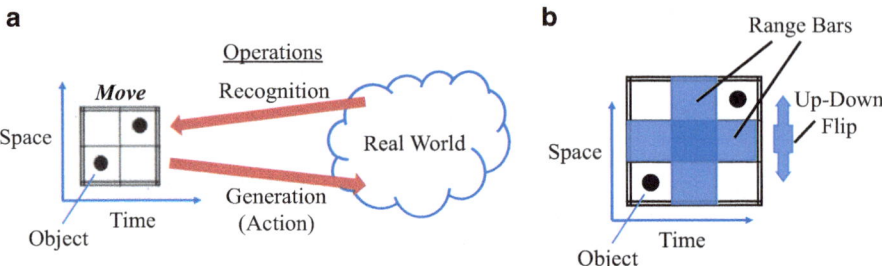

Fig. 4.2 The ground level (symbolic) operational representation of the concept Move. Each *square* represents an elemental discrete spatial or temporal location. (**a**) A *specific* Move in which the Object moves over an elemental spatial location over an elemental time frame and in the upward direction. (**b**) A *general* Move in which the Object moves over *any* amount of spatial location over *any* amount of time and in the *up* or *down* direction

The *Spatial-Location* predicate basically has three arguments and in English, Eq. (4.1) says that "for every *Object*, l and t, if the spatial location of *Object* is at l at time t and the spatial location of *Object* is at $l+\Delta$ at $t+\Delta$, then the Object moved."

A representation that treats space and time on equal footing would be:

$$\forall Object, l, t \quad Spatiotemporal\text{-}Location(Object, (l, t)) \\ \wedge \; Spatiotemporal\text{-}Location(Object, (l + \Delta, t + \Delta)) \quad (4.2) \\ \rightarrow \; Move(Object)$$

where l and t are now like the usual x, y in a 2D spatial coordinate system.

We submit that whether the form of the symbolic representation of Move is in the spatiotemporal form of Fig. 4.2 (pictorial-symbolic form) or the predicate logical form of Eq. (4.1) (word-symbolic form), they are both *ground level* (symbolic) representation of Move. Therefore, the distinction is one between higher-level symbolic representations such as the word symbols "Trunk" and "Branches," and ground level symbolic representations such as that of Fig. 2 and Eq. (4.1).[2]

Figure 4.2 and Eq. (4.1) also clarify that what constitutes a ground level representation is not whether the form of the representation is picture or word, but what the parameters involved are – in this case, both Fig. 4.2 and Eq. (4.1) rely on spatial and temporal locations to define Move which are in turn treated by a

---

[2] This solves a mystery why dictionary definitions are typically not grounded in the sense that we define groundedness here. Because dictionaries tend to be used by human beings who already have some fundamental experience with the world and hence who have already internalized some basic concepts such as Move operationally, the dictionary definitions themselves do not need to elaborate further on this, otherwise the elaborations can be quite complicated, as can be seen in some of the examples of the representations of grounded concepts that we will be considering in the rest of this chapter.

4.1 Ground Level Knowledge Representation and Meaning

Fig. 4.3 Operational representation (the concept of Increase) for motivational states of Happiness, Comfort, Hunger, Pain, etc

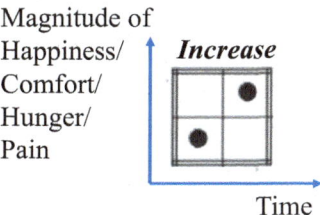

noological system as fundamental concepts to operate with. We provide another example concept below that does not involve spatial parameters.

In Chap. 1, Sect. 1.4 we mentioned that there is an internal ground level in a noological system that involves certain motivational states that do not require further elaboration (states at the bio-noo boundary). For example, a noological agent in normal situations would try to increase the magnitude of its internal states of Happiness, Comfort, etc. while trying to decrease the magnitude of its states of Hunger, Pain, etc. These are states at the bio-noo boundary. The changes in these states can also be represented spatiotemporally at the epistemic ground level as shown in Fig. 4.3.

Figure 4.3 basically shows that a spatiotemporal representation similar to that for changes in spatial location can be used for changes in the various parameters of internal states. Thus, this is the epistemic ground (symbolic) level representation for the concepts of "Increase in Happiness," "Increase in Pain," etc.

Returning to consider the concept Move, we name the form of representation in Fig. 4.2 an *operational representation* because it captures the operation of the concept Move in a direct manner. Move is an *operation* that causes an object to change location over time. Associated with the explicit spatiotemporal representation of Fig. 4.2 is a set of operations or computational processes that use the explicit representation to perform certain tasks, and these tasks are typically performing recognition, effecting action, and carrying out problem solving on the real world, as shown in Fig. 4.2. For recognition operation, suppose a movement occurs in the real world, the operational representational template of Fig. 4.2 can be used to recognize that a movement event has occurred by *matching* the template of Move to the real world event. If the system wishes to *effect* a movement, the same template is used in the "opposite" direction: The template specifies that in the second time frame the object must be at a different location relative to a first time frame and the system emits actions to bring that about. Thus, these operations, together with the explicit spatiotemporal representation, constitute the complete operational representational mechanisms for, in this case, the concept Move.

In the rest of this chapter, we will use the pictorial form of the operational representation of Figs. 4.2 or 4.3 rather than the predicate logical form of Eq. (4.1). As will be seen, the pictorial form is clearer for the concepts that we will be considering.

4.2 An Operational Representational Scheme for Noological Systems

Other than the concept Move above, in this chapter we will identify a set of fundamental operational representations that can describe and capture knowledge at the epistemic ground level – whether it is at the macroscopic level (such as trees, waves, people, cars, rockets, clouds, stars, and galaxies) or the microscopic level (such as microbes, molecules, and atoms). Figure 4.4 provides an overview of the use of the representational scheme for grounding experience and actions.

Events and processes in the environment (external and internal) basically take place over space and time and the noological system observes and learns about these activities and interactions and analyzes and represents them in operational representations. These internal representations may be organized in some symbolic conceptual hierarchy (such as will be discussed in association with Figs. 4.8 and 4.9) and are manipulated by internal noological processes. When the system generates actions and activities in the real world, which are also basically spatio-temporal patterns, it composes these internal operational representations into these actions and activities in a manner directed by the goals or requirements of certain noological tasks.

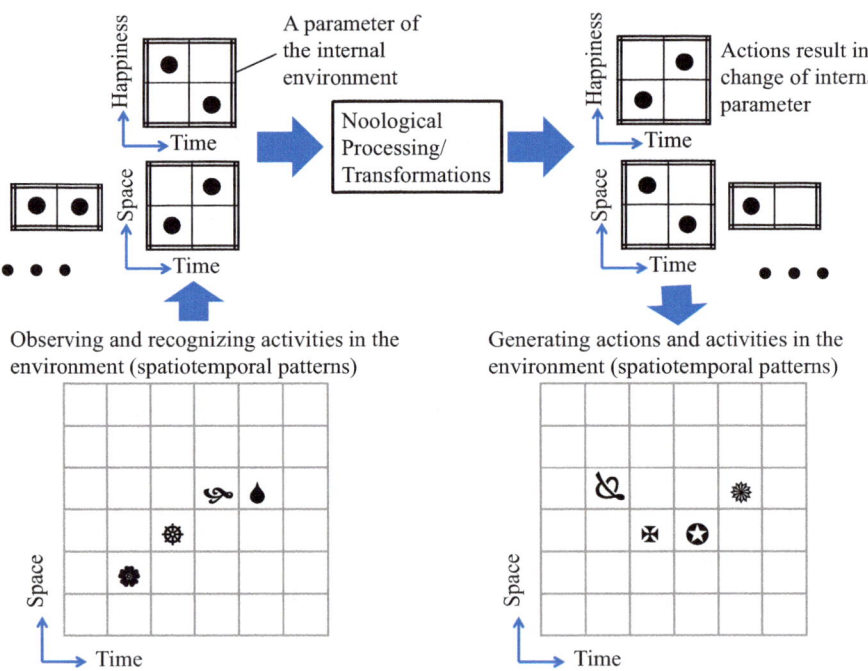

Fig. 4.4 Overview of using operational representations to ground experience and actions. The environment consists of both external and internal environments and a parameter of the internal environment – Happiness – is shown in operational representational form

4.3 Operational Representation   153

Note that, as discussed in Chap. 1, the environment involved consists of both the external as well as internal environments, and in Fig. 4.4 we show a parameter of the internal environment – the state of Happiness – being represented in the form of operational representation. And, like the physical activities in the external environment, its activity is being observed, recognized, and changed by the noological processes as well.

The basic operational representational scheme developed here involving explicit characterization of state changes in the spatial and temporal dimensions would seem intuitive, as space and time are very fundamental in our reality. However, a representational scheme is only meaningful and effective if it can support and enhance noological processes. Therefore, the next chapter explores how the representational scheme can be used in encoding causal rules that can participate in reasoning and problem solving processes.

In developing the basic representational scheme, we focus on considering 1D spatial activities. Even though our reality is 3D space and 1D time, by first considering 1D activities and hence addressing the issues associated with state changes in 1D space and 1D time, we can gain insights into how the temporal dimension enters into the picture. There are some very important issues that arise just by the addition of the temporal dimension, hence these have to be addressed first. Hence, in this chapter and the next, we focus on "depth-wise" investigations – for relatively simple but illustrative 1D situations, the basic representations can be build up to representations of causal rules, representations of extended objects, and finally using these representations for reasoning and problem solving in some interesting 1D situations. We will only touch on 2D and 3D issues briefly and relegate the details to future investigations.

4.3 Operational Representation

In this section we develop the various basic operational representations that constitute a set of "atomic operational representations" that are used to represent events and activities at the epistemic ground level. These representations can then be used to build up to more complex, higher level representations to encode more complex events and activities.

4.3.1 Basic Considerations

We begin by considering the behavior of various complex and simple objects and their attendant activities. Figure 4.5 shows two relatively complex objects – Fig. 4.5a, some fluid, and Fig. 4.5b, a plant with leaves – and one relatively simple object – Fig. 4.5c, a rigid rod. The fluid is amorphous and its shape changes continuously as it flows over, say, some flat surfaces (with mainly 2D motion)

Fig. 4.5 (**a**) Fluid with arbitrarily complex movements on a flat surface. (**b**) Leaves on a plant with potential movements in complex 3D manners. (**c**) Translational and rotational motions of a rigid object and the elemental movements of its subparts (From Ho 2012)

and the leaves on the plant or tree may move in some very complex 3D manner when they are blown by winds. To capture and represent the static visual features in these objects, the approach in traditional computer vision is to try and fit straight lines or simple curves to capture the various subparts of these objects (e.g., Forsyth and Ponce 2011; Shapiro and Stockman 2001; Szeliski 2010, etc.). However, it is instructive to note that not only are there relatively complex subparts in these objects, each of the *points* on these subparts can also potentially move independently in some arbitrarily complex manner. A human being's naturally endowed visual senses may not be able to attend to and capture all these static and dynamic details in a short span of time, but the full perception can be built up over time or recording devices can be used to capture a full characterization of the static and dynamic details involved.

For the case of the rigid object in Fig. 4.5c undergoing translational and rotational motions, the subparts' movements are simpler than that of the fluid and the moving leaves, but they still have to be characterized at the elemental level as every point on the rigid rod has the potential of interacting with other objects. Therefore, this means that we need to provide the mechanisms for the representation of *elemental* physical details and their attendant potential *elemental* movements to capture knowledge at the so-called epistemic ground level. This detailed information will have implications for the various noological processes of conceptualization, reasoning, and problem solving.

Without presuming the kind of visual shapes that may appear in the physical world and how they or their subparts may move, the most general approach would be to provide a representational scheme that can capture the most elemental details, both in space (the static aspect) and time (the dynamic aspect). Our perceptual and conceptual systems work at multiple resolutions – sometimes only the gross aspects of objects are important for a cognitive task (e.g., consider the high-level, symbolic description of a Tree as made up of Leaves, Branched, and Trunk in Fig. 4.1), but when the system "zooms" in onto the details, they are there to be examined by the cognitive system for certain purposes. For example, consider a simple task such as when a person acts to lean on the trunk of a tree. Other than identifying the general shape of the trunk and its spatial relationship to the tree and its surrounding that provide information on the location and orientation of the trunk for the leaning task,

4.3 Operational Representation

one also needs to know if the trunk is comfortable to lean on by examining its textural details. Furthermore, it would be beneficial to represent the temporal dimension explicitly, as the changes in the temporal dimension, just like the changes in the spatial dimension, will have important consequences for the various noological processes.

Often, the spatial features of an object, such as a protrusion, is visually analyzed and represented for the purpose of characterizing the static aspects of the object involved. Similarly, in attempting to characterize the activities and interactions in the physical world, the noological system needs to analyze and represent temporal features to characterize the dynamic aspects of these activities and interactions. For the case of a visual feature such as a protrusion, this would be how its shape changes over time that may result in it interacting with other objects in some manner. Therefore, there is no better way to do this than to represent the temporal dimension explicitly, just like what we normally do for the spatial dimension: A most general representation for capturing the fundamental spatiotemporal features of activities and interactions is an *elemental* representation that captures both the spatial and temporal aspects explicitly as shown in Fig. 4.6.

Figure 4.6 shows elemental spatiotemporal representations for both 2D and 1D that capture an elemental specific movement – i.e., an elemental object (represented by a black disc) occupying a location in space at a specific time instant then occupies a neighboring spatial location in the next time instant.[3] Here we consider only specific movements much like in Fig. 4.2a but the general movement of Fig. 4.2b can likewise be treated. In these representations, both the spatial and temporal aspects are represented explicitly and the temporal aspect has been spatialized. What is important is how these spatiotemporal representations are used. Firstly, as mentioned above in connection with Fig. 4.2, these spatiotemporal representations can be used as templates to recognize instances in the physical world that signify an "elemental movement" event. A matching process is carried out between the template and the activities in the "physical" world. If an elemental object moves from a spatial location to another neighboring spatial location in the physical world over an elemental time (which is a complicated and round-about way of saying "it moves"), the template is triggered, and this signifies that a movement has taken place.[4]

---

[3] Figure 4.6 shows only the 1D and 2D representations but a 3D representation can be likewise defined. Also, later we will highlight that these "elemental objects" occupying elemental locations of space-time need not be physical objects but can be something more ephemeral such as a point of light or something more abstract such as the position of a certain parameter in some internal "mental" space-time representing the magnitude of some mental states such as the state of "pain" or "happiness" that can change over time, much like a physical object can change its position in space over time.

[4] For the case of the 2D Cartesian movement in Fig. 4.6a, there will be a need for eight kinds of templates, one for each of the eight possible neighboring spatial locations, and for the 1D movement in Fig. 4.6b, there will be a need for two kinds of template, one for the "upward" movement – i.e. in the positive-x direction – and one for the "downward" movement – in the negative-x direction.

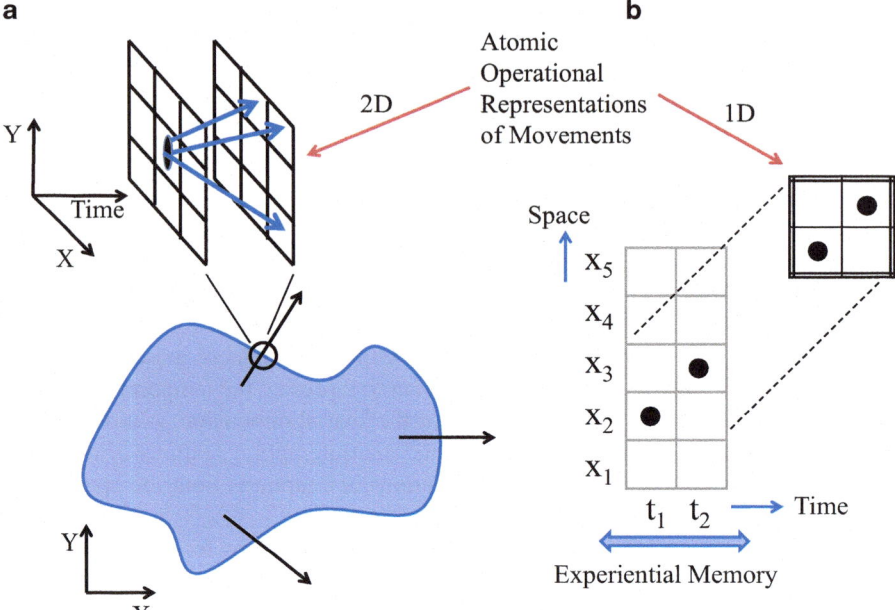

Fig. 4.6 Atomic operational representations for elemental specific movements. (**a**) For 2-dimensional (2D) movements of points, say, on the border or interior of an *amorphous shape*. (**b**) For 1-dimensional (1D) movements of an elemental object (represented by a *black dot*)

What is significant in capturing and representing the concept of "movement" (or "elemental" movement) in this manner is that it need not rely on complex word symbolic or linguistic descriptions – such as a description using predicate logic in Eqs. (4.1) and (4.2). The representation captures the "operation" of movement and in an explicit manner – the transformation in the temporal dimension is spatialized and "laid out" explicitly. That is why we term this spatiotemporal template an "operational representation" or an "operator." Moreover, an elemental movement is "atomic" in the sense that it cannot be divided further, therefore the spatiotemporal templates in Fig. 4.6 are also called "atomic operational representations" or "atomic operators." As mentioned earlier in Sect. 4.1, we submit that these atomic operators in Fig. 4.6 capture the "meaning" of movement at the epistemic ground level.

We would like to highlight that in the simple movement operator in Fig. 4.6 we label the two elemental objects in the two diagonally separated spatiotemporal locations with something that is of the same shape and color. By this, we are stipulating that these are the same elemental object at two different spatiotemporal locations. The convention we are adopting here is that if they are not of the same shape and color, then the operator may be stipulating some other things than a simple object movement – e.g., an object that moves and transmutes into another object. We are assuming that there are some other processes that would determine, say, whether two elemental objects in separate spatiotemporal locations in the

4.3 Operational Representation

physical world are indeed the same object before the movement operator operates on them to determine whether it was just a movement that has taken place instead of a transmutation and a movement. Often, a noological system would assume that if there is no change in the appearance of the object over two spatiotemporal locations then they are indeed the same object.

To allow activities such as movements in general or an elemental movement as described above in particular to be recognized in the physical world using the movement operator, a noological system must have an Experiential Memory – a memory structure that will store perceived information over more than one time instance – to allow the changes in the temporal dimension to be explicitly represented for the movement operator's matching process to take place, as shown in Fig. 4.6b. This is often also known as episodic memory (Gleitman et al. 2010).

Note that the movement representations in Fig. 4.6 capture a certain "speed" of movement – i.e., in a certain elemental time it moves an elemental distance in space. To represent a variety of speeds in operational representations, atomic operators corresponding to different speeds are needed. In Sect. 4.3.7.1 we will discuss this further. However, in most of our subsequent discussions, we will use just one basic "canonical" speed of movement – that which corresponds to the basic movement representation in Fig. 4.6 – while we focus on the applications of the movement operators in building more complex representations (this chapter) and in problem solving situations (next chapter).

As was illustrated and discussed in connection with Fig. 4.2, other than being used to recognize an instance of an elemental movement, the atomic operators of movement in Fig. 4.6 can also be used to "generate" movement. There are two kinds of target output for the generation process. One is it specifies what needs to take place in space and time to effect an elemental movement in the physical world – if an object is at a spatial location in one instance, it needs to be at (or "shifted to") another neighboring spatial location in the next (or neighboring) instance. The other is it directs the generation output to a "mental space" or a "mental array" internal to the noological system for reasoning purposes (e.g., like the mental array of Funt (1980) or Ho (1987)). (The Experiential Memory of Fig. 4.6 which may be implemented as an "array data structure" in a computer program that stores data that can be transformed is also an example of a "mental space.") This is sometimes called "imaging" or "imagining" (Kosslyn 1980, 1994). We will term the use of the operators in both the recognition direction and the generation direction *Recognition-Generation Bidirectionality*.[5]

We would like to clarify the nature of the "elemental blob" (the black circle) in the atomic operators of Fig. 4.6. This blob is an "elemental" bit of occurrence but it

---

[5] We normally think of the recognition process as information starting from the outside world and flowing "into" the noological system to be processed and when the noological system intends to effect a movement in the outside world, the information flows "outward" to the limbs or actuator systems and finally effects something in the physical world. The actions could of course be directed as well toward an internal "mental array" as mentioned above.

need not correspond to any "truly elemental" bit of nature such as the fundamental particles that are deemed to be indivisible (in any case, truly fundamental particles may not exist in nature – physicists seem to be continually discovering particles that are more and more fundamental). It is an elemental bit of occurrence functioning at a level relevant to the noological task involved. E.g., depending on the resolution of the reasoning process involved, the "blob" could be an atom, a person, a vehicle, a point or region of light, or any recognizable points on the human body, an artifact, or a tree. These blobs could refer to something out in the real world or they could be entities in the noological system's internal mental processes.

4.3.2 Existential Atomic Operators

A more primitive activity or operation than movement that can take place in the physical world is the appearance (coming into existence) and disappearance (going out of existence) of an entity at a specific location. Such an entity could be a point of light or an elemental "blob" of substance – an elemental object. (In Sect. 4.3.8 we will discuss how the appearance or disappearance of some mental states/objects – such as the feeling of "pain" or any emotion – could be likewise handled using similar operators.) A point of light could be turned on and off and an elemental object could appear at or disappear from a location. The appearance of an elemental object at a location could be due to the object moving from a neighboring spatial location or due to the materialization of the object "out of thin air" at that location (assuming the technology to effect this exists). The former will be termed *loose* appearance and the latter *strict* appearance.

Similarly, for disappearance, there is *loose* disappearance from a spatial location, in which an elemental object moves away to another neighboring spatial location, or *strict* disappearance, in which an object dematerializes from the location or a point of light at that location is extinguished. The loose appearance of an object at a spatial location is more commonly referred to as the *arrival* of the object at that spatial location from a neighboring spatial location. Similarly, loose disappearance is commonly referred to as *departure*.

Having appeared at a location, the light or object may "stay" at that location for a while. For strict materialization such as the turning on of a point of light, this is often referred as *persisting* at that location and for loose materialization in which an object moves into a neighboring location and stops moving for a while, it is usually referred to as *staying* at that location.

Figure 4.7 shows a point of light or an elemental object appearing at a certain location, persisting for a while, and then disappearing from that location. The atomic operational representations of *Strict-Appear, Loose-Appear, Strict-Disappear, Loose-Disappear,* and *Persist* activities are shown in Fig. 4.7. We show only 1D versions of these operators as the 2D ones or even 3D ones are similar. *Loose-Appear* is a sub-operator of *Strict-Appear*. If an activity triggers *Strict-Appear*, it will trigger *Loose-Appear*, but the reverse is not necessary true. The situation is

4.3 Operational Representation

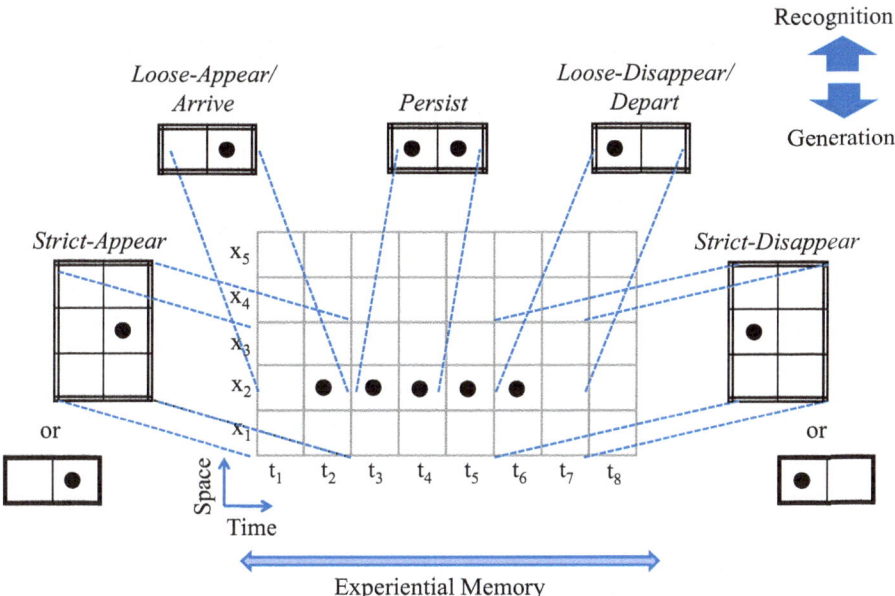

Fig. 4.7 Atomic operational representations (1D) for *Strict-Appear, Loose-Appear/Arrive, Persist, Strict Disappear,* and *Loose Disappear/Depart*

similar for *Loose-Di*sappear and *Strict-Disappear*. In most parts of this chapter, we will be concerned with just *Loose-Appear*ance of physical objects, and often we will refer to it as *Arrive*, and similarly, *Loose-Disappearance, Depart*. For *Strict-Appear* and *Strict-Disappear* we will sometimes shorten them to *Appear* and *Disappear* respectively. In subsequent discussions and depictions, for ease of illustration, we will use an alternative, smaller version of the *Strict-Appear* and *Strict-Disappear* templates that look like the *Loose-Appear* and *Loose-Disappear* templates respectively but with differently marked borders (which are thicker lines) as shown in Fig. 4.7. We submit that these atomic operators in Fig. 4.7 capture the "meaning" of the respective concepts (*Appear, Persist, Disappear,* etc.) at the epistemic ground level.

Figure 4.7 shows, as before, that an Experiential Memory is necessary to hold incoming sensory information over more than one time instance for these operators to detect the corresponding activities. Also, similarly, the operators can be used for recognition as well as generation – e.g., for generation, the *Appear* operator would perform something like a *"Make-Appear"* operation.

It is interesting to note that the concept of "stationarity", whether it be *Stay* associated with movement or *Persist* associated with existence, which connotes the idea of "changelessness," is easily encoded in our operational representational scheme in which time is represented explicitly, thus allowing the encoding of the idea of "no change in position in time" in a direct and fundamental way.

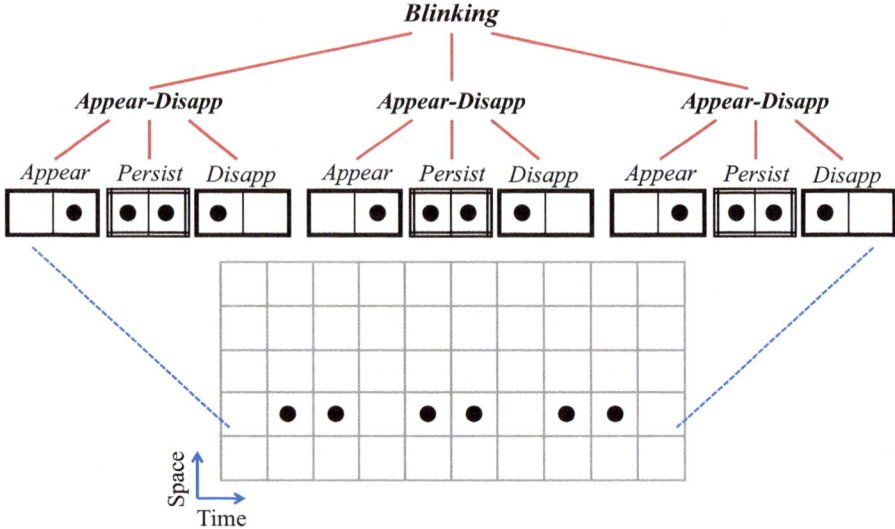

Fig. 4.8 Conceptual hierarchy for *Blinking* (From Ho 2012)

The atomic operational representations can be built upon to represent higher level concepts. Figure 4.8 shows, using a 1D example, how conceptual hierarchies can be built from the atomic operators of Fig. 4.7. In Fig. 4.8, the concept of "blinking" is shown to be built from the atomic operators of *Appear (Strict-Appear)*, *Persist*, and *Disappear (Strict-Disappear)* in a two-level structure. If a point of light (or a blob of substance) keeps appearing and disappearing, it is *Blinking*. The intermediate level concept *Appear-Disapp* captures just the appearance and disappearance aspects of the external event, ignoring the persistence aspect, which means it is not dependent on whether there is a persistent state or how long the persistence is.

It can be seen in Fig. 4.8 that at the bottommost level, the concepts are represented using atomic operators but at the next two levels up, the concepts (Appear-Disappear, Blinking, etc.) are represented word-symbolically (as mentioned in Sect. 4.1, symbols can be words or pictures) but they are ultimately grounded in the epistemic ground of the bottommost level.

4.3.3 Movement-Related Atomic Operators

Figure 4.9 shows a blob of substance materializing at a point in a 1D space, moving up and down in a certain manner, and then disappearing. Likewise, higher level concepts building up to a conceptual hierarchy can be built upon the atomic operators of *Appear, Move-Up, Move-Down, Stay,* and *Disappear*. (As mentioned earlier, the representation of *Stay* is similar to *Persist* – "unchanged over time." The

4.3 Operational Representation

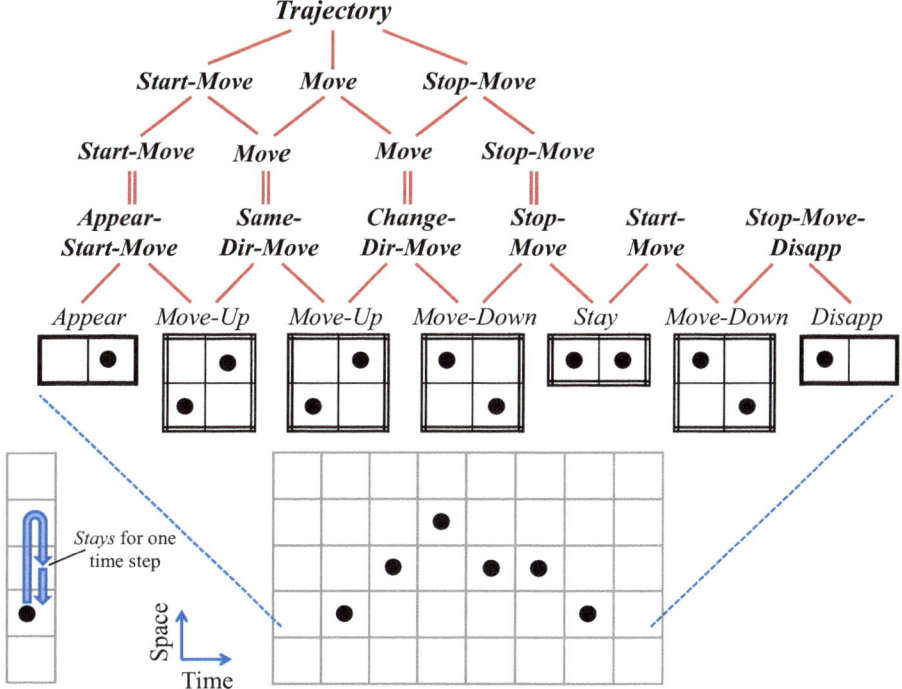

Fig. 4.9 Conceptual hierarchy for *Trajectory* (From Ho 2012)

difference is *Stay* implies that there was a previous motion, whereas *Persist* does not.) The higher level concept of *Start-Move* is *Stay* followed by either *Move-Up* or *Move-Down*. It can also be *Appear-Start-Move* – appearing and starting to move – up or down – right away. The concept of *Change-Dir-Move* (change of movement direction) is *Move-Up* followed by *Move-Down* or vice versa. The concept of *Move* is either *Move-Up*, *Move-Down*, *Same-Dir-Move* or *Change-Dir-Move*. The concept of *Trajectory* is *Start-Move* followed by *Move* followed by *Stop-Move* and it doesn't matter how much movement takes place in between the *Start-Move* and *Stop-Move*. Therefore, there is a higher level concept of *Move* that collapses many sub-instances of *Move*.

Thus, a high level operational concept of *Trajectory* can be built from various lower level operational concepts as well as the atomic level operational concepts. The higher level concepts are thus grounded through the atomic operational concepts. (Another set of operators discussed earlier in association with Fig. 4.7, the *Loose-Appear/Arrive* and the *Loose-Disappear/Depart*, would in principle be triggered at all the spatiotemporal locations as the object moves along the trajectory in Fig. 4.9 but they are not useful in characterizing the conceptual hierarchy illustrated in Fig. 4.9 and are hence not shown.)

In Figs. 4.8 and 4.9 we have indicated only that some atomic operators are triggered in various spatiotemporal locations but have not indicated any other constraints on the exact spatiotemporal locations at which they are triggered. In Sect. 4.3.7 we will make finer distinctions between *Blinking* vs *Continuous-Blinking* and *Trajectory* vs *Continuous-Trajectory* by considering more constraints on the spatiotemporal locations of the triggered atomic operators.

These higher level operational concepts such as *Blinking* or *Trajectory* also exhibit the property of *Recognition-Generation Bidirectionality* in that they can be used in both the recognition direction (i.e., to recognize an activity or event in the physical world being a *Blinking* or *Trajectory* activity or event) and the generation direction (i.e., to generate a *Blinking* or *Trajectory* event or activity). The noological processes would traverse up (for recognition) or down (for generation) the conceptual hierarchies to achieve these noological operations accordingly.

4.3.4 Generation of Novel Instances of Concepts

In the process of generation, based on the generalizations embedded in the conceptual hierarchy, novel combinations of the atomic operators may give rise to novel instances of spatiotemporal processes or activities. Figure 4.10 shows a novel combination of atomic operators that is also a *Blinking* event, but the amount of persistence is different in different stages of the blinking (compare with Fig. 4.8). This is because the concept of *Blinking* ignores the amount of persistence at each *Appear-Disappear* stage, as mentioned above. Which particular instance of a general concept such as *Blinking* or *Trajectory* with its attendant unique combination of atomic operators is generated would depend on the requirement of the noological task that initiates the generation process for specific reasoning or problem solving purposes.

Other than through the conceptual hierarchies, another process by which generation takes place is through the application of causal rules for problem solving that will be discussed in the next chapter.

4.3.5 Example of 2D Movement Operations

Figure 4.11 shows how the changing shape of some amorphous fluid as it flows can be described by operational representations. Built upon 2D atomic movement operators, concepts such as "Almost Stationary Back Portion," "Slowly Developing Side Concavity" and "Rapidly Protruding Front Portion" that describe the time course of change of various regions of the fluid can be created. Similarly, many complex activities and events can be described at various levels of detail for the purpose of cogitation – the complex movements of the leaves on a tree in the wind, the flapping of hanging laundry in the wind, the operations of complex machineries with many parts, etc. These descriptions would be the necessary steps to support other noological processes such as reasoning and problem solving. However, as

4.3 Operational Representation

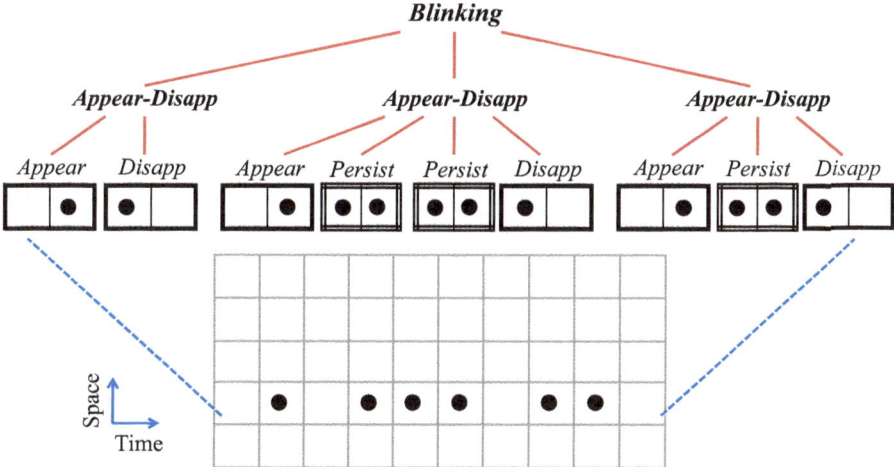

Fig. 4.10 A novel *Blinking* event (From Ho 2012)

mentioned earlier, in the current and the next chapters, we will focus on investigating in the "depth" direction – how the representational scheme can be applied to reasoning and problem solving in relatively simple 1D situations to develop the necessary computational structures and processes that can be later generalized to 2D and 3D, more complex and more general situations. Suffice it to note that the operational representational scheme proposed here has the power and generality to represent any complex activities in the physical world at the ground level.

4.3.6 Characteristics of Operational Representation

Base on the foregoing discussion, the critical characteristics of operational representation in general and atomic operational representation in particular are summarized as follows (Ho 2012):[6]

- *Explicit Temporal Representation* – the temporal dimension is explicitly represented in operational representation – temporal changes are laid out in a spatial extent for simultaneous processing. This requires the noological system utilizing the representation to have an explicit Experiential Memory.
- *Elemental Representation* – atomic operators are elemental in that they represent the smallest discrete changes in the respective dimensions.
- *Cognitive Hierarchy Construction* – higher level concepts can be built directly upon the lower level as well as the atomic level operational representations.

---
[6] These characterizations are succinct and definitive and hence cannot be easily altered so they are taken verbatim from Ho (2012).

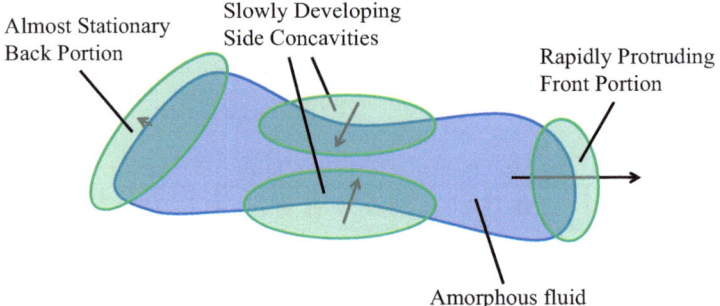

Fig. 4.11 Descriptions of the regions of some amorphous fluid

- *Grounding of Representations* – atomic operational representations are grounded directly in the environment or the mental space, higher level conceptual representations are grounded through the atomic representations.
- *Recognition-Generation Bidirectionality* – operational representation can be used for recognition as well as generation purposes. Recognition and generation can involve things, events, and processes in both the physical environment and the mental space.
- *Noological Manipulatability and Recomposability* – operational representations, atomic or higher level, can be noologically manipulated and recomposed into novel instances in cognitive processes.

4.3.7 Further Issues on Movement Operators

In the previous sections we have used two kinds of operators, the existential and the movement operators, to discuss the motivations and considerations behind the scheme of operational representation. We have ignored some details associated with the operators in order to focus on the general considerations and conceptions that provided the motivations. In this section we formulate further details associated with the existential and movement operators described in the previous sections.

Chief among all the issues concerning operational representations is the way in which the operators is computationally structured and applied to recognize the instances in the physical environment that may trigger the operators. Take, for example, the *Move-Up* operator as shown in Fig. 4.9. An elemental *Move-Up* operation or activity could in principle take place at any spatiotemporal location in the physical world. Therefore, a cognitive system must be ready to detect this operation/activity at any spatiotemporal location. One way that the various operators could be organized to be prepared for detecting any occurrence of the corresponding activity at any spatiotemporal location is as shown in Fig. 4.12. In Fig. 4.12, each of the various operators *Appear, Stay, Disappear, Move-Up, Move-Down,* etc. is shown to populate an Operator Array, each of which containing one

4.3 Operational Representation

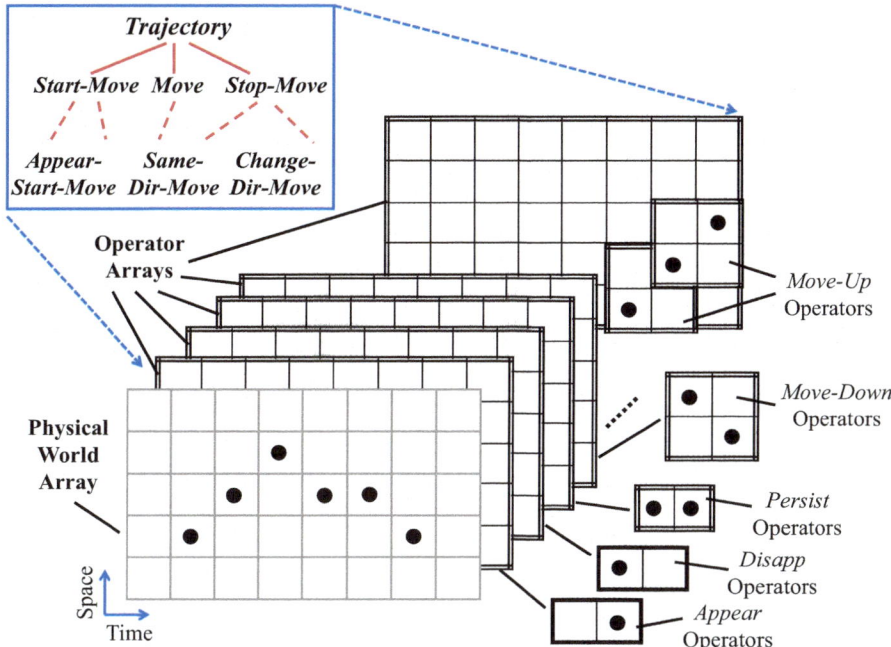

Fig. 4.12 Operator Arrays. *Top left* corner shows a typical conceptual hierarchy "looking at" the Operator Arrays to determine if a certain concept is present in the Physical World Array (The Experiential Memory)

kind of operator. In each of the Operator Arrays, there is an operator assigned to every spatiotemporal location (as an example, the last array, the *Move-Up* operator array, shows two *Move-Up* operators assigned to two different neighboring spatiotemporal locations).[7] Therefore, the operators are ready to capture any activity that may take place at any spatiotemporal location in the Physical World Array (the Experiential Memory).

We mentioned in Sect. 4.3.3 that we did not distinguish between the concepts of *Trajectory* vs *Continuous-Trajectory* or *Blinking* vs *Continuous-Blinking* with the earlier formulation. Now, with the Operator Arrays defining specific spatiotemporal locations for various operators, we can define the concept of *Continuous-Trajectory*. A *Continuous-Trajectory* is one in which all the atomic operators at the ground level are "joined up" spatiotemporally. The definition of spatiotemporally *joined-up* is as follows. Each of the movement or existential operators contains some "subsquares" – for the existential operator, there is a left sub-square (corresponding to an earlier instance) and a right sub-square (corresponding to an elemental "next"

---

[7] There is neuroscientific evidence showing that the brain organizes various kinds of visual features detectors in a topographical manner over the cortical surface much like these Operator Arrays (Nicholls et al. 2011).

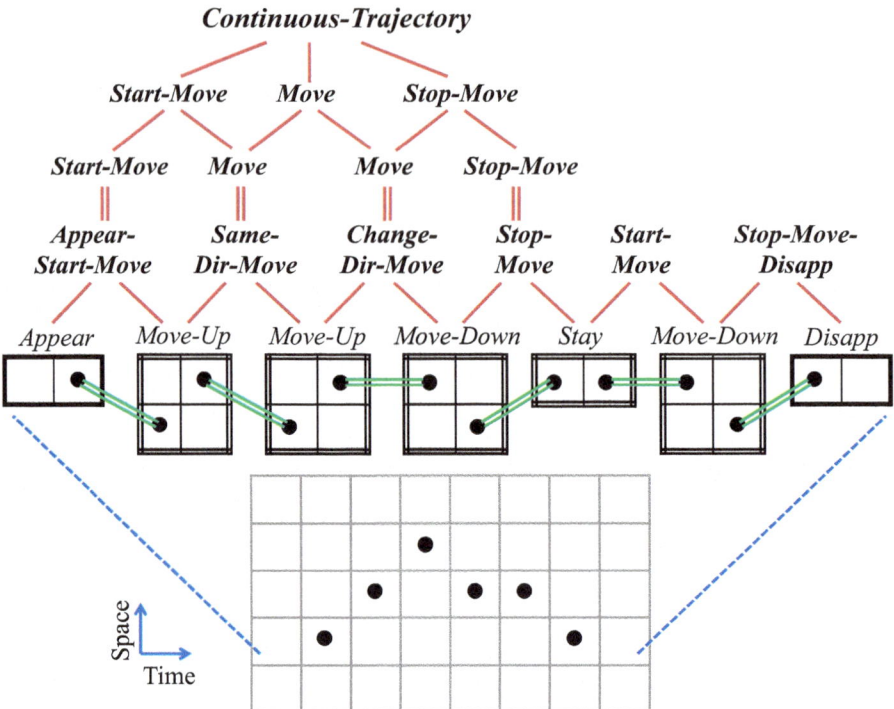

Fig. 4.13 Concept of *Continuous-Trajectory*. The *green double lines* joining the atomic operators stipulate the spatiotemporal *joined-up* condition as defined in the text (From Ho 2012)

instance), and the movement operator likewise has four sub-squares – and we will group them into left ones and right ones. The spatiotemporal joined-up condition is satisfied if a *right* sub-square of an (existential or movement) operator in a given time instance that contains an expected elemental object (a "blob", i.e.) is spatio-temporally coincident with a *left* sub-square of the operator in the subsequent elemental instance that also contains an expected elemental object. This can be determined by having the purported trajectory (say, the one illustrated in the Physical World Array in Fig. 4.12) trigger all the operators in the Operator Arrays in Fig. 4.12 and then the spatiotemporal *joined-up* condition can be checked to see if it holds. Figure 4.13 illustrates the concept of *Continuous-Trajectory* – the green double lines between the atomic operators represent the spatiotemporal coincident, joined-up condition. The top left corner of Fig. 4.12 shows how a typical conceptual hierarchy may "look at" the Operator Arrays to see if a certain concept can be triggered.

Hence even though the same set and sequence of atomic operators as shown in Figs. 4.9 and 4.13 may be triggered but if they are not all *joined-up* spatiotempo-rally as defined above, then the *Trajectory* is not continuous spatiotemporally in the sense that is usually understood (there would appear to be sudden "jumps" at certain spatiotemporal locations). The higher (second) level operators such as *Start-Move* or *Chang-Dir-Move* shown in Fig. 4.13 are meant to be built on *joined-up* atomic operators. If the operators are not *joined-up* spatiotemporally, then a *Move-Up* followed by a *Move-Down* operator may correspond to other kinds of phenomena – e.g., an object moving upward, disappearing, and then appearing somewhere else in space and/or time and moving downward. This will not correspond to an object changing its direction of movement in the way we normally understand it. Hence a "*spatially* continuous" trajectory as we observe it over time is *joined-up spatiotemporally* in the sense defined above.

An important thing to note is that the set of Atomic Operator Arrays in Fig. 4.12 could thus *describe/recognize* or *generate* all possible activities in this restricted universe, activities that are characterized by the combinations of the various kinds of existential and movement atomic operators at various spatiotemporal locations. In a typical noological system, not all possible conceptual hierarchies are created beforehand as there are too many possibilities. The conceptual hierarchies can be constructed as and when activities are encountered, such as the *Continuous-Trajectory* hierarchy shown in Fig. 4.13, or when certain types of events need to be generated in some noological tasks – e.g., in solving certain problems, a discontinuous or special kind of trajectory may be needed. (Problem solving with operational representations is discussed in the next chapter.) One can imagine special types of trajectories such as zig-zaging trajectory (more relevant for 2D space situations), jerky trajectory (starting and stopping often), etc. that can be defined based on the same set of atomic operators. Non-trajectory activities such as *Blinking* also have their own many specialized types of activities.

Many kinds of conceptual hierarchies can be created to represent many kinds of concept. Again, we will not delve any deeper than what we have done with *Trajectory* and *Blinking* but will proceed to refine other important constructs in our continued explorations in the "depth" direction.

4.3.7.1 Generalizations and Specializations of Atomic Movement Operators

In Sect. 4.3.1 we mentioned that the movement operators depicted in Fig. 4.6 and some of the subsequent figures represent a "canonical speed" of movement and they will suffice for most of our subsequent discussions on reasoning and problem solving. A movement operator could also signify either a movement that takes place at a specific location or movements that take place *anywhere* in space and time. The Operator Arrays depicted in Fig. 4.12 allow the representation of both the general and specific aspects of the movement and the existential operations. A *particular* operator in a specific location in the Operator Array, when activated (in the recognition or the generation direction), represents an elemental activity that takes place at those locations, while *any* of the operators in the Operator Array, when activated, represents activity that takes place anywhere/somewhere in spacetime. One can also conceive of grouping these operators in the Operator Array in

168   4 Conceptual Grounding and Operational Representation

such a way as to capture the concept of, say, "movements in this region of space-time" if the concept is of any use in some noological processes.

In this section, we extend the canonical movement operator of Fig. 4.6 to represent other speeds, more general concepts such as ranges of speed, as well as acceleration. Figure 4.14a–d show that different speeds can be represented by specifying different amount of spatial displacement relative to a certain amount of temporal displacement. In Fig. 4.14a, the operator represents 2 units of spatial displacement ("upward") per 1 unit of temporal displacement, and in Fig. 4.14b it is 4 units of spatial displacement ("upward") per 1 unit of temporal displacement (corresponding to 2 units and 4 units of speed upward respectively). These two operators thus represent speeds that are higher than the canonical speed of 1 unit of spatial displacement per unit of temporal displacement (i.e., 1 unit of speed). In Fig. 4.14c, d, the operators represent speeds that are lower than the canonical speed – 1 unit of spatial displacement per 2 units of temporal displacements and per 4 units of temporal displacements respectively (corresponding to ½ and ¼ speed units respectively). These operators can be used in recognition and generation processes in the same way as that of the canonical operator described earlier.

Figure 4.14e–h are movement operators that represent ranges of speed. Figure 4.14e shows a movement operator similar to that of Fig. 4.14a – i.e., an operator that represents 2 speed units (i.e., 2 units of spatial displacement per 1 unit of temporal displacement) – but with two additional blue double parallel lines demarcating the two squares laid out "horizontally" in between the upper and lower elemental objects and a vertical double arrow in between the two parallel lines. We label this operation a "Range Bar" in Fig. 4.14e. The two lines and the arrows are used to represent the fact that the spatial separation between the lower (temporally earlier) and upper (temporally later) elemental objects has a minimum amount of 2 units but it can stretch to any amount that is more than 2 units of separation. That is, this operator represents any speed that is greater than or equal to 2 units.

Figure 4.15a illustrates how this operator might be used for recognition. Consider an event in the physical world that is an elemental object moving at more than 2 speed units, and in the specific case of Fig. 4.15a, it is an elemental object moving at 4 speed units. The operator would begin from the designated minimum spatial ("vertical") separation between the temporally earlier (lower) elemental object and the temporally later (upper) elemental object (and in this case it is 2 units of space) and stretch the separation in subsequent steps, each time increasing, say, the separation by 1 unit, and each time attempting to match the current operator with the event in the physical world. If there is a match at some point then the operator is triggered – i.e., an event that is an object moving at a speed greater than or equal to 2 units is recognized/detected. Similarly, for generation, the operator can be activated in the same way but in the "opposite" direction – i.e., the operator will begin with the minimal spatial separation (in this case, 2 units of space) designated and stretch it step by step, and halt at any step and output a movement – with the locations of the object in the earlier and the later time instances as specified by the transformed operator. This would correspond to moving an elemental object at a speed of more than 2 units. The specific step at which the halting takes place could

4.3 Operational Representation

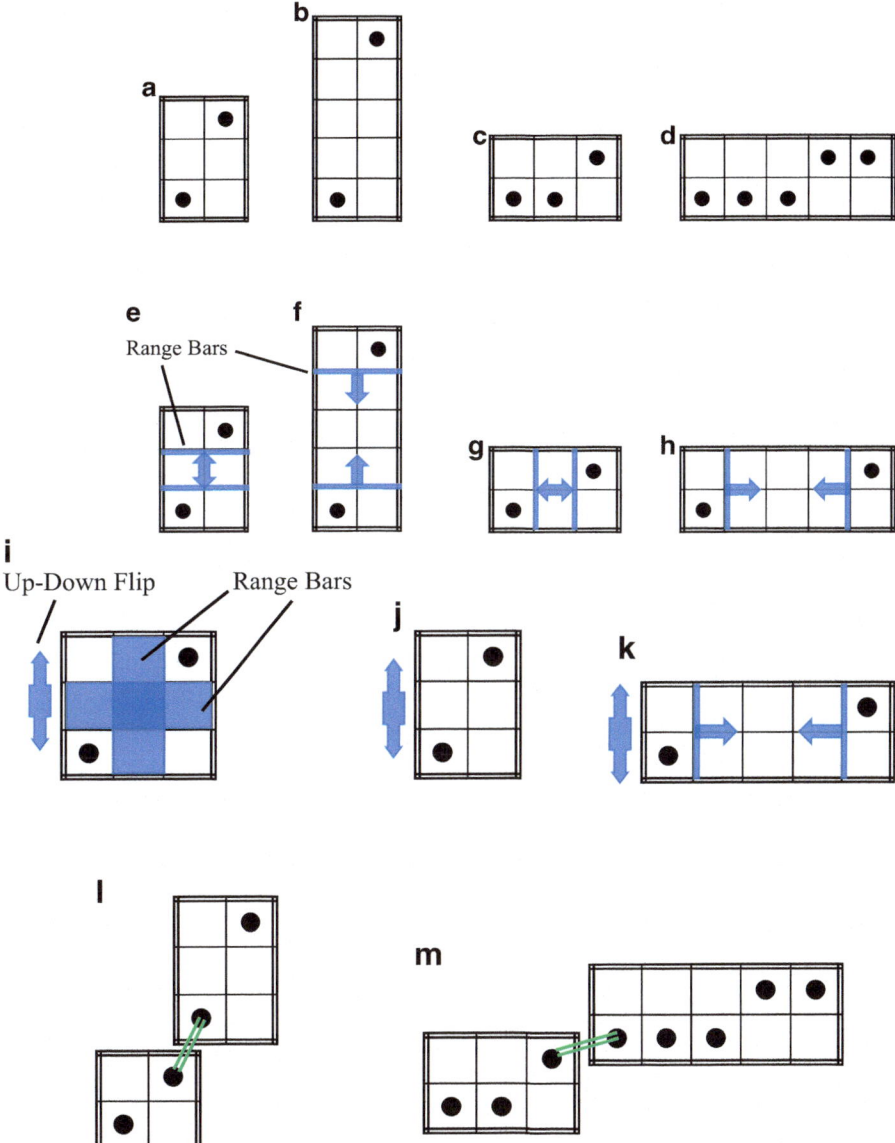

Fig. 4.14 Representations of different kinds of speed. (**a**) 2 units of upward spatial displacement per 1 unit of temporal displacement (2 speed units upward). (**b**) 4 speed units upward. (**c**) ½ speed unit. (**d**) ¼ speed unit. (**e**) Greater or equal (>=) to 2 speed units. (**f**) Smaller than or equal to (<=) 4 speed units but greater than or equal to (>=) 1 speed unit. (**g**) Smaller than or equal to (<=) ½ speed unit. (**h**) Greater than or equal to (>=) ¼ speed unit but smaller than or equal to (<=) 1 speed unit. (**i**) Representation of any direction of movement at any speed. (**j**) Movement upward or downward at 2 speed units. (**k**) Movement upward or downward at greater than or equal to ¼ speed unit but smaller than 1 speed unit. (**l**) Acceleration upward. (**m**) Deceleration upward

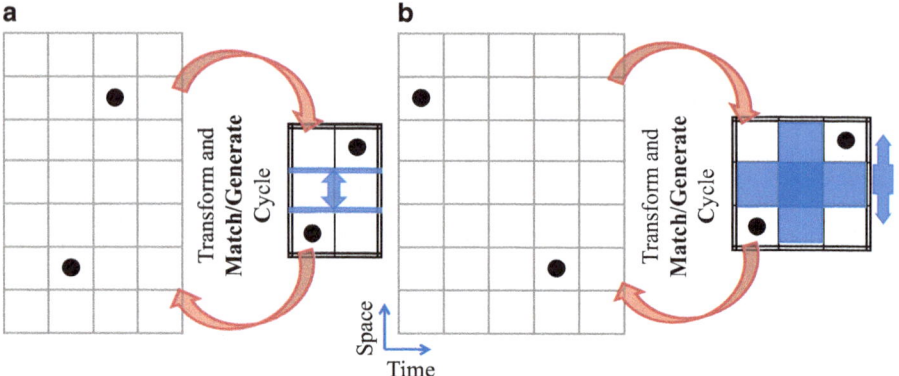

Fig. 4.15 Transform and match process for recognition of general kinds of speed of movement. (**a**) Operator for movement greater or equal to 2 speed units. (**b**) Operator for movement at any speed in any direction

be determined by a random process (i.e., moving by an arbitrary speed greater than 2 units) or by the requirement of other noological processes that may dictate exactly how much greater than 2 units the speed should be.

In a similar vein, the operators in Fig. 4.14f–h correspond to speeds of $>=1$ unit but $<=4$ units, $<=½$ unit, and $>=¼$ unit but $<=1$ unit respectively. For the cases of Fig. 4.14g, h the separation that is variable is in the temporal dimension and for the cases of Fig. 4.14f, h with the "inward pointing" arrows, the operator begins with the designated *maximum* separation and reduces the separation at every step of the recognition or generation process.

Figure 4.14i is an operator than represents "movement at any speed in any direction" or "any movement." The spatiotemporal regions in the operator covered by continuous shading (which are also "Range Bars" but these ones have arbitrary "widths") represent separation or interval (and in this case both spatial and temporal) that can vary from nothing to any arbitrary amount, and the "Up-Down Flip" transformation specification means that the operator can be applied in the configuration shown in Fig. 4.14i for any "upward movement" or one in which the configuration is flipped about a "horizontal" axis – i.e., now the elemental object in the later temporal location in the operator is below the elemental object in the earlier temporal location, thus representing a movement in the "downward" direction. If one needs to represent a more specific movement like "any movement in the *upward* direction" one can remove the "Up-Down Flip" transformation specification in Fig. 4.14i.

Figure 4.15b shows, in a similar vein as in Fig. 4.15a, how the "any movement" operator of Fig. 4.14i operates – as dictated by the Range Bars, it starts from zero separations in the spatial or temporal dimension between the two elemental objects as shown in Fig. 4.15b and increases the separation step-by-step space-wise and time-wise in turn for the recognition or generation for the upward direction movement, and then it flips the operator according to the Up-Down Flip transformation specification to carry out the same process for the downward direction.

Figure 4.14j shows how the Up-Down Flip transformation can be used to represent "upward or downward movement of 2 speed units" and Fig. 4.14k shows a representation of "upward or downward movement of great than or equal to ¼ speed unit but smaller than 1 speed unit." Figure 4.14l shows an acceleration upward – an increase of speed in the following time steps – and Fig. 4.14m shows a deceleration upward – a decrease of speed in the following time steps. Note that the *joined-up* squares on the operators (shown as a green double line between the squares) in Fig. 4.14l, m have a similar meaning as defined in Fig. 4.13 – that they occupy the same spatiotemporal location.

For the representation of real world speeds, grids with much finer resolution than those depicted in Fig. 4.14 would be needed but the basic representational mechanisms would remain the same.

4.3.8 Atomic Operational Representations of Scalar and General Parameters

Movement is the change of spatial location. There are other quantities, physical and mental, that may also change in time. Example of physical quantities that may change elementally are the hardness of substance, mass of objects, brightness of light, etc. In mental space, there are ideas and feelings that have "strengths" that may change elementally. Using similar atomic operators as for movements in the foregoing discussion, the changes in these non-spatial or scalar parameters can be likewise represented as shown in Fig. 4.16. (We use arrays with dotted line borders to represent "mental space.") Examples of the parameters for which their corresponding magnitudes are represented in the "vertical" axis in the figure are physical parameters such as "brightness" or "mass," or mental parameters such as "pain" or "happiness." The magnitude of these parameters *Appears, Increases, Decreases, Persists*, or *Disappears* over time. The atomic operators in Fig. 4.16 thus capture the changes in non-spatial parameters in the same way that the movement and existential operators capture the changes in spatial parameters described in the previous sections. (In Sect. 3.6 we have used a similar temporal representational scheme for the Energy Level of an agent.)

The *Increase* and *Decrease* operators here are the analogs of the *Move-Up* and *Move-Down* operators associated with spatial movement. *Appear, Disappear* and *Persist* operators here are the analogs of the equivalent ones associated with the spatial existential operations. Also, the "change of directions" associated with movement in Fig. 4.9 becomes *Maximum* and *Minimum* in Fig. 4.16. Hence, similarly, conceptual hierarchies can be built upon the atomic operators in the domain of the scalar parameters – e.g., Pain, a scalar parameter measuring an internal mental state, can appear and disappear repeatedly (like "blinking") and

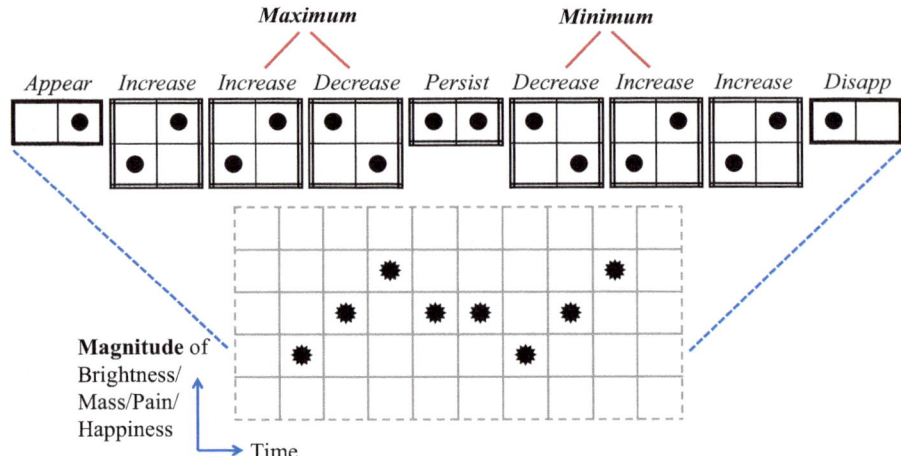

Fig. 4.16 Changes of a scalar parameter over time. Concept of *Maximum* and *Minimum* can be defined accordingly (*Arrays* with *dotted line* borders represent "mental space")

can get worse (*Increase*) at one point, reach a *Maximum* and get better (*Decrease*) subsequently (like "change of direction" in the domain of movement).

As mentioned in Chap. 1, there is a bio-noo boundary at which there are certain built-in needs and motivations for a natural or artificial noological system. At the lower sensory levels there may be certain built-in sensors for saltiness, sweetness, bitterness, etc. if these are necessary for the system's survival. Naturally, there will be changes in the degree of these sensations. These changes would need to be represented internally for the attendant noological processing (e.g., identifying what causes the increase or decrease in these parameters). The system would then need atomic operators for saltiness, sweetness, bitterness, etc. and represent the changes such as that shown in Fig. 4.16. A similar process is required for other "qualia" that manifest themselves to consciousness such as pain, sadness, anxiousness, brightness, darkness, redness, and spaciousness. These qualia have intensities that vary over time but they do not have further internal structures scrutinizable by consciousness. They are atomic in nature and their changes can be likewise represented.

With this representation, one can represent the "trajectory" of an intelligent agent's life in terms of the "trajectory" of some internal mental state parameters such as, say, Happiness – how it changes over time, and this can be represented or captured at multiple levels of temporal resolution – the changes in Happiness can be described as variations over weeks, months, or years, or over finer moments of seconds and minutes.

4.3 Operational Representation

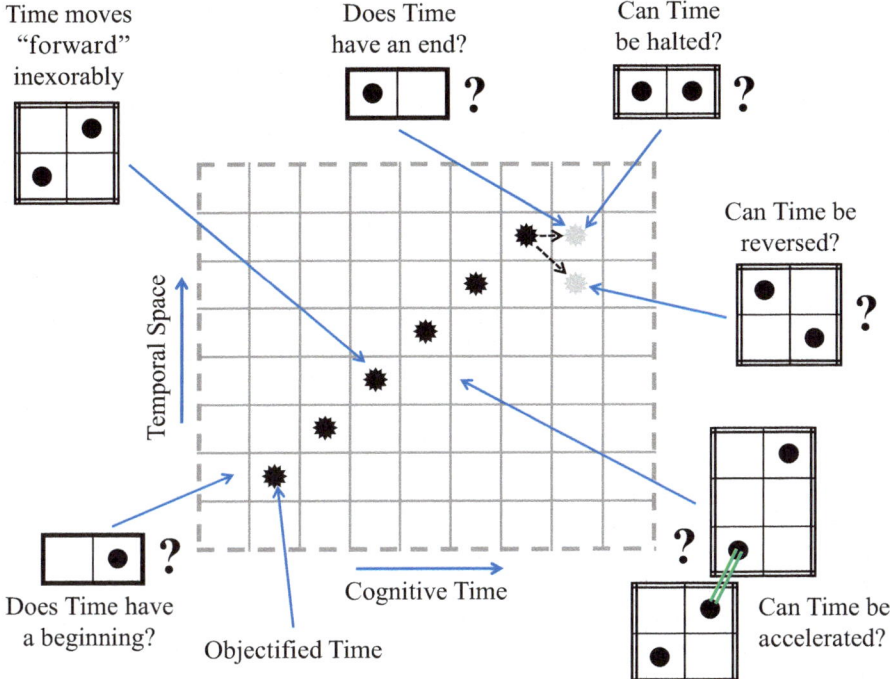

Fig. 4.17 Cogitating about Time. Representations capturing the movement, acceleration, halting, beginning, ending, and reversal of Time for the purpose of cogitation involving Time (From Ho 2012)

4.3.9 Representing and Reasoning About Time

We do not know what happens in the minds of other naturally occurring noological systems but for human beings, we know we are able to think and reason about Time like it is a physical entity. We conceptualize Time as being able to speed up, slow down, or even stopped. We wonder if Time has a beginning or an end (Hawking 1998). Since operational representations represent Time explicitly, it should be able to handle the conception of Time itself.

Interestingly, when we conceive of Time as a physical entity that can come into existence, move at different speed, etc., there must be another kind of time to be used as a reference in order to characterize, say, "coming into existence" – essentially the *Appear* operation as discussed earlier. We call this "Cognitive Time" and we represent it using the horizontal axis as shown in Fig. 4.17.

In Fig. 4.17, Time is objectified into a physical entity (a "blob"). The vertical axis is the "Temporal Space" which is the "space" in which the objectified Time moves. The idea that usually Time moves forward inexorably is represented by a continued movement of this objectified Time, much like an object moving inexorably under its own inertia. Likewise, we can represent the acceleration of Time, the beginning of Time, the end of Time, the halting of Time, the reversal of Time, etc. The usual *Appear*, *Disappear*, and the various movement operators can be used accordingly. There is a difference between Time coming to a halt and time coming to an end. For Time coming to a halt, it would correspond to the *Persist/Stay* operator, and for Time coming to an end, it would correspond to the *Disappear* operator. In physics, the reversal of Time has been used to explain certain phenomena (Feynman 1988) and this corresponds to the *Move-Down* operator as shown (this is reversal in Temporal Space as the conceptualization moves *forward* in Cognitive Time, much like an object moving backward in space while the process moves *forward* in time).

Once Time has been objectified, the cogitation of Time thus employs similar processes as the cogitation of the usual physical processes. This again illustrates the power of operational representation.

4.3.10 Representation of Interactions

When there is more than one causal agent or entity in the environment, interactions arise. A causal agent can be something that is visible and/or tangible such as a point of light or an object, or something that is invisible/intangible such as a "force" (e.g., electric or magnetic force) from somewhere. With interactions, the physical world becomes complex and interesting. In this section we explore various situations in which interactions take place.

4.3.10.1 Propulsion and Materialization

Physical investigations of the past centuries have identified various forms of interactions (such as gravitational and electromagnetic interactions) that act on objects at a distance and cause them to move. In our daily experiences, we identify "forces" as causal agents of movements whatever their origin and physical nature. Often, these forces seem to act in close proximity, requiring a contact event (say, between a human hand exerting a force and an object being pushed) for the force to have an effect (even though, as we now understand from physics, this kind of force is still electromagnetic in nature and there is no "real contact" involved if understood at the microscopic level).

In this chapter, we represent a force and its associated direction in the form of a thick blue arrow that is present in the same spatiotemporal location as the elemental object that the force is influencing as shown in Fig. 4.18. In most cases in this chapter we would consider only forces of 1-unit magnitude that will cause an elemental object to move one elemental spatial location in one elemental time but our representational scheme is general enough to represent forces of all kinds of magnitude (see Chap. 7, Sect. 7.2). Also, in Fig. 4.18 and subsequent figures, what is illustrated in the spatiotemporal frame may be a subsegment of a larger event and

4.3 Operational Representation

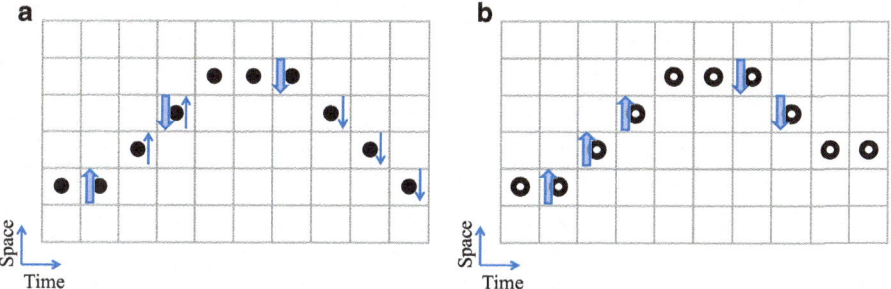

Fig. 4.18 Representation of forces and their effects. (**a**) A force acting on a "light" object. (**b**) A force acting on a "heavy" object. *Thick blue arrow* is force, *thin blue arrow* is momentum (**a** From Ho 2013)

we do not describe what happens to the object outside the spatiotemporal frame that is not of relevance to the discussion.

In Fig. 4.18a we show a kind of "light" object which when acted on by a force will acquire a momentum and continue to move until its movement is counteracted by an opposing force. Two situations result in this kind of dynamics: (1) the object involved is traveling in free space and hence there is no friction to counteract its movement; (2) the object involved is traveling on the ground with friction but the force is large relative to the friction so that the object would seem to be traveling unimpeded for a while after being acted on by the force. In the latter case, the object would still come to a halt after some time. We use a thin blue arrow next to an elemental object in the same spatiotemporal square to represent the presence of momentum on the elemental object involved. When momentum is present, the elemental object would continue to move in the absence of forces.

Figure 4.18a shows that after an "upward" force is applied to the object, it travels upward for two time frames under its momentum until a "downward" force counteracted its momentum and it stops moving one time frame later. Subsequently, another downward force propels it downward.

In Fig. 4.18b we show a "heavy" object that when acted on by a force would travel only for one elemental unit of space over one elemental time step and halt. Continued movement of the object involved would require constant pushing by the force – the presence of a force on the object at every elemental time frame. In the situations of Fig. 4.18a, b we are assuming no other object(s) blocks the immediate movement of the object acted on by the force or the continuing movement of the object with momentum, otherwise the resultant motion will be described by other interaction situations to be described later (e.g., Figs. 4.20, 4.21a, and 4.22, etc.).

Materialization is a concept in which something appears at a location "out of thin air." *Materialization* can involve the switching on of a point of light (as mentioned in Sects. 4.3.1 and 4.3.2) or the appearance of a physical object. The latter is technically impossible currently but is a possibility often explored in science fiction. So, the idea of *Materialization* is shown in Fig. 4.19a in which something is not at a spatial location one time frame before and it appears in that

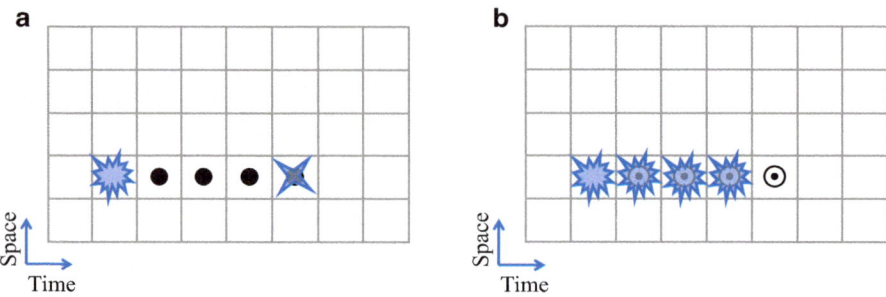

Fig. 4.19 (a) *Materialization* and *Dematerialization* operations. The "starburst" *shape* represents the materialization action and the 4-pronged star *shape* represent dematerialization. (b) Non-persistent materialization – the object represented by a *circle* and a *dot* in the center is non-persistent and requires continued *Materialization* actions to exist ((a) From Ho 2013)

spatial location one time frame after. The operation is represented by a "starburst" icon at the temporal location one time frame before the *Materialization*.

Dematerialization is the opposite of *Materialization* as shown in Fig. 4.19a – the operation, shown as a "four-pronged star," operates on an object at a spatial location at a certain time frame and the object ceases to exist in the next time frame. The switching off of a point of light is also a kind of *Dematerialization*.

In parallel with the case of movement, we distinguish two kinds of *Materialization*. In one kind represented in Fig. 4.19a, the materialized object is persistent and would persist until a *Dematerialization* operation causes it to cease to exist. The other kind of materialized object is non-persistent – it would disappear automatically unless continued to be maintained by an action of *Materialization* at every elemental time frame as shown in Fig. 4.19b. After the last materialization action, the non-persistent object (represented by a circle with a dot in the center) disappears on its own without the need for an explicit dematerialization action, unlike the case in Fig. 4.19a.

4.3.10.2 Reflection, Obstruction, and Penetration

When more than one object exists in an environment, it creates the possibility that these objects may come into contact with each other or influence each other through forces acting from a distance. This would normally result in changes to the original states of the objects involved. We consider first the simpler situation in which one of the two interacting objects is immobile as shown in Fig. 4.20. We use a filled square to represent the immobile object and the mobile object, represented as a filled circle, would reflect off the immobile object in the direction from which it came. Both objects are elastic. Note that the event shown in Fig. 4.20 is a movement and reflection of an elemental object in *1D space* as shown in the small picture on the left of the figure. The spatiotemporal representation of the event is shown on the bottom right. The impinging mobile object is shown to have a momentum toward

4.3 Operational Representation

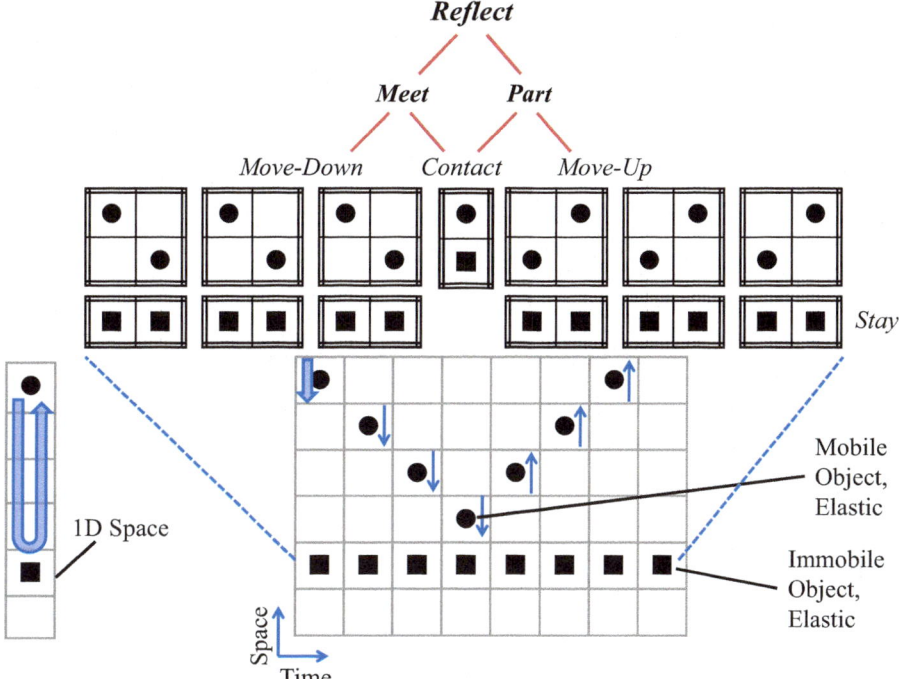

Fig. 4.20 Representation of *Reflection*. The *bottom left* pictures shows the event happening in 1D space. The *bottom right* picture is the spatiotemporal representation of the same event (From Ho 2013)

the stationary object initially and after the contact event, its momentum reverses direction.

In Fig. 4.20 we also show a conceptual hierarchy built up to represent the concept of *Reflect* from the atomic operators. A *Contact* event takes place at the moment when the reflected object changes direction. The *Contact* operator shown in Fig. 4.20 consists of the two kinds of objects placed next to each other spatially. We also illustrate the other atomic operators *Move-Down*, *Move-Up*, and *Stay* that are triggered separately by the incoming object and the immobile object. Intermediate level concepts *Meet* and *Part* are also shown.

In Fig. 4.20 the assumption is that both the impinging and immobile objects are elastic, therefore a reflection takes place for the impinging object when it comes into contact with the immobile object that results in some momentum in the reverse direction after the contact. In the real world, if only one of the impinging or the immobile object is elastic, reflection will also take place. Also, in the real world, reflection is not really immediate (infinitely fast) because of the nature of physical processes – e.g., with certain kinds of elasticity, the reflection may be delayed significantly from the time of contact between the impinging and immobile objects. We do not need very different representational structures to handle these.

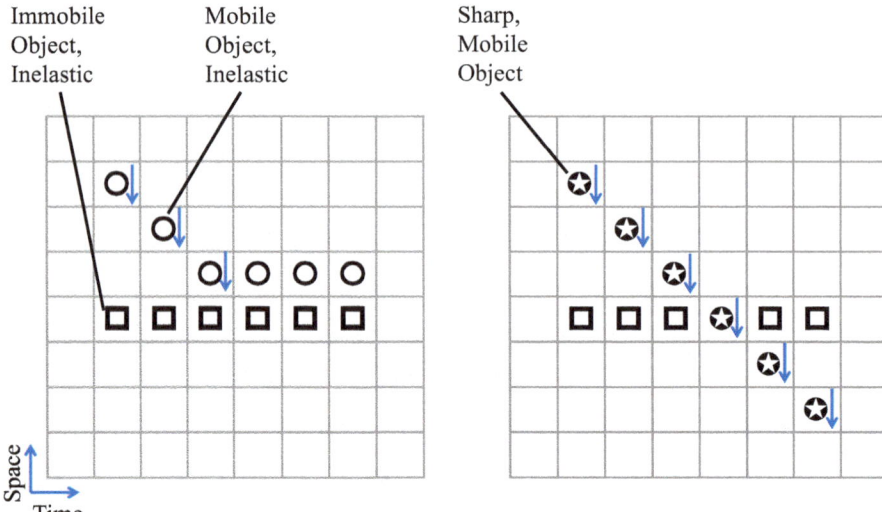

Fig. 4.21 (**a**) Hit and stay, no reflection. (Both impinging object and "obstacle" are inelastic.) (**b**) A *Penetration* situation (From Ho 2013)

Remember that the time steps represented in Fig. 4.20 (and for that matter all the figures before and after) are discrete and correspond to certain time gaps. Given a certain time resolution of interest, if the delay in the reflection is within the span of the time steps, then the same reflection representation as in Fig. 4.20 can be used – i.e., the reflected movement already begins in the following time step as represented in Fig. 4.20. If the reflection delay spans more than one time step, then the impinging object would stay next to the immobile object for some time steps after making contact with it and then a movement in the reverse direction would begin.

Unlike the situation above when one of the objects involved in a contact situation is elastic, there would be a reflection of the impinging object, if both objects – the impinging and the immobile objects – are inelastic, the impinging object would come to a halt next to the immobile object. This situation is illustrated in Fig. 4.21a. (An unfilled shape is used to represent an inelastic object.) In physics, this is understood as the momentum of the impinging object being "absorbed" in the inelastic interaction and normally the energy is converted to other forms such as heat. Figure 4.21a also shows that the impinging object's momentum disappears after the contact event.

In Fig. 4.21b, we illustrate the situation in which an incoming object is able to *Penetrate* the immobile object (shown to be an inelastic object here) because it has some kind of "sharpness" property. We use a star inside a circle to represent a "sharp" penetrating object. In contrast to the situations in Figs. 4.20 and 4.21a, in this penetration situation the impinging object continues its motion pass the

4.3 Operational Representation 179

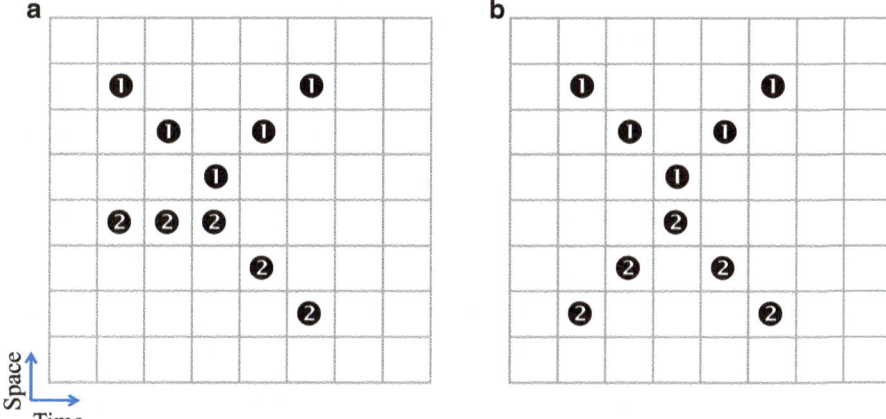

Fig. 4.22 Canonical interactions between two objects that are both mobile that are represented by filled *circles* but separately identified with a number on the *circle*. (**a**) *Object 2* is originally stationary. (**b**) *Object 1* and *Object 2* both move toward each other at the same speed originally

immobile object into the space on its "other side." In Fig. 4.21b we show that the penetrated object is an immobile *inelastic* object. Presumably, the same penetration event will take place even if the immobile object is elastic as typically the effect of a sharp impinging object is independent of the elasticity of the penetrated immobile object.

Figure 4.22 illustrates a situation in which both interacting objects are mobile. Since both objects are of the same nature and are mobile, we use the same filled circle to represent them. However, since they are also separate objects, we use a number on the object to distinguish them. In Fig. 4.22a, *Object 2* is originally stationary and in Fig. 4.22b both *Objects 1* and *2* move toward each other originally. Both objects separate from each other after the interaction. What is depicted here is a "canonical" situation in which the exact velocity, acceleration, and mass of the objects involved are omitted. The effects of different velocity, acceleration, and mass can also be handled in the operational representational framework.

4.3.10.3 Action at a Distance

Sometimes objects do not need to come into contact before they can interact and affect each other's movements. This can happen with electrical or magnetic forces that can act over a distance. Figure 4.23 depicts the situations with objects that are charged electrically with the same polarity as well as with different polarity of charges. In Fig. 4.23a, two objects that are both electrically positively charged are shown to move apart with no other external forces acting on them. Acceleration is omitted here but in the real world, the behavior of charged objects is such that the force between them weaken with distance and they would be moving at a lower

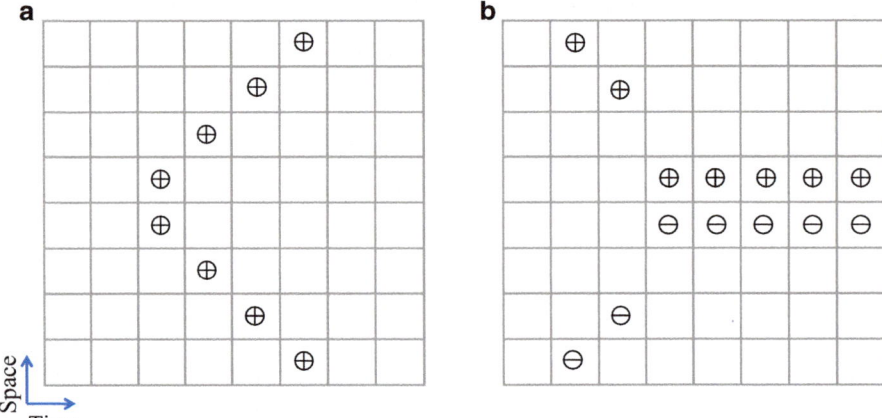

Fig. 4.23 (**a**) Both objects are electrically positively charged. No other external force is needed to begin the movement. No acceleration consideration. (**b**) Objects with opposite charges attract each other, with consideration of acceleration – when the objects are nearer each other, the speeds are higher

speed as they are farther apart. In Fig. 4.23b two objects that are oppositely charged are shown to move toward each other with no other external forces acting on them and some acceleration is shown – when they are closer, they move faster toward each other.

4.3.10.4 Attach, Detach, Push, and Pull

So far we have considered situations that involve elemental objects that move independently of each other. If an operation of *Attach* is applied to two elemental objects, the resultant situation is such that they would move in unison afterward. They also form an *extended object* – object composed of more than one elemental object. This is shown in Fig. 4.24a. We use a small bar to represent an "attach-link" between the objects. This attach-link is just a representation in the same vein as the momentum representation in the form of an arrow next to the object in the foregoing discussion (e.g., Fig. 4.18a). It represents a state of the objects involved. This link may be visible sometimes (e.g., a physical link between two objects) or may not be visible (as in, say, electrostatic attraction). But the consequence of an *Attach* operation is such that the objects involved will pass a "Pull Test" to be described below that can determine if they are indeed attached. *Detach* is the opposite operation of *Attach* as shown in Fig. 4.24b – after the operation the objects involved no longer move in unison and will not pass the "Pull Test." The attach-link is shown to be removed after the operation.

Next we consider the *Push* operation. In Fig. 4.25a, we show a force being applied to a first object (labeled "1") in the direction of a second object (labeled

Fig. 4.24 (a) *Attach* and (b) *Detach* actions. The "attach-link" indicates attachment between the elemental objects (From Ho 2013)

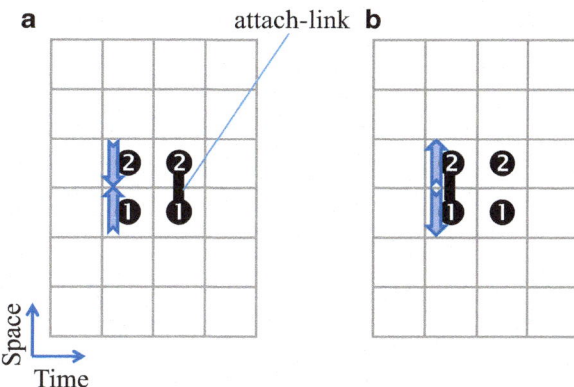

"2"). The first object is in contact with the second object but not necessarily attached to it (the operation applies to both attached and non-attached situations). In the next time frame both objects move an elemental step in the direction of the force. This is a *Push* operation in which the force is transmitted from one object to another object and causes the second object to move in the absence of other direct, explicit force being applied to the second object. Figure 4.25a also shows that both objects acquire a momentum in the direction of the pushing force. The *Move-Up-in-Unison* operator in Fig. 4.25a depicts the movement specification for both elemental objects involved in the *Push* event. We are assuming that these are "light" objects, otherwise they would not acquire any momentum and would move only one elemental spatial location and halt without further movement (the meanings of "light" vs "heavy" objects were discussed in connection with Fig. 4.18 in Sect. 4.3.10.1). And as in Sect. 4.3.10.1, we consider only forces of unit magnitude that will move the objects involved one elemental spatial position over one elemental time step.

The opposite of a *Push* action is a *Pull* action. However, the *Pull* action only applies to objects that are attached to each other as shown in Fig. 4.25b. In this situation, a force is applied to the first object in a direction away from the second object and the second object then follows the first object, and both move one elemental spatial location after one time frame. Momentum is acquired like in the *Push* situation if both objects are light. A corresponding *Move-Down-in-Unison* operator is shown. Similarly, here we are concerned with only pulling forces of unit magnitude as shown. In the events of *Push* and *Pull* described in Fig. 4.25a, b respectively, we are assuming that there are no other objects blocking the movement of the objects involved.

The *Pull* action can be used to perform a "Pull Test" on objects that are seemingly attached to each other. As mentioned earlier, sometimes the attach links between objects may not be visible or that the objects are linked through invisible forces such as electrostatic forces. A Pull Test will reveal whether they are indeed attached.

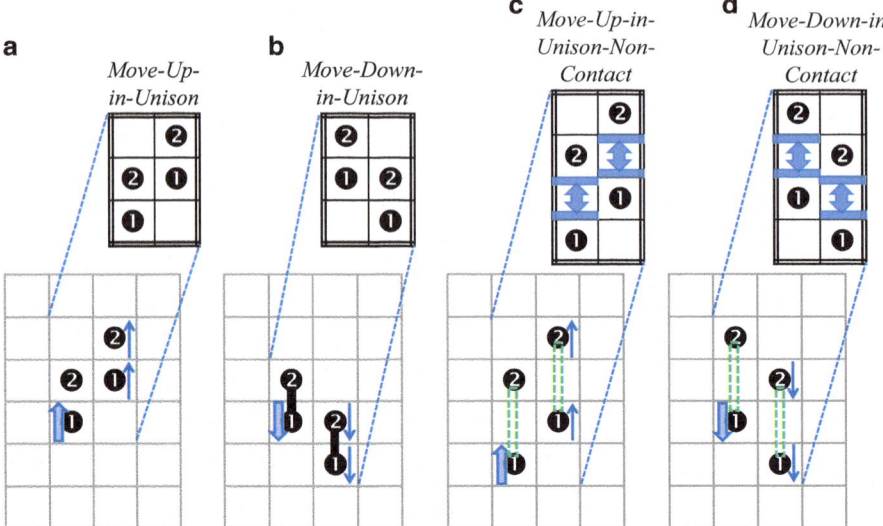

Fig. 4.25 (**a**) *Push* event with a *Move-Up-in-Unison*-operator. (**b**) *Pull* event with a *Move-Down-in-Unison* operator. (**c**) *Distant-Push* (*Move-Up-in-Unison-Non-Contact operator*). (**d**) *Distant-Pull* (*Move-Down-in-Unison-Non-Contact operator*) ((**a**, **b**) from Ho 2013)

In Fig. 4.25c, d we depict events in which objects that are not in immediate contact are being pushed or pulled from a distance and they subsequently move in unison. These are *Distant-Push* and *Distant-Pull* situations. These could be due to action at a distance (due to electrically charged objects, say) or the presence of other intervening, possibly connected elemental objects. The corresponding *Move-Up-in-Unison-Non-Contact* and *Move-Down-in-Unison-Non-Contact* operators have Range Bars (shaded spatial intervals with arrows like those described in Sect. 4.3.7.1, Fig. 4.14) that represent one or more than 1 unit of space in between the objects involved in these situations.

The attach operation sets the stage for the description of *extended* objects in which more than one elemental object is linked together and all the elemental objects involved move in unison. The smallest possible extended object would be those that are depicted in Fig. 4.25 in which *two* elemental objects are linked together. In the next chapter (Sect. 5.2.3) when we discuss problem solving we will see how more than two elemental objects are linked together to form a useful tool of sorts. In Chap. 9, an extended object representation is used to represent an "Agent." Extended objects can also be rigid or elastic. In the former, the elemental objects within the extended objects would all move in perfect unison much like the two objects in Fig. 4.25b. In the latter, some of the elemental objects may lag behind others in their movements. These can also be represented in operational representations in the same vein as the rigid extended object of Fig. 4.25b.

4.4 Issues on Learning

The basic structure employed in operational representations, namely the explicit spatiotemporal structure, is highly amenable to learning, especially unsupervised learning through direct observations made of the activities and interactions in the physical environment. Activities and interactions by their very nature are characterizable as changes of some parameters over time, hence a representational scheme based on direct encoding of spatiotemporal features dovetails very well with the fundamental features of activities and interactions that are to be captured, encoded, and represented.

Figure 4.26 shows an approach called the "cookie cutter" approach that can conceivably be used to extract atomic operators from the environment in an unsupervised manner. This approach is similar to that described by Uhr and Vossler (1981) in their pattern recognition system. Basically, if a pattern is encountered in the environment, the recognition system simply "cuts-out" sub-patterns of the encountered pattern and stores them as potentially useful features for recognizing future occurrences of the same pattern or other patterns which are characterizable as consisting of a number of sub-patterns, including the earlier "cut-out" ones. A "weight" is associated with each sub-pattern/feature to capture its importance in helping to characterize the pattern involved. In subsequent encounters with more patterns, the importance of these sub-patterns/features is then adjusted by adjusting the "weights" associated with them. Sub-patterns/features that are later found to be useless (say, when the weights are reduced to zero or near zero because they are not critical in characterizing the patterns involved) may be removed entirely.

Whereas in Uhr and Vossler (1981), the patterns thus extracted are *spatial* sub-patterns for the purpose of characterizing spatial patterns, what we have in operational representations are *spatiotemporal* sub-patterns that can be used to characterize spatiotemporal patterns, namely activities and interactions. These spatiotemporal sub-patterns could be the atomic operators that we discussed in Sect. 4.3 and illustrated here again in Fig. 4.26. Therefore, an unsupervised learning process to extract these atomic operators can conceivably work as follows. When the system observes the occurrences of some spatiotemporal patterns (i.e., activities and interactions) in the physical environment, it will extract various sub-patterns of spatiotemporal features that occur in those activities by "cutting" them out from the spatiotemporal pattern as shown in Fig. 4.26. These atomic operators are then used to recognize future occurrences of the same spatiotemporal patterns or some other spatiotemporal patterns that are also characterizable using these atomic operators, as we have discussed extensively in this chapter (e.g., recognizing the concept of *Trajectory* in Fig. 4.9 or *Blinking* in Fig. 4.10).

In Sect. 4.3.7.1 we also discussed atomic operators that are more general in that they represent ranges of certain parameter values. For example, the operator in Fig. 4.14e represents greater than or equal to 2 speed units (upward) and the operator in Fig. 4.14i represents movement at any speed in any direction (up or down). This kind of general operators can conceivably be created when various

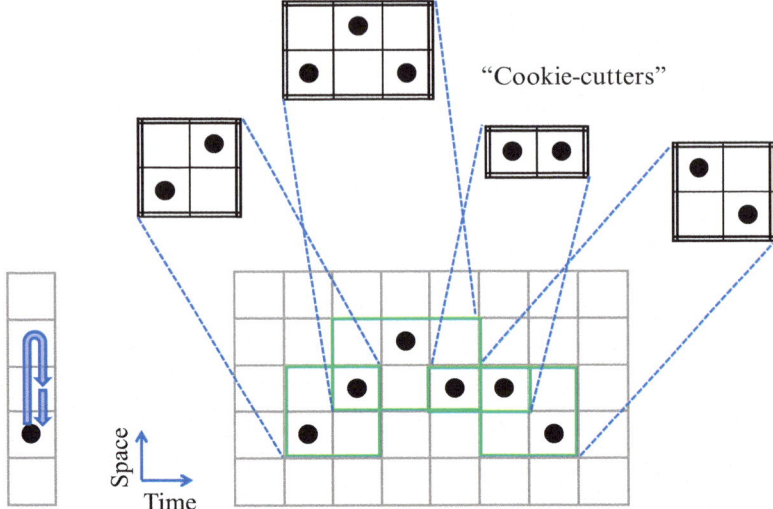

Fig. 4.26 Unsupervised learning of atomic operators – the "cookie-cutter" approach (Uhr and Vossler 1981). The activities here involve an object moving upward and then downward, halting for one time frame, and then moving downward again

speeds have been encountered in the environment and various kinds of specific operators have been created to characterize them. And then, a process of generalization combines them into the more general concepts. Figure 4.27 shows that operators representing speeds of all kinds are combined into a general movement in any direction and at any speed, which is the general speed operator depicted in Fig. 4.14i. Instead of creating these more general operators at random, the system may do so only under certain circumstances, such as when the combination may be useful for certain cognitive tasks. This issue is left to be pursued in future research.

The conceptual hierarchies such as that depicted in Figs. 4.8 and 4.9 that consist of various higher level operators defined in terms of the lower level or atomic operators can also be created in an unsupervised manner. If there are occurrences of some activities or interactions in the environment that trigger a series of atomic operators, these operators can be grouped together in a hierarchical manner guided by certain rules. Similar to how Uhr and Vossler (1981) keep track of the usefulness of various sub-patterns extracted from the environment, the system can likewise observe the usefulness or importance of these higher level operators in subsequent activities and interactions based on some criteria related to their roles in, say, certain noological functions and decide whether to keep them or discard them. Future research should strive to understand how conceptual hierarchies are formed.

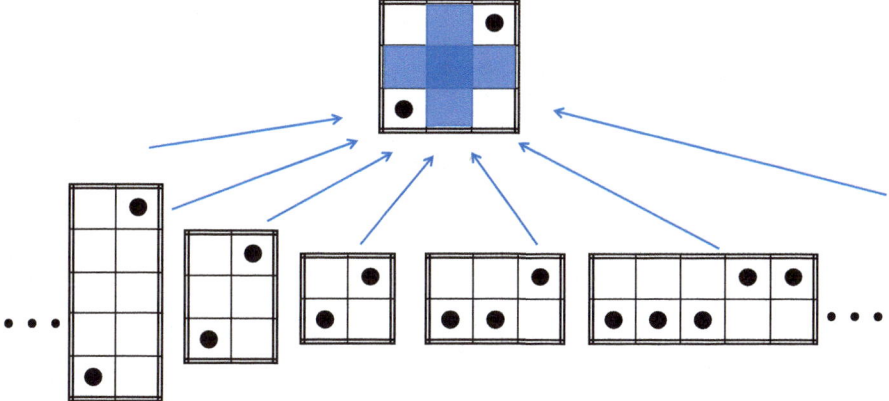

Fig. 4.27 Generalization of movement operators with specific speeds to general operators

4.5 Convergence with Cognitive Linguistics

As mentioned in Sect. 4.1 in the beginning of this chapter, in recent years an area of linguistics, cognitive linguistics, has arisen as a new paradigm for the study of meaning through perceptual and cognitive characterizations (e.g., the work by Croft and Cruse 2009; Evans and Green 2006; Geeraerts 2006; Johnson 1987; Langacker 1987, 1991, 2000, 2002, 2008, 2009; Ungerer and Schmid 2006, etc.). Specifically, because many concepts consist of a temporal component, the cognitive linguistic way of representing them involves a straight forward use of an explicit temporal dimension, much like in operational representations. Figure 4.28 illustrates the concept of Arrive as represented in cognitive linguistics (Langacker 1987) and contrasts that with a typical dictionary-type definition. The figure also shows how each lexical item in the dictionary definition can be mapped onto a part of the spatiotemporal representation of Arrive, thus grounding its meaning.

Figure 4.28 also shows the cognitive linguistic representation of the concept of After. In cognitive linguistics, a distinction is made between a *trajector* (tr) and a *landmark* (lm). This is like the figure/ground distinction in visual perception. The trajector is the focus of the sentence. In the case of the sentential definition of Arrive, if we divide the sentence into two events, the first of which is "to reach a place" and the second is "traveling" then the trajector is "to reach a place." "After" specifies a temporal order of these two events as shown.

In Fig. 4.28, it can be seen that the cognitive linguistic representation of Arrive consists basically of a SPACE and a TIME axes. Along the TIME axis, there is a sequence of rectangles, each containing the picture of a snapshot of the Arrive event at a particular time. The oval in the picture represents a "place." The landmark (lm) resides in the place. The trajectory (tr) starts from some distance away and moves toward the lm and *arrives* at the place after a number of snapshots, each showing that the tr is moving nearer the place. This is the full, spatiotemporal

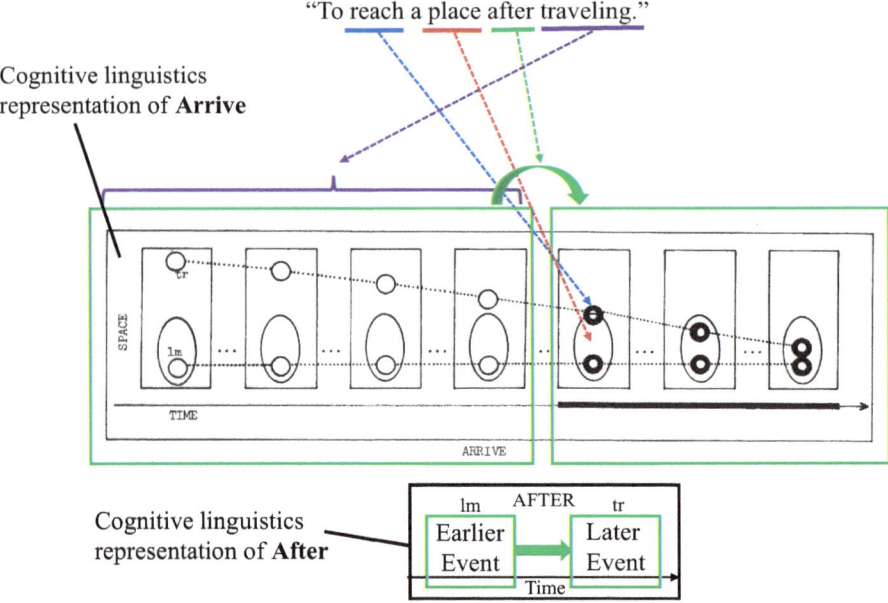

Fig. 4.28 The cognitive linguistic representation of the concepts of Arrive and After (Langacker 1987). The dictionary definition maps directly onto the various parts of the spatiotemporal representation. *lm* landmark, *tr* trajector. ("ARRIVE" figure from Foundations of Cognitive Grammar, VOL. I: THEORETICAL PREREQUISITES by Ronald W. Langacker. Copyright © 1987 by the Board of Trustees of the Leland Stanford Jr. University. All rights reserved, Reprinted by permission of the publisher, Stanford University Press, sup.org)

representation of Arrive in cognitive linguistics, which echoes strongly the operational representational scheme that we have developed in this chapter. It is a fully grounded representation of the event and concept of Arrive. It has the Recognition-Generation Bidirectionality property (Sect. 4.3.6) of operational representation. It can be used to recognize an instance of the Arrive concept, and it can also be used to generate an instance of the Arrive concept.

Now, let us consider the dictionary definition of Arrive at the top of the figure – "to reach a place after traveling." Each of the verbs and nouns in the sentence – reach, place, after, and traveling – can be mapped nicely onto a construct in the spatiotemporal representation as shown. So, if a noological system knows the grounded meanings of all the lexical items – reach, place, after, and traveling – and the rules of putting them together in a sentence, it can generate the sentence "to reach a place after traveling" to explain what Arrive means. Alternatively, if it does not know the meaning of the lexical item Arrive and it takes in the dictionary definition, using the rules of the language it can then construct the grounded spatiotemporal representation of the concept of Arrive accordingly. With that, further deep reasoning and problem solving with the concept of Arrive will be possible.

The representational framework of cognitive linguistics thus dovetails very well with that of operational representation. In the next chapter, after we discuss the application of operational representation to the representation of causal rules and problem solving, we show that the grounded representation of concepts engenders learning to solve problems through language, which is a rapid kind of learning that sets noological systems with language faculty significantly apart from those without, in terms of their survival abilities.

4.6 Discussion

The intention of this chapter is to present a representational framework to encode meaning at the epistemic ground level and the primary constructs are explicit spatiotemporal representations called operational representations. As the world is fundamentally dynamic, activities and interactions have to be captured and represented adequately at the epistemic ground level in order for noological systems to understand the world. We submit that the various operational representations discussed represent the "meaning" of the concepts adequately at the epistemic ground level in the form of explicit spatiotemporal representations. In the next chapter, we will demonstrate how these representations structured into causal rules can be used effectively for reasoning and problem solving.

Simplifications are necessary for exploring the fundamental issues and uncovering fundamental principles that can then be generalized to more complex situations. However, over-simplifications may hinder the discovery of adequate fundamental principles. We began with 1D space + 1D time considerations of activities and interactions to uncover and establish basic principles but we believe by handling the issues with the explicit temporal dimensions in depth and in an adequate fashion, it allows us to uncover the fundamental principles without sacrificing the understanding of certain critical issues.

The chapter therefore leaves many issues for future investigations, chief among which is the issue of 2D and 3D space + 1D time representations to handle real world objects and situations. Do these require new and distinctly different principles? What about the representations of mass and other physical parameters? (We will discuss some issues concerning mass in Chap. 7, specifically Sect. 7.2.) What about the roles of "mental" parameters such as pain, the basic representations of which were presented but not pursued further? (We will discuss in some detail the functional role of pain and anxiousness in Chap. 8.) What about complex physical objects such as fluid and trees? We have started discussions on them but have not presented complete treatments on them.

In the next chapter, we turn to the use of operational representations for encoding causal rules and the application of the causal rules in reasoning and problem solving. We will see that the use of operational representation in the explicit spatiotemporal form (i.e., Fig. 4.2 vs the predicate logic form of Eq. 4.1) is straightforward and natural for reasoning and problem solving tasks.

Problems

1. Figure 4.25 shows that when two elemental objects are attached together, they move in unison. If many elemental objects are attached together likewise, they form an *extended* object, and if they all move in exact unison as in Fig. 4.25 (i.e., all the parts move in the same manner at the same time), they constitute a "rigid" extended object. Show how the same representational framework can be used to represent non-rigid, *elastic* extended objects (i.e., not all elemental parts move in the same manner at the same time).
2. Extend operational representation as discussed in this chapter to 2D and 3D scenarios. One major difference in terms of movement possibility between 1D representation on the one hand and 2D and 3D representations on the other is the existence of rotational movement in 2D and 3D space.

References

Croft, W., & Cruse, D. A. (2009). *Cognitive linguistics*. Cambridge: Cambridge University Press.
Evans, V., & Green, M. (2006). *Cognitive linguistics: An introduction*. Mahwah: Lawrence Erlbaum Associates.
Ferrucci, D., Brown, E., Chu-Carroll, J., Fan, J., Gondek, D., Kalyanpur, A. A., Lally, A., Murdock, J. W., Nyberg, E., Prager, J., Schlaefer, N., & Welty, C. (2010). Building Watson: An overview of the DeepQA project. *AI Magazine, 31*(3), 59–79.
Feynman, R. P. (1988). *QED: The strange theory of light and matter*. Princeton: Princeton University Press.
Forsyth, D. A., & Ponce, J. (2011). *Computer vision: A modern approach* (2nd ed.). Englewood Cliffs: Prentice Hall.
Funt, B. V. (1980). Problem solving with diagrammatic representations. *Artificial Intelligence, 13*, 201–230.
Geeraerts, D. (2006). *Cognitive linguistics*. Berlin: Mouton de Gruyter.
Gleitman, H., Gross, J., & Reisberg, D. (2010). *Psychology* (8th ed.). New York: W. W. Norton & Company.
Hawking, S. (1998). *A brief history of time*. New York: Bantam.
Ho, S.-B. (1987). *Representing and using functional definitions for visual recognition*. Ph.D. thesis, University of Wisconsin-Madison.
Ho, S.-B. (2012). The atoms of cognition: A theory of ground epistemics. In *Proceedings of the 34th annual meeting of the cognitive science society*, Sapporo Japan (pp. 1685–1690). Austin: Cognitive Science Society.
Ho, S.-B. (2013). Operational representation – A unifying representation for activity learning and problem solving. In *AAAI 2013 fall symposium technical reports-FS-13-02*, Arlington, Virginia (pp. 34–40). Palo Alto: AAAI.
Johnson, M. (1987). *The body in the mind*. Chicago: The University of Chicago Press.
Kosslyn, S. M. (1980). *Image and mind*. Cambridge, MA: Harvard University Press.
Kosslyn, S. M. (1994). *Image and brain*. Cambridge, MA: MIT Press.

References

Langacker, R. W. (1987). *Foundation of cognitive grammar* (Vol. I). Stanford: Stanford University Press.
Langacker, R. W. (1991). *Foundation of cognitive grammar* (Vol. II). Stanford: Stanford University Press.
Langacker, R. W. (2000). *Grammar and conceptualizations*. Berlin: Mouton de Gruyter.
Langacker, R. W. (2002). *Concept, image, and symbol: The cognitive basis of grammar*. Berlin: Mouton de Gruyter.
Langacker, R. W. (2008). *Cognitive grammar: A basic introduction*. Oxford: Oxford University Press.
Langacker, R. W. (2009). *Investigations in cognitive grammar*. Berlin: Mouton de Gruyter.
Miller, G. A., & Johnson-Laird, P. N. (1976). *Language and perception*. Cambridge, MA: Harvard University Press.
Nicholls, J. G., Martin, R., & Fuchs, P. A. (2011). *From neuron to brain* (5th ed.). Sunderland: Sinauer Associates.
Russell, S., & Norvig, P. (2010). *Artificial intelligence: A modern approach*. Upper Saddle River: Prentice Hall.
Shapiro, L. G., & Stockman, G. C. (2001). *Computer vision*. Upper Saddle River: Prentice Hall.
Szeliski, R. (2010). *Computer vision: Algorithms and applications*. New York: Springer.
Uhr, L., & Vossler, C. (1981). A pattern-recognition program that generates, evaluates, and adjusts its own operators. In E. A. Feigenbaum & J. Feldman (Eds.), *Computers and thought*. Malabar: Robert E. Krieger Publishing Company.
Ungerer, F., & Schmid, H.-J. (2006). *An introduction to cognitive linguistics*. Harlow: Pearson.

Chapter 5
Causal Rules, Problem Solving, and Operational Representation

Abstract Building on the conceptual grounding framework set up by the previous chapter, this chapter illustrates the application of grounded operational representations for the formulation of causal rules. Causal rules are noologically efficacious entities that enable effective problem solving. Causal rules for effecting movement, propulsion, reflection, obstruction, penetration, attachment, etc. are described. The mechanism of incremental chunking for problem solving is also described using these causal rules. It is shown how a complex problem solving process can employ these rules built on grounded representations and how, as a result, a noological system can learn rapidly through language instructions because concepts are properly represented and understood at the ground level.

Keywords Causal rule • Conceptual grounding • Semantic grounding • Meaning • Problem solving • Incremental chunking • Monkey-and-bananas problem • Learning through language

In the previous chapter we have described the use of operational representation to encode epistemic ground level knowledge. As mentioned in Chap. 1, the primary noological processing backbone is problem solving and as discussed in the previous chapter, grounded representation of knowledge can facilitate problem solving at the ground level. For a noological agent to use its knowledge for reasoning and problem solving, the knowledge would have to be encoded in generalizations in the form of rules and in the case here, *causal rules* – how various causal agents affect and determine the outcome of physical or mental events and processes. In this chapter we formulate the causal rules based on the physical properties of the 1D world described in the previous chapter and show how these rules can be used for reasoning and problem solving.

5.1 Representation of Causal Rules

In keeping with the operational representational framework, the causal rules are represented using basically the same spatiotemporal representations with the temporal dimension explicitly represented. A causal rule consists of basically three portions – a *Start*, *Action*, and *Outcome* portions. (Similar to the script structure discussed in Chap. 3). The *Action* is what *causes* the *Start* state to be transformed into the *Outcome*. The rules therefore specify the transformations on spatiotemporal patterns by various causes. The spatiotemporal patterns that exist in the environment together with the actions can be matched directly to the spatiotemporal patterns specified in the *Start* state and *Action* portions of the rules to "trigger" the rules – i.e., to reason what would happen given a certain environmental situation. Similarly, if there is a certain desired goal spatiotemporal representation, it can be matched directly to the spatiotemporal pattern specified in the *Outcome* portion, and the matched rule may be used in a backward reasoning problem solving process to discover what is needed to achieve the *Outcome*. Also, the most basic causal rules described in this section are "atomic" in the sense that they specify transformations over 1 elemental time step. These are described in Sects. 5.1.1, 5.1.2, and 5.1.3. There are other more complex causal rules that can be built upon the basic atomic causal rules in a problem solving process and these will be discussed in Sects. 5.2.2 and 5.2.3.

Note that not all the physical behaviors described in Chap. 4, Sect. 4.3 are encapsulated in the causal rules described in this chapter but the causal rules described here are sufficient for our subsequent discussion in Sect. 5.2 on reasoning and problem solving.

5.1.1 *Materialization, Force, and Movement*

We begin with describing a few fundamental causal rules associated with materialization, force, and movement. Figure 5.1a depicts a rule for materialization (*Rule 1*) – at a particular instance of time t_2, a *Materialize* action is performed at a spatial location and in the next instance t_3 an elemental object appears at that spatial location. At time t_1, which is the Start state, there is nothing in the location. This is encoding the operation of *Materialization* as described in Fig. 4.19a in Sect. 4.3.10.1 of Chap. 4. What the causal rule specifies is that if the *Action* is performed, the situation in the *Start* state will transform to that in the *Outcome*. In this diagram and subsequent diagrams for causal rules, what is shown in each of the *Start*, *Action*, and *Outcome* portions is basically a "fragment" of the physical world that contains the event of interest represented in the usual form of the spatiotemporal arrays that we have been using so far. These fragments in all three portions have exactly the same dimensions and they are meant to be the exact same segment of space-time.

All the causal rules described in this chapter can be learned through a rapid unsupervised causal learning process described in Chap. 2 – an event (a change of state) that takes place across an earlier temporal transition t_1 to t_2 causes a later event/change of state that takes place across a subsequent temporal transition t_2 to t_3.

Most of the empty spatiotemporal locations in the causal rule representation indicate a "don't care" condition – i.e., it doesn't matter whether other objects are

5.1 Representation of Causal Rules

Fig. 5.1 Causal rules associated with *Materialization*. (**a**) *Rule 1 – Materialize*. (**b**) *Rule 2 – Dematerialize*. (**c**) *Rule 3 – Stay*. The *Start*, *Action*, and *Outcome* portions of the rules that are specified in spatiotemporal forms can be matched directly to the spatiotemporal patterns in the physical world or a goal specification to effect reasoning processes. The *vertical blue rectangles* highlight the spatiotemporal slice of concern in each portion. All rules are translationally and rotationally invariant. As in Chap. 4, a *black round dot* represents an elemental object (which could also be, say, a point of light)

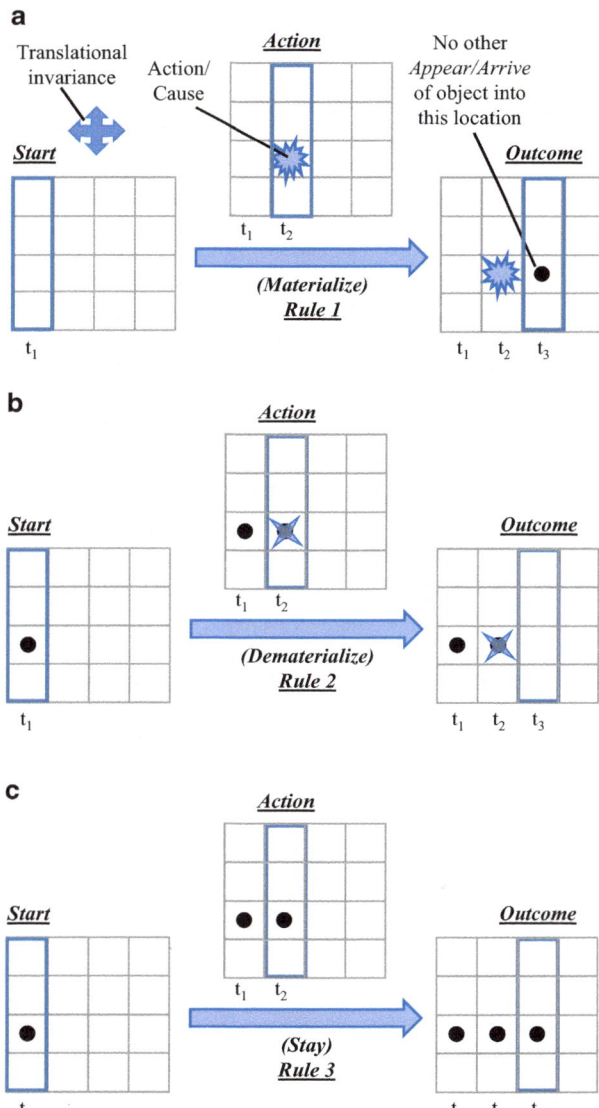

present there in matching the *Start* or *Outcome* situation to the physical world in the process of reasoning or problem solving. The only exception is when a spatiotemporal location in the *Outcome* situation receives an "appearance" or "arrival" of an elemental object as a result of the transformation. There must not be an appearance or arrival of another elemental object at the same spatiotemporal location, otherwise there will be a conflict situation that needs to be resolved separately. This is indicated in Fig. 5.1a as "No other *Appear/Arrive* of object into this location" in the spatiotemporal location of concern.

In the *Action* portion of the rule, the *Action/Cause* is specifically pointed out in Fig. 5.1a but we will omit this in the subsequent figures as it is quite obvious which entity in the figure is the action – the *Action/Cause* is the entity that is present in the *Action* portion but not in the *Start* portion. In this figure, we also indicate that the rule is spatiotemporal location invariant with a symbol on the top left corner. This means this rule applies to anywhere in space and time (i.e., any translational displacement in space-time). Subsequent rules are likewise invariant and we will omit the symbol.

Figure 5.1b shows the causal rule for *Dematerialization (Rule 2)* – at a particular time step, a "de-materialize" action is performed at a spatial location on an elemental object and in the next time step the elemental object disappears from that spatial location. This is capturing the *Dematerialization* operation as described in Fig. 4.19b in Sect. 4.3.10.1 of Chap. 4. Figure 5.1c shows a *Stay* rule *(Rule 3)* – no action is necessary for the elemental object to stay in its earlier spatial location. (This operation is described in Fig. 4.9).

Figure 5.2a shows a *Propel a Light Object* rule *(Rule 4)* – after the application of a force (indicated as a thick arrow) of a unit magnitude on an elemental object, it moves one elemental spatial location in one elemental time step and acquires a momentum (indicated as a thin arrow). This is the same as the operation as described in Fig. 4.18a. The cause of the movement is the force and there is a similar conflict situation as in *Materialization* above – no other objects should *Appear/Arrive* at the same location that the elemental object is moving into. There is a similar rule to *Rule 4*, which we refer to as *Rule 4'* (not shown in the figure), which is the "Up-Down Flip" version of *Rule 4* – i.e., the force, the consequential direction of movement, and momentum acquired are all in opposite direction to those in *Rule 4*. Many of the rules to be described below have similar "Up-Down Flip" versions.

The learning of the concept of momentum in *Rule 4* that is marked in the t_3 time frame is dependent on having learned and encoded the next rule – the *Momentum Continuation* rule first. Figure 5.2b shows the *Momentum Continuation* rule *(Rule 5)* – an object that has a momentum in a time step continues to carry the momentum in the next time step (the operation as described in Fig. 4.18a). No explicit action is necessary for this transformation. Note that similar to Fig. 4.18a, a thin arrow next to the elemental object is used to represent the presence of momentum in the object involved.

The learning of the concept of momentum through causal learning depends on the observation that there is an earlier change of state, which is the change of location from t_1 to t_2, followed by a later change of state – a change in location from t_2 to t_3. Thus, the cause of the later change of state is the earlier change of state. There are no other explicit causes. We thus first add a thin upward arrow to the spatiotemporal location where the object is at t_2 to make the cause explicit – that is, the object has a property called *momentum* at that time frame and that causes the movement from t_2 to t_3. However, if one observes one more time frame further, one would see that the object continues to have the property of momentum that causes yet further movements. Therefore, the object has the property of momentum in all

5.1 Representation of Causal Rules

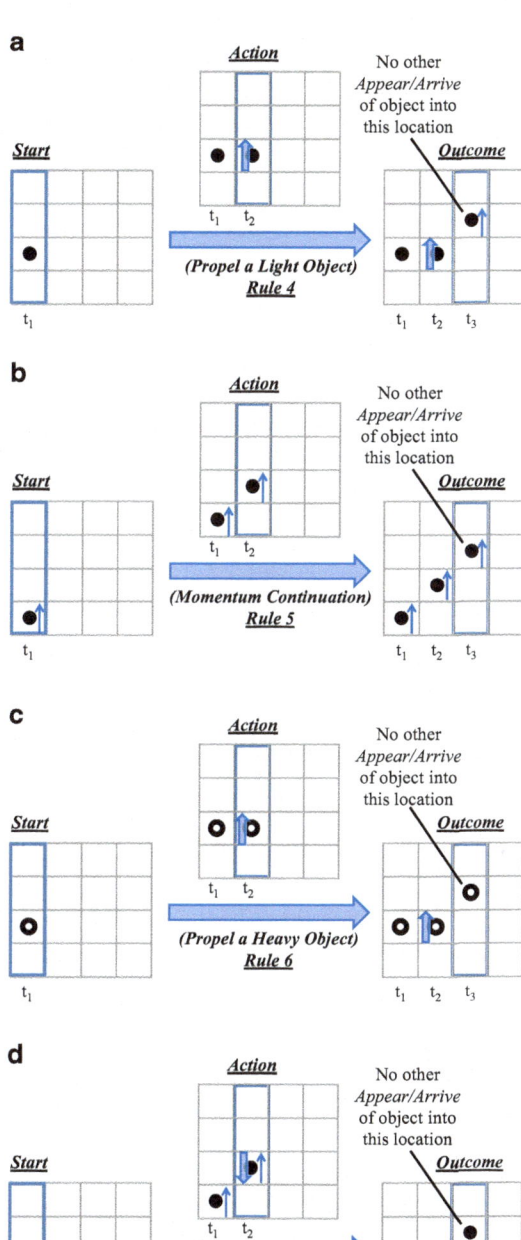

Fig. 5.2 Causal rules associated with Force and Movement. (**a**) *Rule 4 – Propel a Light Object*. (**b**) *Rule 5 – Momentum Continuation*. (**c**) *Rule 6 – Propel a Heavy Object*. (**d**) *Rule 7 – Momentum Cancellation*. The Start, Action, and Consequence portions of the rules that are specified in spatiotemporal forms can be matched directly to the spatiotemporal patterns in the physical world or a goal specification to effect reasoning processes. A *thick arrow* represents a force and a *thin arrow* represents a momentum

the time frames from t_1 to t_3 in *Rule 5*. The property of momentum is then similarly labeled in the t_3 time frame of *Rule 4*.

Figure 5.2c shows a *Propel a Heavy Object* rule (*Rule 6*). This rule is based on the operation described in Fig. 4.18b and the propelled object moves one unit space in one unit time after a unit force is applied to it and it does not acquire any momentum. Figure 5.2d shows a *Momentum Cancellation* rule (*Rule 7*). This rule is based on the operation described in Fig. 4.18a – an opposing force at a time step cancels the momentum of an object as it arrives at the next spatial location in the next time step.

5.1.2 Reflection on Immobile Object, Obstruction, and Penetration

In this section we describe a set of rules associated with interactions involving one immobile object and another mobile object that were described in Figs. 4.20 and 4.21 in Sect. 4.3.10.2. Figure 5.3a is a causal rule for an elastic mobile object reflecting off an immobile elastic object (*Rule 8*), encapsulating the event described in Fig. 4.20. In this event, the impinging object carries a momentum when it makes contact with the immobile object and in the next time step it moves away from the immobile object with a momentum in the opposite direction from the original one.

Figure 5.3b shows a rule describing the situation in which a reflection off an immobile elastic object is caused by a force applied to an elastic object that is originally stationary and contacting an immobile elastic object and the applied force is in the direction of the immobile object (*Rule 9*). Figure 5.3c captures the situation in which an inelastic impinging object collides with an immobile inelastic object (*Rule 10*) as described in Fig. 4.21a.

In Fig. 5.3a, c we are assuming that to match the rules to the real world to deduce whether *Rules 8* or *10* is applicable, we need to know whether the impinging and immobile objects involved, represented by shapes with different visual features (which are cues for a noological system to predict the properties of the shapes involved), are elastic or inelastic. An intelligent system can rely on visual features such as the surface textures of the impinging and immobile objects to determine whether it is elastic or inelastic if it has some prior experience with the connection between textures and elasticity. If the property of elasticity is indeterminate or determined wrongly, correct reasoning would not take place initially. Instead, the system could then infer the elasticity of the objects involved by observing the outcome of some collision events – i.e., whether the outcome matches the *Outcome* of *Rule 8* or *Rule 10*. If the former, then at least one of the impinging and immobile object is elastic, otherwise both are inelastic. The system would then label the objects accordingly for future accurate reasoning and prediction.

Figure 5.3d is a causal rule (*Rule 11*) that encapsulates the *Penetrate* action described in Fig. 4.21b. The penetrating "sharp" object moves past the immobile

5.1 Representation of Causal Rules

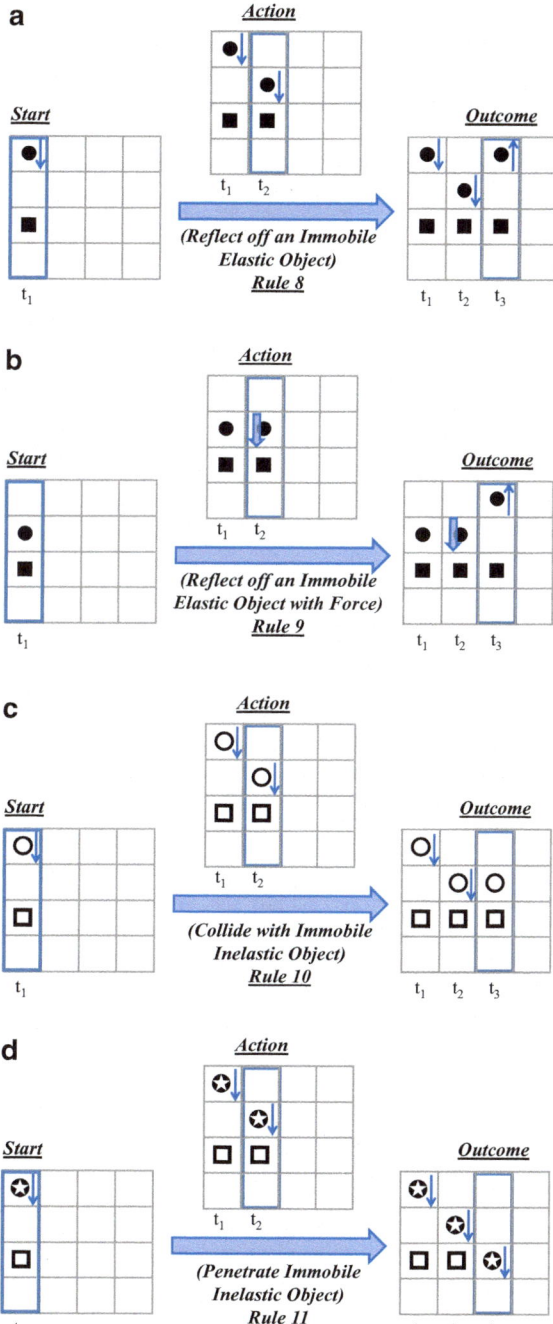

Fig. 5.3 Rules encoding the interactions between one immobile and one mobile objects. (**a**) An elastic object with a momentum makes contact with an immobile object which is also elastic (*Rule 8*). (**b**) A force applied to an elastic mobile object in the direction of an immobile object with which it makes contact (*Rule 9*). (**c**) Inelastic impinging object contacting inelastic immobile object (*Rule 10*) and after the "collision" the impinging object loses its momentum entirely and stays next to the immobile object. (**d**) A "sharp" object penetrating an inelastic object (*Rule 11*)

object that it penetrates with a momentum. Here the immobile object is labeled an inelastic object. Presumably, there is another similar rule for an immobile *elastic* object as typically the penetration effect of a "sharp" object is independent of the elasticity of the penetrated object.

5.1.3 Attach, Detach, Push, and Pull

This section describes various causal rules that encapsulate the *Attach*, *Detach*, *Push*, and *Pull* actions described in Figs. 4.24 and 4.25. Figure 5.4a is an *Attach* rule based on the *Attach* operation described in Figs. 4.24a and 5.4b is a *Detach* rule based on the *Detach* operation described in Fig. 4.24b. For the *Push* operation shown in Fig. 4.25a, we show the corresponding rule in Fig. 5.4c and there is a similar rule that is not shown that involves *Pushing* one object against another one that is attached to the first one. In that situation, both objects will move much like in Fig. 5.4c. In Fig. 5.4d, a *Pull* situation is encoded in the rule (based on Fig. 4.25b).

Similar to the situations discussed above concerning "elastic" and "inelastic" immobile objects in Sect. 5.1.2 that give rise to different interaction effects, if the existence of a link between seemingly attached objects cannot be determined or is determined wrongly, the reasoning process may not arrive at the correct prediction of the outcome of certain actions, but based on the outcome, the system will then be able to label the existence of the link accurately for subsequent accurate reasoning – e.g., a "Pull Test" (Sect. 4.3.10.4) will reveal whether two elemental objects are indeed linked (Fig. 5.4d).

Therefore, the labeling of a link in *Rule 12*'s *Outcome* portion relies on the separate experience with the presumably linked objects through *Rule 15* – the Pull Test. If an attempted attach action on a pair of objects (such as the objects in *Rule 12*) is followed by the objects successfully passing the Pull Test, then the causal rule – *Rule 12* – is formed/learned accordingly, showing the link between the objects in the *Outcome* portion. This is similar to the labeling of the momentum property of an object after a force application on it in Fig. 5.2a above (*Rule 4*).

5.2 Reasoning and Problem Solving

Having described a number of causal rules in Sect. 5.1, we now describe how these rules are used in reasoning and problem solving. In Fig. 5.5, we show the three major processing modules that are involved in this process in a reasoning system. First, there is the Physical World Array that holds the state of the physical world over time. The state of the physical world includes the various objects at various spatiotemporal locations and their attendant types (i.e. represented by their shapes as to whether they are mobile, immobile, elastic, inelastic, etc.), the forces present

5.2 Reasoning and Problem Solving

Fig. 5.4 Various interaction rules. (**a**) *Attach* (*Rule 12*). (**b**) *Detach* (*Rule 13*). (**c**) *Push* (*Rule 14*). (**d**) *Pull* (*Rule 15*) – Pull Test

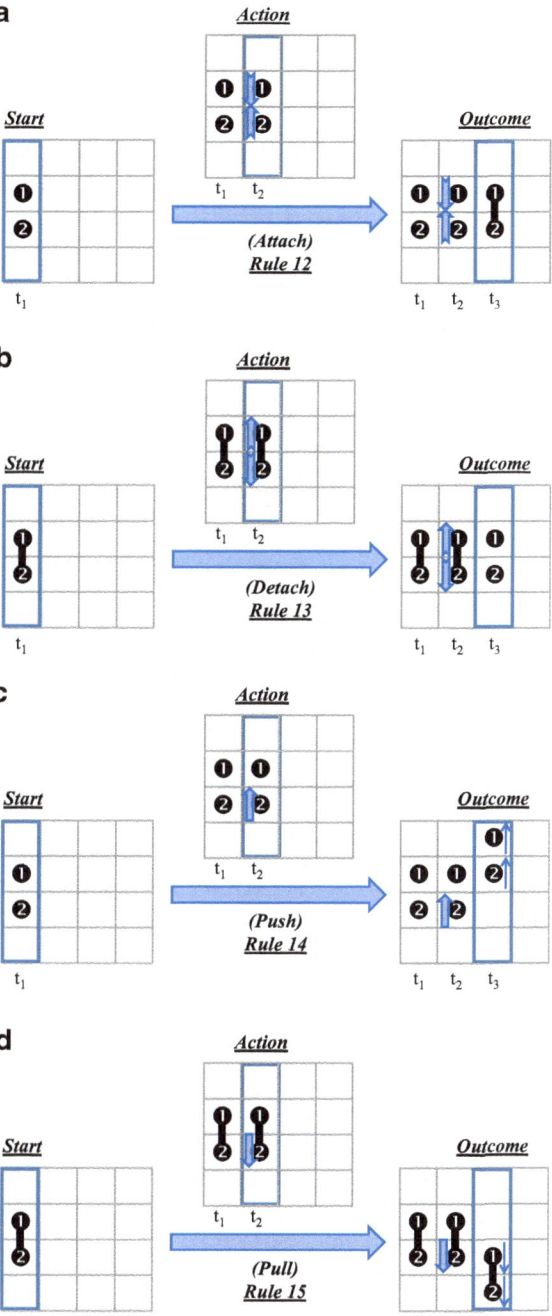

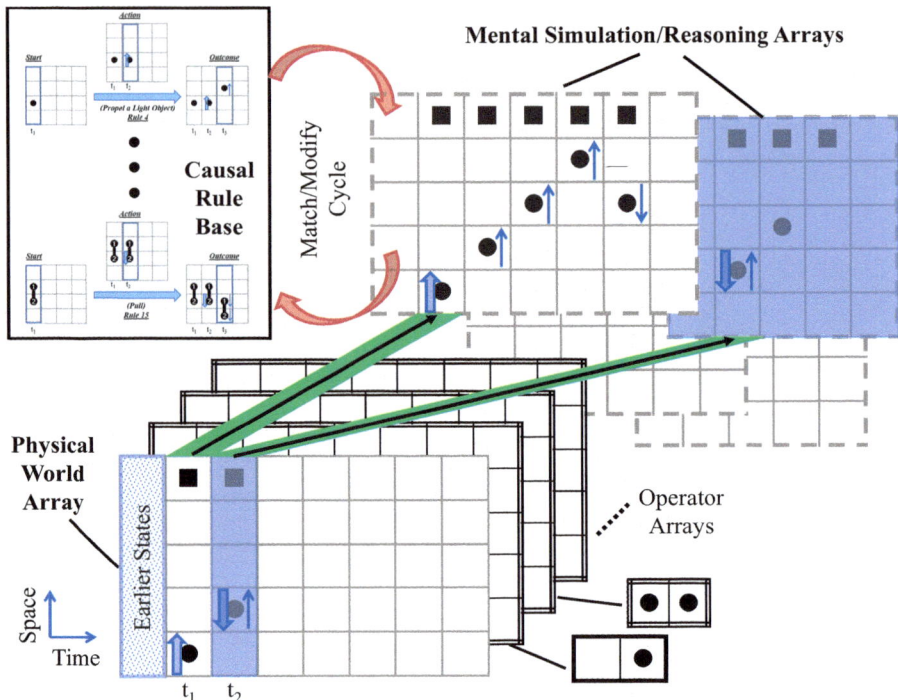

Fig. 5.5 A reasoning system with three major components: a Physical World Array, a Causal Rule Base, and a set of Mental Simulation/Reasoning Arrays. Mental space arrays have *dotted line* borders. The Operator Arrays are like those described in Fig. 4.12

at certain spatiotemporal locations, the momentum associated with the various objects, etc. The Operator Arrays at the back of the Physical World Array are like those described in Fig. 4.12. Second, there is a Causal Rule Base that holds the causal rules, such as those described in Sect. 5.1. Third, there are some Mental simulation/Reasoning Arrays that contain the results of "mental simulations" – given a certain external configuration of objects and causal agents observed in the "physical world," mental simulations determine the potential consequences of their interactions. (Consistent with the notation used in Figs. 4.16 and 4.17, these "mental space arrays" have dotted line borders). Refer also to the discussion on mental simulation in connection with Fig. 1.14 in Sect. 1.7, Chap. 1.

In a forward reasoning process, the physical configuration in the Physical World Array is matched to the Causal Rule Base to determine future consequences of interactions. The Physical World Array holds more than one time step of information, but usually, what is of interest in forward reasoning are the consequences in the future given the current configuration. In Fig. 5.5, t_1 is the time slice for the "current" physical configuration. It shows a force applied to a mobile object in the "upward" direction with an immobile object some distance away in the upward direction. This current configuration is matched to the Causal Rule Base and rules are invoked in the forward direction (i.e., matched to *Start* and *Action*, implying certain *Outcome*) to transform the configuration to "imagine" what may happen in the Mental Simulation Array. In the Mental Simulation Array it is shown that given the current configuration, it is expected that as time goes on, the mobile object would contact the immobile object and would be reflected by it, based on the "knowledge" encoded in the Causal Rule Base. Note that the rules in the Causal Rule Base are applied to all physical objects in the current configuration simultaneously – therefore the immobile object stays immobile as dictated by the rules.

However, in the next time step, the physical world may change in such a way that the earlier expected event does not occur. Figure 5.5 shows that in the next time step, t_2 (time slice highlighted in blue), a "downward" force is applied to the upward moving mobile object (carrying an upward momentum). According to *Rule 7* in Fig. 5.2d, the upward momentum would be canceled by the downward force, and the mobile object would move one more spatial step and then would lose its momentum and come to a halt after that. It is not expected to come into contact with the immobile object unless further changes in the physical world make that happen. This is simulated in another Mental Simulation Array (highlighted in blue).

In a backward reasoning process (Barr and Feigenbaum 1981), the reasoning system begins with a goal, matches it to the *Outcome* portions of causal rules in the Causal Rule Base, and then applies the causal rules backward to find a linked sequence of rules, the final one of which has a *Start* state that matches the current physical configuration. This represents a solution – should the sequence of rules be applied forward from the current *Start* state, the physical configuration will be transformed into the goal configuration.

5.2.1 Examples of Problem Solving Processes

Figure 5.6 illustrates two examples of how problems are represented in our framework. In Fig. 5.6a the problem is to find a solution for a mobile object to move three spatial steps upward in five time steps and in the end state the mobile object must stop there – i.e., it does not possess any momentum. In Fig. 5.6b the problem is to find a solution for a mobile object to move three time steps upward in five or *fewer* time steps, and the mobile object continues to possess a momentum at the final location. The START state is labeled with a solid-line rectangle box and the GOAL state is labeled with a dashed-line rectangular box. In subsequent figures, to reduce clutter, we will only label the objects themselves in the START and GOAL states with square boxes and not the entire physical state space with the vertical rectangles.

The examples shown in Fig. 5.6 are problems that are physical in nature – they involve the movement of a certain physical object toward a certain physical spatiotemporal location. Our framework is also extendable to problems that involve a system attempting to reach, say, a certain affective state. As described in Sect.

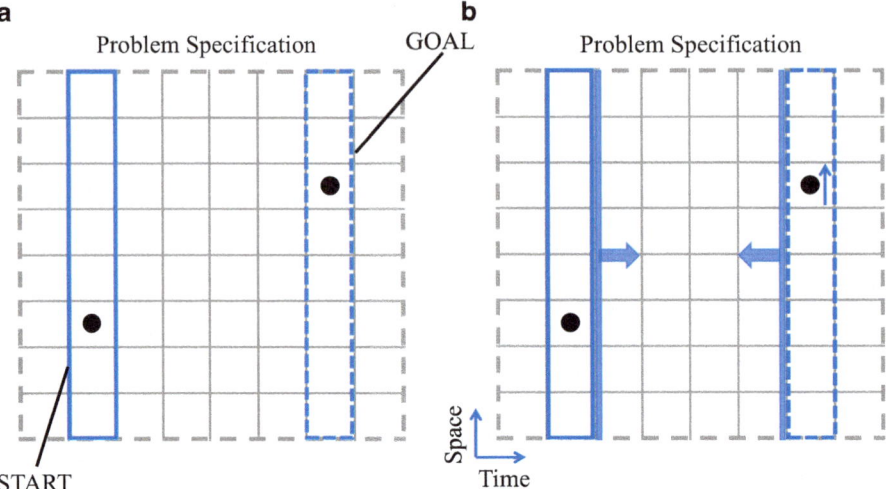

Fig. 5.6 Problem representations. (**a**) A problem to be solved within a fix time frame. (**b**) A problem to be solved within a time frame with an upper limit. The START state is labeled with a *solid-line rectangle box* and the GOAL state is labeled with a *dashed-line rectangle box*

4.3.8, affective states such as happiness and pain can be represented in operational representations and they can also be specified accordingly in some problem representations similar to those described in Fig. 5.6. If a system contains rules that transform mental as well as physical states, then a solution to a problem that includes mental as well as physical state specifications can be found through a similar process as that described for solving physical problems below. The EX-IN causal rule of Sect. 3.6 in Chap. 3 is just such an example of "mental state" related rule.

There could be trivial problems that could be represented using the problem representational method depicted in Fig. 5.6, such as one that involves the transformation of a START state (e.g., a mobile object at a certain location) to an GOAL state in a single step (e.g., moving the object to a neighboring location). For such problems, ready solutions may be found in the Causal Rule Base of Fig. 5.5 (e.g., *Rule 4* – Fig. 5.2a – that specifies that a force can be used to propel an object and move it to a neighboring spatial location). In general, more complex problems will require reasoning processes in the forward or backward direction using the rules encoded in the Causal Rule Base.

Figure 5.7 is a basic backward reasoning algorithm (Barr and Feigenbaum 1981) and Fig. 5.8 shows part of the backward reasoning process when the causal rules described in Figs. 5.1, 5.2, 5.3, and 5.4 are applied to the problem in Fig. 5.6a. Starting from the GOAL state in the problem specification of Fig. 5.6a, there may be more than one rule that matches the GOAL state on their *Outcome* portions and for each of the matches more branching may take place. Figure 5.8 shows only part of one of the backward reasoning branches.

5.2 Reasoning and Problem Solving

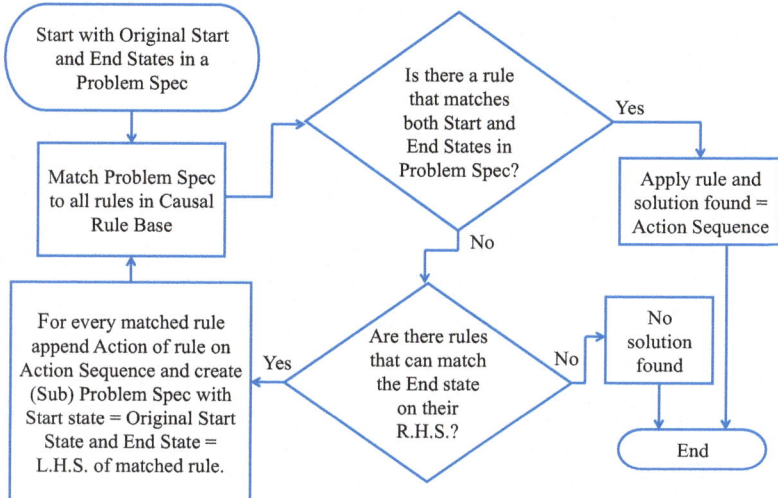

Fig. 5.7 A basic backward reasoning process (Barr and Feigenbaum 1981)

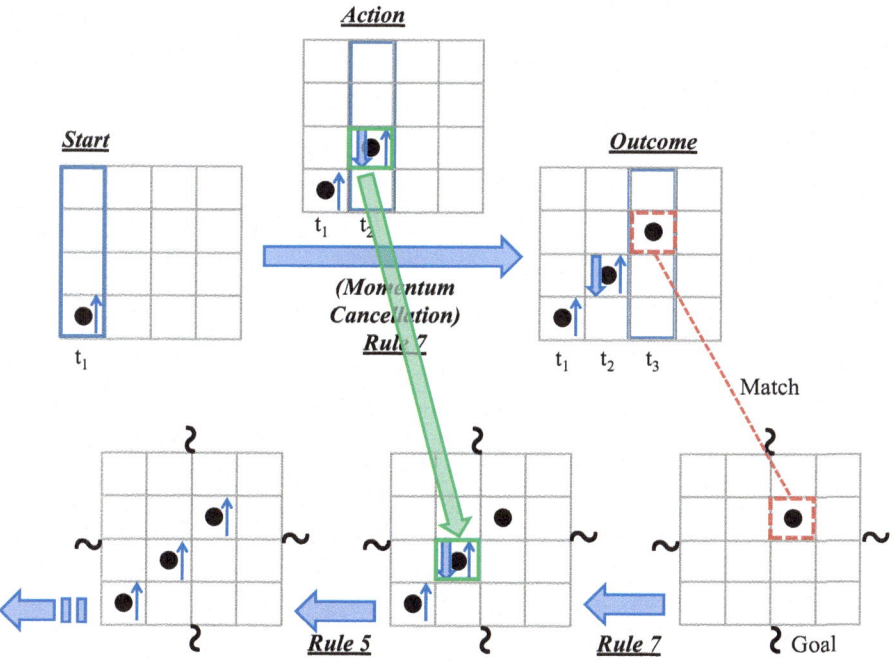

Fig. 5.8 Partial trace of one branch of the backward reasoning process of the problem in Fig. 5.6a

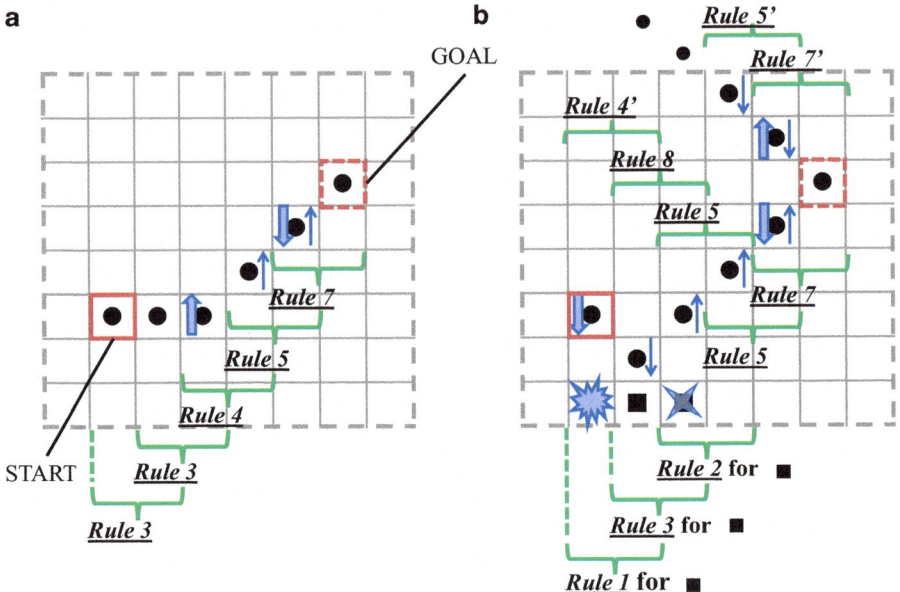

Fig. 5.9 Possible solutions for the problem of Fig. 5.6a. (**a**) A complete trace of one of the branches of the backward reasoning process leading to a solution. (**b**) Two other possible backward reasoning branches with one of them leading to a solution but not the other. The rules with the primes (e.g., *Rule 5'*) are same as those corresponding ones without primes but are "Up-Down Flip" versions

Figure 5.9a provides a complete trace for one of the backward reasoning branches that gives rise to a solution and shows the rules that are involved. Figure 5.9b illustrates two other possible backward reasoning branches. One, the lower one, gives rise to a solution that is different from that described in Fig. 5.9a. It involves the materialization of an immobile object that reflects the mobile object that is originally propelled to move in a direction (downward) that is opposite to the direction (upward) of the GOAL location relative to the START location. This reflection action sends the mobile object back toward the desired end location and results in it arriving at the GOAL location at the prescribed time. Whether this solution is acceptable depends on other constraints, for example, whether the materialization (or in the real world, the construction) of an immobile object at that specific location is allowed. The other backward reasoning branch, the upper one, does not give rise to a solution as it is headed in the "wrong direction." Solutions to problems can also be found from combining both forward and backward reasoning (Barr and Feigenbaum 1981).

5.2.2 Incremental Chunking

Despite the fact that the physical world that we have defined in Chap. 4 and this chapter so far contains relatively few elements, such as the relatively small number of objects in a physical configuration and the relatively small number of types of causal agents (moving objects, forces, materialization actions, etc.), that can be present at any given time, the search processes directed by the backward reasoning algorithm depicted in Fig. 5.7 operating in conjunction with the set of causal rules described in Figs. 5.1, 5.2, 5.3, and 5.4 can still potentially give rise to a large search space.

One way this search space can be reduced is by a process we term "Incremental Chunking of Causal Rules" as discussed in Chap. 3, Sect. 3.5. The basic idea consists of two parts. One part is "Chunking of Causal Rules." As we know, currently the causal rules as described in Figs. 5.1, 5.2, 5.3, and 5.4 are all fundamental/atomic rules that describe operations involving only one time step. Now, if the system has encountered a problem earlier and found a solution which consists of multiple time steps of operations, this "multi-step causal rule" can be added to the Causal Rule Base to aid future problem solving. Thus, causal rules involving a smaller number of steps can be chunked into more complex rules involving more steps.

The other part is the aspect of "Incremental Chunking." Though chunking in itself can assist future problem solving by providing a "known solution," if the original problem encountered is a complex one that involves multiple steps, such as that depicted in Fig. 5.6a, the system still needs to explore a potentially large search space to solve the problem first to arrive at a chunked, multi-step solution to be encoded as a multi-step rule. However, if simpler problems are first encountered and solved relatively easily and quickly, and subsequently each problem encountered has *incrementally* increased complexity, then each time the search space explored is not as large. This idea has been described in detail in Chap. 3, Sect. 3.5 in connection with a pure spatial movement problem. Here, we illustrate the idea again using our operational representational framework with the problem specification discussed currently in Fig. 5.6a.

Figure 5.10 illustrates a problem simpler than the one depicted in Fig. 5.6a. It consists of the desired movement of a mobile object by two locations in two time steps. Consider the two rules – *Rule 7* and *Rule 7'* (the "Up-Down Flip" inverse of *Rule 7*) – that can both match the GOAL state. We assume that the backward reasoning process is a parallel one in which both potential directions of search are pursued "simultaneously" – expanding one step backward in each direction before expanding the next step, etc. In the situation depicted in Fig. 5.10, a worst case of 4 rule applications would take place – *Rule 7'*, *Rule 7*, *Rule 4'* and then *Rule 4* – before the solution is found (and the wrong branch of search discarded). The solution of two time steps – application of an upward force followed by the application of a downward force – is then encoded as a two-step causal rule (*Rule 16*) as shown in Fig. 5.11 and added to the Causal Rule Base.

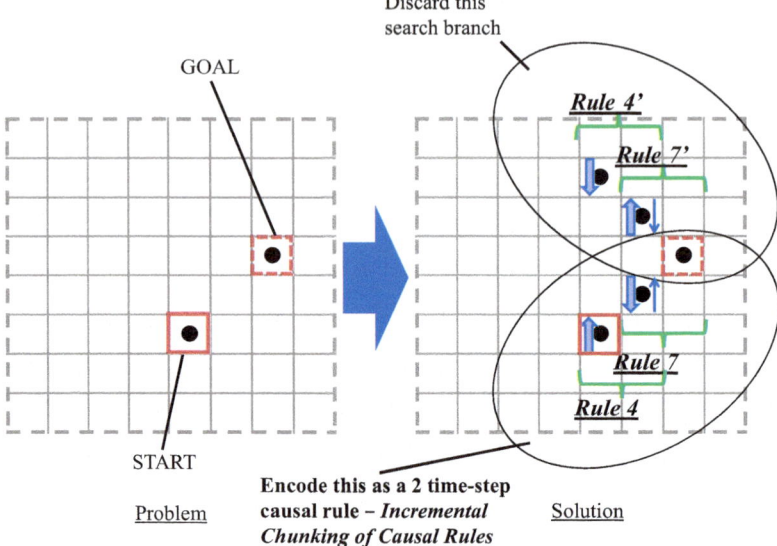

Fig. 5.10 An example of incremental chunking of causal rules

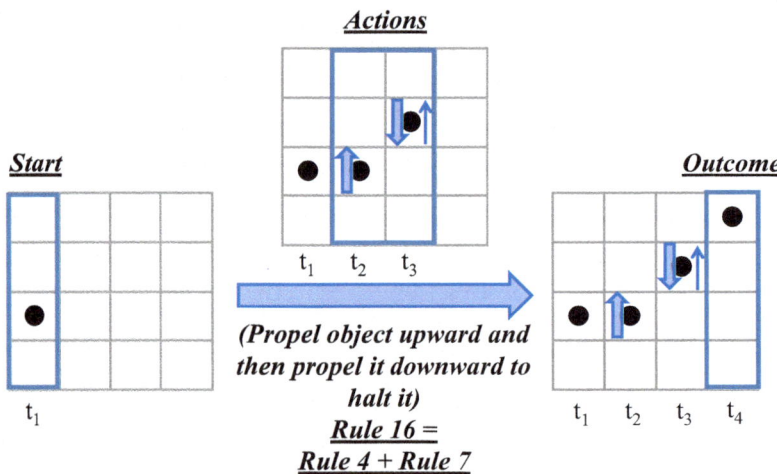

Fig. 5.11 An example of a two time-step causal rule – *Rule 16*

Suppose now the system is given a problem that is slightly more complex than the one in Fig. 5.10 which potentially requires a three time-step solution as shown on the left side of Fig. 5.12. Furthermore, we employ a heuristic that gives priority to rules that have been successfully used and encoded in similar situations. On the right hand side of Fig. 5.12 is shown some of the earlier encountered problems (that have been solved with attendant solutions). Of the two problems shown, the one

5.2 Reasoning and Problem Solving

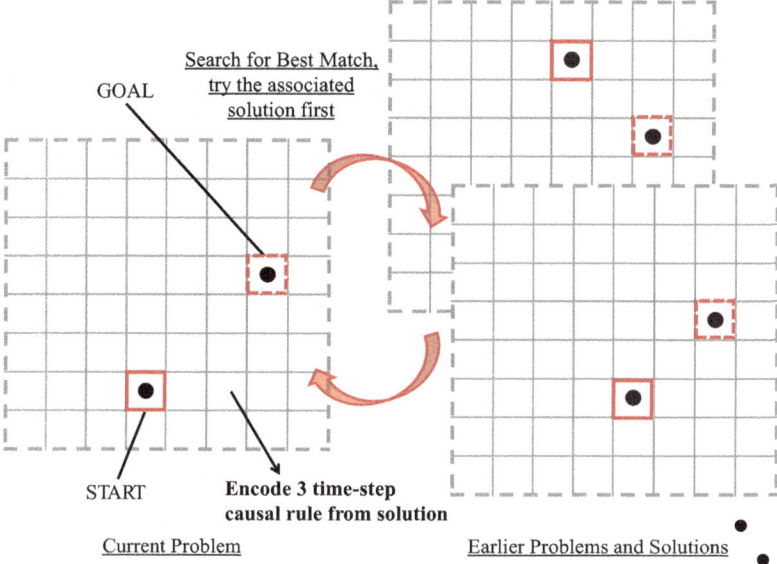

Fig. 5.12 A slightly more complicated Current Problem than that in Fig. 5.10 and two similar problems encountered earlier with multi-step solutions stored in Causal Rule Base

below (same as the problem in Fig. 5.10) would be more similar to the current problem and its corresponding solution rules would be considered first. (These two earlier problems would already have their multi-step solutions encoded and stored in the Causal Rule Base.) A possible solution trace is given on the right hand side of Fig. 5.13 engaging *Rule 16* that is a two time-step chunked rule from the earlier simpler problem. Compare that solution with the one on the left hand side of the same figure with no chunked rule engaged (derived in a similar manner as that in Fig. 5.10). What is shown on the left hand side is actually only part of the potential search space as there are possibly side branches of the search process that are not shown. But it represents a *lower bound* on the amount of search needed if no chunking is used. The solution on the right hand side hence achieves some savings in the size of the search space compared to that on the left hand side. As problem complexity increases, various chunked rules from earlier problem situations could potentially reduce the search space significantly. Also, in this case of Fig. 5.12's Current Problem, a three time-step rule would be added to the Causal Rule Base after the successful discovery of the solution.

The Incremental Causal Rule Chunking process described here is a microcosm of the learning process that takes place often in the real world with noological agents. Complex organisms such as humans go through an extensive "childhood" phase in which simpler tasks and concepts are learned incrementally before engaging in more complex tasks and concepts in adulthood. Even in adulthood, during which time most organisms would have already acquired enough knowledge for fundamental survival needs, incremental learning is always essential for tackling

Fig. 5.13 Solution steps for the Current Problem in Fig. 5.12, with and without the use of chunked rules

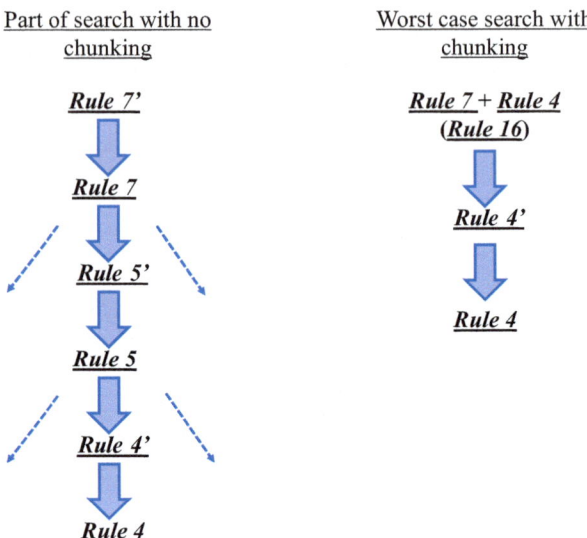

ever bigger and more complex problems. The process of incremental chunking of rules must lie at the core of adaptive noological systems.

Note that incremental rule chunking as described above can also take place in the absence of environmental input. A system can begin with whatever problems that have known solutions and conjure up similar problems and solve them "in its head" before even encountering them in the real world. This would be like the "thought roaming" process that human beings often engage in – we often imagine problem situations before encountering them and come up with solutions "mentally" – and that can greatly speed up our process of adaptation to the environment.

5.2.3 A More Complex Problem

Figure 5.14 illustrates a more complex problem and its solution (similar to the problem discussed in Ho (2013)). The problem statement is shown on the left side of the figure in the form of problem representation as discussed in Sect. 5.2.1. It depicts a starting physical configuration consisting of two mobile objects (*Objects 1* and *2*) and an intervening immobile inelastic object (represented as an open square, labeled as Wall). After an indefinite interval (represented by the longer and thicker Range Bar) *Object 1* and *Object 2* acquire some downward momentum, and while *Object 2* is specified to be at a certain final location (a location that is at an opposite side of the Wall from its original location), *Object 1* can be in any position specified by the Range Bar under it. *Object 1* can be thought of as an "agent" used to retrieve *Object 2* from "behind" the

5.2 Reasoning and Problem Solving

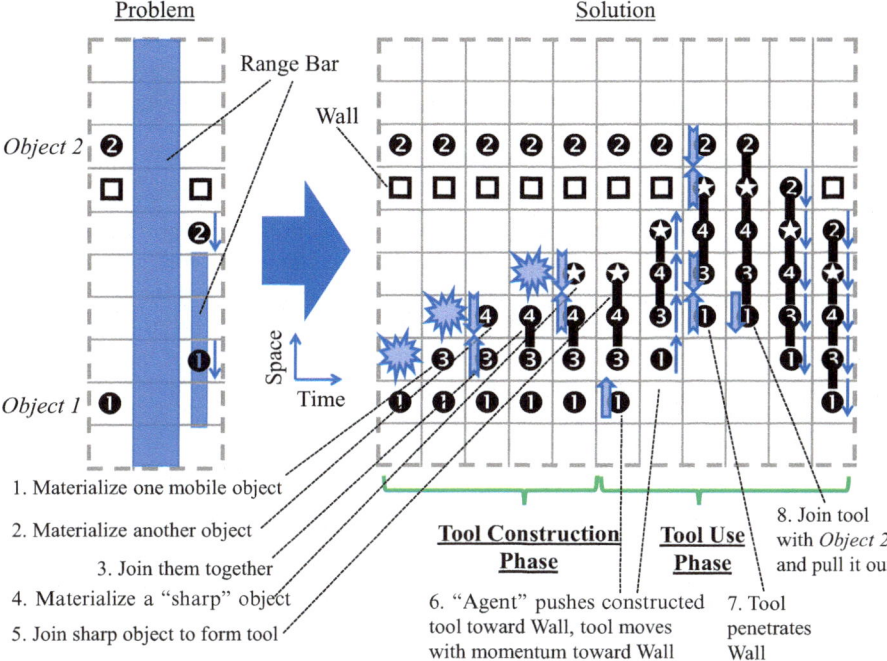

Fig. 5.14 A more complicated problem than in Fig. 5.12 and a possible solution. The problem is to retrieve *Object 2* from "behind" the Wall from the position of *Object 1*. In order to make *Object 1* behave like an active "agent" and other objects passive, we add the constraint that a force can only act on *Object 1*. The *longer and thicker Range Bar* indicates an indefinite temporal interval and the *shorter* one indicates a range of allowable spatial locations

Wall. In order to make *Object 1* behave like an active agent and other objects passive, we add the constraint that a force can only act on *Object 1*.

A backward reasoning process using the rules we have discussed can conceivably provide a solution as shown on the right side of Fig. 5.14. The solution shows that a tool is first constructed by materializing two mobile objects and one mobile "sharp" object (as discussed and defined in Sect. 4.3.10.2 and Fig. 4.21b), and attaching them together in such a way that the sharp end of the tool can then be used to penetrate the immobile object and get itself attached to *Object 2*.[1] Following that, *Object 2* can be *pulled* downward (by *Object 1*, after it has been attached to the tool), achieving a momentum as specified in the problem specification. This is basically a process of constructing a tool and then using it to retrieve some objects behind an obstacle.

---

[1] Here we show the sharp object coming to a halt immediately after penetrating the immobile object (Wall), whereas the sharp object in Fig. 4.21b continues to move after penetration. This version here is a slight variation of the earlier-depicted penetrating behavior.

The solution can be divided into a "Tool Construction Phase" and a "Tool Use Phase" but the solution process does not distinguish this explicitly – the backward reasoning process just assembles the steps of the solution together continuously through the rule matching process as described in Sect. 5.2.1. Therefore, there is a nice continuity between tool construction and tool use in the problem solving process. We can identify this two phases of solution in many real world problem solving processes as well – e.g., building a car (constructing a "tool") and driving it to somewhere to pick up something (using the tool). In the real world, "materialization" of various objects necessary for the construction of the tool as depicted in Fig. 5.14 can mean bringing parts from other places to the scene of the construction of the tool. That process would often require another sub-problem to be solved.

Obviously, incremental chunking is necessary for the solution of the problem in Fig. 5.14, otherwise the search space will be prohibitively large.

5.3 Elemental Objects in 2D: Representation and Problem Solving

In this section, we digress a little and briefly discuss the activities and interactions associated with *elemental* objects in 2D. Note that this is to be distinguished from 2D objects that are extended – i.e., objects that are composed of many elemental parts and that have dimensions in the "X" and "Y" directions (the usual "length" and "width" directions). Here we extend the previous considerations of elemental objects (objects without dimensions, in a sense) that move in a 1D space to those that move in a 2D space (i.e., now there are two degrees of translational freedom – the "X" and "Y" directions, as are conventionally labeled).

Despite the increase in translational degrees of freedom, the activities and interactions of 2D elemental objects can be characterized by similar computational structures as those for 1D elemental objects discussed above. Figure 5.15 illustrates a 2D operator that represents "*Reflection off an Immobile Object with Momentum*" – much like the situation described in Sect. 4.3.10.2, illustrated in Fig. 4.20, and encapsulated in *Rule 8* (Fig. 5.3a) for the 1D case, except now that the operator is "3D" – 2D space with "X" and "Y" axes and 1D time. These dimensions are needed for the characterization of the reflection interaction in the 2D space physical world (the 2D operator has coordinates X' and Y' while the real world coordinates are X and Y). A 2D Rotational Transformation for the operator shown in the figure specifies that the operator can be matched to any reflection situation in any orientation. Specialized reflection operators can also be defined for certain subset of specific orientations.

Figure 5.16 shows how the *Push*, *Momentum Continuation*, and *Non-Penetration* operations described for the 1D situation (corresponding to *Rules 14*, *5* and *10* respectively) can be extended naturally to the 2D situation. The object shown in Fig. 5.16 is some amorphous fluid spreading over a 2D surface. Even though strictly

5.3 Elemental Objects in 2D: Representation and Problem Solving

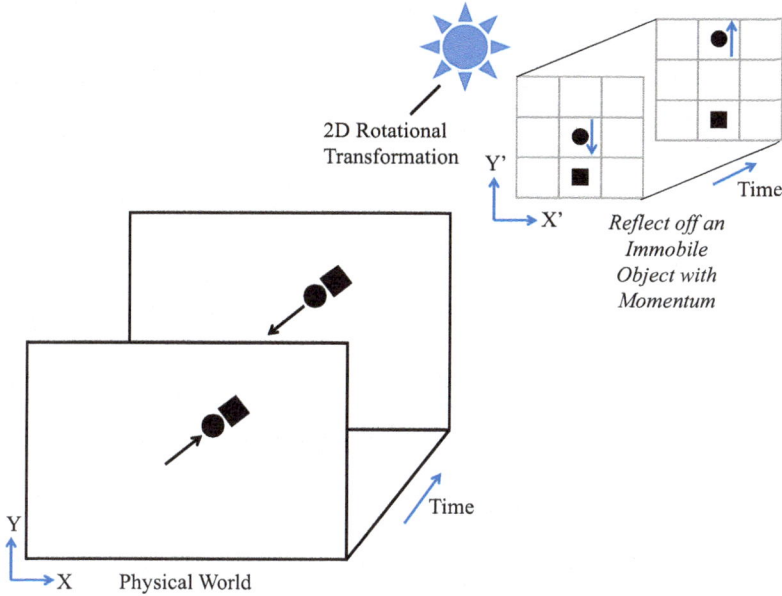

Fig. 5.15 *Collision and Reflection* operator for 2D elemental object with rotational symmetry

speaking the amorphous fluid is a 2D extended object – implying that the elemental objects composing it do not move independently of each other, for a short span of time, the behavior of the elemental objects comprising it can be approximately characterized as independent elemental objects moving in 2D. We therefore apply the *Push*, *Momentum Continuation*, and *Non-Penetration* operators to characterize, respectively, the "pushing" of some of the elemental parts of the fluid by other parts of the fluid (say something from the source of the fluid), the continuation of the momentum at the advancing "front" of the fluid, and the behavior of the fluid at locations at which it is halted by an immobile non-elastic barrier. (In the discussion section, Sect. 5.5 below, we contrast our method of representation of the behavior of liquid with that in qualitative physics – e.g., Hobbs and Moore 1985).

With some further extensions to the current rules, the problem solving situation involving some fluid depicted in Fig. 5.17 can also conceivably be handled. With the representation and understanding of the concept of "*Non-Penetration*," barriers can be constructed much like the construction of the "penetration-tool" in Fig. 5.14 to achieve a certain goal – such as to confine the fluid to certain areas. A goal specification similar to those that we have described above, but now for the desired changes of the fluid shape with respect to time,[2] could be used to drive reasoning processes that provide the construction (in a "Tool Construction" phase much like

---

[2] And for places where we wish the fluid to stop flowing further and beyond we specify no further shape changes in time – i.e., a "halt" in further motion.

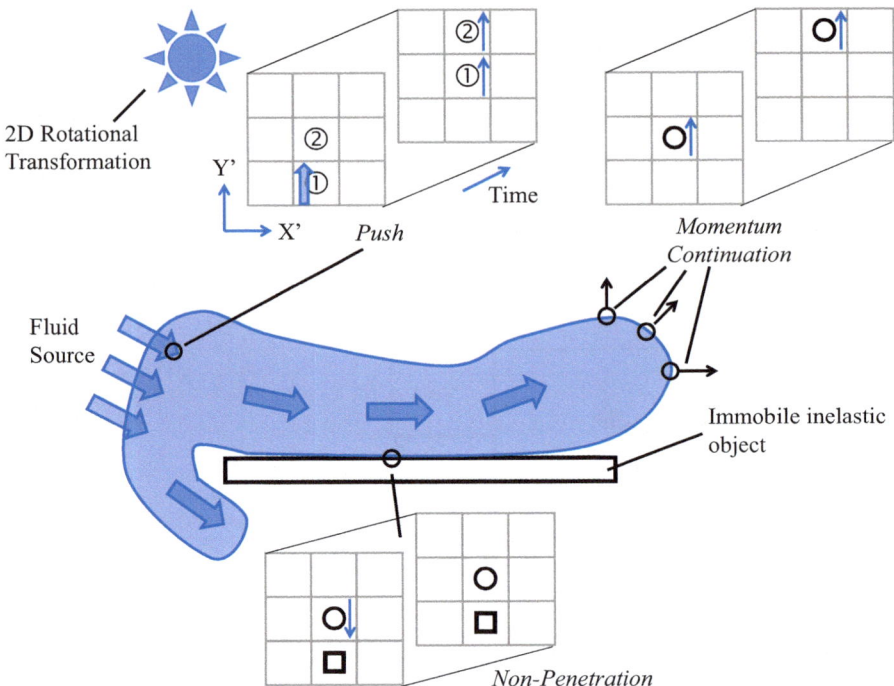

Fig. 5.16 2D fluid – approximated by a collection of independent elemental objects moving in 2D directions. Operators involved are *Push*, *Momentum Continuation*, and *Non-Penetration*

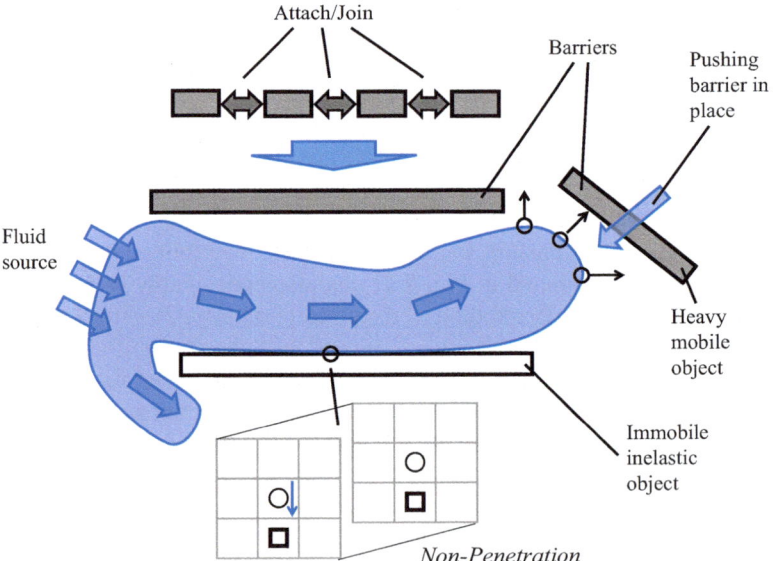

Fig. 5.17 Constructing and placing barriers to control fluid flow

that in Fig. 5.14) and placement (in a "Tool Use" phase) of barriers in certain places in certain manners as a solution. Figure 5.17 shows barriers being constructed and pushed into certain places to control the fluid flow.

Hence the operational representational scheme allows the straightforward and natural representation of knowledge for encoding physical behavior of even physically complex objects such as fluids and generating solutions to certain desired configurations of the objects involved. In a 3D world and under the constraint of gravity, various kinds of containers for the containment of liquids can be conceived and constructed by a noological system using similar representational constructs as in the example for 2D fluid in Fig. 5.17.

In figures such as Figs. 4.8 and 4.9 in the previous chapter, we showed the idea of conceptual hierarchies. Conceptual hierarchies such as semantic networks are constructs often used in AI for reasoning (e.g., Russell and Norvig 2010) and in psychology for the modeling of human knowledge structure (e.g., Gleitman et al. 2010). What are the roles of conceptual hierarchies in problem solving? One possibility can be seen by considering Figs. 4.11 and 5.17. In Fig. 4.11 we suggested how some higher level concepts regarding the fluid – such as "rapidly protruding front portion" – can be built on the lower level atomic operators and this can potentially direct a problem solving process to provide some solutions to manage certain regions of the liquid accordingly to achieve certain goals. The ensuing problem solving process illustrated in Fig. 5.17 provides the detailed and specific physical constructs built from elemental parts – i.e., the various barriers – for the management of the flowing fluid. In general, concepts and conceptual hierarchies help to reduce the search space involved in problem solving processes and make these processes more efficient.

5.4 Natural Language, Semantic Grounding, and Learning to Solve Problem Through Language

We began the discussion on operational representation in Chap. 4 and ended with the discussion of using operational representation for causal rule encoding, reasoning, and problem solving in this chapter. In Chap. 4 we claimed that the operational representations of various concepts such as *Move*, *Appear*, *Stay*, *Attach*, and *Push* that we have discussed represent the epistemic ground level knowledge and we argued for the case by using some examples in which operational representations engender certain "ground level" problem solving that is otherwise not possible (e.g., knowing the spatiotemporal behaviors of leaves on branches allows a person to imagine or use "leaves on branches" as a tool to sweep floor or fan oneself – Sect. 4.1). We submitted that this ground level representation allows the noological system to "really understand" the concepts involved through an appropriate encoding of the meaning of the concepts. In this chapter above, we further illustrated how causal rules represented in the form of operational representations can be used for effective reasoning and problem solving. These causal rules encode conceptual generalizations of the concepts involved (such as *Push, Momentum*, and *Attach*). Therefore, in a sense what we are showing is that *knowing* (the concepts involved) is *knowing how to act* (with the concepts).

We have shown in Fig. 5.14 how a tool could be constructed to solve a certain problem and the process of tool construction and application was worked out by the system through a backward chaining problem solving process with causal rules that have been learned earlier. Now, suppose a system (or a person) has worked out this solution before. Could the system use natural language to communicate to another system the method to solve the problem without the second system having to undergo a potentially involved problem solving process? In fact, a vast proportion of human progress was achieved through learning from others through natural human language about all kinds of solutions to all kinds of problems. Of course, this is only possible because the recipients of the instructions in natural language form *really understands* what the strings of symbols emitted by the instructors mean.

Using our framework, we illustrate how instructions though language is possible because our representational framework captures meaning appropriately. In Fig. 5.18 we show an example, derived from the same problem as that in Fig. 5.14, of using English language instructions from an "instructor" to teach a "student" how to solve the problem involved. One could roughly characterize the meaning of a sentence as existing at two levels – the syntactic level and the lexical level (e.g., Saeed 2009). For this example, we assume that there are some syntactic transformational rules that are peculiar to each language (in this case, English) that convert the syntactic structures to some deeper representations that capture the syntactic level meaning of the sentences involved. This aspect of meaning processing is complex in itself and we will not delve into the details there. What we would like to illustrate with Fig. 5.18 is how the individual lexical items (the "words") can be associated with some ground level representations that provide the meaning of the lexical items involved and this *understanding* allows the recipient of the language instructions to carry out tool construction and application, illustrating that the recipient *really understands* the strings of symbols emitted by the instructor through language.

On the right side of Fig. 5.18 is the solution to the problem in Fig. 5.14 and on the left side is a sequence of instructions in English instructing how a tool could be constructed and used to solve the problem. To avoid clutter, we extract two sentences from the left panel and place them right below the Solution picture and show how each of the words in the sentences can be mapped onto an item, an activity, or an action in the Solution. Again, to avoid clutter, we show only some of the mappings. Also, there are linguistic devices such as the article "the" whose function cannot be clearly identified with an external object or event and we omit the mappings for these as well. Other than the words corresponding to some of the concepts we have discussed above, we also show how the concepts of "first," "after," and "then" can be mapped onto the temporal order of events in the Solution

5.5 Discussion

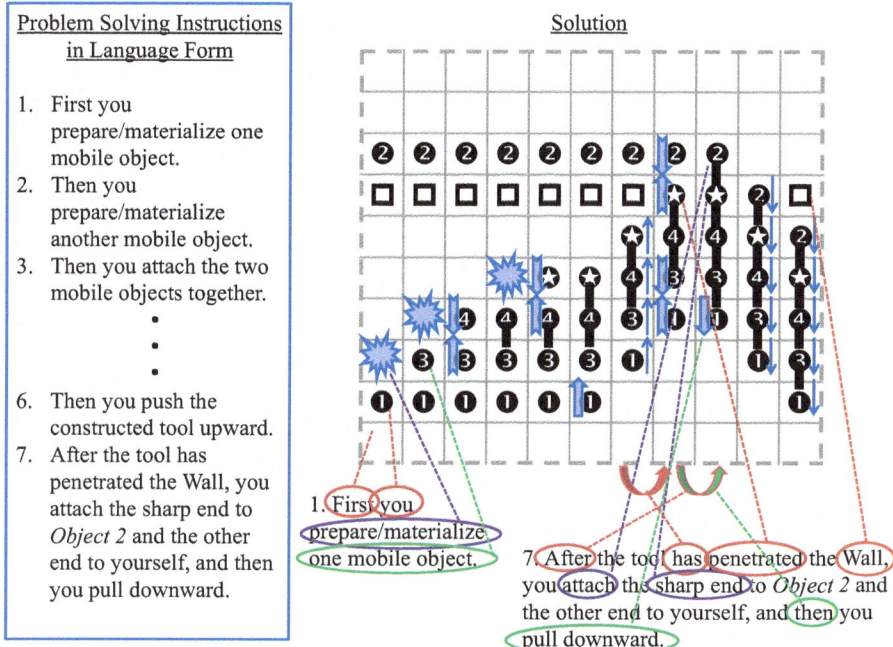

Fig. 5.18 Semantic grounding of natural language sentences through grounding of the meaning of the lexical items in the sentences. This translates into the actions carried out according to the natural language instructions, illustrating the idea of *knowing* (some concepts) is *knowing how to act* (with the concepts)

picture. Through an appropriate "meaning construction" process using the grounded understanding of each of the lexical items in the sentences and the syntactic structure of the sentences, it is then possible for the recipient of the instructions in natural language to carry out the appropriate actions accordingly. We characterize this process of learning from language instruction as *learning of grounded concepts for rapid problem solving through language* which constitutes the bulk of learning for noological systems that have the language faculty, such as human beings. The great strides made in the human community in problem solving, as distinct from that of other animals, critically depend on this kind of learning.

5.5 Discussion

We would like to contrast our operational representation paradigm described in both the previous chapter and this chapter with standard AI approaches in two major aspects. One aspect concerns the representation of the temporal aspects of the world. In many AI and neuroscience investigations and methods, it is deemed that to capture temporal information, a recurrent neural network or a Markov chain (Arbib 2003; Sutton and Barto 1998) would be the most natural choice as changes in the state of the world from one time frame to another can be embedded in these structures both for recognition and generation purposes. In neural networks, a succession of the physical states of the world which change over time could be learned and stored and later used to recognize novel instances with some generalizations (Arbib 2003). Recurrent neural networks can also generate a succession of states representing action sequences to be taken in the world.

In Markov chains, the temporal nature of the physical world can be captured in state transition probabilities linking one physical state to another, representing how likely the transitions have been observed in the real world. These transition probabilities can direct an intelligent system to output action sequences to achieve certain goals. Even though in some sense the temporal aspects of the world are captured in these representations, as these representations can recognize and generate states over time, the temporal aspect is implicit. Take, for example, the basic movements of *Move-Up* and *Move-Down* as discussed in Chap. 4, Sect. 4.3.3. State transition representations would link the state in the earlier time step to the later time step, and perhaps assign a transition probability to the transition, but they do not really capture the physical nature of the movements. The fact that, say, *Move-Up* and *Move-Down* are actually intimately physically related to each other and related in various manners to other operations, is best represented and revealed through the spatiotemporal representations in which the temporal dimension is explicitly represented such as that we have advocated here – in the form of operational representations. The participation of these operational representations, in the construction of conceptual hierarchies and in various reasoning and problem solving processes, becomes very natural and straightforward, as can be seen in various examples described in both chapters. It would be difficult to see how a neural network or Markov chain representation can support relatively complex problem solving processes of Fig. 5.14 and also allow explicit linguistic descriptions and instructions such as that illustrated in Fig. 5.18.

Another aspect concerns a field of AI investigation known as qualitative or naïve physics (Hobbs and Moore 1985; Weld and de Kleer 1990). In qualitative physics, knowledge of the physical behavior of objects and causal agents is also encoded and represented for reasoning and problem solving purposes. But the various methods advanced are often haphazard and no standard methods exist for handling various situations. Many methods are proposed to handle temporal information but none of them is explicit and principled like what we propose here. Moreover, certain approaches seem convoluted, unnatural, and difficult to extend to more general domains, such as the representation of liquids using predicate logic as described by Hayes (1985). Comparing Hayes' scheme with what we have outlined for representing fluids in the operational representational framework (Figs. 4.5, 4.11, 5.16, and 5.17), the current method is more straightforward, natural, and effective, with the possibility that the knowledge for the physical behavior of fluids being learnable by observing and interacting with the environment directly through causal learning as discussed in Chap. 2. The area of qualitative physics (Hobbs and Moore

5.5 Discussion

1985; Weld and de Kleer 1990) does not address the issue of learning of causal knowledge.

A note should be made about the convergence of our method here and an area of research known as cognitive linguistics as discussed in Chap. 4, Sect. 4.5 (e.g., the work by Croft and Cruse 2009; Evans and Green 2006; Geeraerts 2006; Johnson 1987; Langacker 1987, 1991, 2000, 2002, 2008, 2009; Ungerer and Schmid 2006, etc.). In cognitive linguistics, many concepts such as the concept of *Arrive* are represented in a form with an explicit temporal axis such as illustrated in Fig. 4.28 (Langacker 1987). From the point of view of the cognitive linguists, the pictorial representation is an adequate representation of the *meaning* of the concepts involved. What we have achieved here goes beyond that. We provide a fully computationalized account of a representational scheme that is based on explicit temporal representation together with the successful demonstration of the power and effectiveness of the representational scheme in noological tasks of learning, reasoning, and problem solving. However, the convergence with the cognitive linguistic approach further strengthens our belief that our representational scheme does indeed provide the foundation for capturing the meaning of concepts adequately for noological processing.

A further point to note is that in a full-blown AI system that employs operational representations to handle, in an epistemically grounded manner, various real world objects, events, and processes that are 3D and of very high resolution, a massive amount of information processing is expected. However, there is no short-cut for creating truly intelligent systems. Human beings deal with the full complexity of the physical world in a massively parallel processing manner with their large number of neurons connected mostly in parallel in their nervous systems to achieve rapid processing for the purpose of survival. If epistemically grounded knowledge and characterization of the world is necessary for intelligent actions and survival, there is no escape from addressing the massive nature of information processing required. The structured and principled methods outlined in our operational representational scheme have the generality and power to scale-up to drive noological processing that addresses physically realistic situations in the real world.

In Sect. 5.4 we pointed out that a critical kind of learning, learning of grounded concepts for problem solving through language, constitutes the most important kind of learning for noological systems with the language faculty, which engenders rapid learning of a huge amount of knowledge for survival. This critical mechanism resulted in the rapid advancement of the human community compared to that of the other animals. This kind of learning is only possible through grounded conceptual representations such as the various operational representations discussed in the past two chapters.

We would like to highlight the significance of the spatialization of time from the point of view of another discipline – physics. As we know, one major breakthroughs in physics in the past century was the discovery of the theory of relativity that depends on the spatialization of time to achieve a more general formulation of physical laws and hence a deeper understanding of reality (Lorentz et al. 1923). Hence our explicit temporal representation of some grounded concepts here may have the same significance to noology as the spatialization of time has to physics.

Problem

The problem in Fig. 5.18 is analogous to the "monkey-and-bananas" problem in traditional AI (McCarthy 1968; Nilsson 1980) – a monkey/robot cannot reach some bananas attached to a high ceiling and hence has to use a tool (such as a long stick) or climb up one or more boxes to reach the bananas. The scenario in Fig. 5.18 could be interpreted as a monkey constructing a tool (with a sharp end) and jumping up and using the tool to poke through the ceiling to spear at and retrieve the banana(s) on the other side of the ceiling. In contrast to the approach in traditional AI, when a problem such as this (the left side of Fig. 5.18) is posed, a noologistically realistic approach, as has been demonstrated in this chapter, is to tackle all the issues involved from learning (of physical laws in this case) to problem solving. This includes incremental chunking of knowledge (Sects. 3.5 and 5.2.2) and learning of heuristics (Sect. 2.6.1). Using the 2D/3D representational scheme devised in the problem at the end of Chap. 4, formulate a real/3D-world monkey-and-bananas problem and its solution in the spirit of the noologistically realistic approach as expounded in the foregoing discussions.

References

Arbib, M. A. (2003). *The handbook of brain theory and neural networks* (2nd ed.). Cambridge, MA: MIT Press.
Barr, A., & Feigenbaum, E. A. (Eds.). (1981). *The handbook of artificial intelligence: Volume I*. Los Altos: William Kaufmann.
Croft, W., & Cruse, D. A. (2009). *Cognitive linguistics*. Cambridge: Cambridge University Press.
Evans, V., & Green, M. (2006). *Cognitive linguistics: An introduction*. Mahwah: Lawrence Erlbaum Associates.
Geeraerts, D. (2006). *Cognitive linguistics*. Berlin: Mouton de Gruyter.
Gleitman, H., Gross, J., & Reisberg, D. (2010). *Psychology* (8th ed.). New York: W. W. Norton & Company.
Hayes, P. (1985). Naïve physics I: Ontology for liquids. In J. R. Hobbs & R. C. Moore (Eds.), *Formal theories of the commonsense world*. Norwood: Alex Publishing.
Ho, S.-B. (2013). Operational representation – A unifying representation for activity learning and problem solving. In *AAAI 2013 Fall Symposium Technical Reports-FS-13-02*, Arlington, Virginia (pp. 34–40). Palo Alto: AAAI.
Hobbs, J. R., & Moore, R. C. (Eds.). (1985). *Formal theories of the commonsense world*. Norwood: Alex Publishing.
Johnson, M. (1987). *The body in the mind*. Chicago: The University of Chicago Press.
Langacker, R. W. (1987). *Foundation of cognitive grammar* (Vol. I). Stanford: Stanford University Press.
Langacker, R. W. (1991). *Foundation of cognitive grammar* (Vol. II). Stanford: Stanford University Press.
Langacker, R. W. (2000). *Grammar and conceptualizations*. Berlin: Mouton de Gruyter.
Langacker, R. W. (2002). *Concept, image, and symbol: The cognitive basis of grammar*. Berlin: Mouton de Gruyter.
Langacker, R. W. (2008). *Cognitive grammar: A basic introduction*. Oxford: Oxford University Press.

Langacker, R. W. (2009). *Investigations in cognitive grammar*. Berlin: Mouton de Gruyter.
Lorentz, H. A., Einstein, A., Minkowski, H., & Weyl, H. (1923). *The principle of relativity: A collection of original memoirs on the special and general relativity*. London: Constable.
McCarthy, J. (1968). Program with commonsense. In M. Minsky (Ed.), *Semantic information processing*. Cambridge, MA: MIT Press.
Nilsson, N. J. (1980). *Principles of artificial intelligence*. Los Angeles: Morgan Kaufmann.
Russell, S., & Norvig, P. (2010). *Artificial intelligence: A modern approach*. Upper Saddle River: Prentice Hall.
Saeed, J. (2009). *Semantics*. Malden: Wiley-Blackwell.
Sutton, R. S., & Barto, A. G. (1998). *Reinforcement learning: An introduction*. Cambridge, MA: MIT Press.
Ungerer, F., & Schmid, H.-J. (2006). *An introduction to cognitive linguistics*. Harlow: Pearson.
Weld, D. S., & de Kleer, J. (Eds.). (1990). *Readings in qualitative reasoning about physical systems*. San Mateo: Morgan Kaufmann Publishers, Inc.

Chapter 6
The Causal Role of Sensory Information

Abstract This chapter investigates in a deep and principled manner how sensory information contributes to a noological system's learning and formulation of causal rules that enables it to solve problems rapidly and effectively. The spatial movement to goal with obstacle problem is used to illustrate this process. The process is one of deep thinking and quick learning – the causal learning process requires few training instances, but extensive use is made of the many causal rules learned to reason out a solution to the problem involved. Heuristic learning and generalization emerge in the process as an aid to further accelerate problem solving.

Keywords Sensory information • Problem solving • Causal rule • Heuristic • Heuristic generalization • Learning of heuristic • Script • Recovery from thwarted script • Deep thinking • Quick learning

In Chap. 1, Sect. 1.2, using a simple and yet illustrative example, we showed how sensory information contributes to the problem solving abilities of a noological system. The role of sensory information is that of a "service function" to the problem solving process as illustrated in Fig. 1.7. In Chap. 2, Sect. 2.6.1, we demonstrated how certain parameters sensed and provided by the sensory system allow an agent to learn useful heuristics that guide and vastly reduce the problem solving search effort in the SMG problem.

Even though computer vision systems are improving in their visual abilities (e.g., Forsyth and Ponce 2002; Shapiro and Stockman 2001; Szeliski 2010, etc.), by far the most successful visual systems are found in biological systems. Humans and primates lie at the apex of these abilities, with well-endowed visual cortices. What do these visual systems do and what information can they provide for an agent to assist with problem solving? Marr (1982) characterizes the visual system's stages of processing as including the stages of primal sketch, 2.5D sketch, and 3D model representation. The 3D model representation is an object centric representation that requires the use of some top-down models to construct as typically our visual system cannot see the "backside" of objects from a particular vantage point. Therefore, the information a visual system can derive from a particular vantage point on a scene with object(s) inside it is a 2.5D sketch – a depiction of the scene in terms of the surface orientation at every point of the surfaces in the scene, the distance to every point on the surfaces, the discontinuities in the surface orientation,

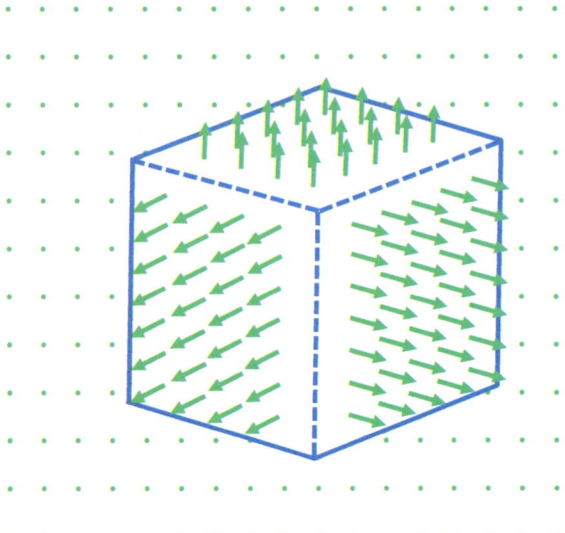

Fig. 6.1 A 2.5D sketch of a scene with an object. Scene consists of a cubicle object floating in front of a "wall." Information includes local surface orientation (shown as *arrows*, and the background wall has *arrows* pointing toward the reader), distances of points on surface from the viewer, discontinuities in surface orientation (shown as *dotted lines*), and the discontinuities in depth of surface points (shown as *solid lines*) (Marr 1982)

and the discontinuities in the depth of the surfaces as viewed from the vantage point (Fig. 6.1). In this 2.5D sketch the backside of various objects in the scene are occluded and cannot be seen.

The contour information in a 2.5D sketch basically implies that the agent has a full map of the relative distances of every point in the visual scene from its position. From this information, it is implied that the agent would also be able to compute the relative distances between any two points in the scene. This level of characterization is necessary before the characterization of full 3D information, such as a 3D model of an object in the scene, which would lead to the next step of characterization which would be the category of the object involved. Constructing 3D information from 2.5D information requires some pre-knowledge as may be gained from earlier experiences (e.g., having seen the object involved from a different direction). Sometimes, a novel object or an object of no interest may not have been assigned a label or category, but their 2.5D or 3D information can still be made available from the visual input.

As we shall see in our subsequent sections in this chapter as well as in the rest of the book, this 2.5D and/or 3D information is an important foundation upon which a noological system builds its causal knowledge about the world which it then applies to problem solving situations.

6.1 Information on Material Points in an Environment

Let us now consider the kind of rudimentary information that is captured by a visual system that is then presented to the higher level causal learning processes. As the concern of this book is not with vision per se, we shall use an idealized "super-

6.1 Information on Material Points in an Environment

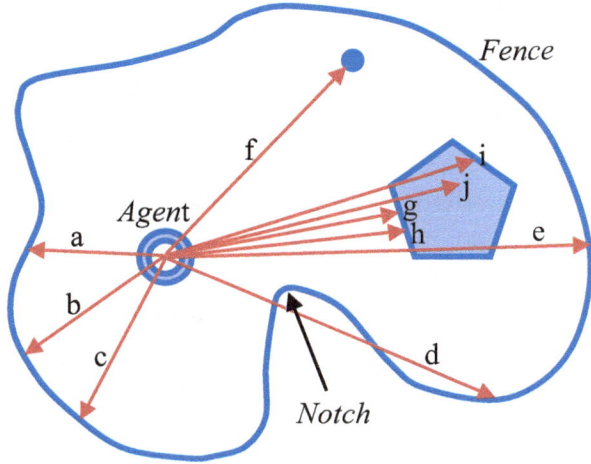

Fig. 6.2 A super-visual perception of the environment

visual" system, a visual system that provides more information than that of a natural visual system or an artificial computer vision system. As will be described below, what this super-visual system could "perceive" at any instance is not only based on the information from one vantage point.

Figure 6.2 shows an Agent embedded in an environment with an arbitrarily shaped *Fence* and a couple of objects. We assume that the visual system of the agent is "super-visual" and "all-sensing" in that as long as a material points exists, it can sense it. Therefore, occlusion does not matter, and the Agent can see in all 360° directions simultaneously – i.e., it has an unlimited viewing angle. In a realistic visual system, points that are occluded can also be uncovered through exploration and points that are currently not in view can be brought into view by the change of direction of placement of the visual sensory organs – i.e., often by turning the body or the head on which the organs rest, or by turning the sensory organ within its supporting structure (e.g., turning an eye in an eye socket). Of course, if some special technology is available, such as some kind of "penetrating radar," an artificial system may be able to perceive all this information about the objects in the environment at one instance.

Therefore, in Fig. 6.2, arrows a, b, c, d, e represent relative distances to points on the *Fence* that are perceived, whether they are occluded or not (The "*Notch*" would occlude d in a normal visual situation). The *Fence* does not have any thickness so there is only a linear series of material points on its "surface." We assume there is a finite number of points on a seemingly "continuous" object such as the *Fence* that can be perceived and registered. Each point can be thought of as an elemental object from which the *Fence* is made. (If there are material points that exist beyond the *Fence*, they can also be sensed.) Arrow f is the relative distance to an elemental object (characterized by a singular point). Arrows g and h are relative distances to two elemental points on the surface of another object, from which the local shape of the object can be constructed. We also assume that any occluded point (arrow i) and any interior point (arrow j) of any object can be perceived. In natural visual systems such as the human visual system, the occluded points cannot be perceived directly but can be known cognitively based on earlier perception that is currently available in some sort of visual memory. Similarly, interior points of non-transparent objects are typically not perceived directly by natural visual systems but can be known cognitively through inferences made based on the surface points perceived and/or earlier perceptual experiences with the objects involved.

In our real world, there are three ways an object that is originally not perceived by a visual sensory organ that has a limited viewing angle can come into perceptual existence. One way is, the object that is currently not within the viewing angle of the sensory organ can be brought into view through the rotation of the body or the head on which the sensory organ rests or rotation of the sensory organ within its supporting structure (e.g., rotation of an eye in its socket). A second way is, the object can come into perceptual existence through its movement or the movement of the agent if it had earlier been occluded by other objects. The third way is, the object had originally not existed and it comes into existence through materialization. (In our current real world, materialization is not common other than in particle accelerators and the closest analog of materialization is the switching on of some light.)

In any case, when an object comes into existence, some processes are triggered in the visual system. Firstly, the existence of the object is registered. In a computer system, this could correspond to the creation of a new entry in the memory system that contains the following predicate logical representation: $Exist(Object(X))$. The representation must of course be accompanied by the corresponding processes that interprets the meaning of "*Exist*" and we posit that this is one of the fundamental built-in processes of a noological system. (The concept *Exist* is grounded in the spatiotemporal pattern as discussed in Chap. 4.) As a result of the existence of the object, the relative distances to the surface points of the object are computed by the visual system, and in our super-visual system, the distances to the interior points also become available. As discussed in Chap. 4, an extended object is made up of a number of material points existing at different locations. This process is shown in Fig. 6.3.

In Fig. 6.3, the predicate $Exist(Object(X), S = \{x\})$ means that an $Object(X)$ is seen by the noological *Agent* in the environment that occupies the set of locations, $S = \{x\}$. For simplicity, we use x to represent the usual x, y coordinate tuple $[x, y]$ for 2D space. When that happens, assuming that the object is made up of a collection of material *Points*, it implies that there is a material *Point* at every one of the locations $\{x\}$. The following rule states this situation:

$$Exist(Object(X), S = \{x\}) \rightarrow (\forall x\, x \in S \rightarrow Exist(Point(x))) \quad (6.1)$$

Note that this logical assertion asserts the internal mental representations that are created within the agent's "mental" processing machineries. Now, if a material point exists at a certain location x, it also means that the visual system of the *Agent* would compute a relative distance, *RD*, to it:

6.1 Information on Material Points in an Environment

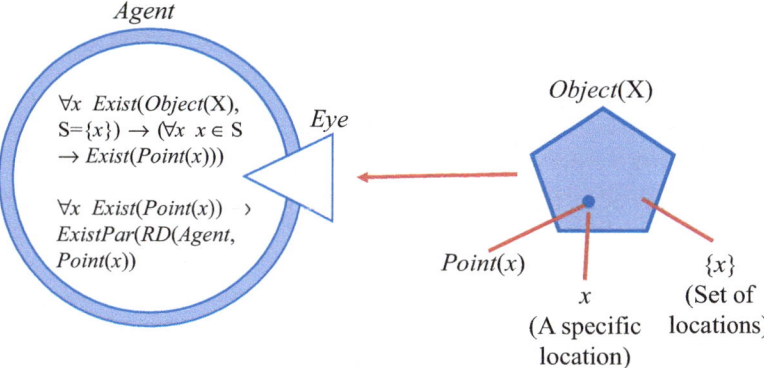

Fig. 6.3 Internal "mental" events of a noological system that take place when an *Object*(X) is observed by a visual system. For simplicity, x is used to represent the usual x, y coordinate tuple $[x, y]$ for 2D space

$$\forall x \, Exist(Point(x)) \rightarrow ExistPar\big(RD\big(Agent, Point(x)\big)\big) \quad (6.2)$$

ExistPar is a process by which the parameter concerned, in this case the relative distance measure, *RD*, becomes available to the *Agent*, provided for by the visual system. Note that this logical assertion asserts the internal noological processes that take place within the *Agent's* mental processing machineries. The "*Exist*" operator is used not only to describe the entities that exist "out there" in the external environment but also entities that are created and in existence in the noological system's "mind" as a result of sensory processing and representations.

And we assume that the visual system can also provide the relative distance information, *RD*, between any two material points, whether they belong to the same extended object or not:

$$\begin{aligned}\forall x, y \; Exist(Point(x)) \wedge \; Exist(Point(y)) \\ \rightarrow ExistPar(RD(Point(x), Point(y)))\end{aligned} \quad (6.3)$$

The role of the visual system is to provide information on possible causal agents (inanimate or animate objects) in the environment for the noological system to take advantage of for its problem solving purposes. As the simple example in Chap. 1, Sect. 1.2 shows, if the physical existence of a certain kind of object in the environment can provide alleviation to the agent's hunger state, and the agent has access to information on the parameters (such as location) of the object(s) involved, it can take advantage of this information to quickly solve its possibly urgent problem of hunger alleviation before its death state is reached.

Also, other than the visual system, there are other kinds of sensory systems (such as one that operates with laser or sound waves, in robots or animals such as bats) that can inform the noological system about relative distance information. Other than relative distance information, typically there might also be information on absolute location, relative to some larger reference frame, obtained through the visual or other senses, that is available to a noological system (e.g., Chap. 2, Fig. 2.15).

There is an interesting relationship between absolute location and relative distances. Suppose the absolute location is defined with respect to a visible frame of reference, such as the *Fence* in Fig. 6.2, there exists a unique set of relative distances to the visible frame of reference that would correspond to a particular absolute location. (In general, in 2D space, we need only two relative distances from two points on the reference frame to uniquely define an absolute location, due to the unique nature of the space.) However, it is possible that an absolute location information can be defined with respect to an invisible frame of reference – the agent/organism simply receives information telling it where it is in terms of absolute location (e.g., GPS signals transmitted from afar.). Therefore, we keep both absolute and relative distance parameters separate and assume both can be available at the same time.

6.2 Visual Information Provides Preconditions for Causal Rules

Recall that in Chap. 2, Sect. 2.4, in the discussion concerning effective causal learning, we considered the case of the viability of the gravitational law in various contexts. These contexts form the synchronic preconditions for the diachronic cause-effect (namely the gravitational effect on objects) to take place. For a noological system to be able to make use of this contextual information to formulate cause-effect rules, it has to observe and identify these contextual preconditions. If its sensory system, such as the visual system, could provide such information, it will be advantageous and it should be fully exploited.

Now we consider a simple situation depicted in Fig. 6.4a in which the existence of material points as detected by a visual system provides information for forming the causal preconditions for a certain cause and effect situation. The cause and effect situation in Fig. 6.4a is none other than the commonly encountered situation of a force causing an object to move – we term the rule governing this the *force-movement rule*. In the real world that we inhabit, it is typically the case that the force-movement rule applies ubiquitously in most locations and contexts. Exceptions might include situations in which one is at a location with an obstacle that obstructs the movement of the object on which the force is applied, or in which the object is electrically charged and there are distal electrical fields that exert a counteracting force on the object's movement. In other universes or other physical realities, the rules governing force-movement could be different. In this section, we show how effective causal learning could be applied to learn the rule in the general situation.

We consider an object, a force, and a "$Fence(F_1)$" in Fig. 6.4a. For simplicity, we assume that the object, *Object*(X), is a point object (even though we exaggerate its

6.2 Visual Information Provides Preconditions for Causal Rules

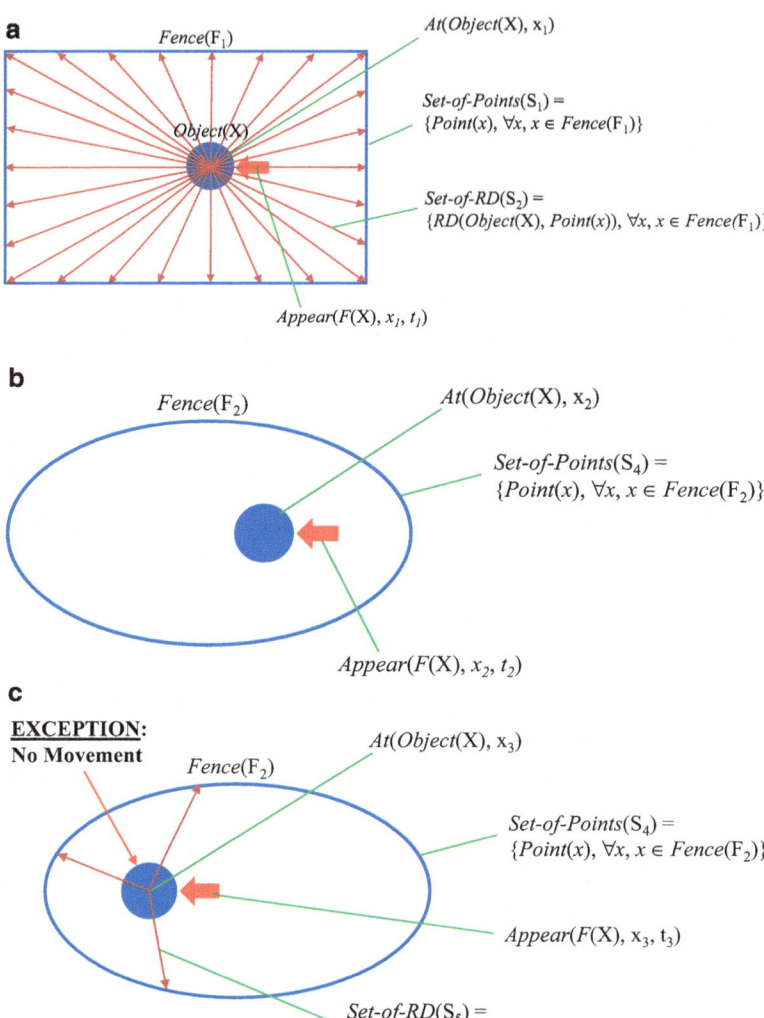

Fig. 6.4 (a) A force $F(X)$ acting on an $Object(X)$ at a certain location x_1 and time t_1, given a certain context – $Fence(F_1)$. (b) A force $F(X)$ acting on the same $Object(X)$ at a different location x_2 and time t_2 and different context – $Fence(F_2)$. (c) At yet another third location, x_3, $Object(X)$ does not move as a result of a force $F(X)$. $Object(X)$ is a point object but its size is exaggerated here

dimension in the figure that makes it look like an extended object), and the relevant parameters describing the state of the object are its absolute location, x, and its state of movement, $Move(Object(X))$.

Figure 6.4a shows the starting situation in which an agent observes an appearance of a force $F(x)$ at a certain absolute location x_1 at time t_1 followed by an elemental movement of the $Object(X)$. (The force could also have been generated by the agent itself). The basic diachronic causal relation is

$Appear(F(X), x_1, t_1) \rightarrow Move(Object(X), t_1 + \Delta)$, which means after force $F(x)$ is applied to the object $Object(X)$ at time t_1, $Object(X)$ will move an elemental time later, at $t_1 + \Delta$. The temporal parameter encodes the fact that the effect of an action (cause) is not instantaneous. (Unlike the full characterization of force and movement in terms of magnitude and directions as will be discussed in Chap. 7, Sect. 7.2, here we ignore the magnitude and direction parameters in order to focus our discussion on the relevant issues to be discussed here.) However, at this time there is a $Fence(F_1)$ of material points that exist and there are hence certain relative distances from $Object(X)$ to the various material points that constitute $Fence(F_1)$. Suppose there is also absolute location information, x, available. The causal rule incorporating the SYN causes at this first encounter would be:

$$At(Object(X), x_1)[\textbf{SYN}] \wedge$$
$$\exists Set\text{-}of\text{-}Points(S_1) = \{Point(x), \forall x\, x \in Fence(F_1)\}\,[\textbf{SYN}] \wedge$$
$$\exists Set\text{-}of\text{-}RD(S_2) = \{RD(Object(X), Point(x)), \forall x\, x \in Fence(F_1)\}\,[\textbf{SYN}] \wedge \quad (6.4)$$
$$Appear(F(X), x_1, t_1)[\textbf{DIA}] \rightarrow Move(Object(X), t_1 + \Delta)$$

Strictly speaking, "*Fence*" should be "*Locations on Fence*" but we short-formed it for convenience. [SYN] and [DIA] are the synchronic and diachronic conditions respectively as discussed in Chap. 2, Sect. 2.4. Also, $\exists Set\text{-}of\text{-}Points\ldots$ is the same as $Exist\,(Set\text{-}of\text{-}Points\ldots)$. In English, this equation says:

$Object(X)$ being at absolute location $x_1\,[\textbf{SYN}] \wedge$
There exists the set of points that makes up $Fence(F_1)\,[\textbf{SYN}] \wedge$
There exists the set of relative distances from $Object(X)$ to all the points on $\quad\quad(6.4')$
$\quad Fence(F_1)\,[\textbf{SYN}] \wedge$
The force $F(X)$ appears at absolute location x_1 at time t_1 [**DIA**]
$\rightarrow$ (implies) $Object(X)$ moves at time $t_1 + \Delta$

There are three main SYN causal conditions that are associated with this rule and they are illustrated in Fig. 6.4a. These are the absolute location, x_1, of the Object (X), the existence of all the material points on $Fence(F_1)$ at their respective locations, $Set\text{-}of\text{-}Points(S_1)$, and the relative distances to all the points on $Fence(F_1)$, $Set\text{-}of\text{-}RD(S_2)$, as stated above in Eq. (6.4) and in Fig. 6.4a. The agent at this stage has to assume that these are all necessary preconditions. The agent needs at least a second experience to determine whether some of these conditions can be relaxed based on, say, dual instance generalization as discussed in Chap. 2, Sects. 2.4 and 2.5. It could be that $Object(X)$ can move only when at a certain absolute location, or when it has a certain combination of relative distances to all the points on the $Fence(F_1)$, or it may not require both these conditions at all.[1]

---

[1] As mentioned above, in principle only two relative distances are needed in a 2D plane to uniquely determine the set of all other relative distances, but this is the unique feature of 2D space. In the interest of generality, we keep all the relative distances for the synchronic condition. Partly, this is also to avoid having to identify any two particular points to characterize the relative distances.

6.2 Visual Information Provides Preconditions for Causal Rules

Suppose the force is applied again (at a later time, t_2) and the object moves elementally again (absolute location is now x_2 and relative distances have changed so now it is $Set\text{-}of\text{-}RD(S_3)$):

$$At(Object(X), x_2)[\mathbf{SYN}] \land \\ \exists Set\text{-}of\text{-}Points(S_1) = \{Point(x), \forall x\, x \in Fence(F_1)\}[\mathbf{SYN}] \land \\ \exists Set\text{-}of\text{-}RD(S_3) = \{RD(Object(X), Point(x)), \forall x\, x \in Fence(F_1)\}[\mathbf{SYN}] \land \\ Appear(F(X), x_2, t_2)[\mathbf{DIA}] \to Move(Object(X), t_2 + \Delta)) \quad (6.5)$$

Since $x_1 \neq x_2$, $S_2 \neq S_3$, and $t_1 \neq t_2$, applying dual instance generalization as described in Chap. 2, Sect. 2.5 (suppose the situation corresponds to "medium desperation", otherwise, more instances may be needed to achieve the same generalization), the rules are generalized to:

$$At(Object(X), x_{ANY})[\mathbf{SYN}] \land \\ \exists Set\text{-}of\text{-}Points(S_1) = \{Point(x), \forall x\, x \in Fence(F_1)\}[\mathbf{SYN}] \land \\ Appear(F(X), x_{ANY}, t_{ANY})[\mathbf{DIA}] \to Move(Object(X), t_{ANY} + \Delta) \quad (6.6)$$

Note that we remove the $Set\text{-}of\text{-}RD$ (...) but keep the "$At(Object(X)...)$" precondition with an "ANY" subscripted for the x parameter. In principle both preconditions can be removed but we keep the "$At(Object(X)...$" part for the purpose of keeping the description of the existence of the object involved somewhere in the rule. Hence, at this stage, the existence of the $Fence(F_1)$ with a combination of points at their corresponding *absolute locations* remains as a SYN condition, but the set of *relative distances* from the object to the points on the $Fence$ (F_1) has been removed as a SYN condition. The only way the $Fence(F_1)$ can be removed as a SYN condition is if another $Fence(F_2)$ constituting a set of points with a different set of locations as shown in Fig. 6.4b is the context of the force rule. Then:

$$At(Object(X), x_{ANY})[\mathbf{SYN}] \land \\ \exists Set\text{-}of\text{-}Points(S_4) = \{Point(x), \forall x\, x \in Fence(F_2)\}[\mathbf{SYN}] \land \\ Appear(F(X), x_{ANY}, t_{ANY})[\mathbf{DIA}] \to Move(Object(X), t_{ANY} + \Delta) \quad (6.7)$$

Since $S_1 \neq S_4$, combining both Eqs. (6.6) and (6.7) we get:

$$At(Object(X), x_{ANY})[\mathbf{SYN}] \land \\ Appear(F(X), x_{ANY}, t_{ANY})[\mathbf{DIA}] \to Move(Object(X), t_{ANY} + \Delta) \quad (6.8)$$

Hence this is the most general force rule that says that you only need a force to move something, and that can take place *anywhere*. It is like the generality of the law of gravity everywhere on Earth in *any context* before the discovery of some contexts within which the law does not apply (Chap. 2, Sect. 2.4). A closer example is the learning of the force law of Fig. 2.7 in the same section.

However, there is a possibility that there is an exception to this rule (much like the case of gravity and force as discussed in Sect. 2.4). Suppose at some specific absolute location with a corresponding set of relative distances to some *Fence*, it is found that the general force law does not work as shown in Fig. 6.4c. At this stage, we can activate a retroactive restoration process (Sect. 2.4) and restore all the conditions that have been removed, much like the case of gravity we discussed in Chap. 2. However, suppose the number of situations involving Rule (6.8) far outnumber those of the exception, instead of retroactively restoring a large number of the rules with SYN conditions (i.e., now instead of x_{ANY} for the argument of the absolute location parameter for Rule 6.8, we need to list all the locations that the rule works except the *one* location that it does not), it is more space efficient to store the rule in an exception form as follows:

$At(Object(X), x_{ANY})[\mathbf{SYN}] \wedge$
$\quad\quad Appear(F(X), x_{ANY}, t_{ANY})[\mathbf{DIA}] \rightarrow Move(Object(X), t_{ANY} + \Delta)$
EXCEPT
$At(Object(X), x_3)[\mathbf{SYN}] \wedge$
$\exists Set\text{-}of\text{-}Points(S_4) = \{Point(x), \forall xx \in Fence(F_2)\}[\mathbf{SYN}] \wedge$
$\exists Set\text{-}of\text{-}RD(S_5) = \{RD(Object(X), Point(x)), \forall xx \in Fence(F_2)\}[\mathbf{SYN}] \wedge$
$\quad\quad Appear(F(X), x_3, t_3)[\mathbf{DIA}] \rightarrow Not(Move(Object(X), t_3 + \Delta))$

(6.9)

6.3 Inductive Competition for Rule Generalization

Now, an interesting situation arises. Suppose now the agent encounters yet another location (say, x_4) in this same $Fence(F_2)$ environment at which the movement is impeded.[2] If we apply dual instance generalization of this instance in conjunction with the EXCEPTION portion of Rule (6.9), we would have a general rule that says that absolute location and shape of the *Fence*(*Set-of-Points*...) or any combination of relative distances to the *Fence*(*Set-of-RD*...) is *not* a necessary SYN condition for the *non-movement* rule. This is clearly in conflict with the original rule such as

---

[2] One may ask the question if the agent's movement had been impeded in the first instance, would it not get stuck there forever and then it will never be able to move to another location to experience the second instance of impediment? There are a few possible scenarios. One is, the agent may be able to dematerialize itself when it encounters the non-movement and re-materialize itself somewhere else. Another is, the agent can "hop" out of the 2-D plane in the third dimension and land in another spot of the environment. Yet another is, the movement impediment is only in certain directions so the agent can "back-up" and continue its exploration – but we have stated in the beginning of this thread of discussion that we will ignore the consideration of direction to focus the discussion on relevant issues.

6.3 Inductive Competition for Rule Generalization

Rule (6.8) that states the precise opposite (which states that *Set-of-Points*... or *Set-of-RD*... is not a necessary SYN condition for the *movement* rule).

The conflict arises as follows. Firstly, the two rules that give rise to the non-movement are coded as follows:

$At(Object(X), x_3)[\textbf{SYN}] \wedge$
$\exists Set\text{-}of\text{-}Points(S_4) = \{Point(x), \forall x \, x \in Fence(F_2)\}[\textbf{SYN}] \wedge$
$\exists Set\text{-}of\text{-}RD(S_5) = \{RD(Object(X), Point(x)), \forall x \, x \in Fence(F_2)\}[\textbf{SYN}] \wedge$
$\qquad Appear(F(X), x_3, t_3)[\textbf{DIA}] \rightarrow Not(Move(Object(X), t_3 + \Delta))$
OR
$At(Object(X), x_4)[\textbf{SYN}] \wedge$
$\exists Set\text{-}of\text{-}Points(S_4) = \{Point(x), \forall x \, x \in Fence(F_2)\}[\textbf{SYN}] \wedge$
$\exists Set\text{-}of\text{-}RD(S_6) = \{RD(Object(X), Point(x)), \forall x \, x \in Fence(F_2)\}[\textbf{SYN}] \wedge$
$\qquad Appear(F(X), x_4, t_4)[\textbf{DIA}] \rightarrow Not(Move(Object(X), t_4 + \Delta))$

(6.10)

Then, applying our dual instance generalization mechanisms and removing the different pre-conditions, we obtain:

$At(Object(X), x_{ANY})[\textbf{SYN}] \wedge$
$\exists Set\text{-}of\text{-}Points(S_4) = \{Point(x), \forall x \, x \in Fence(F_2)\}[\textbf{SYN}] \wedge$
$\qquad Appear(F(X), x_{ANY}, t_{ANY})[\textbf{DIA}] \rightarrow Not(Move(Object(X), t_{ANY} + \Delta))$

(6.11)

Even though this rule has an additional condition "*Set-of-Points*$(S_4) = \{Point(x), \forall x, x \in Fence(F_2)\}$ [SYN]" compared to Rule (6.8), they are still in conflict because Rule (6.8) states that no matter where you are in the environment or what kind of *Fence* you are surrounded by, you are always able to move the object when you apply a force to it. Moreover, suppose there is a third situation in which the agent is at yet another different absolute location and is surrounded by, say, the "rectangular fence" – *Fence*(F_1) of Fig. 6.4a – and the non-movement situation is still present, then a further generalization will result in the rule:

$At(Object(X), x_{ANY})[\textbf{SYN}] \wedge$
$\qquad Appear(F(X), x_{ANY}, t_{ANY})[\textbf{DIA}] \rightarrow Not(Move(Object(X), t_{ANY} + \Delta))$

(6.12)

Which is clearly in conflict with Rule (6.8) because both (6.8) and (6.12) share the same preconditions but have totally different consequences.

So, should the agent generalize (completely, removing all *Set-of-Points* and *Set-of-RD* preconditions) on the movement rule or the non-movement rule and make the other rule an exception? It will be instructive now to review the purpose of dual instance generalization as spelled out in Chap. 2, Sect. 2.5. Dual instance generalization is an inductive process and we have mentioned in Sect. 2.5 that the number of instances one needs before one "tentatively" confirms a generalized rule and applies it depends on a measure called "desperation." This is a measure of how desperate the agent is in applying the rule. If the agent is not desperate, it can wait for many more instances to confirm the rule. Or, if the application of the rule results in disastrous consequences, it will also wait for more instances before confirming the rule. Now, in the beginning, when the agent has first experienced the movement rule, at a level of moderate desperation, it generalizes after two instances. As the number of instances grows, the agent becomes more confident about the rule. So, by the time that the agent encounters the non-movement rule, since it has already gone way beyond more than two instances for the movement rule, the situation provides a strong competition for inductive generalization. In other words, the system would keep the generalization for the *movement* rule and the two rules corresponding to the two locations at which it encounters *non-movement* despite the application of a force are coded as special case disjunctive conditions as follows:

$$At(Object(X), x_{ANY})[\mathbf{SYN}] \wedge$$
$$\qquad Appear(F(X), x_{ANY}, t_{ANY})[\mathbf{DIA}] \rightarrow (Move(Object(X), t_{ANY} + \Delta))$$
$$\underline{\text{EXCEPT}}$$
$$At(Object(X), x_3)[\mathbf{SYN}] \wedge$$
$$\qquad Appear(F(X), x_3, t_3)[\mathbf{DIA}] \rightarrow Not(Move(Object(X), t_3 + \Delta))$$
$$\underline{\text{OR}}$$
$$At(Object(X), x_4)[\mathbf{SYN}] \wedge$$
$$\qquad Appear(F(X), x_4, t_4)[\mathbf{DIA}] \rightarrow Not(Move(Object(X), t_4 + \Delta))$$
$$(6.13)$$

Despite the fact that Rule (6.13) is not as general compared to Rule (6.8), it still has some good predictive value. If the agent is planning to take actions to move objects in various absolution locations in the environment at which it has no prior experience with in terms of the force-movement situation, and if it is confident about Rule (6.8), it can plan for that to be possible anywhere. If it is confident about Rule (6.13), it can still plan for that to be possible almost anywhere except for a small number of locations. The problem solving abilities of an agent is dependent on the ability to apply these rules in as general a way as possible.

As the agent moves around further and encounters more and more of these specialized rules of non-movement, the length of the disjunction will grow. Unlike the most general Rule (6.8) or relative general Rule (6.13), as the number of locations increases at which the movement situation does not apply, a rule will emerge in which the number of non-movement locations begins to approach the number of locations at which movement is possible on the application of a force on an object. Unless there are some predictable patterns in the locations at which these rules are applicable (e.g., left vs right side of the environment), these rules then become less useful in allowing the agent to predict and plan for what may happen in various locations in the environment and will negatively impact upon its problem solving abilities.

6.3 Inductive Competition for Rule Generalization

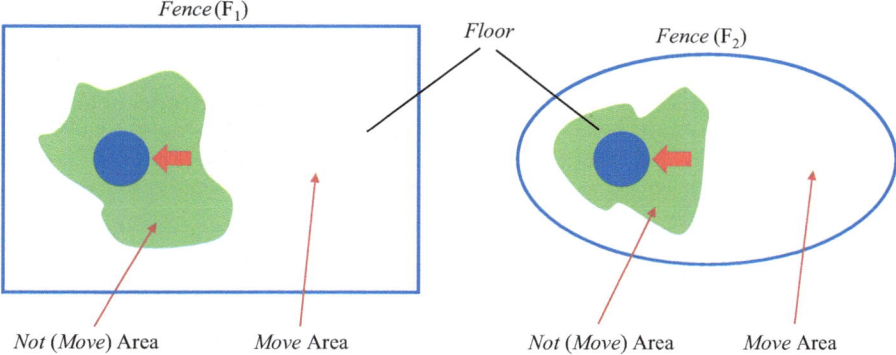

Fig. 6.5 Environments with floor colors distinguishing *Move* and *Not(Move)* areas which correspond to *WHITE* and *GREEN* areas respectively

However, if there are other extra visual information/cue in the environment, the agent can potentially encode more general rules using this information. Figure 6.5 shows an example situation in which there is an additional type of visual information – the color of the "floor" on which the agent rests. This is encoded as *Color* (*Floor(x)*, *y*) which means the color of the floor at absolute location x is y. Also, suppose the physical properties are such that an object can be moved when a force is applied to it on the WHITE area but not on the GREEN area as shown in Fig. 6.5.

Now, suppose there are two instances at different absolute locations in which *Object*(X) is moved successfully by a force in the WHITE area and two instances at different absolute locations in the GREEN area in which *Object*(X) could not be moved despite the application of a force, and these four situations take place in both the *Fence*(F_1) as well as the *Fence*(F_2) environment as shown in Fig. 6.5. Going through the same dual instance generalization process as that described above, the two general rules that would be encoded would be as follows:

$$At(Object(X), x_{ANY}))[\textbf{SYN}] \wedge Color(Floor(x_{ANY}), \text{WHITE})[\textbf{SYN}]$$
$$Appear(F(X), x_{ANY}, t_{ANY})[\textbf{DIA}] \rightarrow Move(Object(X), t_{ANY} + \Delta) \quad (6.14)$$

$$At(Object(X), x_{ANY}))[\textbf{SYN}] \wedge Color(Floor(x_{ANY}), \text{GREEN})[\textbf{SYN}]$$
$$Appear(F(X), x_{ANY}, t_{ANY})[\textbf{DIA}] \rightarrow Not(Move(Object(X), t_{ANY} + \Delta)) \quad (6.15)$$

Note that the second and third x_{ANY} in each of the rules is the "same" x_{ANY} as the first one – i.e., they can be of *any* value, but they are of the *same* value – just like the second t_{ANY} is the same as the first t_{ANY}. Rule (6.15) is not in conflict with Rule (6.14) as they have distinct preconditions. Therefore, unlike in the case of Rule (6.13), in which because there was no visual cue to distinguish between the *Move* and *Not(Move)* situations, the *Not(Move)* situation has to be encoded as a special case, here the *Not(Move)* situation is encoded as a general rule with a specific precondition – the color of the floor on which *Not(Move)*-ment takes place is GREEN.

With the two general rules (6.14) and (6.15), the agent will be better able to plan its actions – it can better predict where an object is expected to be movable by a force and where it is not, thus enhancing its problem solving capabilities. Thus these two rules have better utility than Rule (6.13).

In the real-world physical environment, it is often the case that different friction on the ground will give rise to different resistance to movement given the application of a force on an object (e.g., if you are on a wet surface you will move differently than on a dry surface). And these situations may be distinguished by visual cues. Causal learning and causal rule encoding processes such as that of the above could allow knowledge such as this to be learned and encoded.

6.4 Meta-level Inductive Heuristic Generalization

Note that there are other synchronic preconditions that have been considered and removed in the process of arriving at Rules (6.14) and (6.15) in addition to those involving the *Fence* in the earlier considerations. When the floor acquires a visual property of color, one can think of it as having a material existence like the *Fence*. The elemental bits of the floor would play the same role as the elemental bits of the *Fence* earlier, presenting relative distances and existential context for the movement and non-movement rules. So the two patches of WHITE and GREEN floor provide similar *Set-of-Points* and *Set-of-RD* synchronic conditions:

$$Set\text{-}of\text{-}Points(S_7) = \\ \{Point(x), \forall x\, x \in Floor(x) \text{ s.t. } Color(Floor(x), \text{WHITE})\}$$

and

$$Set\text{-}of\text{-}RD(S_8) = \\ \{RD(Object(X), Point(x)), \forall x\, x \in Floor(x) \text{ s.t. } Color(Floor(x), \text{WHITE})\}$$

for the White area of the floor and:

$$Set\text{-}of\text{-}Points(S_9) = \\ \{Point(x), \forall x\, x \in Floor(x) \text{ s.t. } Color(Floor(x), \text{GREEN})\}$$

and

$$Set\text{-}of\text{-}RD(S_{10}) = \\ \{RD(Object(X), Point(x)), \forall x\, x \in Floor(x) \text{ s.t. } Color(Floor(x), \text{GREEN})\}$$

6.4 Meta-level Inductive Heuristic Generalization

for the GREEN area of the floor. There is a set each of these four conditions for the *Fence*(F1) and *Fence*(F2) environments. However, in a similar process as the elimination of the corresponding conditions for the *Fence*, these conditions are also removed in the generalization process (i.e., based on $S_7 \neq S_9$, $S_8 \neq S_{10}$).

In the discussion so far including the earlier derivation of the general force-movement rules such as Rules (6.8), (6.14), and (6.15), it has often been the case that the *Set-of-Points* (surrounding physical context) and the *Set-of-RD* (relative distances to surrounding physical context) conditions are removed in the process of generalization. In the final formulation of the *Move* vs *Not*(*Move*) rules of (6.14) and (6.15), it is the *Color* condition that is the distinguishing condition rather than the *Set-of-Points* or *Set-of-RD*. It is typically true in our physical reality that for most of the physical phenomena that we observe, whether it is a simple force causing a translational movement or a rotational movement, gravity acting on objects, the force involved is large or small, or the objects involved are heavy, light, or of any particular shape, the surrounding context, as well as the relative distances from the point of physical interaction to the surrounding context do not participate in the synchronic preconditions of the causal rules involved. There are exceptions, of course, such as when the surrounding context is electrically or magnetically charged and the objects involved are likewise charged as shown in Fig. 6.6.

In the case of Fig. 6.6, the object – *Object*(X) – involved and the *Fence* contain positive electrical charges as shown. If the law of electricity is such as that obtains in our current reality, then there is a repulsive force with a magnitude that obeys an inverse square law of distance between every elemental positive charge on the *Fence* on the one hand, and *Object*(X) on the other, and the total effect on *Object*(X) is the sum of all the forces from all the elemental charges on the *Fence*. The resultant magnitude and direction of the force on *Object*(X) is dependent on the location of *Object*(X) in a complex manner and without a scientific understanding of the inverse square law and the ability to carry out mathematical integration to

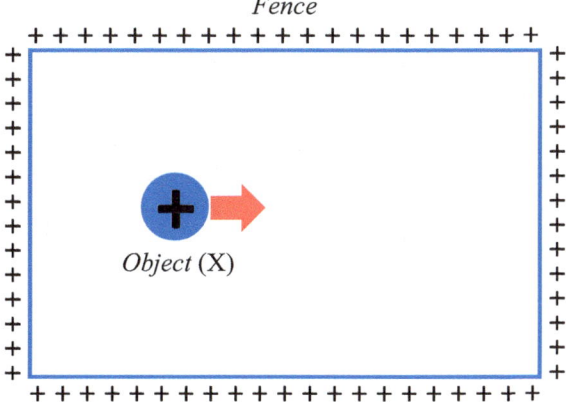

Fig. 6.6 A situation with an "electrically charged" fence around an electrically charged object

derive a general and quantitative equation to describe the force at every location, an agent can only rely on its direct experience to encode the relation between location and force experienced. In this situation, what the agent would discover for the synchronic conditions that affect the magnitude and direction of the force on *Object* (X) would be as follows:

- The absolute location (relative to a larger frame of reference within which the *Fence* can be positioned) of *Object*(X) is not a necessary precondition for the force on it.
- The *Set-of-RD* from *Object*(X) to all the points on the *Fence* is a necessary precondition for the force on it.
- The *Set-of-Points* – i.e., the shape of the *Fence* – is a necessary precondition.
- For a particularly shaped *Fence*, other than some symmetrical locations relative to the *Fence*, the forces at all the locations are different and specialized rules that are functions of *Set-of-RD* are needed to describe them.

Now, suppose the agent, in its earlier experiences with the environment, did not encounter electrically or magnetically charged objects and so in most of its experiences, *Set-of-Points* of any surrounding objects/contexts and *Set-of-RD* to any surrounding objects/contexts do not feature in the physical phenomena that it encountered, then a set of meta-level heuristic rules could be formulated as such:

$$\forall x, y \; Generally(Rule(x) \wedge \textit{Rule-For}(x, object(y))$$
$$\wedge \textit{Property}(Object(y), \text{PHYSICAL}) \wedge \textit{Rule-Type}(x, \text{PHYSICAL})$$
$$\rightarrow \textit{Not-Important}(Parameter(\textit{Set-of-Points}), SYN\textit{-Precondition}(Rule(x)))) \tag{6.16}$$

$$\forall x, y \; Generally(Rule(x) \wedge \textit{Rule-For}(x, object(y))$$
$$\wedge \textit{Property}(Object(y), \text{PHYSICAL}) \wedge \textit{Rule-Type}(x, \text{PHYSICAL})$$
$$\rightarrow \textit{Not-Important}(Parameter(\textit{Set-of-RD}), SYN\textit{-Precondition}(Rule(x)))) \tag{6.17}$$

In English, these say:
Generally it is the case that:

<u>If</u>
x is a *Rule*
And *x* is a *Rule-For Object*(y)
And *Object*(y) is a PHYSICAL object
And *x* is of PHYSICAL *Rule-Type*
<u>Then</u>
The *Set-of-Points Parameter* is *Not-Important*
    as a *SYNchronic-Precondition* for *Rule*(x)

(6.16′)

6.4 Meta-level Inductive Heuristic Generalization

A similar rule is obtained for the Parameter *Set-of-RD*.[3] Before the agent encounters electrically or magnetically charged objects, these heuristic rules can help to expedite the formulation of various physical causal rules. We should think of these heuristics as "Meta-Rules" – rules that can help to direct the formulation of the "base-level" causal rules (e.g., Rules 6.14 and 6.15). There could also be degrees of confidence associated with these Meta-Rules as a result of experience such as the degrees of confidence associated with the base-level causal rules as discussed in association with Fig. 2.4 in Chap. 2, Sect. 2.3. The predicate *Generally* reflects a level of confidence with the Meta-Rule involved.

Consider how these heuristic rules may be used. For example, in our earlier discussion such as that pertaining to Fig. 6.4, we could remove *Set-of-Points* as a synchronic precondition only after observing the causal phenomenon in two separate environments with differently shaped *Fences*. If there is a certain level of desperation in getting the rule formulated and used, then Meta-Rules (6.16) and (6.17) can be used to direct the formulation of the general causal rule – Rule (6.8) – involved without having to experience the second environment – i.e., the oval shaped *Fence* – $Fence(F_2)$ – in Fig. 6.4b. Having the first experience with the rectangular *Fence*, $Fence(F_1)$, is sufficient.

Now, later, suppose electrically charged objects and fences are encountered such as that depicted in Fig. 6.6, the following specific heuristic rules can be formulated:

$\forall x, y \; Generally(Rule(x) \land Rule\text{-}For(x, object(y)))$
$\quad \land Property(Object(y), \text{PHYSICAL}) \land Property(object(y), \text{CHARGED})$
$\quad \land Property(\text{Environment}, \text{CHARGED})$
$\quad \land Rule\text{-}Type(x, \text{PHYSICAL}))$
$\quad \rightarrow Important(Parameter(Set\text{-}of\text{-}Points), SYN\text{-}Precondition(Rule(x))))$

(6.18)

$\forall x, y \; Generally(Rule(x) \land Rule\text{-}For(x, object(y)))$
$\quad \land Property(Object(y), \text{PHYSICAL}) \land Property(object(y), \text{CHARGED})$
$\quad \land Property(\text{Environment}, \text{CHARGED})$
$\quad \land Rule\text{-}Type(x, \text{PHYSICAL}))$
$\quad \rightarrow Important(Parameter(Set\text{-}of\text{-}RD), SYN\text{-}Precondition(Rule(x))))$

(6.19)

In Chap. 2, Sect. 2.6.1 we discussed the learning of a simple heuristic. The learning of these more complicated meta-level heuristic rules will be discussed in Chap. 7.

---

[3] An example of a non-physical type rule applicable to a physical object could be "If you believe the object is valuable, you will sell it at a high price at the market" – this is the situation in which the object involved is physical, but the action taken on it is not and the context in which the rule is applicable matters.

6.5 Application to the Spatial Movement to Goal with Obstacle (SMGO) Problem

The SMG (spatial movement to goal) problem is one of the most basic problems encountered by a noological agent in its process of interacting with the environment for the purpose of satisfying its needs. In Sect. 2.6.1 we considered the simple situation in which there are no obstacles between the starting location and the goal location of the movement process. We used that as a platform to establish the basic principles behind effective causal learning augmented search. In this section we consider the more general situation in which there are obstacles in the form of a "*Wall*" between the start and goal locations as shown in Fig. 6.7. The dotted line in Fig. 6.7 shows that if the *Wall* were not there as an obstacle, the *Agent*, or an *Object* moved by a force or an *Agent*, would have moved straight to the GOAL from the START location. We call this spatial movement to goal with obstacle problem (SMGO). This is a scenario often used in AI research in the areas of problem solving search (Russell and Norvig 2010) and reinforcement learning (Sutton and Barto 1998). In Fig. 6.7 we have exaggerated the size of the *Agent/Object* involved as an extended object for the sake of visual clarity but in the following discussion it is meant to be a point object. Some extensions are needed to handle situations involving extended objects and these will be discussed later in Chap. 7.

A typical general method applied to solve the SMGO problem in traditional AI is the same as that used for the SMG problem – the A* algorithm (Hart et al. 1968). This algorithm has been proven to be able to discover the optimal path for any problem. However, as in the earlier example of using A* to tackle the SMG problem discussed in Chap. 2, Sect. 2.6.1, the search process does not keep track of and learn certain causal information that is very useful in drastically cutting down the amount of search needed for the problem solving process. The A* search when applied to a SMGO problem typically expands a large number of nodes from the beginning to the end of search, much like in the case for the SMG problem. As laid out as a basic principle of noological learning in Chap. 2, Sect. 2.7 and Chap. 3,

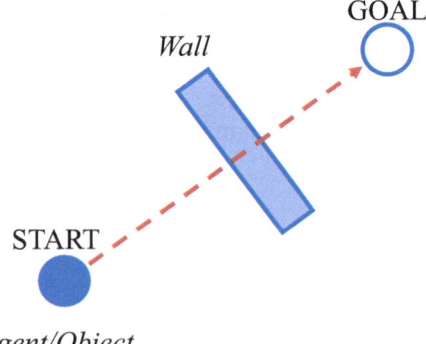

Fig. 6.7 The spatial movement to goal with obstacle (SMGO) problem. The *dotted line* is the path used if the *Wall* were not there. Size of the *Agent/Object* is exaggerated

6.5 Application to the Spatial Movement to Goal with Obstacle (SMGO) Problem

in any problem solving situation, a small degree of blind search/exploration is necessary, and then the system must quickly exploit causal information discovered in this process of blind search/exploration and encode powerful and general causal rules that will obviate or minimize any further blind search.

What we expect a noological system, whether natural or artificial, to be able to do in this obstacle situation is to use its sensory information as well as causal learning abilities to quickly solve this obstacle problem "intelligently." Figure 6.8a shows that even if this is the first time an *Agent* encounters the *Wall* or whatever object and finds it "impenetrable" on its way to the GOAL, it should be able to quickly understand that the *Wall*, as visually identified, is the *cause* of its inability to get to the other side, and therefore it should head toward an area where the *Wall* does not exist. A small amount of trial and error may be needed at the location of the first encounter with the *Wall* but it should discover some general rules quickly. And, among the many possible paths of movement including those that lead "backward"/ away from the *Wall* initially, it should be able to determine that the best and minimal distance to the GOAL path is to first head straight to one of the "corners" of the *Wall*, go around the corner, and subsequently head straight toward the GOAL, as indicated by the dark red arrows in Fig. 6.8a. Having learned this in a first encounter with the problem, in subsequent encounters if the noological system were to start from the same START position, it should follow the path indicated by the blue arrow – head straight to one of the corners of the *Wall* from START. If the problem solving search is directed by an A* algorithm (A* Search Algorithm, Wikipedia 2015), there will be a lot of trial-and-error move exploration (shown as the green area in Fig. 6.8a).

Figure 6.8b, c show situations in which the visual cue to the presence of an obstacle is absent, and the *Agent* would have to resort to more trial and error explorations. Suppose the obstacle is in the form of a totally *Transparent Wall*, in the sense that there is no visual cue to indicate its existence or shape – this can happen if the reflective index of the *Wall* is the same as that of the surrounding medium so that it does not reflect any light to indicate its existence, or the obstacle is some kind of "*Force Field*" (as often portrayed in science fiction movies). The *Agent* will then not be able to use any earlier experiences to predict where else it might encounter impenetrability and plan its path accordingly, without having to keep testing for impenetrability, unlike in Fig. 6.8a. In the situation of Fig. 6.8b, because the sides of the *Transparent Wall* are straight, an Agent may use an inductive rule to predict to some extent the locations at which it may encounter the impediment to its movement toward the GOAL, and it may be able to skip some of the testing actions as it moves along the *Wall* toward a possible turning point. However, because there is no visual cue to the corner of the *Transparent Wall*, the Agent would still have to keep testing for impenetrability all the way to a corner.

For the case of Fig. 6.8c, in which the *Transparent Wall* is totally irregular in shape, an inductive prediction of where the parts of the *Wall* may be situated would be difficult, and the *Agent* would have to keep feeling its way along the *Wall* to reach a turning point toward the GOAL. Of course, having learned the shape of these obstacles through one round of laborious exploration, the *Agent* would be able

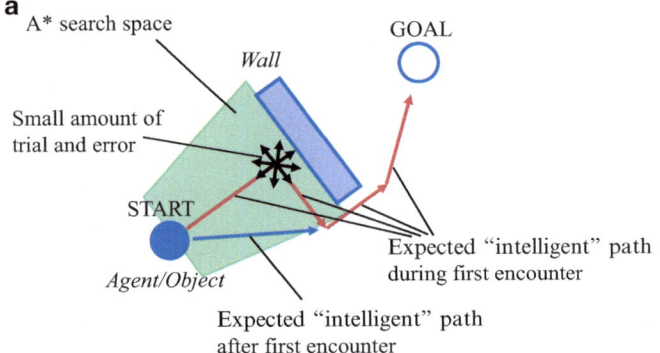

Fig. 6.8 (**a**) Expected "intelligent" solution to the SMGO problem. When the problem is encountered the first time, the expected path is the *dark red arrows*' path. Having learned the general idea of impediment of the *Wall*, the *blue arrow's* path will be taken afterward. *Green* area is what an A* algorithm will have to search through. (**b**), (**c**) If the obstacle is invisible, such as a force field, there is a need to feel about along the "*Wall*" (the *black arrows*) as the extent of the invisible force field is not known or the force field could have some arbitrary invisible shapes

to head straight for a corner or a turning point the next time. (This is provided that these obstacles do not "disappear" or change shapes and locations.) When we contrast the situations in Fig. 6.8b, c with the situation in Fig. 6.8a, we can appreciate the causal role of visual information – that it can provide valuable information for generalization and hence help an agent to reduce problem solving search greatly, and also help it to better plan actions in other localities at which it has no direct prior experience. This situation is similar to the floor-color situation discussed in Sect. 6.3 and Fig. 6.5 in which the extra visual cue of floor color allows the formulation and encoding of more general rules that help the agent to better plan for actions at places in the environment where it does not have direct experience.

In the SMGO problem, we assume a super-visual system is available to the *Agent* (Sect. 6.1), so that even though the GOAL may not be directly visible from certain locations given a normal visual system, in this case the *Agent* knows where the

6.5 Application to the Spatial Movement to Goal with Obstacle (SMGO) Problem   241

GOAL is throughout the problem solving process. We also assume that the GOAL is stationary.

In the ensuing discussion, we will describe in detail the causal learning processes that allow the situations of Fig. 6.8a to be handled in a noologically satisfying manner as we have specified above – i.e., based on a small number of experiences and based on the availability of visual cues, the *Agent* encodes the causal nature of the obstacles involved and then computes a reasonably optimal path of actions to reach the GOAL from the START location. There are a few extra causal learning devices that are not necessary for the SMG process in Chap. 2 but are necessary to tackle the SMGO problem, and these will be described. However, these extra devices are general and not just applicable to the SMGO problem.

6.5.1 Script and Thwarting of Script

In Chap. 2, Sect. 2.6.2 we mentioned the use of a knowledge structure called *script* to encapsulate a sequence of actions leading to a certain consequence/outcome. We provided the example of a MOVEMENT-SCRIPT (Fig. 2.18) consisting of specifications for the START state, the ACTIONS, and the OUTCOME related to a movement activity. The script is learned and encoded based on earlier instances in which the agent had an experience of solving the basic SMG problem and finding and executing a solution as shown in Figs. 2.15 and 2.16. (Another source of script encoding comes from observing other agents or objects executing activities – see Fig. 2.19 – and this will be discussed in more detail in Chap. 7.) This encoded script is a chunked piece of knowledge that can be used in future problem solving situations and can also be used to predict expected courses of actions emitted by agents or objects. (The "incremental chunking" method for knowledge acquisition and problem solving was discussed in Chap. 3, Sect. 3.5.)

The MOVEMENT-SCRIPT of Fig. 2.18 is a *general* movement script that specifies how to get from a START location to a GOAL location for *any* START and GOAL locations. (Basically it says to keep moving in a direction pointing toward the GOAL.) This can be used in the situation in which an agent is applying a force on an object to move it, or an agent is generating a force internally to move itself. In the subsequent discussion, we will discuss just the situation of the movement of an *Object*(X) by an external force/agent and the same discussion/solution will apply to both situations.

Consider the situation of Fig. 6.9a which is the typical SMG problem except that we have included a "*Wall*" in dotted lines to contrast it with the situation in Fig. 6.9b. The problem specification (an "OUTCOME" specification) requires *Object*(X) be moved from the START location to a GOAL location which implies that there is a change in its absolute location (i.e., $+\Delta AL$, using the notation of Fig. 2.16), and its relative location (RD) to the GOAL should change from some

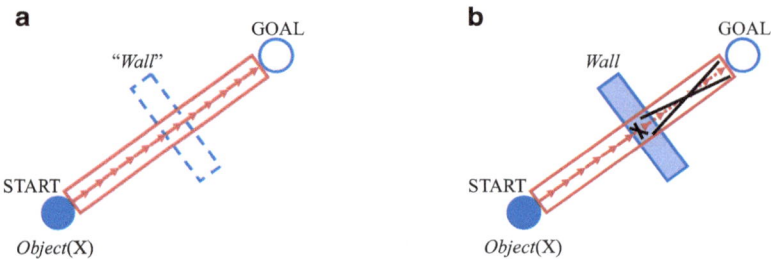

Fig. 6.9 (**a**) A typical SMG problem and its solution – a *"Wall"* in *dotted line* is shown for the purpose of contrasting it with (**b**). (**b**) An SMGO problem with a thwarted solution

value to zero. This PROBLEM SPECIFICATION ($+\Delta$AL, RD $= 0$) is used to query the Script Base of a noological system by matching the OUTCOME portion of the scripts with the PROBLEM SPECIFICATION, and the general MOVEMENT-SCRIPT of Fig. 2.18 is retrieved. Figure 6.10 illustrates this process (also see Chap. 2, Sect. 2.6.2, Fig. 2.21 and associated detailed discussion).

If the START state of the script retrieved matches the current START state of the object involved, the ACTIONS portion of the script is then the solution to the problem. If the START state does not match the current START state of the object involved, then a backward chaining process is activated to look for another script (if available) or execute a problem solving process to transform the current START state of the object to the START state of the retrieved script (Chap. 2, Sect. 2.6.2). In the case of the situation in Fig. 6.9a, the MOVEMENT-SCRIPT is successfully matched and executed as the solution to the problem, and this is shown as a dark red rectangle containing a sequence of prescribed elemental movements of *Object* (X) from the START to the GOAL locations.

Now, suppose *Object*(X) is in the situation as depicted in Fig. 6.9b in which everything else is the same as in Fig. 6.9a except for the obstacle in the form of a *"Wall."* And suppose now the agent is given a similar problem – to move *Object* (X) to reach the GOAL or to reduce the relative distance of *Object*(X) to the GOAL to zero. In searching through the Script Base for a solution, the best match is still the MOVEMENT-SCRIPT of Fig. 2.18. The reason why this is the "best" match but not the perfect match is that there is an obstacle in between the START and GOAL locations. But, before having learned a good way to handle the SMGO, there is no script currently in the Script Base that provides a perfect-match script solution to the SMGO problem. Therefore, it retrieves the MOVEMENT-SCRIPT of Fig. 2.18 to solve the problem and the script then prescribes a sequence of actions. (A corresponding situation in the real world may be a situation in which a person is experiencing a glass wall the first time. She is able to see the glass and is able to see the goal location she wishes to get to beyond the glass, and her best first try

6.5 Application to the Spatial Movement to Goal with Obstacle (SMGO) Problem

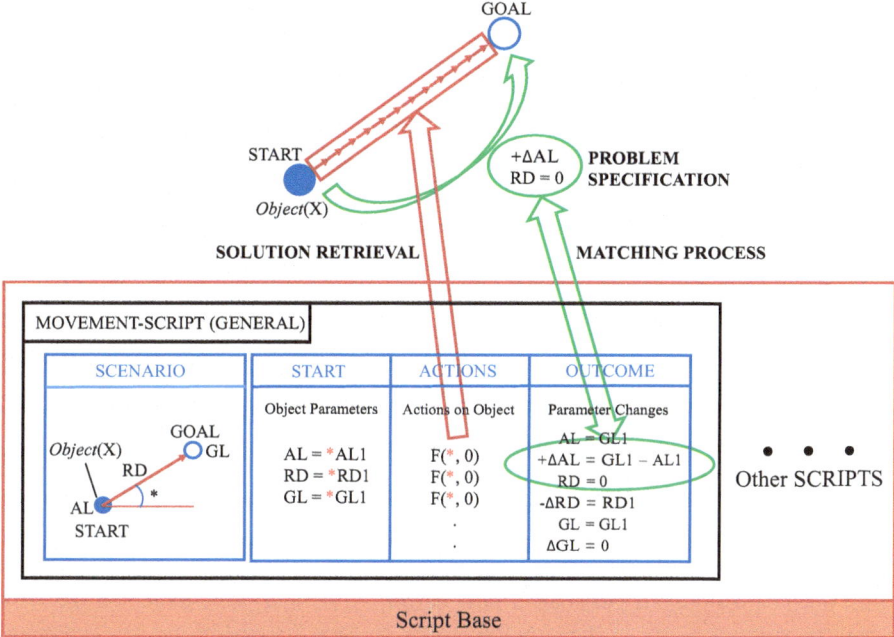

Fig. 6.10 Script matching and solution retrieval in a backward chaining process. (Similar to Fig. 2.21)

would be to head straight toward the goal, as she normally would do, and then in the process she discovers that the glass is an impenetrable obstacle.)

In the process of executing the solution, before reaching the *Wall* object, the agent applies the usual force in the goal direction and at each step manages to effect an elemental movement of *Object*(X) toward the goal. This further confirms the generality of the force rule learned earlier, much like in the situation discussed in Sect. 6.2 (Fig. 6.4) in which the existence of and relative distances to other objects in the environment do not figure in the synchronic causal conditions of the "Force Rule" (6.8). Now, at some point the agent finds that despite the fact that it applies a force to *Object*(X) toward the goal, it does not lead to an elemental movement of *Object*(X) toward the goal, as shown in Fig. 6.9b with a small "black cross," which is not what is expected from the usual Force Rule (6.8). This is when *Object* (X) "touches" the *Wall*. As a result, the rest of the script is thwarted, and all subsequent actions cannot take place, which is shown with a big "black cross" in the figure. In our daily commonsensical environment, we know that this is due to the fact that the *Wall* has the impenetrability property that halts the movement of the agent. However, at this stage, the agent has just encountered this concept and has not fully learned its most general form yet.

6.5.2 Recovery from Script Thwarting Through Causal Reasoning

When an agent encounters an impediment (a thwarting of its plan), it takes one of two actions to attempt to overcome the impediment to fulfill its plan – the *Thwart-Handling Process* depicted in Fig. 6.11. One is, it searches the Script Base (Sect. 2.6.2, Fig. 2.21 or Fig. 6.10) to see if there is an alternative plan it can use to restore its plan and satisfy its current PROBLEM SPECIFICATION (Step ③' of Fig. 6.11) – the Script Base may contain a script that matches the current impediment situation exactly or closely and provide a counter-thwarting plan. (In Fig. 6.11, Step ① is Script Execution and Step ② is testing for Script Thwarting.) In fact, had a similar situation been encountered and solved before, the Script Base would contain just such a counter-thwarting script. The other is, it carries out causal reasoning to identify the cause(s) of the thwarting of the plan (such as discussed above in connection with Fig. 6.8a), as well as exploratory actions (typically a heuristics-guided search process) to discover new scripts that can overcome the impediment (Step ③" of Fig. 6.11 – *Counter-Thwarting Script Discover Process*). In this process, it consults the Script Base to see if there are other scripts that can assist with the re-planning to generate a new script. After that, it re-executes the new script to attempt to solve the problem (back to Step ① of Fig. 6.11). These are general methods of script matching and retrieval, causal reasoning, and exploration applied to overcome the thwarting of a plan and are hence not restricted to the application to the current SMGO problem only.

Let us suppose it is the first time this thwarting situation is encountered and no alternative plans are available in the Script Base to overcome the impediment. The location of *Object*(X) is represented by its center, and at this point of impediment it is x_{11} as shown in Fig. 6.12. We will also call this the *Original Position* (of impediment) of *Object*(X). The point of contact between *Object*(X) and the *Wall* is $x_{11} + \delta$.[4] Before taking any random exploratory actions, the agent first carries out causal reasoning (Step③" in Fig. 6.11) to identify the cause(s) of the thwarting of the script.

Step③" in Fig. 6.11 actually consists of a number of sub-steps. We will discuss these with respect to the scenario of the SMGO problem and then present a general algorithm that is applicable to other problems later.

Based on visual information, because the *Wall* consists of many material points, there are a number of possible synchronic causes. Similar to the situation depicted in Fig. 6.4c and Rule (6.9), the agent now encodes a rule that says that at a certain absolute location (x_{11}) of *Object*(X), and at a certain combination of distances from all the points on the *Wall*, the force rule does not work and the script is thwarted:

---

[4] Here, δ is a vector with "x, y" components that adds to the "x, y" coordinates of x_{11} to reach the point of contact on the *Wall* – remember we had earlier mentioned that we use x to stand for the spatial coordinates in a 2D space, [x, y], to avoid clutter in the equations/rules. If *Object*(X) is a point object, δ becomes 0.

6.5 Application to the Spatial Movement to Goal with Obstacle (SMGO) Problem 245

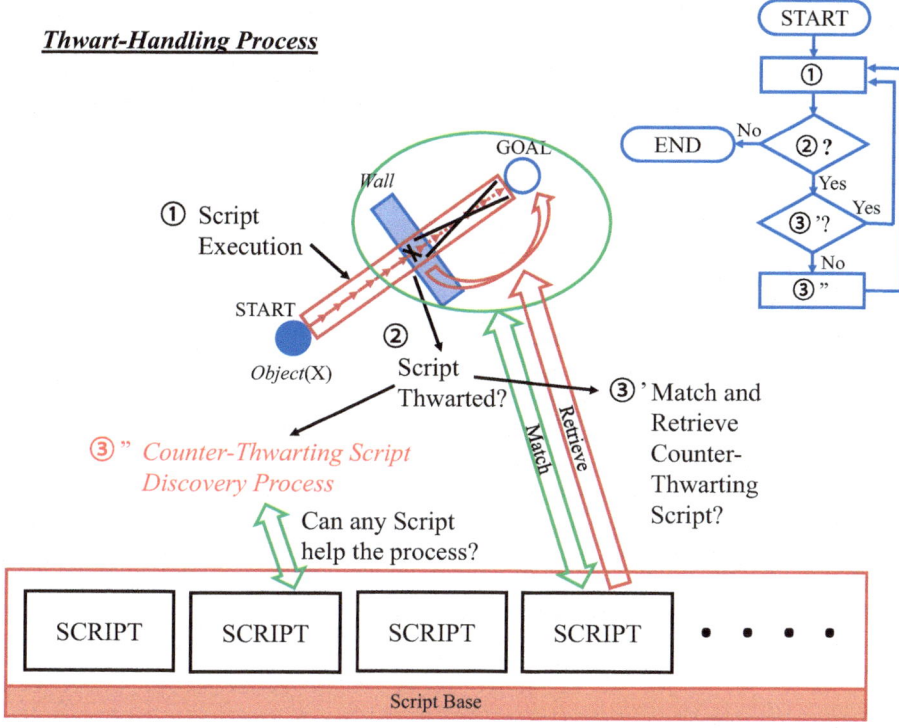

Fig. 6.11 The *Thwart-Handling Process* to overcome the thwarting of a script

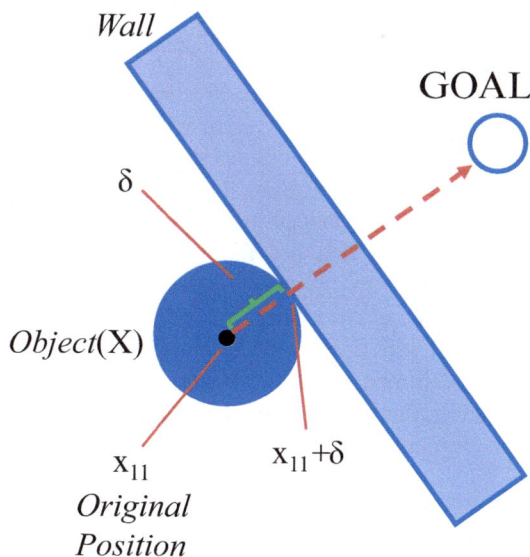

Fig. 6.12 The location of *Object*(X) at the first point of impediment as it heads toward the GOAL. This position is called the Original Position (of impediment) and it is represented by the center of *Object*(X), x_{11}. The point of contact of *Object*(X) and the *Wall* is $x_{11} + \delta$

$At(Object(X), x_{11})[\mathbf{SYN}] \wedge$
$\exists Set\text{-}of\text{-}Points(S_{11}) = \{Point(x), \forall x \, x \in Wall\}[\mathbf{SYN}] \wedge$
$\exists Set\text{-}of\text{-}RD(S_{12}) = \{RD(Object(X), Point(x)), \forall x \, x \in Wall\}[\mathbf{SYN}] \wedge \quad (6.20)$
$\quad Appear(F(X, RA = 0), x_{11}, t_1)[\mathbf{DIA}]$
$\quad \rightarrow Not(Move(Object(X), RA = 0, t_1 + \Delta))$

The "RA=0" parameter specifies that the direction of the force application when the force is thwarted is directed toward the GOAL (recall that RA is the relative direction as defined in Fig. 2.15), which is what the MOVEMENT-SCRIPT specifies for the elemental actions. Therefore, unlike in the earlier rules, the force F and the movement predicate *Move* now have an argument for direction.

Now, since the current situation is that the *Wall* is visible and presumably has an identifiable color, a similar situation to that of Fig. 6.5 is obtained in which the floor color helps to provide a useful synchronic condition to the causal rule involved. Here, by analogy with the $At(Object(X), x)) \wedge Color(Floor(x), COLOR)$ synchronic condition used in conjunction with Fig. 6.5, we add a $Contact(Object(X), Wall(x)) \wedge Color(Wall(x), COLOR))$ synchronic condition to Rule (6.20):

$At(Object(X), x_{11})[\mathbf{SYN}] \wedge$
$\exists Set\text{-}of\text{-}Points(S_{11}) = \{Point(x), \forall x \, x \in Wall\}[\mathbf{SYN}] \wedge$
$\exists Set\text{-}of\text{-}RD(S_{12}) = \{RD(Object(X), Point(x)), \forall x \, x \in Wall\}[\mathbf{SYN}] \wedge$
$Contact(Object(X), Wall(x_{11} + \delta))[\mathbf{SYN}] \wedge Color(Wall(x_{11} + \delta), BLUE)[\mathbf{SYN}] \wedge$
$\quad Appear(F(X, RA = 0), x_{11}, t_1)[\mathbf{DIA}]$
$\quad \rightarrow \quad Not(Move(Object(X), RA = 0, t_1 + \Delta))$

(6.21)

Assuming that the *Contact* point on the *Wall* is δ "distance" away from the location of the center of *Object*(X), which is x_{11}, as shown in Fig. 6.12, and that the color of the *Wall* is BLUE. We are also assuming that *Contact* is a basic predicate supplied by the visual system when a contact situation is observed. Therefore, at this stage, the agent concludes that it is the entire combination of *Object*(X)'s current absolute location, the existence of the set of points on the *Wall*, *Object*(X)'s relative distances to the points on the *Wall*, as well as the contact it makes with the *Wall* that are the synchronic preconditions to the impediment to its movement.

Even though in reality, much like in our physical world, it is only the contact of the agent with the immediate point on the *Wall* adjacent to it at this moment that is the real impediment, the agent is not able to discover this yet. In the general situation, there is no reason for a general noological system to accord the *Contact* situation any particular status unless it has interesting causal consequences. In our physical universe, making contact with something or someone (metaphorically or physically) can usually initiate a chain of consequences. For example, we constantly encounter situations in which contacts with heavy or immobile objects

6.5 Application to the Spatial Movement to Goal with Obstacle (SMGO) Problem

impede our onward movements, and the location of the contact involved (visually identified) is the location of the immobility. However, there are also forces that act at a distance, such as gravitational and electromagnetic forces, and in these cases, interesting causal consequences can arise when the relative distances between the objects of cause and objects of effect may not be 0 – i.e. RD $\neq$ 0.

Continuing the discussion, since Rule (6.8) was true earlier (corresponding to the situations in Fig. 6.4a and in Fig. 6.4b before the *Not(Move)* is experienced), it is repeated here for ease of comparison:

$$At(Object(X), x_{ANY})[\mathbf{SYN}] \wedge$$
$$Appear(F(X), x_{ANY}, t_{ANY})[\mathbf{DIA}] \rightarrow Move(Object(X), t_{ANY} + \Delta)$$
(6.8)

and now Rule (6.21) s is true (also repeated from above for ease of comparison):

$$At(Object(X), x_{11})[\mathbf{SYN}] \wedge$$
$$\exists Set\text{-}of\text{-}Points(S_{11}) = \{Point(x), \forall x x \in Wall\}[\mathbf{SYN}] \wedge$$
$$\exists Set\text{-}of\text{-}RD(S_{12}) = \{RD(Object(X), Point(x)), \forall x x \in Wall\}[\mathbf{SYN}] \wedge$$
$$Contact(Object(X), Wall(x_{11} + \delta))[\mathbf{SYN}] \wedge Color(Wall(x_{11} + \delta), BLUE)[\mathbf{SYN}] \wedge$$
$$Appear(F(X, RA = 0), x_{11}, t_1)[\mathbf{DIA}]$$
$$\rightarrow Not(Move(Object(X), RA = 0, t_1 + \Delta))$$
(6.21)

Rule (6.21) can be added to Rule (6.8) as an exception for the situation in Fig. 6.9b much like in the case of Rule (6.9). What causes the change from *Move* to *Not(Move* (...)) must be the extra SYN conditions. (This is counter-factual reasoning: had these factors been not there – comparing Fig. 6.9a, b – the world would have been different. Therefore, it must be these factors that are the causes of the current situation.)

Therefore, to get *Object*(X) to move at x_{11}:

$$Appear(F(X, RA = 0), x_{11}, t_{ANY}) \rightarrow Move(Object(X), RA = 0, t_{ANY} + \Delta)$$
(6.22)

We consider activating the following:

$Not(At(Object(X), x_{11})) \vee$
$Not(\exists Set\text{-}of\text{-}Points(S_{11}) = \{Point(x), \forall x\, x \in Wall\}) \vee$
$Not(\exists Set\text{-}of\text{-}Points(S_{12}) = \{RD(Object(X), Point(x)), \forall x\, x \in Wall\}) \vee$ (6.23)
$Not(Contact(Object(X), Wall(x_{11} + \delta))) \vee$
$Not(Color(Wall(x_{11} + \delta), BLUE))$

That is, if one of the conjunctive synchronic preconditions is now not true, then the force *may* be able to move the object. This is like the usual contra-positive operation in logic. In words, these mean the following, disjunctively:

(1) Change the current absolute location.
(2) Make all the points on the *Wall* disappear → in short, dematerialize the *Wall* (Chap. 4) (In principle, one needs only to dematerialize a thin "tunnel" connecting one side to the opposite side of the *Wall* but at this point the agent is not able to reason this out yet.)
(3) Change the relative distances to all the points on the *Wall*.
(4) Make the object not contact the *Wall* at the current absolute location.
(5) Make the color of the *Wall* not BLUE at the current location. Let us assume that the color of the *Wall* cannot be changed arbitrarily but there could be other parts of the Wall at which the color is not BLUE. Then, combining (4) and (5), the action would be "Make the object not contact the *Wall* at the current absolute location at which the color of the *Wall* is BLUE."

All these actions *may* remove the impediment to the movement but it is not guaranteed that they necessary would, because there may be other conditions of impediment that are consistent with these actions – e.g., action (1) suggests to change the absolute location of *Object*(X), but there may be another absolute location at which the object's movement in the direction of the GOAL is still impeded. However, trying them is a good bet.

Among the three options, action (2) – dematerializing the *Wall* – is the most expedient if the noological system indeed has this ability. (As of the writing of this book, this method is usually not available to an average agent but is often the stuff of fantasy and science fiction.) It searches its Script Base or Causal Rule Base and if this rule is found, it can be used to generate a new plan and the thwarted plan can be resumed – according to the *Thwart-Handling Process* in Fig. 6.11 (the rule could look like *Rule 2 – Dematerialize* – of Fig. 5.1b, Chap. 5). As for options (1) and (3), the agent would carry out actions to change the current location of *Object*(X). Given the availability of the earlier force-movement rule, this option would be available in its Script/Causal Rule repertoire. Option (4) could be a result of options (1) and (3) or other actions.

Another way to realize option (4) would be to move the *Wall* away from *Object*(X) in some manner while keeping *Object*(X) stationary if the agent has this ability. The reasoning process for moving the *Wall* in a certain manner to allow the unimpeded movement of *Object*(X) is similar to the process of an agent avoiding an oncoming projectile to be discussed in Chap. 9. Therefore, we will not discuss this option in this chapter and will relegate the discussion to Chap. 9.

Suppose the agent tries to change the location of *Object*(X) first, thus attempting to achieve the first negated condition (1) above – $Not(At(Object(X), x_{11}))$. If the Script Base already has rules that can convert $At(Object(X), x_{11})$ to $Not(At(Object(X), x_{11}))$ (such as the ELEMENTAL-MOVEMENT-SCRIPT of Fig. 2.23), these will be used to guide physical experiments to discover the optimal

6.5 Application to the Spatial Movement to Goal with Obstacle (SMGO) Problem   249

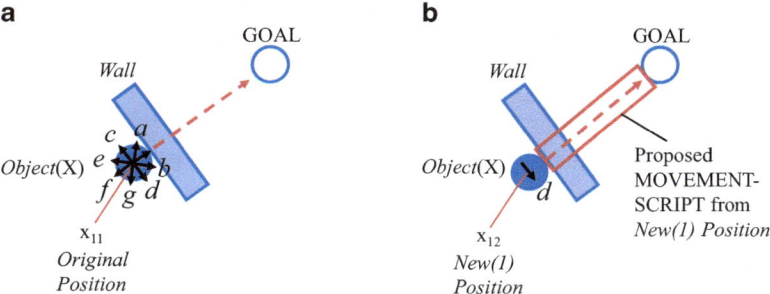

Fig. 6.13 (a) Agent trying out various movement options for *Object*(X) at the *Original Position* (x_{11}) of contact with the *Wall*. (b) Given the currently available movement options and given that the various outcome measures of goodness are known from the counterfactual information, a "best" movement direction is selected and a new movement script is proposed. The elemental movement in this direction brings *Object*(X) to a *New(1) Position* of x_{12}

rule. Otherwise, the agent uses whatever actions are available to it to achieve *Not* (*At*(*Object*(X), x_{11})). Figure 6.13a shows the agent trying to change *Object*(X)'s location by applying a force on it in several possible directions. The original "forward" direction toward the goal obviously leads to non-movement, as has just been experienced by the agent. For directions *a* and *b* that are slightly oblique to the "forward" direction, assuming that the *Wall* has a smooth surface, the resultant motion is one that is parallel to the *Wall*'s surface, similar to what would happen if the agent had tried to move *Object*(X) in directions *c* and *d* respectively. If the full law of physics is taken into consideration, the difference between applying a force and trying to move *Object*(X) in direction *a*, compared to that of applying the same magnitude of force in the direction *c*, would be that the magnitude of movement parallel to the *Wall*'s surface would be less in the case of *a* compared to the case of *c*, because the component of force parallel to the *Wall*'s surface in the case of *a* is smaller. However, in our current discussion, we will ignore the issue of magnitude and just consider the direction of movement. Then there are directions *e*, *f*, and *g* that all satisfy the condition of *Not*(*At*(*Object*(X), x_{11})).

In fact, as we shall see in the following discussion, moving elementally in the directions *c* and *d* may correspond to achieving the most progress toward the GOAL, but if *Object*(X) were to move in these directions it will also encounter movement impediment immediately when attempting to move toward the GOAL again. Directions *e*, *f*, and *g* are "moving *away* from the GOAL" hence they are non-optimal, but *Object*(X)'s immediate subsequent move toward the GOAL is not impeded as it has moved away from the surface, even though in one more attempt to move toward the GOAL, it will still end up contacting the surface again and its movement will be impeded, not unlike the consequence of moving in the *c* or *d* direction first except with an additional step of futile "backward" move.

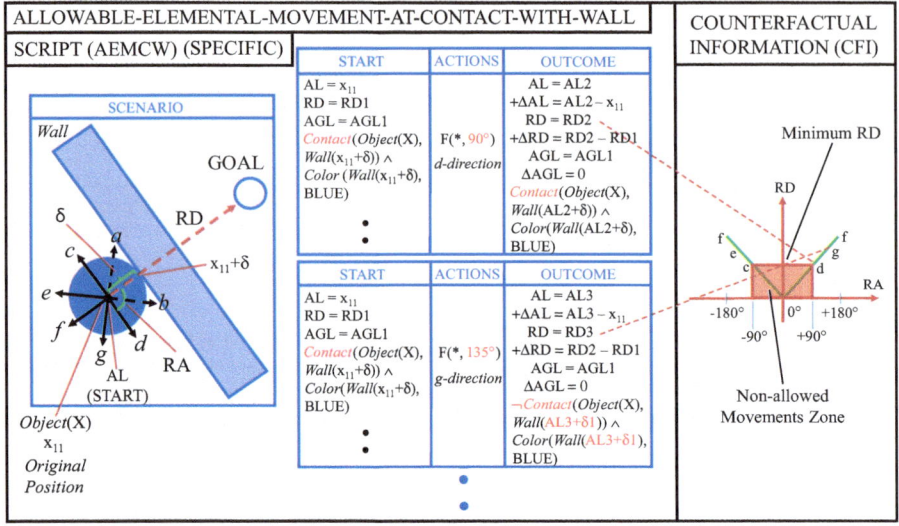

Fig. 6.14 A script (short-formed AEMCW-SCRIPT) that encodes the allowable elemental movements at the contact with the *Wall*. Note that AL3 is the new location of the center of *Object* (X) after being displaced elementally in the *g* direction and δ1 is a vector that is parallel to δ as defined in footnote 4. Therefore, AL3+δ1 (highlighted in *red*) is a point on the *Wall* that *Object* (X) would have contacted if *Object*(X) had moved elementally along only the component of *g* that is parallel to the *Wall*'s surface

After attempting the various directions of elemental movement of *Object*(X), the agent discovers that directions *c*, *d*, *e*, *f*, *g* are all possible movement directions. Hence, the consequences of these actions are recorded in a counterfactual script – the ALLOWABLE – ELEMENTAL – MOVEMENT – AT – CONTACT – WITH – WALL – SCRIPT (AEMCW-SCRIPT) – as shown in Fig. 6.14. This script can be thought of as a special case of the ELEMENTAL-MOVEMENT-SCRIPT of Fig. 2.23, and the GOAL is used as a reference point to measure the relative angle (RA) and relative distance (RD). This script is a SPECIFIC script as it describes what happens specifically at the current location of contact, x_{11}. In fact, it is only at this specific location of contact with the *Wall*, and with the orientation of the long axis of the *Wall* being perpendicular to the direction to the GOAL, that the COUNTERFACTUAL INFORMATION (CFI) portion of Fig. 6.14 has these specific values on the RA axis of the graph for the various directions, *c*, *d*, *e*, *f*, and *g*.

Now, having experienced the physical exploration process and recording the consequences as encapsulated in the script of Fig. 6.14, two general rules are extracted as follows:

6.5 Application to the Spatial Movement to Goal with Obstacle (SMGO) Problem

$At(Object(X), x_{11})[\mathbf{SYN}] \wedge$
$\exists Set\text{-}of\text{-}points(S_{11}) = \{Point(x), \forall x\ x \in Wall\}\ [\mathbf{SYN}] \wedge$
$\exists Set\text{-}of\text{-}RD(S_{12}) = \{RD(Object(X), Point(x)), \forall x\ x \in Wall\}[\mathbf{SYN}] \wedge$
$Contact(Object(X), Wall(x_{11} + \delta))[\mathbf{SYN}] \wedge Color(Wall(x_{11} + \delta), BLUE)[\mathbf{SYN}] \wedge$
$Appear(F(X, 270° < RA < 90°\ \text{relative to GOAL}), x_{11}, t_{ANY})[\mathbf{DIA}]$
$\rightarrow Not(Move(Object(X,\ RA = Same), t_{ANY} + \Delta))$

(6.24)

$At(Object(X), x_{11})[\mathbf{SYN}] \wedge$
$\exists Set\text{-}of\text{-}points(S_{11}) = \{Point(x), \forall x\ x \in Wall\}[\mathbf{SYN}] \wedge$
$\exists Set\text{-}of\text{-}RD(S_{12}) = \{RD(Object(X), Point(x)), \forall x\ x \in Wall\}[\mathbf{SYN}] \wedge$
$Contact(Object(X), Wall(x_{11} + \delta))[\mathbf{SYN}] \wedge Color(Wall(x_{11} + \delta), BLUE)[\mathbf{SYN}] \wedge$
$Appear(F(X,\ 90° \leq RA \leq 270°\ \text{relative to GOAL}), x_{11}, t_{ANY})[\mathbf{DIA}]$
$\rightarrow Move(Object(X,\ RA = Same), t_{ANY} + \Delta)$

(6.25)

This is assuming that there are enough data points of exploratory movements with different values of RA that allow the continuous value conditions for RA to be established.[5] RA = Same in the diachronic effect means it has the same value as that in the diachronic cause. Rule (6.24) basically encapsulates the situation depicted in the SCENARIO portion of Fig. 6.14 that as long as the direction of the force, RA, on *Object*(X) is pointed at a part of the *Wall* in the angular range of 270° < RA < 90°, and *Object*(X) is in contact with the *Wall* at a part of the *Wall* that is BLUE (there could be other parts that are not), then it cannot move *Object*(X) in the same direction as the force. This is a generalization over the situations encapsulated in Rule (6.21). In Rule (6.21) it is encoded that the impediment only takes place when RA = 0.

Rules (6.24) and (6.25) have a specific condition – that *Object*(X) must be at absolute location x_{11} because all these elemental movement explorations take place at that specific location. Later, where similar impediments/non-impediments of movement take place at a different absolute location for *Object*(X), the absolute location x_{11} will be generalized to x_{ANY} and the *Set-of-RD*(S_{12}) condition will be removed. This will be discussed in Sect. 6.5.2.2.1.

Now, based on the CFI portion of the AEMCW-SCRIPT, the "best-action" – the one that gives rise to a minimum distance from the GOAL (minimum RD), under the current movement constraint, is in the direction *c* or *d*.

Hence, compared with the situation in the ELEMENTAL-MOVEMENT-SCRIPT of Fig. 2.23, in which there is no impediment, and in which the best action is in a direction of RA = 0 – i.e., headed directly toward the GOAL – under the current constraint, the best possible directions that correspond to a minimum RD to GOAL are different.

---

[5] We will not delve into the detailed mechanism for learning and establishing parameter ranges in this book but suffice it to note that it is a general mechanism that applies to all kinds of parameters.

A typical A* algorithm would also end up picking either *c* or *d* directions for the next move from the first point of impediment with the aid of the minimum distance to GOAL heuristic. However, so far in works in robotics or other related AI domains in which A* is used, the idea of impediment under certain conditions (such as the *Contact* condition) is built-in (and therefore only applicable in certain known situations) and not learned from a general causal learning mechanism such as that described above, and there is no systematic and general way to record earlier counterfactual experiences such as that captured in the AEMCW-SCRIPT of Fig. 6.14, that can then be used to select the best possible direction of movement. We will see later that this process is a crucial learning process that can vastly improve the problem solving abilities of a nooligical agent. The crucial different between the process described here and that in a typical A* situation is that we are basically making explicit the complete step-by-step noological processes that allow an agent to learn from experiences and apply the learned knowledge to subsequent novel problem solving situations. On the other hand, in A*, the human scientist/programmer has already carried out part of these thinking processes, leaving some parts – such as the laborious node expansion process – to the machine.

Now, between the directions *c* and *d*, which give rise to the same minimum distance to the GOAL,[6] suppose the agent picks direction *d* as shown in Fig. 6.13b. This new elementally displaced location is called *New(1) Position* and it corresponds to an absolute location x_{12} of *Object*(X) (represented by the location of its center) as shown in Fig. 6.13b. From this new elementally displaced location, another MOVEMENT-SCRIPT similar to that activated in Fig. 6.9a is now activated and the agent attempts to execute a solution to reach the GOAL. Recall that script matching, retrieval, and execution to satisfy the PROBLEM SPECIFICATION is a constant process in a noological system such as depicted in Figs. 6.10 and 6.11 and discussed in Chap. 1. When a script is thwarted, the systems takes some actions to extricate itself from the immediate impediment, and then it immediately returns to attempt to solve the problem again through the script matching and retrieval process of Fig. 6.10. In so doing, the agent may be confronted with a special or a general situation, and these will be discussed in the following two sections.

Before that, let us summarize Step ③" of Fig. 6.11 – *Counter-Thwarting Script Discover Process*– in detail in Fig. 6.15, which constitutes the bulk of the discussion of the current section. What we show in Fig. 6.11 is the Basic *Counter-Thwarting Script Discover Process*. Later, this process is enhanced into a *Complete Counter-Thwarting Script Discovery Process* (Fig. 6.21).

In Fig. 6.15, we show the path – in the form of a green path – of the processes up to the current point. Firstly, because there is no script that can immediately implement one of the negated conditions of Eq. (6.18), some exploration is carried

---

[6] This is true only if the current point of contact of *Object*(X) on the *Wall* is such that the line joining the current location to the GOAL is perpendicular to the surface of the *Wall* on the side of the contact point.

6.5 Application to the Spatial Movement to Goal with Obstacle (SMGO) Problem   253

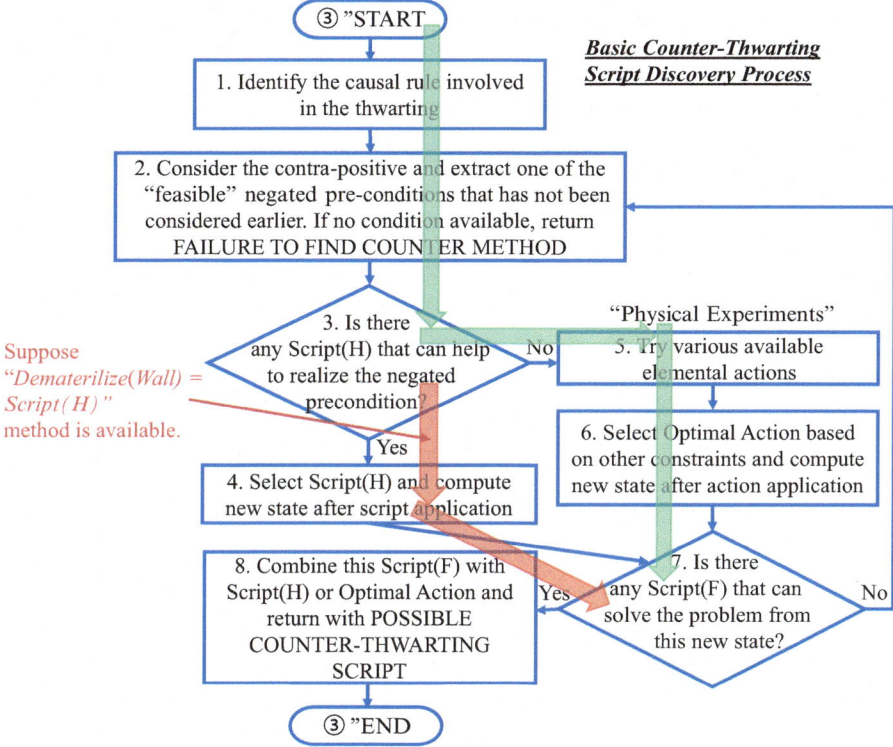

Fig. 6.15 The *Basic Counter-Thwarting Script Discovery Process*, which is the *basic* detailed processes of Step ③" of Fig. 6.11. The *green path* shows the process when there is no Script (H) that can realize one of the negated preconditions. The *dark red path* shows the process when there is a Script(H) that can realize one of the negated preconditions (such as *Dematerialization* of the *Wall*)

out to see if one of the negated conditions can be met. We did this by trying out various available elemental actions at the point of thwarting and an optimal action – an elemental movement in direction d – was picked based on the script of Fig. 6.14 which encodes the results of the exploration. We are currently at the situation testing step (Step 7): *Is there any Script(F) that can solve the problem from this new state?* This is when we consider the special and the general situations from this point onward.

We also indicate in the form of a dark red path on Fig. 6.15 that suppose a *Dematerialize(Wall)* = $Not(Set\text{-}of\text{-}Points(S_{11}) = \{Point(x), \forall x\ x \in Wall\})$ method is available (Script(H)), that action will be taken to get the situation to arrive at a new state (in this case, the absence of the *Wall*), and then since the *Wall* is now no longer present, the agent will now use the basic MOVEMENT-SCRIPT (Script(F)) that would allow *Object*(X) to move all the way to the GOAL (that is why in Step 8 it is a combination of Script(F) *with* Script(H) *or* Optimal Action).

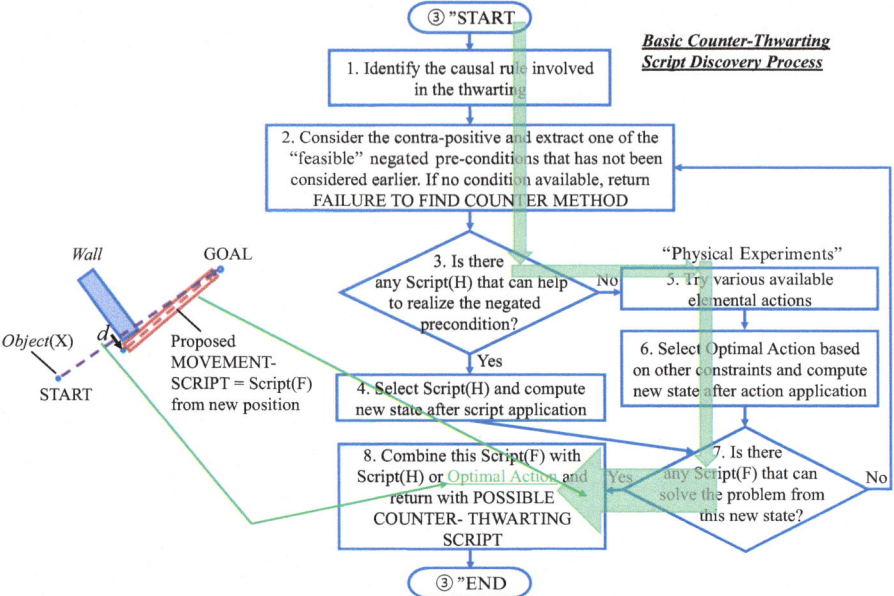

Fig. 6.16 A special situation in which the obstacle of the SMGO problem is such that *Object*(X) first touches the *Wall* right next to one of its corners. After selecting an elemental optimal action d, the agent can move *Object*(X) straight to the GOAL using a basic MOVEMENT-SCRIPT. *Object*(X) is now shown as a small, point object

6.5.2.1 A Special Situation: Next to a Wall Corner

In Fig. 6.16 we show a special situation in which the positioning of the START and GOAL relative to the *Wall* is such that when *Object*(X) first touches the *Wall*, it is right next to one corner of the *Wall*. After having moved an elemental step in the d direction, there is a clear path from that new position to the GOAL. The usual MOVEMENT-SCRIPT (Fig. 2.18) can be used to solve the problem from the new position and the problem is trivially solved. To return a total solution, the *optimal* elemental movement/action in the direction d is combined with the MOVEMENT-SCRIPT (Script(F)) and the process is shown in the flowchart of Fig. 6.16 in the form of a broad green arrow completing the earlier green arrow path in the flowchart in Fig. 6.15.

6.5.2.2 The General Situation

In the general situation, when *Object*(X) touches the *Wall*, it is not right next to a corner of the *Wall*, and the problem situation at this point would be like that shown in Fig. 6.13b. In this problem solving situation of Fig. 6.13b, the entire earlier problem solving process of Fig. 6.9b is repeated and the agent's effort is again

6.5 Application to the Spatial Movement to Goal with Obstacle (SMGO) Problem

thwarted – $Object(X)$ encounters the same impediment in the attempted movement in the direction toward the GOAL. Hence, the entire process of handling the thwarting situation as depicted in Fig. 6.11 is repeated. Again, assuming that there are no readily available counter-thwarting scripts, the agent then enters the step of *Counter-Thwarting Script Discovery Process* (Step ③" of Fig. 6.11) to discover a counter-thwarting script.

In the process of causal reasoning (Step 1 of Fig. 6.15), first a causal rule that gives rise to the current thwarted situation is extracted, though, this time, the location of impediment is slightly different (as shown in Fig. 6.13b, $Object(X)$ is now at location x_{12}), and a rule similar to that of Rule (6.21) but with the absolute and relative locations of $Object(X)$ changed is extracted as follows:

$$At(Object(X), x_{12})[\textbf{SYN}] \wedge$$
$$\exists Set\text{-}of\text{-}points(S_{11}) = \{Point(x), \forall x\, x \in Wall\}[\textbf{SYN}] \wedge$$
$$\exists Set\text{-}of\text{-}RD(S_{13}) = \{RD(Object(X), Point(x)), \forall x\, x \in Wall\}[\textbf{SYN}] \wedge$$
$$Contact(Object(X), Wall(x_{12} + \delta))[\textbf{SYN}] \wedge Color(Wall(x_{12} + \delta), BLUE)[\textbf{SYN}] \wedge$$
$$Appear(F(X, RA = 0), x_{12}, t_1)[\textbf{DIA}]$$
$$\rightarrow Not(Move(Object(X), RA = 0, t_1 + \Delta))$$
(6.26)

Comparing Rules (6.21) and (6.26), since $x_{11} \neq x_{12}$ and $S_{12} \neq S_{13}$, we obtain, through dual-instance generalization:

$$At(Object(X), x_{ANY})[\textbf{SYN}] \wedge$$
$$\exists Set\text{-}of\text{-}Points(S_{11}) = \{Point(x), \forall x\, x \in Wall\}[\textbf{SYN}] \wedge$$
$$Contact(Object(X), Wall(x_{12} + \delta))[\textbf{SYN}] \wedge$$
$$Color(Wall(x_{ANY} + \delta), BLUE)[\textbf{SYN}] \wedge$$
$$Appear(F(X, RA = 0), x_{ANY}, t_1)\ [\textbf{DIA}]$$
$$\rightarrow Not(Move(Object(X), RA = 0, t_1 + \Delta))$$
(6.27)

In Rules (6.27), "ANY" represents *any* value but variables that are labelled ANY are of *any* but the *same* value. This rule is more general than either Rules (6.21) or (6.26) because they state that as long as $Object(X)$ is in contact with the part of the *Wall* that is BLUE (no matter at what absolute location), and as long as the Wall exists, there is an impediment in the direction of the GOAL. Therefore, the agent is now able to discover, through only a small number of training instances, the general causal conditions for the movement impediment of $Object(X)$ that is akin to our typical commonsensical understanding of the causes involved in this situation, and that it was not able to discover at the stage of the learning of Rule (6.21) earlier.

Figure 6.17 shows that this knowledge can now be encoded in the form of an "ANTI-MOVEMENT-SCRIPT" that is derived from the MOVEMENT-SCRIPT of Fig. 2.18. In this script, SCENARIO is shown in pictorial, analogical form with a *Wall* positioned in between an $Object(X)$ and a GOAL, and an associated condition $Contact(Object(X), Wall(x_{ANY} + \delta)) \wedge Color(Wall(x_{ANY} + \delta), BLUE)$. ($x_{ANY}$ is the location of $Object(X)$.) Earlier a more specific ANTI-MOVEMENT-SCRIPT

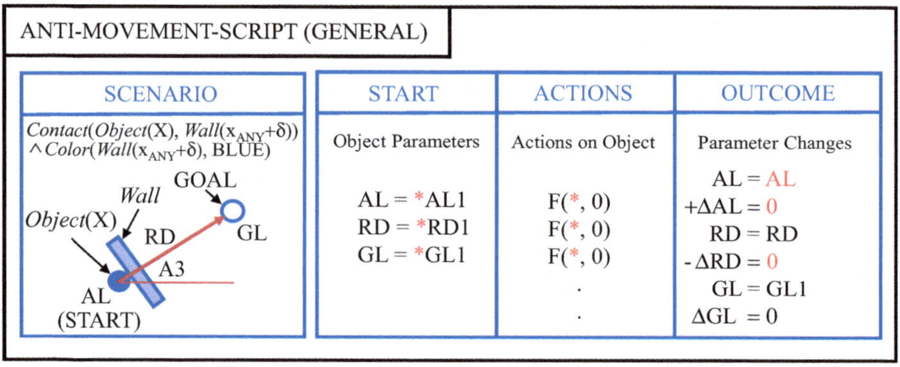

ANTI-MOVEMENT-SCRIPT (GENERAL)			
SCENARIO	START	ACTIONS	OUTCOME
$Contact(Object(X), Wall(x_{ANY}+\delta))$ $\wedge Color(Wall(x_{ANY}+\delta), BLUE)$ [diagram: Wall, Object(X), GOAL, RD, GL, A3, AL (START)]	Object Parameters AL = *AL1 RD = *RD1 GL = *GL1	Actions on Object F(*, 0) F(*, 0) F(*, 0) . .	Parameter Changes AL = AL +ΔAL = 0 RD = RD - ΔRD = 0 GL = GL1 ΔGL = 0

Fig. 6.17 An ANTI-MOVEMENT-SCRIPT learned

corresponding to the SCENARIO of Fig. 6.9b was also encoded (not discussed nor shown earlier) but the current one is more general as the contact point between *Object*(X) and the *Wall* can be at any point on the *Wall*, provided that the part of the *Wall* is BLUE. In the ACTIONS portion it is shown that a force F is repeatedly applied to *Object*(X) much like in the MOVEMENT-SCRIPT of Fig. 2.18. However, in the OUTCOME portion it is shown that both the absolute location of the *Object*(X) (AL) and its relative distance to the GOAL (RD) are not changed as a result. This script can also be used for problem solving in a backward chaining process – if the desired OUTCOME is the *non-movement* of an object with the application of a force, one solution is to create a *Wall* to block its movement.

Similar to the case of exceptional sub-rules in Rule (6.13), this script can be linked to the MOVEMENT-SCRIPT of Fig. 2.18 as an exceptional sub-script.

Now, according to Rule (6.27), there are four synchronic preconditions, and as with the contra-positive argument above (after Eq. 6.15), the negation of *any* one of them *may* allow movement to proceed (i.e., a change from *Not*(*Move*) to *Move*) in the direction of the GOAL.[7] Negating the first condition – *Not*(*At*(*Object*(X), x_{ANY})) – results in *Object*(X) not existing anywhere, which seems to be a queer requirement. Logically this is not wrong as removing *Object*(X) from existence is certainly one way to remove the situation of impediment. However, we ignore this first negated condition as it implies we cannot use some practically implementable change of the absolute location of *Object*(X) to counter the thwarted situation.

---

[7] As discussed earlier, the situation is a "may" situation because after removing the current cause(s) of the impediment by, say, moving *Object*(X) to a new location, there may be yet undiscovered causes of impediments. One possibility is that if the *Wall* were to extend farther in both directions parallel to the direction *d* but with a different color and it can still impede movement, which is typical of the physics of our reality – i.e., color of the *Wall* is typically not material to its property of impenetrability. Then, making *Object*(X) not in *Contact* with the BLUE area of the *Wall* does not remove the impediment. However, it would still be a good try before any heuristics in this regard have been encoded that may advise against it.

6.5 Application to the Spatial Movement to Goal with Obstacle (SMGO) Problem   257

The second negated condition, *Not(Set-of-Points*(S_{11})...), basically negates the existence of the *Wall* and we have said above that if the dematerialization of the *Wall* is an available option to the agent, it is the most expedient way to remove the impediment. Let us assume that this option is not available. Then there is the third and fourth negated conditions combined, *Not*(*Contact*(*Object*(X), *Wall*($x_{ANY} + \delta$)) $\wedge$ *Color*(*Wall*($x_{ANY} + \delta$), BLUE)),[8] which can be achieved by either moving *Object*(X) in such a way that the *Contact* condition on a BLUE part of the *Wall* no longer holds while keeping the *Wall* stationary or vice versa – by moving the *Wall* and keeping *Object*(X) stationary. As mentioned above, we will not consider the option of moving the *Wall* in this chapter as it is similar to another problem – that of moving an agent to avoid an oncoming projectile to be discussed in Chap. 9.

In summary, the three negated conditions under consideration above are:

$$Not(At(Object(X), x_{ANY})) \vee$$
$$Not(\exists Set\text{-}of\text{-}Points(s_{11}) = \{Point(x), \forall x \, x \in Wall\}) \vee$$
$$Not\,(Contact(Object(X), Wall(x_{ANY} + \delta)) \wedge Color\,(Wall(x_{ANY} + \delta), \text{BLUE}))$$

(6.28)

And the third negated condition – the "not contact with part of *Wall* that has a BLUE color" – is picked (compared to Eq. 6.18 where instead it was the "change location of *Object*(X)" condition that was picked). This *Not*(*Contact*(*Object*(X), *Wall*($x_{ANY} + \delta$)) $\wedge$ *Color*(*Wall*($x_{ANY} + \delta$), BLUE)) condition now becomes the sub-goal to achieve to get out of the thwarted situation and the SCRIPT is again queried to see if there is any available step that can assist with this (Step 3 in Fig. 6.15).[9] If not, the agent enters an exploration phase in which actions available to it to move *Object*(X) are physically attempted to see if the *Not* (*Contact*(...)...) condition can be achieved (Step 5 in Fig. 6.15).

Figure 6.18 shows the available directions of elemental movement of *Object* (X) that will result in the *Not*(*Contact*(...)...) condition being achieved. Again, like the earlier situation in Fig. 6.13a, these are achieved through random exploration (conducting "physical experiments"), and we assume that a reasonable number of different directions of movement is attempted to get a good sample of various representative directions. Directions of movement that lead to the *Not*(*Contact* (...)...) situation are gathered to be encoded in a counterfactual script and the rest are ignored. Directions c' and d' are similar to directions c and d respectively in Fig. 6.13a except that they are pointing elementally farther away from the *Wall*. I.e., for directions c' and d' respectively, their associated RA's are not 270° and 90° but

---

[8] Similar to the discussion after Eq. (6.18) above under point (5), let us assume that the color of the *Wall* cannot be changed arbitrarily. Then, the third and fourth conditions are combined conjunctively and the negated action would be "Make *Object*(X) not contact the *Wall* at an absolute location at which the color of the *Wall* is BLUE."

[9] Specifically, as *Object*(X) is currently at x_{12}, the agent needs to achieve *Not*(*Contact*(*Object*(X), *Wall*($x_{12}+\delta$)) $\wedge$ *Color*(*Wall*($x_{12}+\delta$), BLUE)) first (note that δ is as defined in Fig. 6.12).

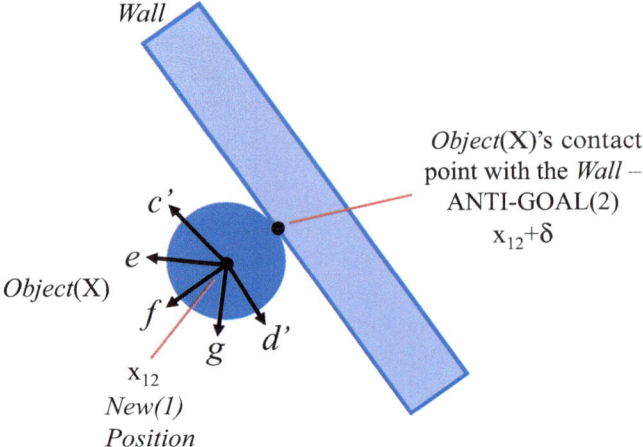

Fig. 6.18 Directions of movement leading to non-contact of the *Wall*

are $270° - \Delta$ and $90° + \Delta$ respectively, where Δ is a small angle. This means any movement in these directions will result in *Object*(X) moving an elemental distance away from the *Wall*.

Figure 6.19 shows the associated counterfactual script learned and encoded for these actions – the ELEMENTAL-MOVEMENT-TO-AVOID-WALL-ANTI-GOAL-SCRIPT (EMAWAG-SCRIPT). As mentioned in footnote 9 above, as *Object*(X) is currently at x_{12}, the agent needs to achieve $Not(Contact(Object(X), Wall(x_{12} + \delta)) \wedge Color(Wall(x_{12} + \delta), BLUE))$ first. The contact point, $x_{12}+\delta$, between *Object*(X) and the *Wall* hence becomes an "ANTI-GOAL" (specifically ANTI-GOAL(2)) – something to be avoided. In general, whatever or wherever is the cause of something undesirable is formulated as an ANTI-GOAL (the point of contact between *Object*(X) and the *Wall* in Fig. 6.13a or 6.14, $x_{11}+\delta$, would also be an ANTI-GOAL, labeled ANTI-GOAL(1)). Note that our ANTI-GOAL here has a different property compared to the ANTI-GOAL illustrated in, say, Fig. 3.10b which could be a physical object that is to be avoided. Here, the ANTI-GOAL is more abstract as it is a *Contact* location. (In Fig. 9.8, Chap. 9, a similar situation occurs.)

In Fig. 6.19, the consequence of each "avoidance" action results in a different relative distance (RD) from the ANTI-GOAL(2) on the *Wall* and a graph capturing these are encoded in the CFI portion of the script. And, as discussed in Chap. 2, Sect. 2.6.4, if the *maximum* distance heuristic were to be applied to this situation, the direction f would be chosen to be the optimal direction of "escape." However, in the current situation, there is another goal beckoning – that of the overall GOAL to which the agent had intended to move *Object*(X) in the first place. Therefore, in this case the maximum distance away from the ANTI-GOAL(2) is not selected, instead, the *minimum* distance from the GOAL that is allowed by the current movement constraints, namely directions c' or d', is selected, much in the same spirit as the

6.5 Application to the Spatial Movement to Goal with Obstacle (SMGO) Problem

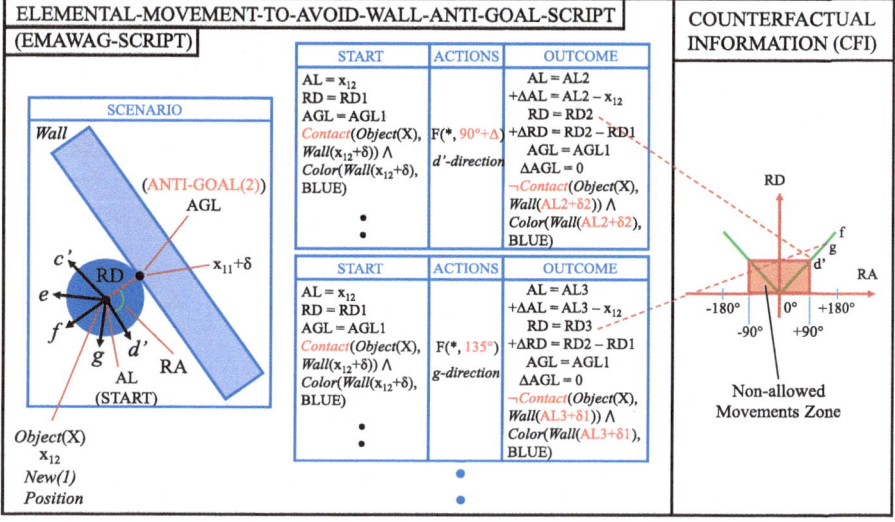

Fig. 6.19 A script that encodes the consequences of various elemental movements away from the ANTI-GOAL(2) on the *Wall*. Note that AL3 is the new location of the center of *Object*(X) after being displaced elementally in the *g* direction and δ1 is a vector that is parallel to δ as defined in footnote 4. Therefore, AL3 + δ1 (highlighted in *red*) is a point on the *Wall* that *Object*(X) would have contacted if *Object*(X) had moved elementally along only the component of *g* that is parallel to the *Wall*'s surface. Similarly, AL3 + δ2 (also highlighter in *red*) is the corresponding point for the movement in the *d'* direction

process associated with Figs. 6.13 and 6.14.[10] We assume that this is a built-in process for the selection of the optimal direction.

Now, suppose direction *d'* is selected,[11] the agent now moves *Object*(X) slightly away from the *Wall* to result in a *Not*(*Contact*(...)...) situation as shown in Fig. 6.20a and arrive at the *New(2) Position*. According to the algorithm that attempts to discover a counter-thwarting script as laid out in Fig. 6.15, the agent now attempts to solve the SMGO problem again by invoking the standard MOVEMENT-SCRIPT (Script(F) in Step 7 of Fig. 6.15) as shown in Fig. 6.20a. This is the combined "Optimal Action and Script(F)" – Step 8 – in Fig. 6.15,

---

[10] The situation is similar to one in which a "catching" game includes also a reward of picking up an item somewhere in the space in which the catching game is carried out. If the simple goal for someone being chased is just to avoid being caught by the chaser, then he would deploy the maximum distance heuristic to stay maximally far away from the chaser. If, on the other hand, the ultimate purpose of the game is to pick up the reward item while avoiding been "caught" in the process, then at some points of the process the person being chased would deviate from the prescription of the maximum distance heuristic and take the ultimate goal – the reward item – into consideration for optimal action.

[11] A justification for this is because *Object*(X) had just come in the *d* direction from its earlier x_{11} location. Taking the *c'* direction will result in *Object*(X) returning back to somewhere near x_{11}. This can be built-in as a heuristic specifying a situation to avoid.

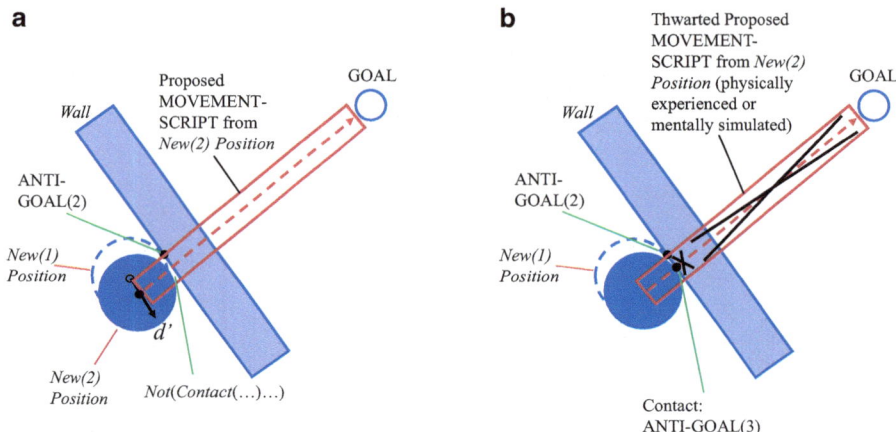

Fig. 6.20 (a) Agent first moves *Object*(X) in d' direction to move it away from the *Wall*, arriving at the *New(2) Position*, and then it attempts to apply the usual MOVEMENT-SCRIPT toward the GOAL. (b) *Object*(X) is again thwarted in its movement

providing a POSSIBLE COUNTER-THWARTING SCRIPT. However, as in the similar situation in Fig. 6.13b, its movements are thwarted again, as shown in Fig. 6.20b. This new point of thwarting is labeled ANTI-GOAL(3).

There are two ways at this stage the agent can find out if *Object*(X), in executing the MOVEMENT-SCRIPT, will meet an impediment. One way is to carry out the actions physically as prescribed by the script and experience the consequence physically, which corresponds to returning from the Step ③" END step of the *Basic Counter-Thwarting Script Discovery Process* of Fig. 6.15 to Step ①of Fig. 6.11 and then discovering that the script is thwarted at Step ②. However, the agent is now armed with the earlier learned knowledge of Rule (6.27) which says that if the agent tries to apply a force on *Object*(X) in the direction of the GOAL while *Object*(X) is in *Contact* with the *Wall* at ANY location with a BLUE color given the scenario of, say, Fig. 6.13b, then *Object*(X) does not move. With this knowledge, the agent can hence instead activate a *mental simulation* to infer the consequence of executing the proposed MOVEMENT-SCRIPT from this *New (2) Position*. (See Fig. 5.5 of Chap. 5 for the overall noological processing architecture, including a Causal Rule Base/Script Base and Mental Simulation/Reasoning Arrays that can support mental simulations.) There is also the ANTI-MOVEMENT-SCRIPT of Fig. 6.17 that the Agent can draw on for the mental simulation. Therefore, after one step of mental simulation using the proposed MOVEMENT-SCRIPT to move *Object*(X) toward the GOAL, the *Contact* event takes place, and either Rule (6.27) or the learned ANTI-MOVEMENT-SCRIPT can be used to infer that the movement will be thwarted.

In general, if mental simulation is carried out and the result is such that there is no further thwarting of the plan, then the process can return to Step ① of the

6.5 Application to the Spatial Movement to Goal with Obstacle (SMGO) Problem

Thwart-Handling Process (Fig. 6.11) to execute the proposed script.[12] Otherwise, further reasoning and mental simulation can be carried out to discover a potentially viable counter-thwarting plan before proceeding to physically executing it. This corresponds to adding some extra steps at the end of the ③" algorithm of Fig. 6.15. This is shown in the modified algorithm – The *Complete Counter-Thwarting Script Discovery Process*- in Fig. 6.21. The *Further Reasoning and Mental Simulation to Discover Possible Counter-Thwarting Script* step is shown as Step 12 in Fig. 6.21.[13] The details of the *Further Reasoning and Mental Simulation...* module will be discussed in Sect. 6.5.2.2.2 (Fig. 6.25).

6.5.2.2.1 The Learning of the General AEMCW-SCRIPT

Before proceeding to discuss the *Further Reasoning and Mental Simulation...*module (Step 12 of Fig. 6.21), we would like to discuss some other rules that were learned along with the exploration process to discover the *Not (Contact(...)...)* directions in Figs. 6.18 and 6.19. These rules will be very useful for further reasoning and mental simulation to discover a viable counter-thwarting script.

In the process of discovering the directions that resulted in *Not(Contact(...)...)* in which a force was physically applied to *Object*(X) in the various directions, two consequences were observed – *Move* or *Not(Move)* of *Object*(X) – much like in the earlier situation in the encoding of the AEMCW script of Fig. 6.14. Now, however, the absolute location of *Object*(X) is x_{12} (the *New(1) Position* – Figs. 6.13a and 6.18) and the focus of movement is relative to the ANTI-GOAL(2), at $x_{12} + \delta$), and hence these *Move* and *Not(Move)* rules are encoded as follows (see Fig. 6.22 for the pictorial details, and compare these rules below to Rules 6.24 and 6.25):

---

[12] This can happen in the situation of Fig. 6.13b because at that moment the system has not encoded a rule that says that impediment can happen anywhere along the *Wall* (the impediment so far is at a specific location of $x_{11} + \delta$, corresponding to *Object*(X) being at the *Original Position*), so the result of the mental simulation is such that the proposed solution, consisting of an elemental movement in the *d* direction combined with a straight-line movement toward the GOAL, can work. On returning to Step ① of Fig. 6.11 to execute the solution physically, then the agent discovers that the solution does not work.

[13] This new algorithm covers the case of the earlier, "old" situation of Fig. 6.13(a) as well (i.e., after arriving at x_{11}, the *Original Position*). There are two situations. One is, the agent has not encoded any general rule about the *Wall* being able to impede the movement of *Object*(X) anywhere along the *Wall*, and the situation will be covered by footnote 12 above – the process will proceed along Steps 9, 10, and then 11 of Fig. 6.21 and return to Step ①of Fig. 6.11, executing the POSSIBLE COUNTER-THWARTING SCRIPT physically. The other is, the agent has learned the general rule in some earlier experiences. Then the process will proceed along Steps 9, 10, and 12, and in the *Further Reasoning and Mental Simulation to Discover Possible Counter-Thwarting Script* module of Step 12 (Fig. 6.25), it will start composing the "proposed long action sequence" according to the process of Fig. 6.25.

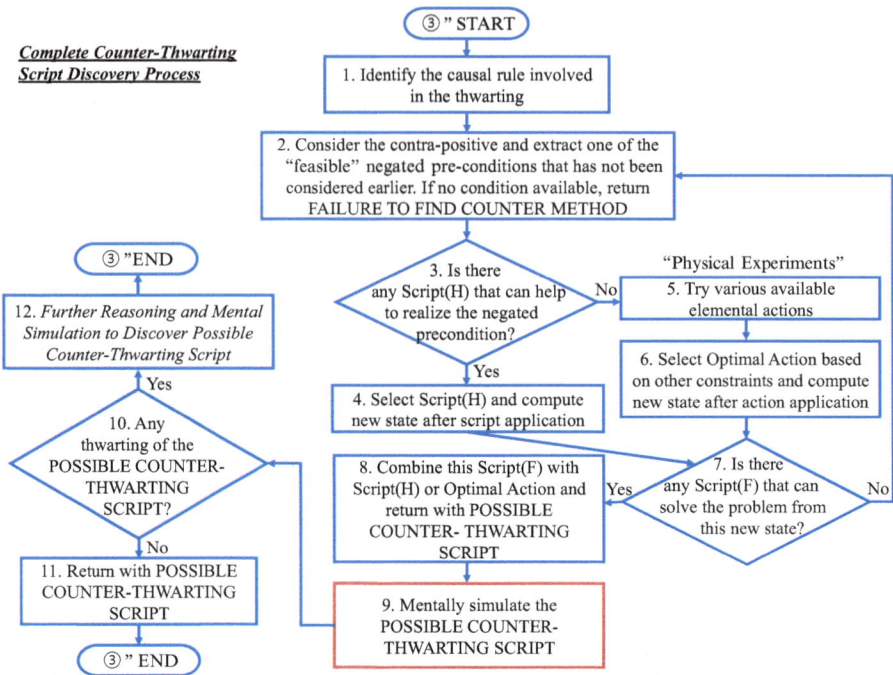

Fig. 6.21 The *Complete Counter-Thwarting Script Discovery Process* – modified *Basic Counter-Thwarting Script Discovery Process* of Fig. 6.15 to include mental simulation of possible counter-thwarting solution (Step 9) and the subsequent handling (Step 12) of the failure of the POSSIBLE COUNTER-THWARTING SCRIPT (failing the test in Step 10). The details of Step 12 – *Further Reasoning and Mental Simulation to Discover Possible Counter-Thwarting Script* – is shown in Fig. 6.25

$At(Object(X), x_{12})$ [**SYN**]∧
$\exists Set\text{-}of\text{-}Points(S_{11}) = \{Point(x), \forall x \; x \in Wall\}$ [**SYN**]∧
$\exists Set\text{-}of\text{-}RD(S_{13}) = \{RD(Object(X), Point(X)), \forall x \; x \in Wall\}$ [**SYN**]∧
$Contact(Object(X), Wall(x_{12} + \delta))$ [**SYN**] ∧ $Color(Wall(x_{12} + \delta), BLUE)$ [**SYN**]∧
$\quad Appear(F(X, 270° < RA < 90°$ relative to $x_{12} + \delta), x_{12}, t_{ANY})$ [**DIA**]
$\quad \rightarrow Not(Move(Object(X, RA = Same), t_{ANY} + \Delta))$

(6.29)

$At(Object(X), x_{12})$ [**SYN**]∧
$\exists Set\text{-}of\text{-}Points(S_{11}) = \{Point(x), \forall x \; x \in Wall\}$ [**SYN**]∧
$\exists Set\text{-}of\text{-}RD(S_{13}) = \{RD(Object(X), Point(X)), \forall x \; x \in Wall\}$ [**SYN**]∧
$Contact(Object(X), Wall(x_{12} + \delta))$ [**SYN**] ∧ $Color(Wall(x_{12} + \delta), BLUE)$ [**SYN**]∧
$\quad Appear(F(X, 90° \leq RA \leq 270°$ relative to $x_{12} + \delta), x_{12}, t_{ANY})$ [**DIA**]
$\quad \rightarrow (Move(Object(X, RA = Same), t_{ANY} + \Delta))$

(6.30)

6.5 Application to the Spatial Movement to Goal with Obstacle (SMGO) Problem

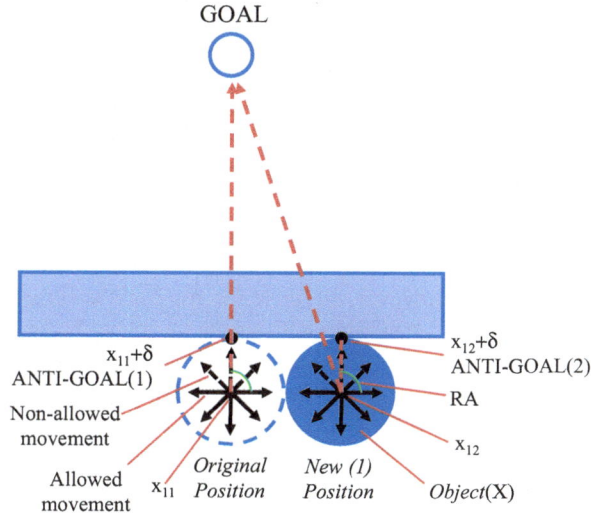

Fig. 6.22 The process of learning the general AEMCW script – the experience and rules learned from the earlier *Original Position* and the later *New(1) Position* are compared. *Solid arrows* indicate directions in which movements are allowed; *dashed arrows*, not allowed

In Fig. 6.22 it is shown that there is the *Original Position* of *Object*(X) and the *New(1) Position* of *Object*(X) (distance between them exaggerated for clarity and refer to Figs. 6.13 and 6.18 for these positions). $x_{12}+\delta$ is ANTI-GOAL(2). Now, had RA been measured with respect to the GOAL, not the $x_{12}+\delta$ ANTI-GOAL(2), from this new location x_{12} of *Object*(X), the rules would have looked like (these rules are the same as Rules (6.29) and (6.30) except for the *Appear*... portion):

$At(Object(X), x_{12})[\textbf{SYN}] \wedge$
$\exists Set\text{-}of\text{-}Points(S_{11}) = \{Point(x), \forall x\, x \in Wall\}[\textbf{SYN}] \wedge$
$\exists Set\text{-}of\text{-}RD(S_{13}) = \{RD(Object(X), Point(x)), \forall x\; x \in Wall\}[\textbf{SYN}] \wedge$
$Contact(Object(X), Wall(x_{12}+\delta))[\textbf{SYN}] \wedge Color(Wall(x_{12}+\delta), BLUE)[\textbf{SYN}] \wedge$
$Appear(F(X, 270° + \Delta < RA < 90° + \Delta\; \text{relative to GOAL}), x_{12}, t_{ANY})[\textbf{DIA}]$
$\quad \rightarrow Not(Move(Object(X, RA = Same), t_{ANY} + \Delta))$

(6.31)

$At(Object(X), x_{12})[\textbf{SYN}] \wedge$
$\exists Set\text{-}of\text{-}Points(S_{11}) = \{Point(x), \forall x\, x \in Wall\}[\textbf{SYN}] \wedge$
$\exists Set\text{-}of\text{-}RD(S_{13}) = \{RD(Object(X), Point(X)), \forall x\; x \in Wall\}[\textbf{SYN}] \wedge$
$Contact(Object(X), Wall(x_{12}+\delta))[\textbf{SYN}] \wedge Color(Wall(x_{12}+\delta), BLUE)[\textbf{SYN}] \wedge$
$Appear(F(X, 90° + \Delta \leq RA \leq 270° + \Delta\; \text{relative to GOAL}), x_{12}, t_{ANY})[\textbf{DIA}]$
$\quad \rightarrow Move(Object(X, RA = Same), t_{ANY} + \Delta)$

(6.32)

One can see from Fig. 6.22 that because RA, as measured (clockwise) with respect to the line joining the center of *Object*(X) to the GOAL, is different from

RA as measured with respect to the line joining the center of *Object*(X) to ANTI-GOAL(2) at $x_{12} + \delta$, the same direction of elemental movement will have a slightly larger value of RA: above in Rules (6.31) and (6.32), it is shown that $270°+\Delta <$ RA $< 90°+\Delta$ relative to GOAL, compared with $270° <$ RA $<90°$ relative to $x_{12}+\delta$, ANTI-GOAL (2), in Rules (6.29) and (6.30).

If we compare Rules (6.31) and (6.32) on the one hand and Rules (6.24) and (6.25) on the other, even though they look very similar, we cannot carry out dual instance generalization between them and remove some of the synchronic preconditions because the diachronic preconditions are different due to the RA range differences. However, suppose we replace the direction toward the GOAL in the earlier Fig. 6.14 SCENARIO in the script with the direction toward the absolute location $x_{11} + \delta$, which is ANTI-GOAL (1) in Fig. 6.22 associated with the *Original Position*, which happens to lie in the same direction toward the GOAL, Rules (6.24) and (6.25) can be reformulated as[14]:

$At(Object(X), x_{11})[\mathbf{SYN}] \wedge$
$\exists Set\text{-}of\text{-}Points(S_{11}) = \{Point(x), \forall x \, x \in Wall\}[\mathbf{SYN}] \wedge$
$\exists Set\text{-}of\text{-}RD(S_{12}) = \{RD(Object(X), Point(x)), \forall x \, x \in Wall\}[\mathbf{SYN}] \wedge$
$Contact(Object(X), Wall(x_{11} + \delta))[\mathbf{SYN}] \wedge Color(Wall(x_{11} + \delta), BLUE)[\mathbf{SYN}] \wedge$
$\quad Appear(F(X, 270° < \text{RA} < 90° \, \text{relative to} \, x_{11} + \delta), x_{11}, t_{ANY})[\mathbf{DIA}]$
$\quad \rightarrow Not(Move(Object(X, \, \text{RA} = \text{Same}), t_{ANY} + \Delta))$

(6.33)

$At(Object(X), x_{11})[\mathbf{SYN}] \wedge$
$\exists Set\text{-}of\text{-}Points(S_{11}) = \{Point(x), \forall x \, x \in Wall\}[\mathbf{SYN}] \wedge$
$\exists Set\text{-}of\text{-}RD(S_{12}) = \{RD(Object(X), Point(x)), \forall x \, x \in Wall\}[\mathbf{SYN}] \wedge$
$Contact(Object(X), Wall(x_{11} + \delta))[\mathbf{SYN}] \wedge Color(Wall(x_{11} + \delta), BLUE)[\mathbf{SYN}] \wedge$
$\quad Appear(F(X, 90° \leq \text{RA} \leq 270° \, \text{relative to} \, x_{11} + \delta), x_{11}, t_{ANY})[\mathbf{DIA}]$
$\quad \rightarrow Move(Object(X, \, \text{RA} = \text{Same}), t_{ANY} + \Delta)$

(6.34)

Then these can be dual-instance generalized with Rules (6.29) and (6.30) respectively as follows:

---

[14] We will not delve into the computational detail of the automatic and general process that underlies this replacement process. Suffice it to note that the system would search for ways to effect generalizations between the rules that it has in the Script Base and try a small number (not large scale search) of possibilities by replacing reference points for various angular or distance measurements. In this case, since an ANTI-GOAL has been identified, it is tried as a possibility. If that does not work, other reference points such as some turning points on the *Wall* or other objects may be tried. If nothing can be found after a small amount of search, it will abandon the process and move on to other methods to solve the counter-thwarting problem.

6.5 Application to the Spatial Movement to Goal with Obstacle (SMGO) Problem

$At(Object(X), x_{ANY})[\textbf{SYN}] \wedge$
$\exists Set\text{-}of\text{-}Points(S_{11}) = \{Point(x), \forall x \, x \in Wall\}[\textbf{SYN}] \wedge$
$Contact(Object(X), Wall(x_{ANY} + \delta))[\textbf{SYN}] \wedge$
$Color(Wall(x_{ANY} + \delta), \text{BLUE})[\textbf{SYN}] \wedge$
$\quad Appear(F(X, 270° < RA < 90° \text{ relative to } x_{ANY} + \delta), x_{ANY}, t_{ANY})[\textbf{DIA}]$
$\quad \rightarrow Not(Move(Object(X, RA = Same), t_{ANY} + \Delta))$

(6.35)

$At(Object(X), x_{ANY})[\textbf{SYN}] \wedge$
$\exists Set\text{-}of\text{-}Points(S_{11}) = \{Point(x), \forall x \, x \in Wall\}[\textbf{SYN}] \wedge$
$Contact(Object(X), Wall(x_{ANY} + \delta))[\textbf{SYN}] \wedge$
$Color(Wall(x_{ANY} + \delta), \text{BLUE})[\textbf{SYN}] \wedge$
$\quad Appear(F(X, 90° \leq RA \leq 270° \text{ relative to } x_{ANY} + \delta), x_{ANY}, t_{ANY})[\textbf{DIA}]$
$\quad \rightarrow Move(Object(X, RA = Same), t_{ANY} + \Delta)$

(6.36)

Rules (6.35) and (6.36) are powerful generalizations. Basically it says that no matter where it is along the *Wall*, if *Object*(X) is *Contact*ing it at a certain point (the ANTI-GOAL), its movement will be impeded in certain directions relative to the *Contact* point. Pictorially, Fig. 6.23 summarizes Rules (6.35) and (6.36).

Figure 6.23 also shows that the thwarted directions toward the GOAL always fall in the range of RA within which the movement of *Object*(X) is impeded. If one of the synchronic preconditions of Rules (6.35) or (6.36) is not met, then the rules may not apply. For example, if there is a patch on the *Wall* that is not BLUE, then the *Not* (*Move*(...)) consequence may not obtain. This does not mean that the *Move* consequence *will* obtain, but it may be worth the agent's effort to give it a try to move to that patch of *Wall* in a new process of physical exploration. Also, if there is *no Contact* with any part of the *Wall*, or at a position at which the *Wall* does not exist at the ANTI-GOAL location of *Object*(X), the *Not*(*Move*(...)) consequence may not obtain.

The knowledge as depicted in Fig. 6.23 and encoded in Rules (6.35) and (6.36) can be incorporated in a script much like that of Fig. 6.14 except that it would be a AEMCW (GENERAL) script. This is shown in Fig. 6.24.

The main difference between Figs. 6.14 and 6.24 is in the ANY value for the START position of Object(X) – AL = x_{ANY} – and the associated *RD1 (recall that "*" means "don't care," which also implies "any" value). Due to the restriction of space, in Fig. 6.24 we did not show the details of the OUTCOME in the situation when there is a force in the direction $270° < RA < 90°$. In this situation, in the OUTCOME portion, *Object*(X) will remain in contact with the Wall but the change in AL will be zero. I.e., AL(OUTCOME) = x_{ANY} and $+\Delta AL = x_{ANY} - x_{ANY} = 0$. This is much like what is shown in the OUTCOME portion of Fig. 6.17.

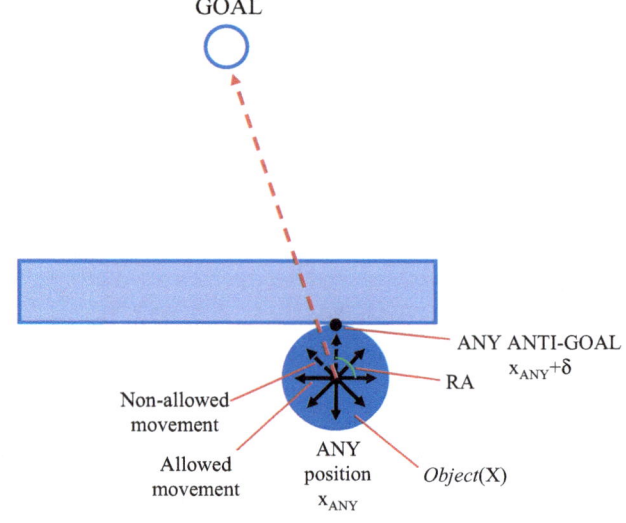

Fig. 6.23 The pictorial interpretation of Rules (6.35) and (6.36) that forms the basis for the general AEMCW script in Fig. 6.24. "Non-allowed movements" are the thwarted directions of movement toward the GOAL

6.5.2.2.2 Further Causal Reasoning with Mental Experiments and Simulations

In the algorithm of Fig. 6.21, termed the *Complete Counter-Thwarting Script Discovery Process*, attempts are made to effect some basic actions (on the thwarted object, *Object*(X), itself) in the physical world or to look for possible scripts (that may contain a chain of actions) in the Script Base that may assist partially or fully with the current situation. The attempt may generate a viable solution for some situations, such as the specific situation depicted in Fig. 6.16. If the attempt fails, then the process enters the next phase of *Further Reasoning and Mental Simulation to Discover Possible Counter-Thwarting Script* (Step 12 of Fig. 6.21). Figure 6.25 shows the process for this phase. We will discuss the various steps in turn.

6.5.2.2.2.1 Counterfactual Script for Possible Solutions

In Step 1 of the *Further Reasoning and Mental Simulation...* process (Fig. 6.25), we check if there are two or more possible solutions for the same problem arising from Step 5 of Fig. 6.21 (e.g., Figs. 6.13 and 6.20 are both attempted solutions toward the GOAL, one moving in the d direction first and the other in the d' direction first, and these d and d' elemental actions were discovered in Step 5 of Fig. 6.21). If not, the process proceeds to Step 3 to compose and propose a longer action sequence based on the only possible optimal elemental action (such as possibly a movement in the d direction first). Otherwise, the process proceeds to Step 2. In Step 2 we construct a counterfactual script based on the earlier attempts (say the processes described in connection with Figs. 6.13 and 6.20). Even though those attempts might not have provided viable counter-thwarting scripts, they might still contain useful information for further reasoning to find a viable script.

6.5 Application to the Spatial Movement to Goal with Obstacle (SMGO) Problem

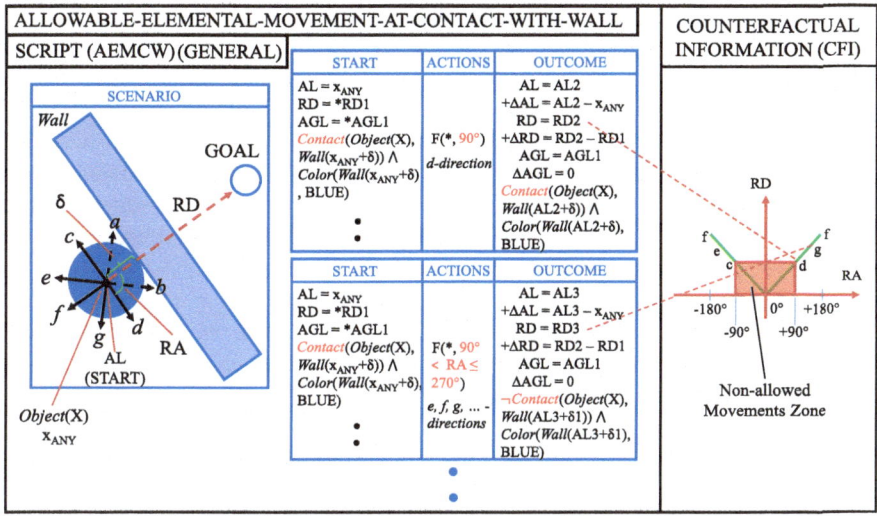

Fig. 6.24 The general AEMCW script – in contrast with that of Fig. 6.14, which is specific

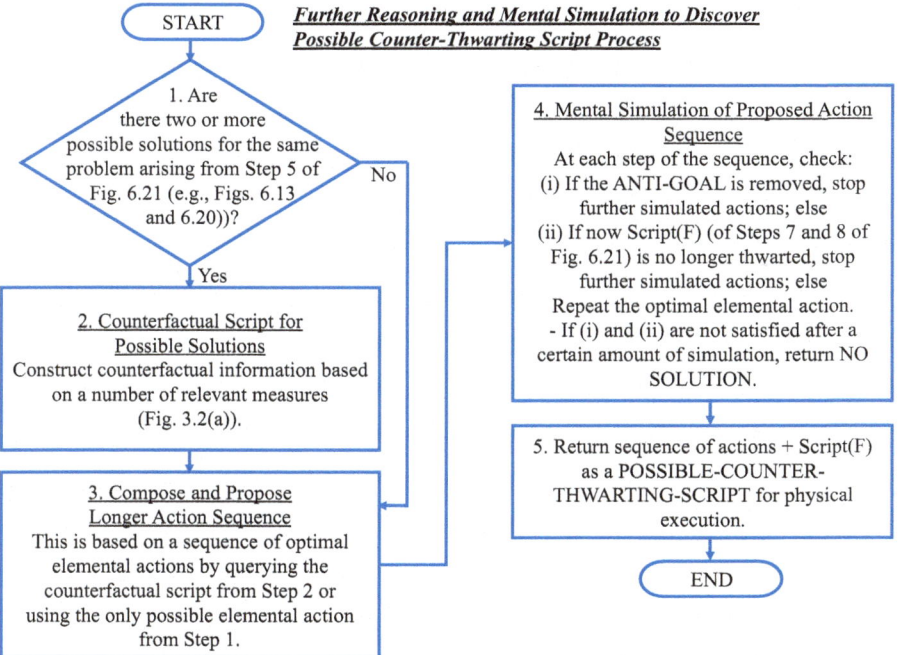

Fig. 6.25 The *Further Reasoning and Mental Simulation to Discover Possible Counter-Thwarting Script process* – Step 12 of the *Basic Counter-Thwarting Script Discovery Process* of Fig. 6.21

Whereas what was learned in the processes of Fig. 6.13a, b is the general condition that being in contact with the (BLUE part of the) *Wall* and at the same time trying to move in the direction of the *Wall* at ANY contact point along the *Wall* will result in the movement being thwarted, as encapsulated in the script of Fig. 6.14, what was learned in the processes of Fig. 6.20a, b is that by removing the existing *Contact* condition through elemental actions to some new position does not help with the removal of the thwarting situation, because the subsequent attempt to move in the direction of the GOAL from the new position is still thwarted (except in some special situations such as that in Fig. 6.16 as mentioned above). Also, in the processes of Fig. 6.20, the reasoning (that the *Wall* can thwart the movement of *Object*(X) in certain directions) can be carried out through *mental* simulation based on the knowledge learned in the processes of Fig. 6.13 through *physical* experience. In both the proposed solutions in Figs. 6.13 and 6.20, there are two MOVEMENT-SCRIPTS involved, one is to move away, using an elemental action (a one-step MOVEMENT-SCRIPT), from the movement thwarted location, identified as ANTI-GOAL(1), followed by an attempted movement toward the GOAL (a second MOVEMENT-SCRIPT that is multi-step). The two attempted solutions of Figs. 6.13 and 6.20 are reproduced in Fig. 6.26a, b respectively for ease of comparison. Even though in Fig. 6.20, the "original position" of *Object*(X) is labeled "*New (1) Position*," we label both "original positions" of Fig. 6.26a, b as *Original Position* for ease of comparison.

At this point, before proceeding further to find a way to remove ANTI-GOAL (1), the system considers what could be learned from the previous two schemes of movement. We summarize again the two schemes of movement: One is to move along the *Wall* in the *d* direction and then attempt to move toward the GOAL, and encountering the impediment (Fig. 6.26a). The other is to move away from the *Wall*, in the *d'* direction, immediately achieving the *Not*(*Contact*(...)...) situation,

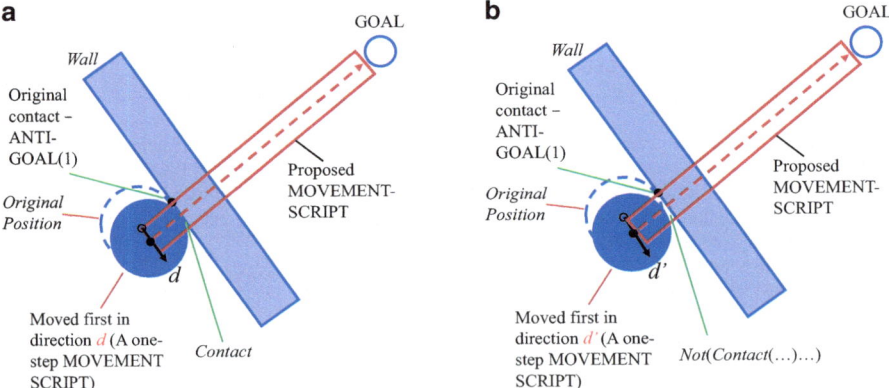

Fig. 6.26 Proposed solutions for counter-thwarting using a DUAL MOVEMENT-SCRIPT (**a**) Reproduced proposed solution from Fig. 6.13. (**b**) Reproduced proposed solution from Fig. 6.20. Both "original positions" in (**a**) and (**b**) are labeled *Original Position*

6.5 Application to the Spatial Movement to Goal with Obstacle (SMGO) Problem

and then attempt to move toward the GOAL, and also encountering the impediment (Fig. 6.26b).

Much like in the COUNTERFACTUAL SCRIPT CONSTRUCTION process of Fig. 3.2 that is responsible for the creation of the counterfactual information part of the MOVEMENT-SCRIPT (with CFI) of Fig. 3.1, the moment there is more than one solution to the same scenario, the system would construct a script with the attendant counterfactual information for the purpose of comparing the two solutions, even though, as mentioned above, in this case the solutions are only possible solutions that were found to be unworkable based on subsequent physical experience and/or mental simulation. This is Step 2 in the process of Fig. 6.25. The purpose of this is so that even though in these attempted and failed solutions a thwarting of the script is met and an ANTI-GOAL is created, any heuristics learned in this process may be able to guide future attempts to remove the ANTI-GOAL. This process encodes the idea "had these solutions been successful, which would have been a better solution?" The POSSIBLE-SMGO-SOLUTION-SCRIPT (PSS-SCRIPT) is shown in Fig. 6.27. Some details in the script are omitted for clarity. The "goodness" measures of the possible solutions include the energy expanded – ΔE – to Goal and to ANTI-GOAL, among possibly others. Two solutions – SOLUTION1 and SOLUTION2 (corresponding to those in Fig. 6.26a, b respectively) are shown in the script.

Now for the current experiences of Figs. 6.13 and 6.20, they cannot be included in the same counterfactual script directly because they have slightly different starting locations. We add in a built-in routine that says if there are two observed solutions to a problem that are close to each other, carry out a mental experiment that align them exactly – e.g., if the difference is in the starting position, then align the starting position. Hence in this case the system would carry out a mental experiment with the scheme of moving in the d' direction first while starting from the *Original Position* (Fig. 6.13a), in order to align it with the action sequence created in Fig. 6.13a, which results from moving from the *Original Position* in the d direction first. That way both solutions have the same starting positions and the same GOAL positions, and they can be compared for their consequence on relevant parameters such as energy expanded for the movement sequence – ΔE – as shown in Fig. 6.27. The system can also carry out more mental experiments and try out the other directions such as g and f, and it will find out that they do not give rise to a smaller ΔE than the direction d. Also, direction c would give the same value as d, so either one can be picked as the minimal option.

At this moment the script of Fig. 6.27 is still specific in the sense that the starting position for *Object*(X) is the specific *Original Position* as labeled in Fig. 6.13a. If another set of mental experiments is carried out from the *New(1) Position* (Figs. 6.13b or 6.20), including trying out at least directions d and d' and perhaps others as well, then a general script can be constructed (based on dual instance generalization) as a general version of Fig. 6.27. The counterfactual information in the general script would encode the knowledge that had there been an attempt to move to the GOAL or ANTI-GOAL using the DUAL MOVEMENT-SCRIPT scheme (Fig. 6.26) from *any* position along the *Wall*, then it would have been

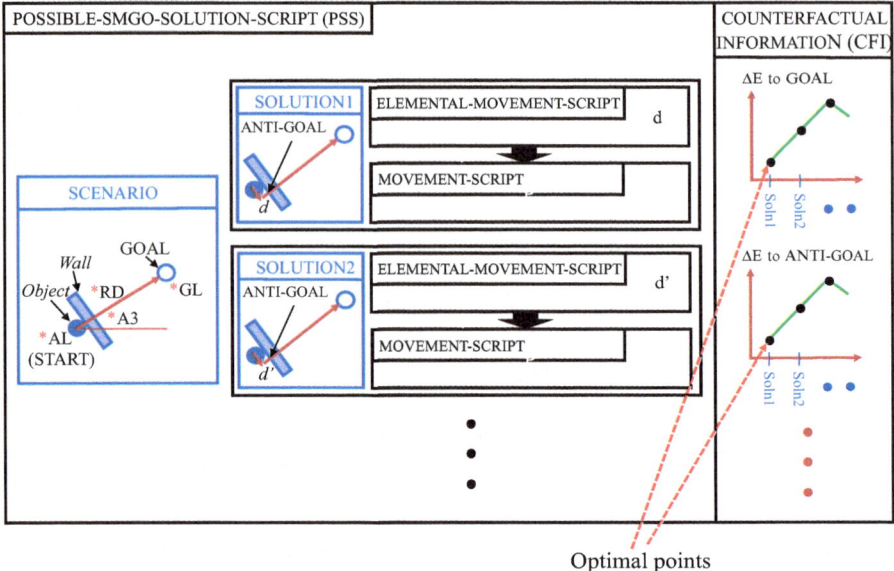

Fig. 6.27 Script containing counterfactual information on energy expanded – ΔE – for the various possible counter-thwarting solutions. SOLUTION 1 results in a lowest ΔE

better off moving *Object*(X) in the direction d rather than d' first – the attempt to first remove the *Contact* situation through moving in the direction d' is futile.

So, from the CFI portion of Fig. 6.27 (the general version), an optimality query (Fig. 3.2) would extract "SOLUTION1" – the one with the starting direction of movement being d – as the optimal solution to achieve minimal ΔE. Note that all this knowledge is inductive in the sense that there is no way at this stage, given a limited exploration of the environment, to know that perhaps by moving in a direction other than d, *Object*(X) may encounter some novel physical phenomena that will result in it being able to reach the GOAL or ANTI-GOAL with even less effort. One scenario often depicted in science fiction is that there is warped space-time that does not behave like our normal space-time. This may give rise to "worm hole" locations that can teleport an object across space instantaneously to another location with no effort or very little effort. If that is the case, there is always a possibility that perhaps moving in the direction d' from a specific location will result in an encounter with a worm hole, resulting in much less effort to get to the GOAL or ANTI-GOAL. Suppose such an experience is encountered in some exploration process, then it will be recorded as a special case to the general script such as that of Fig. 6.24.

6.5.2.2.2.2 Composing and Proposing Longer Action Sequence

Since in the previous attempts, elemental actions (the processes in Figs. 6.13 and 6.20, and Step 5 of Fig. 6.21) were attempted from around the thwarted location to

6.5 Application to the Spatial Movement to Goal with Obstacle (SMGO) Problem

no avail, the next attempt will look at constructing longer sequences of actions from the thwarted location that can potentially counter the thwarting effect of the *Wall*. This is Step 3 of Fig. 6.25.

To do this the system looks for actions in the earlier problem solving attempts that were executed *prior* to encountering the ANTI-GOAL (and thus further actions were thwarted) and tries and put them together to create a longer optimal sequence of actions.

In SOLUTION1 in Fig. 6.27, the action executed prior to encountering the ANTI-GOAL was an elemental action in the direction d. In SOLUTION2 of the same figure, the actions executed prior to encountering the ANTI-GOAL was an elemental action in the direction d' *followed by* an elemental action in the direction of the GOAL. Therefore, for SOLUTION1 there is one action while for SOLUTION2 there are two actions before encountering the ANTI-GOAL. In our case, the optimal action(s) before encountering an ANTI-GOAL happens to come from SOLUTION1 (Fig. 6.27) consisting of just one elemental action while in general, the optimal actions before encountering the ANTI-GOAL may consist of more than one elemental action.

To compose a longer sequence of optimal actions, the following piece of built-in knowledge is used:

> An optimal sequence of actions consisting of more than one elemental action can be constructed from a sequence of optimal elemental actions or optimal sub-sequences consisting of more than one elemental action each (6.37).

Given the script of Fig. 6.27, the system retrieves elemental action in the direction d from the script (SOLUTION1) and composes a sequence of actions consisting of a series of elemental actions in the direction d to attempt to solve the problem.

Recall that in the discussion in the beginning of Sect. 6.5.2.2.2.1, we mentioned that if there is just one elemental action (say, movement in direction d) available from the *Original Position* and hence Fig. 6.27 has only "one SOLUTION," then that elemental action is used to compose a sequence of actions (failing the test in Step 1 of Fig. 6.25).

The above processes (Steps 2 and 3 in Fig. 6.25) are not only specially crafted for the SMGO problem but are actually very general processes that can be applied to a variety of situations. Let us consider a different situation depicted in Fig. 6.28.

In Fig. 6.28 we show a situation that involves a Person trying to get to the other side of the Door (GOAL) but he finds that the Door is locked (there is an identifiable Lock on the Door) and he does not have the key. The Lock is identified as the ANTI-GOAL. There are two kinds of objects in the room that can be picked up by the Person – Light-and-Sharp (LS) objects and Heavy-and-Round (HR) objects. In a way, the situation is similar to the SMGO problem – there is an impediment to movement to another side of a "barrier." Suppose now the Person does not have the choice of "going around" the door because she is trapped in a room, unlike in the case of the SMGO problem. There is something analogous to the SMGO problem in terms of "elemental actions" available to the Person. The Person can pick up an LS

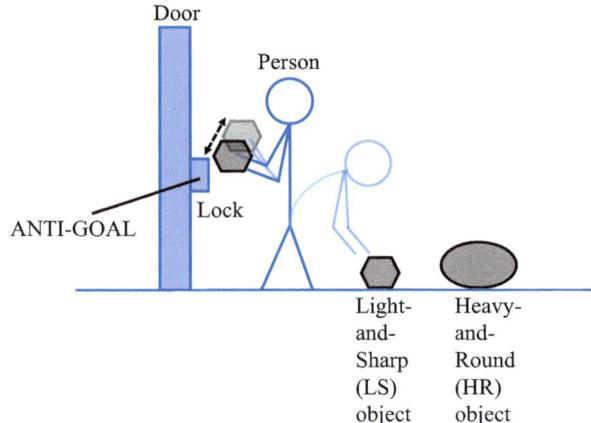

Fig. 6.28 Processes of Fig. 6.25 applied to another scenario – picking up Light-and-Sharp (LS) or Heavy-and-Round (HR) objects to break Lock in order to open Door

or a HR object and hit at the Lock. Let us simplify the situation and consider each "picking-up-and-hitting" action at the Lock with either kind of object an *elemental* action (in reality, it is made up of many more-elemental actions of picking up the object, moving it to the Lock, hitting the Lock with it, etc.)

Now suppose the Person, in an exploration phase to see what "elemental actions" are available to him, first picks up an LS object. Then, she puts down the LS object and picks up a HR object. This would correspond to the step of "Try various available elemental actions" (Step 5) in the *Basic Counter-Thwarting-Script Discovery Process* in Fig. 6.21. Between picking up an LS or an HR object, she selects an LS object because it is lighter so less effort is needed. This corresponds to the step of "Select Optimal Action" (Step 6) in Fig. 6.21. (Strictly speaking, the measure of action optimality includes the effect of the action on the Lock – how much it can damage the Lock but for simplicity in the current case we assume both actions lead to about the same effect on the Lock). Therefore, up to this point the Person has carried out these processes:

- Standing by the Door and encountering a Lock (Thwarted state)
- Picking up an LS or an HR object (Try various elemental actions)
- Selecting an LS object (Select Optimal Action, because an LS object is lighter)

Then she *mentally* simulates what she plans to do and expects will happen:

- Hitting the Lock with the LS object (one-time "elemental action")
- Expecting the Lock to break (impediment removed)
- Opening the Door and leaving the room (final sequence of actions to achieve GOAL. This is Script(F) of Steps 7 and 8 of Fig. 6.21 and is analogous to the final "Proposed MOVEMENT-SCRIPT from *New(1) Position*" of Fig. 6.13b)

This would correspond to the entire procedure on the right side of Fig. 6.21 – the *Basic Counter-Thwarting Script Discovery Process* of Fig. 6.11. Now, suppose the Person executes the entire sequence of actions above physically (Step ① of the

6.5 Application to the Spatial Movement to Goal with Obstacle (SMGO) Problem 273

procedure in Fig. 6.11) and is thwarted in her attempt as one-time hitting of the Lock does not damage it enough to allow the Door to be opened.[15] At this stage the Person can try to see what an HR object instead would do to the Lock with one elemental action – hitting the Lock one time. She tries it and the Lock sustains similar damage as before but more effort is needed to pick up the HR object.

Now the Person can build an internal counterfactual script much like that of Fig. 6.27 listing the "cost" of various attempted possible solutions. This would be executing Step 2 in the process of Fig. 6.25. What would be compared would be the cost of each of these actions *before* encountering the ANTI-GOAL – in this case the Lock – because the ANTI-GOAL for each of these actions is not removed by the actions, and the system is attempting to measure the cost of each of these thwarted actions to form a basis for building a longer chain of actions to see if it could help. This cost measure is analogous to the "ΔE to ANTI-GOAL" measure of Fig. 6.27.

Through this example one can understand better why built-in knowledge Rule (6.37) above refers to elemental actions as well as sub-sequences of elemental actions. For the case of the SMGO discussed above, the choice of the d direction constitutes one elemental action before the ANTI-GOAL is met and had there been no ANTI-GOAL, *Object*(X) then heads toward the GOAL. The choice of d' constitutes two elemental actions – one moving in the d' direction, the second one moving elementally one time in the GOAL direction – before the ANTI-GOAL is met. In the current example of Fig. 6.28, the actions of picking up an LS or HR object and hitting the Lock can also be thought of as constituting a series of elemental actions, but what matters is the effort involved in the entire sequence of actions before encountering the ANTI-GOAL. And in the current example, as in the SMGO example, the most optimal way of composing a sequence of actions is to join up sub-sequences of actions that are each optimal.

The next step in the procedure of Fig. 6.25, Step 3, is about composing a longer sequence of actions to see if it can assist with the desired thwarting of the ANTI-GOAL – in this case, the removal/destruction of the Lock. The corresponding counterfactual script (not shown) of Fig. 6.27 for the current example of Fig. 6.28 is consulted and picking up and hitting with the LH object is the optimal choice. This "elemental sequence" of actions is then composed into a longer sequence of actions, resulting in the composed long action sequence of "repeatedly hitting the Lock with the LH object" (the picking up action has to be carried out only once since the LH object would still be in the hand of the Person after the first hitting action).

---

[15] In our current example, we do not have the situation that corresponds exactly to the situations in Figs. 6.13 and 6.15. In those situations, the two instances of the movement being impeded by the *Wall* give rise to certain generalization about the movement impeding property of the *Wall*, and that is obtained everywhere along the *Wall*. There is also no situation here that corresponds exactly to the "attempting to move away from the *Wall*" possible solution – there is nothing in the synchronic preconditions of the thwarted actions involved here that would give rise to precisely this recommended action after causal reasoning processes such as those based on Rule 6.23, say, are carried out.

The Person carrying out this repeated hitting action may also make an intelligent observation in the process: Are these actions leading to increasing deterioration of the Lock, thus leading to the possibility of it being finally destroyed? If not, some other actions should be attempted rather than blindly continuing with the repeated hitting. Therefore, the composed longer sequence of actions is actually a *proposed* longer sequence of actions composing the repeated application of the optimal elemental step or sub-sequence of steps. This would be an *open-ended script* for physical or mental experiments to see if the sequence of actions can lead to any desirable results. An open-ended script specifies the START conditions and the sequence of actions but its OUTCOME portion is instead used as a check to see if the sequence of actions can ultimately lead to it. This is different from the usual script in which the OUTCOME portion specifies what is known to be the consequence of the actions involved and that can be used in a backward chaining process. Therefore, the OUTCOME portion for the script currently applied to the situation in Fig. 6.28 is something like "*Broken*(Lock)?", meaning "Is the Lock broken yet?" The number of steps in the sequence of actions is also not finite as it is not known beforehand – i.e., it is not known beforehand how many repetitions of the hitting action will lead to the Lock being broken.

In our current situation of the SMGO problem, this corresponds to the agent looking ahead to see if the composed sequence of actions, in this case the continued series of elemental movements in the d direction, would lead to any possibility of overcoming the current thwarted attempt to reach the GOAL.

6.5.2.2.2.3 Mental Simulation of Proposed Action Sequence

Having composed and proposed a longer sequence of actions in Step 3 of Fig. 6.25 based on the earlier learned optimal movement direction d, Fig. 6.29a shows the process of mentally simulating the proposed sequence of actions to see if the ANTI-GOAL can be removed. This is Step 4, Part (i) of Fig. 6.25. Compared to a typical heuristic search process such as A* in which at every step a number of possibilities are generated and evaluated and one course of action – the one that the heuristic measure deems best – is attempted, in this case the process is effectively skipping generation of various possibilities and heuristic evaluation at every step and simply attempts a specific action at every step, based on earlier determination that it is the optimal action to try out. The process, of course, does evaluate whether the GOAL is reached or the removal of the ANTI-GOAL is achieved at every step, just like in a typical search process. But a large amount of computation is obviated compared with the A* process. The reason this is achievable is that the optimal elemental actions are based on some generalization through causal learning that has already been determined earlier in steps that did involve some exploration and evaluation, but subsequently the system then does not keep generating actions and evaluating them with respect to some heuristics until the GOAL is reached or the ANTI-GOAL is removed. This is in the same spirit as the problem solving process of the SMG problem discussed in Chap. 2, Sect. 2.6.1.

6.5 Application to the Spatial Movement to Goal with Obstacle (SMGO) Problem

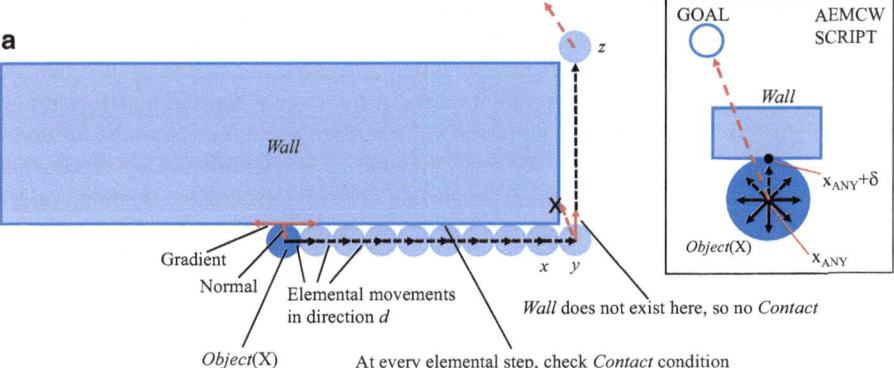

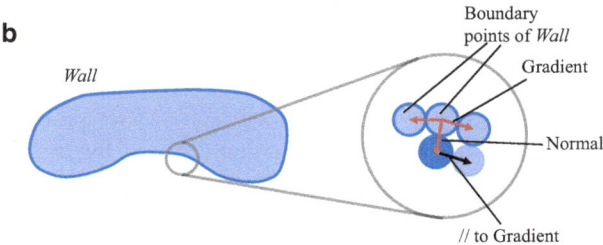

Fig. 6.29 Mental simulation of proposed long action sequence. Assuming no super-visual ability. (**a**) A rectangular Wall. (**b**) A curved Wall

For the ease of consideration, we have depicted the *Wall* to be relative large compared to *Object*(X) in Fig. 6.29a. After the last movement in the direction *d* from position *x*, *Object*(X) arrives at position *y*, which is a *Not*(*Contact*(...)...) situation. This means the ANTI-GOAL has been removed (Part (i) of Step 4 of Fig. 6.25) and the process returns the sequence of actions in direction *d* plus "Script (F)" (straight-line movement from position *y* to GOAL) as a POSSIBLE-COUNTER-THWARTING-SCRIPT (Step 5 of Fig. 6.25) for physical execution. If the Wall has a significant thickness as shown in Fig. 6.29a, then an impediment to its movement happens (marked with an "X") and the problem solving process would repeat the same whole process as discussed above – this is a new but same SMGO problem as in Fig. 6.7 with *Object*(X) starting from location *y*. This is assuming that the agent does not have super-visual ability and is hence not able to see the thickness of the *Wall* during the simulation. After repeating the problem solving process all over from location *y*, it would then bring *Object*(X) to position *z* and from that position Script(F) (straight-line movement toward the GOAL) will presumably not be impeded in the physical execution.

Now we digress and consider the optimal movement direction *d*. Direction *d* has been defined as a direction that is 90° relative to the ANTI-GOAL (Fig. 6.26).

However, there is another way to define direction d. Given a typical visual system, a usual piece of useful information that is available is the depth map of every point of a surface from an observer. From the depth map, there are two pieces of information that is typically derived from any point on the surface – the Normal at that point on the surface and the Gradient, which is how the surface's distance from the observer changes away from that point. We can characterize the Gradient as a vector that represents the direction along which the surface "moves" away from the point under consideration as shown with the red solid arrows in Fig. 6.29a.

Now, in the case of the *Wall* in Fig. 6.29a, the direction d happens to be always 90° from the direction of the ANTI-GOAL as one moves along the surface as shown but in general this may not be the case. Consider the situation of Fig. 6.29b in which the *Wall*'s surface is curved as shown. We enlarge the region around one of the boundary points on the *Wall* and show that actually in this case the direction of movement parallel to the Gradient at that point is not 90° away from the direction of the ANTI-GOAL. However, the most optimal movable direction for *Object*(X) is always along the same direction as that of the Gradient (and in the case of this curved *Wall* this fact can be learned and discovered through the same exploration process as that described for the straight *Wall* of our current SMGO problem above). This fact is the same for both the straight and curved *Walls* of Fig. 6.29a, b respectively. Moreover, for the curved *Wall*, the Gradient changes direction as one moves along the boundary of the *Wall*, and the optimal movement direction for *Object*(X) is not only not 90° away from the direction toward the ANTI-GOAL, it is also not constant relative to that direction. Therefore, in general the optimal movement direction for *Object*(X), "*d*", is the direction of the Gradient at every point on the *Wall*'s boundary. This knowledge/rule can be learned in a similar causal learning process as discussed above for the straight *Wall*, but we will not go into the details here.

Since this Gradient information is available from the visual system, it will be included as a parameter for consideration much like the absolute location (AL) and relative distance (RD), and rules can be learned and encoded using this parameter for specifying the desired movement of *Object*(X).

Let us return to the earlier discussion where we considered the simulation proceeding along direction d and discuss in further detail how Step 4 is executed. Now, as the mental simulation process moves *Object*(X) along the *Wall* (Fig. 6.29a), its impediment conditions are governed by the AEMCW-GENERAL script of Fig. 6.24 or Rules (6.35) and (6.36). A pictorial representing of the gist of the AEMCW script is shown on the right side of Fig. 6.29a which basically encapsulates Rules (6.35) and (6.36). Rules (6.35) and (6.36) basically say that if *Object*(X) is at location x_{ANY} and it is *Contact*ing the *Wall* at location $x_{ANY} + \delta$, there are certain directions along which you could apply a force on *Object*(X) and it will move and there are certain directions along which that will not happen.

Now, recall that we are currently executing Step 4 of the process in Fig. 6.25. As the mental simulation process moves *Object*(X) along the *Wall* in the direction d, it checks if $Not(Contact(Object(X), Wall(x_{ANY}+\delta)) \wedge Color(Wall(x_{ANY}+\delta), BLUE))$ is true, if it is, the ANTI-GOAL is removed. This relates to the precondition of Rule

6.5 Application to the Spatial Movement to Goal with Obstacle (SMGO) Problem

(6.35). This can happen in two situations. One is, *Object*(X) is physically contacting the *Wall* but is not contacting a BLUE part of the *Wall* (so far the experience of the agent with the *Wall* is always with the BLUE part of the *Wall*). The other is, there is no *Wall* at location $x_{ANY}+\delta$ (locations *y* and *z* in Fig. 6.29a). Remember that the removal of the current ANTI-GOAL may mean that the situation is improved but it does not guarantee that the impediment is removed as there may be other new physical effects not experienced and encoded. However, once a situation is discovered through mental simulation that the ANTI-GOAL is removed, it is worth converting the mental simulation to physical actions to see if it does improve the current impeded situation.

At the same time, the conditions of Rule (6.36) are also checked. This is Part (ii) of Step 4 of Fig. 6.25 – "Check if now Script(F) of Steps 7 and 8 of Fig. 6.21 is no longer thwarted." After the most recent elemental movement, if the GOAL is now in the direction which satisfies Rule (6.36), Script(F) is no longer thwarted. In the current SMGO scenario, Script(F) is the MOVEMENT-SCRIPT toward the GOAL from the current location of *Object*(X). In the case of the scenario of Fig. 6.28, it would be attempting to open the door after each hitting of the lock with the object.

Now, let us consider the situation in which the agent does have super-visual ability and hence the entire *Wall*'s constituent parts' locations are all known to the agent, along with all the corresponding boundary gradients. The mental simulation will continue with *Object*(X) following the boundary gradients in contact with the *Wall* as discussed above in connection with Fig. 6.29a, b. Since the shape of the boundary of the *Wall* beyond the right bottom corner is known, *Object*(X) will move "around" the right bottom corner of the *Wall* and move "upward." This process is shown in Fig. 6.30 – one can see that if the boundary gradient is followed along the *Wall*'s boundary, there is no *Object*(X)'s location in the mental simulation at which the *Contact* condition becomes false. Instead, we show the last few locations of *Object*(X) as *x*, *y*, and *z* in Fig. 6.30, and somewhere between locations *y* and *z*, the direction of the GOAL falls within one of the allowed movements' directions based on Rule (6.36) while *Object*(X) is still in *Contact* with the *Wall*. (In Fig. 6.30, for the sake of clarity, not all the allowed movements' directions are shown in the final two locations, *y* and *z*.) As soon as that location is reached, according to Step 4 Part (ii) of Fig. 6.25, the multiple optimal elemental movement process will stop as Script(F) – in this case the straight-line MOVEMENT-SCRIPT to GOAL – is no longer thwarted from this location.

Then, in Step 5, the long sequence of movements from the original thwarted position to this location between *y* and *z* is joined with Script(F) (the straight-line movement toward the GOAL) and this POSSIBLE-COUNTER-THWARTING-SCRIPT is returned. This brings the process back to Step ① of Fig. 6.11 in which this returned script is physically executed. The problem is solved as there is no further thwarting of the straight-line movement from the "upper right" corner of the *Wall* as shown in Fig. 6.30.

This process above replicates closely a typical human mental process in "thinking about" and solving this problem. Recall in Fig. 6.8a in which a series of dark red

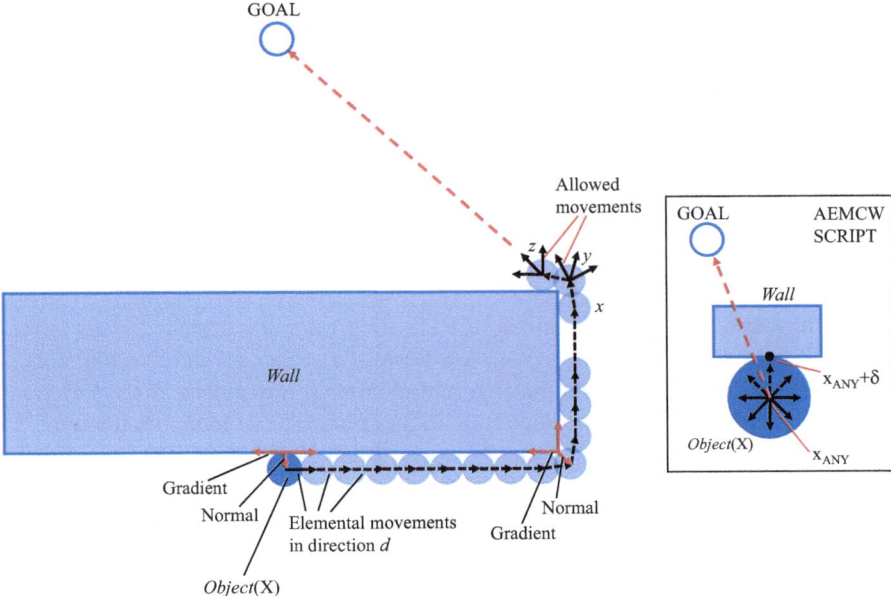

Fig. 6.30 Mental simulation in which *Object*(X) continues to move around the *Wall* because the agent has super-visual ability and can see the full thickness of the *Wall* from the starting position of the simulation. For the sake of clarity, we do not show all the directions of allowable movements for the last two locations of *Object*(X) – y and z

arrows are used to indicate the mental or physical path along which a "truly intelligent" agent would travel in solving this problem. Starting from the first thwarted position where the agent or agent pushing *Object*(X) encounters the *Wall* in the original straight-line path toward the GOAL from START, and after having determined that the *Wall* is impenetrable, the agent looks in the direction along the surface of the *Wall* (direction "*d*") to see if there is a "break" on the *Wall*'s surface at which the impenetrable condition does not hold. Therefore, before an attempt at physical movement to execute a plausible solution, it "mentally simulates" a path along the same direction and next to the surface of the *Wall* until it sees a change in the condition – in this case, a point along the *Wall* at which it is possible to head toward the GOAL unimpeded. For an agent with super-visual ability, this "point of movement freedom toward the GOAL" is between location y and z in Fig. 6.30. For an agent that does not have super-visual ability, this point is at the first corner of the *Wall* encountered along the surface in direction *d* (location y in Fig. 6.29a). The processes we discuss here replicate both these situations respectively in computational detail.

Suppose the *Wall* is huge relative to the agent (e.g., a typical wall encountered in our daily lives relative to an ant, or a "super large wall" relative to a human being), and the corners or other possible transitional points along the *Wall* are not visible from the thwarted position (e.g., the *Original Position* in Fig. 6.13), then the mental

6.5 Application to the Spatial Movement to Goal with Obstacle (SMGO) Problem

simulation of the long sequence of action terminates, as indicated at the end of Step 4 of Fig. 6.25, at which time a NO SOLUTION is returned. This is like "running out of energy or information to mentally simulate a solution" in a natural noological system.

In Fig. 6.8a it is also shown that having worked out the dark red arrows path in the first encounter with the *Wall*, if the agent encounters the SMGO problem again from the same START position, it should go along the blue arrow straight to one of the corners of the *Wall* to execute a more optimal path. This requires the knowledge of something akin to the Pythagoras' theorem that could have been learned earlier in some other situations. This is something worth investigating in future research.

6.5.2.2.2.4 The Non Super-Visual Situation

Now, in the real world situation, the visual system is not super-visual as we have defined it in Sect. 6.1, and as we have discussed above in connection with Fig. 6.29a. We now discuss further details concerning this situation. We assume the agent knows the location of the GOAL from earlier visual information or some other sources of information, which is the basis of the SMGO problem, but it does not have the information of the entire shape of the *Wall* or its "thickness." It sees the *Wall* only from its current starting location, which is the location labeled *Current Starting Position* as shown in Fig. 6.31. When it simulates the movement of *Object* (X) to location *x*, which is the most distant visible portion of the *Wall* at this juncture, it does not have any information on what happens to the *Wall* beyond that point "around the corner." Therefore, we use a built-in heuristic rule that says if the ongoing *Wall*'s boundary gradient is unknown, use the last known gradient to guide the next elemental movement in mental simulation. This then brings *Object*

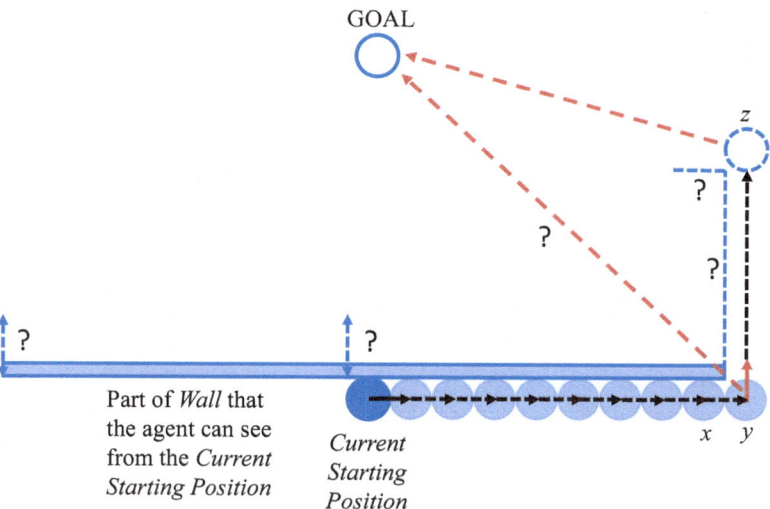

Fig. 6.31 The mental simulation process if the agent does not have super-visual ability

(X) to location *y*, at which point the *Not*(*Contact*(...)...) condition holds and there is a possibility that the straight-line movement toward the GOAL from that location is not thwarted. Hopefully, more information on the *Wall* can be observed from the new position after a physical movement to that location is made. According to Step 4 Part (i) of Fig. 6.25, the mental simulation for the long sequence of optimal elemental movements stops at this point – the ANTI-GOAL is removed because *Object*(X) no longer *Contact* the *Wall* at location *y*. And then, in Step 5 it returns the sequence of actions up to this point plus Script(F) (the straight-line movement to the GOAL from this point) and the agent is expected to use this POSSIBLE-COUNTER-THWARTING-SCRIPT to execute it physically to see if the solution works.

In the above discussion, we always assume that the agent stays with *Object*(X) as it explores the physical property of the *Wall* with *Object*(X) and observes the environment visually. If the agent and *Object*(X) can be separated, the agent can move around the environment and observe more of the *Wall* and then plan movements for *Object*(X) and come back to move *Object*(X) accordingly. This will be like in the case of the agent with a super-visual system as we have defined above in Sect. 6.1.

Similar to that discussed in connection with Fig. 6.29a above, Fig. 6.31 shows further what happens after the mental simulation of *Object*(X) moving to location *y*. After this, the agent (or agent and *Object*(X), if we assume they are the same entity or inseparable), receiving the returned POSSIBLE-COUNTER-THWARTING-SCRIPT from Step 5 of Fig. 6.25, will execute the script and will first move *Object*(X) to location *y* *physically* using the long sequence of recommended elemental movements along the *Wall*. Then it will make an attempt to move it toward the GOAL from location *y* in a straight-line (Step ① of Fig. 6.11). If the *Wall* is negligibly thin, the problem is solved. If the *Wall* extends vertically as shown (and this is now visible to the agent from location *y*), then the entire process above has to be repeated from Steps ②and③" of Fig. 6.11 because the attempted movement toward the GOAL from location *y* is now thwarted. Another location *z* will be derived accordingly to be the next physical location to move to, at which point an assessment will be made further to see if the problem can be solved.

If the *Wall* at the location *z* turns "backward" (Fig. 6.31) in such a way that it does not impede the straight-line path from location *z* to the GOAL, then the problem is solved.

6.6 A Deep Thinking and Quick Learning Paradigm

In this chapter we showed how, given available sensory information about the environment, a noological system can learn quickly about the physical properties of objects and encode useful generalizations that can drastically speed up problem solving processes, without the use of relatively "blind" and "unintelligent" search processes such as the A* process used in traditional AI (Hart et al. 1968). The bulk of this chapter was devoted to showing, in computational detail, how this can be done for the SMGO problem, which is typically handled using the A* algorithm in traditional AI.

It can be seen that the problem solving process discussed in Sect. 6.5 for the SMGO problem involves two critical mechanisms: one is the rapid effective causal learning mechanism of Chap. 2, Sect. 2.6.1 which requires only a small number of training instances (and in all the examples shown in this chapter, it requires two instances, at a medium desperation setting), and the other is a set of relatively involved but general reasoning processes (Figs. 6.21 and 6.25). Therefore, the paradigm can be characterized as one that involves *deep thinking* and *quick learning*. This is in contrast with the deep learning and reinforcement learning paradigms that are relatively slow (requiring a huge number of training instances – e.g., LeCun et al. 2015; Sutton and Barto 1998), and the old AI paradigm of relatively simple problem solving ("thinking") processes without any learning (such as the A* search algorithm of Hart et al. 1968).

It can be seen that both the solutions of Figs. 6.30 and 6.31 satisfy the original requirement as discussed in connection with Fig. 6.8, which we have pointed out would be what is expected of the behavior of a "truly intelligent" system. Both the deep learning paradigm and the traditional AI search processes, including reinforcement learning, are not able to provide quick problem solving processes that involve quick exploration and learning like what we have demonstrated in this chapter. Moreover, the deep learning paradigm is not able to provide explicit symbolic representations of the concepts learned. Explicit symbolic representations support complex reasoning and provides the foundation for communication between noological systems through language that greatly speeds up learning (e.g., learning of problem solving methods through language such as the tool construction and application example of Fig. 5.18, Chap. 5). The deep thinking and quick learning paradigm we have established represents an important fundamental principle for noological systems.

There is one more requirement that was discussed in connection with Fig. 6.8 that has not been addressed. We mentioned that having solved the SMGO problem the first time, subsequently the "intelligent" thing to do would be not to approach the *Wall* again and instead head straight to one of the corners of the *Wall* as shown with the blue arrow in Fig. 6.8. Currently, at the end of the problem solving process in Sect. 6.5, we obtained a sequence of scripts corresponding to the sequence of dark red arrows in Fig. 6.8 and this solution, being a chunked solution of the problem encountered, would be stored in the Causal Rules/Script Base of Fig. 2.28 for future quick deployment. More reasoning mechanisms are needed to replace the two dark red arrows, one representing the movement from the START location to the *Wall*, and the other representing the movement along the *Wall* to one corner of the *Wall*, with the blue arrow of Fig. 6.8. This requires the knowledge of something akin to the Pythagoras' theorem that could have been learned earlier in some other situations. We relegate this to future research.

We would like to make a final note concerning the "thinking and reasoning processes" of Figs. 6.21 and 6.25: they are also a kind of "problem solving script" which is amenable to learning. We also relegate this to future investigations.

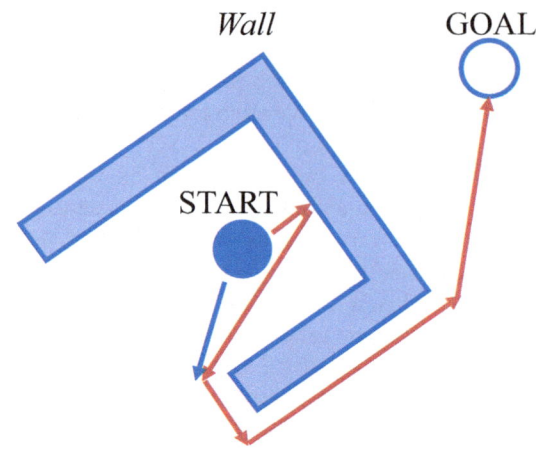

Fig. 6.32 A more complicated SMGO problem. The meanings of the *dark red* and *blue arrows* are the same as that in Fig. 6.8a

Problem

Consider the SMGO problem in Fig. 6.32. This is slightly more complicated than the SMGO problem discussed in this chapter (Fig. 6.7). Can the same mechanisms discussed in this chapter be used to solve this? If not, devise reasonable extensions to the method discussed in this chapter to solve it

References

A* Search Algorithm, Wikipedia (2015). https://en.wikipedia.org/wiki/A*_search_algorithm.
Forsyth, D. A., & Ponce, J. (2002). *Computer vision: A modern approach*. Englewood Cliffs: Prentice Hall.
Hart, P. E., Nilsson, N. J., & Raphael, B. (1968). A formal basis for the heuristic determination of minimum cost paths. *IEEE Transactions on Systems Science and Cybernetics SSC4, 4*(2), 100–107.
LeCun, Y., Bengio, Y., & Hinton, G. E. (2015). Deep learning. *Nature, 521*, 436–444.
Marr, D. (1982). *Vision*. San Francisco: W. H. Freeman and Company.
Russell, S., & Norvig, P. (2010). *Artificial intelligence: A modern approach*. Upper Saddle River: Prentice Hall.
Shapiro, L. G., & Stockman, G. C. (2001). *Computer vision*. Upper Saddle River: Prentice Hall.
Sutton, R. S., & Barto, A. G. (1998). *Reinforcement learning: An introduction*. Cambridge, MA: MIT Press.
Szeliski, R. (2010). *Computer vision: Algorithms and applications*. Berlin: Springer.

Chapter 7
Application to the StarCraft Game Environment

Abstract The StarCraft game environment provides an ideal computational platform to test and illustrate the various principles of noology that have been described in the previous chapters. In this chapter we describe an implemented AI program that plays against the StarCraft built-in game engine. Causal learning is applied successfully to rapidly learn the causal rules to engage and attack enemy agents. Scripts are learned along the way to accelerate problem solving. Counterfactual information associated with scripts that were alluded to in previous chapters are shown here to play a critical role in providing information for the planning of battle strategies. Affective competition is implemented as a high-level goal prioritizing mechanism for the agent involved. As in the previous chapter, the learning of heuristics is shown here to assist in reducing the search space needed for problem solving. Also, as in Chap. 5, it is illustrated here how the grounded conceptual representations used enable the system to learn problem solving methods rapidly through language.

Keywords StarCraft • Causal learning • Script • Counterfactual information • Affective computing • Affective competition • Heuristic • Heuristic generalization • Learning of heuristic • Problem solving • Battle strategy • Learning through language

In this chapter, we use a commercially available computer game environment called StarCraft II (2015) to test and illustrate the various noological principles that have been described so far in this book (Ho and Liausvia 2014). The StarCraft environment consists basically of simulated battle scenes with various fighting agents (the "soldiers") and equipment that supports the battles. The usual mode is to have human players play against the built-in simulator in which the human players would control the fighting agents and specify the setting up of various resources to support the battles. The built-in StarCraft simulator will generate the agents, the equipment, and the agents' tactics and strategies to challenge the human players.

StarCraft provides an API to allow another software to play the game in place of the human players. We therefore developed an "AI" system based on the foregoing principles of noology that we have developed to play against the StarCraft simulator. The AI system would begin with no knowledge of how to play or win in the games against the simulator and would learn through experience (training instances). In contrast to many other efforts that basically employ reinforcement learning (Sutton and Barto 1998) to learn battle tactics and strategies to play against the StarCraft simulator (e.g., Wender and Watson 2012 and a host of others), which usually requires many training instances, we employ causal learning as developed in this book that requires a small number of training instances. In the process, many of the devices developed in the foregoing chapters were employed, such as rapid causal learning, script learning, counterfactual script construction and query, affective competition, knowledge chunking, and learning of heuristics.

In addition to quick learning from direct experience, another advantage of our method is that the tactics and strategies learned are explicitly represented and therefore they can be extracted and described in symbolic language form for ease of communication with humans and other AI systems. These systems may then learn, through language, valuable problem solving methods from the other systems that have in turn learned through earlier direct experience of playing the game. This is "learning to solve problems through language" which is an extremely rapid form of learning which set human beings apart from all other animals (discussed also in Sect. 5.4, Chap. 5).

The following principles/methodologies are employed in the system that will be discussed in this chapter:

- Effective Causal Learning (Chap. 2)
- Learning of Script (Chaps. 2 and 6)
- Learning and Application of Counterfactual Script (Chaps. 2 and 6)
- Knowledge Chunking (Chaps. 2 and 3)
- Learning of Heuristics (Chaps. 2 and 6)
- Affective competition (Chap. 3)

As will be seen, the StarCraft game environment provides the opportunity to address the various noological processing aspects from perceptual processes to episodic and semantic memory processes, motivational and affective processes, conceptual processes, goal formation processes, action planning processes, and learning processes as depicted in Figs. 1.2 and 1.17 in Chap. 1.

7.1 The StarCraft Game Environment

The StarCraft game environment consists of many kinds of agents and many kinds of facilities (such as factories) that can generate battle-related resources such as ammunition, vehicles, and various kinds of structures. However, for our purpose here, we focus on just a few types of agents and we do not involve any of the facilities. Our focus here is on how the agents first learn the basic rules to engage/ attack individual enemies and then learn some good strategies to defeat a group of enemies. Figure 7.1 captures a typical scene in StarCraft consisting of some agents

7.1 The StarCraft Game Environment

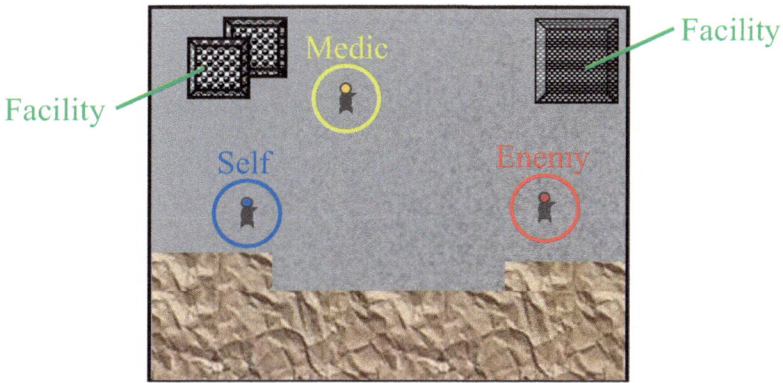

Fig. 7.1 A typical StarCraft scene with agents and facilities (There will be quite a number of "screenshots" from the StarCraft game environment illustrated in this chapter. Due to copyright reason, we did not capture the exact screenshots and illustrate them here. Instead, the same visual contents are redrawn and re-represented. To have an idea of what the original visual output looks like, please follow some of the hyperlinks provided in the rest of this chapter)

and facilities. For all the activities to be described in this chapter, the facilities are ignored.

We name the agent fighting on "our" side – controlled by our AI system – Self agent (circled in blue in Fig. 7.1). The fighting agent controlled by the StarCraft simulator is called Enemy agent (circled in red). Another type of agents, called the Medic agent (circled in yellow), is for "repairing" the fighting agents and restoring their "health state" to a higher level. In StarCraft, Medic agents are available for both the Self side and the Enemy side. We are concerned only with the Medic agent for the Self side, therefore the Medic agent circled in yellow in Fig. 7.1 is the Medic agent for the Self side.

Figure 7.2 shows a temporal log of ten parameters in the StarCraft environment. These parameters are associated with the agents present at any given time and together they define the state of the system. There are two agents present currently. These parameters change with time as the states of the agents change. In Fig. 7.2, the Time parameter is shown in the leftmost column (TIME) and the numbers in the column are the time steps. The two columns immediately to the right are the agents' absolute orientation, one containing the values for the Self agent (AG_1_0) and the other for the Enemy agent (AG_2_0). The numbers "1" and "2" in the names of the parameters indicate the agent involved – agent "1" or "2".

The other parameters are as follows: ATT is the Attack State of the agent, DIST is the distance between the agents (there are two values, one between agents 1 and 2 and the other between agents 2 and 1, which are of exactly the same values). HEAL is the Heal State of the agent which is 1 if the agent is in the process of being healed by the Medic. HP is the Health Point of the agent which reflects a level of "health" and which will be reduced as soon as the agent is "shot" by another agent. LAGL is the agent's relative orientation to another agent. MOV indicates whether

TIME	Agents' Absolute Orientation		Attack State		Distance between Agents		Heal State		Health Point	
	AG_1_0_	AG_2_0_	ATT_1_0	ATT_2_0	DIST_1_0	DIST_2_0	HEAL_1_	HEAL_2_	HP_1_0_	HP_2_0_
142	2	4	0	0	136	136	0	0	80	80
143	2	4	0	0	132	132	0	0	80	80
144	2	4	0	0	128	128	0	0	80	80
145	2	3	0	0	124	124	0	0	80	80
146	2	3	1	0	124	124	0	0	80	80
147	1	3	1	0	124	124	0	0	80	80
148	1	3	1	0	124	124	0	0	80	74
149	1	3	1	0	124	124	0	0	80	74

Agent's Relative Orientation to Another Agent		Move State		Magic Point		Absolute Location of Agent			
LAGL_1_	LAGL_2_	MOV_1_0	MOV_2_0	MP_1_0_	MP_2_0_	X_1_0_1	X_2_0_1	Y_1_0_1	Y_2_0_1
24	44	1	0	0	0	721	671	252	378
25	43	1	0	0	0	719	671	255	378
25	43	1	0	0	0	716	671	258	378
26	64	1	0	0	0	713	671	261	378
1	64	0	0	0	0	713	671	261	378
1	64	0	0	0	0	713	671	261	378
1	87	0	0	0	0	713	671	261	378
1	87	0	0	0	0	713	671	261	378

Fig. 7.2 The StarCraft parameters. See text for explanation (©2014 IEEE. Reprinted, with permission, from Ho, S. B. and Liausvia, F., "A Rapid Learning and Problem Solving Method," Proceedings of the IEEE Symposium on Computational Intelligence for Human-like Intelligence, Page 112, Fig. 2)

the agent is moving. MP is Magic Point which is applicable to Medic agent only – it reflects how much "power" the agent has to heal/charge up the fighting agents. X and Y are the co-ordinates for the absolute location of the agent.

7.1.1 The Basic Scripts of the StarCraft Environment

Figure 7.3 shows a comparison between the MOVEMENT-SCRIPT (GENERAL) that was discussed earlier (from Fig. 2.18, replicated at the top of Fig. 7.3) and the STARCRAFT-MOVEMENT-SCRIPT (GENERAL) that was learned and encoded in StarCraft (shown at the bottom of Fig. 7.3). Both movement scripts describe a straight-line movement from one point to another. The chief difference between the two scripts is that while in the idealized situation of Fig. 2.18, there are only three parameters associated with the script – the AL, RD, and GL parameters, in the more

7.1 The StarCraft Game Environment

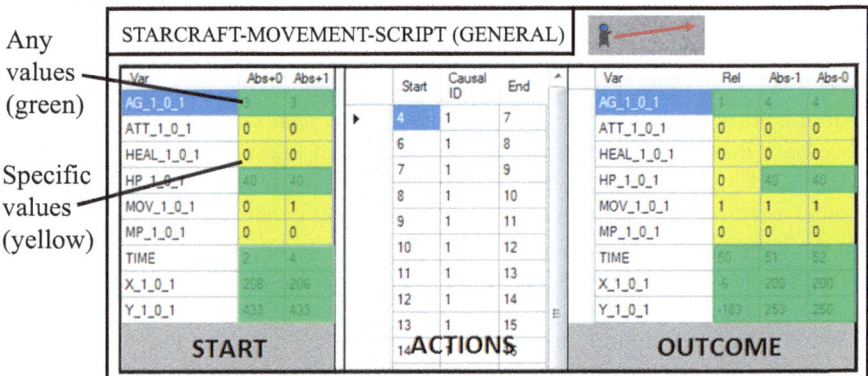

Fig. 7.3 Comparison between the MOVEMENT-SCRIPT (GENERAL) (From Fig. 2.18) and STARCRAFT-MOVEMENT-SCRIPT (GENERAL). The details of the script are explained in the text. Parameters in *yellow* remain constant across all instances of activities associated with the script ("specific values"). Parameters in *green* are those that change across instances ("any values"). The Var column contains the StarCraft parameters as explained in connection with Fig. 7.2. Abs+0, Abs+1, Abs−1, Abs−0 are the first two and the last two time steps. Rel is the difference between Abs−0 and Abs+0. Causal ID is the causal rule ID assigned to the particular action. Start and End are the starting and ending time steps of the rule involved (Bottom illustration above: ©2014 IEEE. Reprinted, with permission, from Ho, S.B. and Liausvia, F., "A Rapid Learning and Problem Solving Method," Proceedings of the IEEE Symposium on Computational Intelligence for Human-like Intelligence, Page 116, Fig. 7)

realistic situation of the STARCRAFT-MOVEMENT-SCRIPT, there are many more parameters associated with the environment and the agent involved. In the simulation, all the ten parameters described in Fig. 7.2 were used. However, in the interest of space, only a few of them are shown in Fig. 7.3.

Another difference is that while the START portion of the earlier MOVEMENT-SCRIPT contains only one value for each of the parameters, the START portion of the STARCRAFT-MOVEMENT-SCRIPT contains two sets of values, one at time "0" (Abs+0) and another at time "+1" (Abs+1). This is to allow the encoding of a possible starting condition that is diachronic – i.e., a certain change in a parameter always happen or is necessary for the event to proceed. In the case of the STARCRAFT-MOVEMENT-SCRIPT in Fig. 7.3, except for the TIME parameter, most of the two sets of values happen to be the same, but in the subsequent scripts that we will encounter (e.g., Figs 7.15 and 7.17, etc.), many of them are different.

The STARCRAFT-MOVEMENT-SCRIPT (GENERAL) was learned and encoded in a basic learning phase in which straight line movements with different starting and ending locations of one or more than one agent was activated. This learning phase would be similar to the Second Infant Learning Phase as discussed in Chap. 3, Sect. 3.5.1. As mentioned in Chap. 2, Sect. 2.6.2, one way these scripts are encoded would be the result of a problem solving process during which a START and a GOAL locations are specified and during which the problem solving process generates, in this case, a straight-line solution. As with the process of creating the MOVEMENT-SCRIPT (GENERAL) (Fig. 2.18) from a first instance MOVEMENT-SCRIPT (SPECIFIC) (Fig. 2.17a), the various instances of straight-line movements from different START and GOAL locations are grouped into the STARCRAFT-MOVEMENT-SCRIPT (GENERAL).

Among the various instances of movements, some parameters remained constant across these instances and they are colored yellow in Fig. 7.3, including the ATT, HEAL, MOV and MP parameters. These parameters remained constant across instances because the various instances of movement events either did not change them or changed them in a similar manner. Take for example, the ATT parameter. This Attack parameter remains 0 from the beginning of the movement to the end of the movement because there is no attack involved in a movement event. As for the MOV parameter, it started as 0 but changed to 1, indicating that there was a movement, and this change was the same across all the instances.

The parameters that are colored in green are parameters that are changed across instances. These include AG, HP, TIME, X, and Y. These parameters have the same properties as those labeled with a "*" in the MOVEMENT-SCRIPT (GENERAL) (Fig. 2.18). Take, for example, the TIME parameter. As the movement event can begin at any time, there is no reason why this parameter will be the same from one movement instance to another. Similarly, for the parameter HP, which is the Health Point of the agent, it can be of any value from instance to instance. Both these values can be set by a higher level action of initiating an agent with an arbitrary value of HP, and then setting the agent in motion at an arbitrary TIME.

Figure 7.4 shows the original situation from which the STARCRAFT-MOVEMENT-SCRIPT was learned. The agent "11" was moved from one location L1, to another location, L6. The Movement Event is bounded by the start of the change of location to the end of the change of location, indicated by the blue box. On the left hand side, we indicate how the three parameters TIME, LOC, and MOV change (we use L1, say, to encapsulate both coordinates, say, X1 and Y1 of LOC). AGENT and FORCE are not parameters provided by the StarCraft environment but are placed there for ease of understanding the process. In the StarCraft environment, a command can be given to move the agent in a certain direction and over a certain distance. The command is equivalent to a FORCE. The other parameters such as ATT and HP are not shown.

When the force F is applied at a certain time frame, say, t2, the change of location is effected in the next time frame, say t3. This can be seen in the change of

7.1 The StarCraft Game Environment

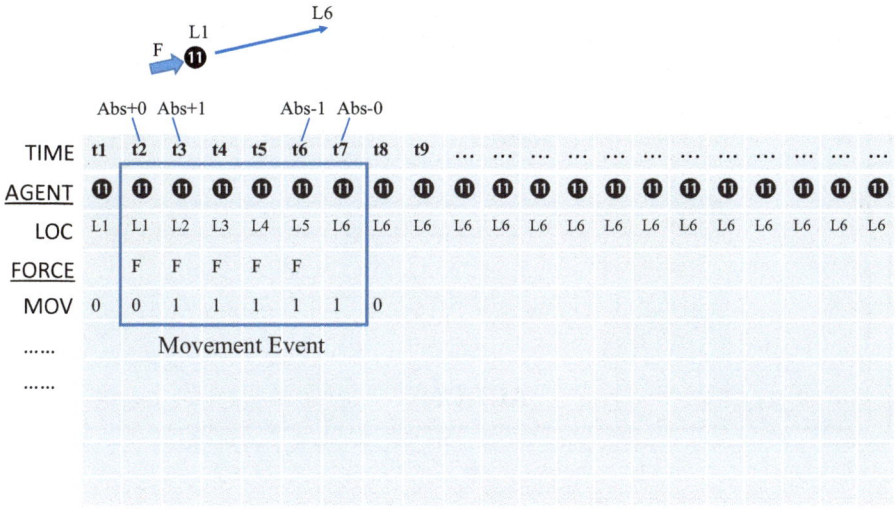

Fig. 7.4 The learning of the STARCRAFT-MOVEMENT-SCRIPT. Abs+0, Abs+1, Abs−1, Abs−0 are the first two and the last two time steps

L1 to L2 from t2 to t3. The MOV state is changed from 0 to 1 at t3, and then it stays at 1 until after the last change of location at t7.

In Fig. 7.4 we also indicate the definitions of the Abs+0, Abs+1, Abs−1, Abs−0 time frames relative to the event that are also used in the script in Fig. 7.3. Basically these are the first two and the last two time steps of the event involved. In Fig. 7.3, the Rel value ("Relative value") at the OUTCOME portion of the script is the difference between the values at Abs−0 and Abs+0 – i.e., the change from the beginning of the script to the end of the script.

The causal rule (indicated as "Causal ID = 1" in Fig. 7.3) that corresponds to the application of the force "F" (Fig. 7.4) occupies three time frames. It begins one time frame before the force is applied to one time frame after the force is applied. The Start and End in the ACTIONS portion of the STARCRAFT-MOVEMENT-SCRIPT in Fig. 7.3 basically indicate the starting and ending time frames of the rule.

Note that the parameters such as AG, HP, TIME, X, and Y are settable in StarCraft. That is, one can choose the TIME to initiate the movement, the Locations (X and Y) to start and end the movement, the Orientation (AG) of the agent involved, the Health Point (HP) of the agent, and the Magic Point (MP) of the agent (MP is only relevant to the Medic agent). The other parameters are generated by StarCraft in response to the situation, such as ATT, HEAL, and MOV.

Using the example of the STARCRAFT-MOVEMENT-SCRIPT, Fig. 7.5 shows all the possibilities of how the parameter values change across instances and within instances of scripts. There are four possibilities as shown in the figure. Consider the first one that says "Value starts with a *specific* value and does not change over the event" in connection with the ATT parameter. Recall that in the discussion above

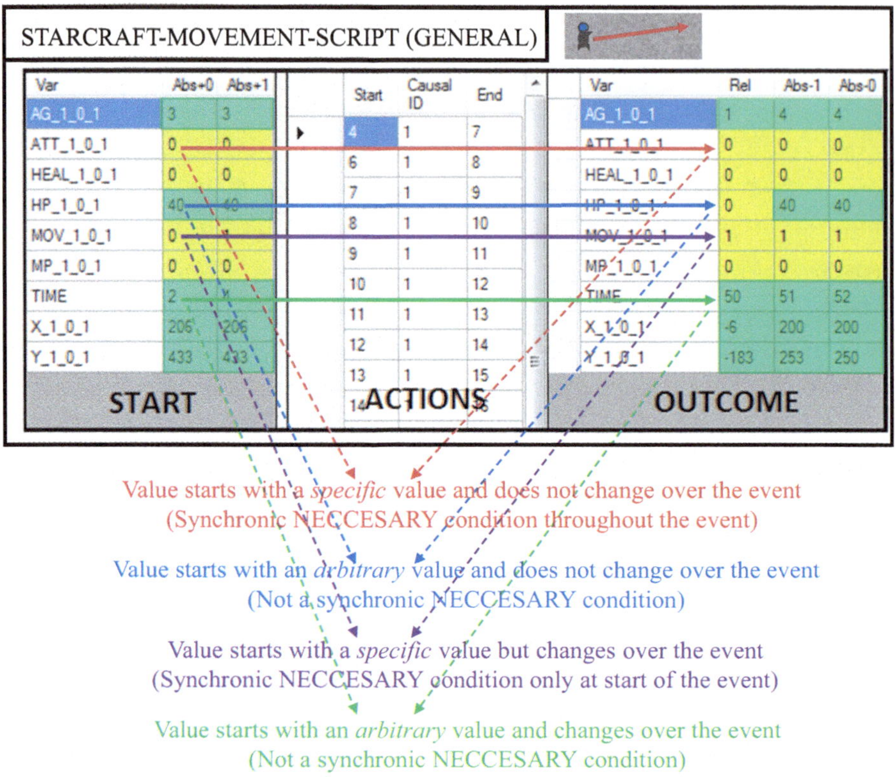

Fig. 7.5 Script parameter value changes across and within instances of scripts. (©2014 IEEE. Reprinted, with permission, from Ho, S.-B. and Liausvia, F., "A Rapid Learning and Problem Solving Method," Proceedings of the IEEE Symposium on Computational Intelligence for Human-like Intelligence, Page 116, Fig. 7)

we mentioned that the ATT parameter does not change from one instance of movement to another, not only for its starting value at the beginning of the event, but also throughout the entire event. Therefore, this parameter's value starts with a specific value and does not change over the event, and it is characterized as a "synchronic NECESSARY condition throughout the event." Recall from Chap. 2 that a synchronic necessary condition is one that has to be present in the precondition of a rule for the rule to be applicable. In this case, it means that for a movement event to take place, the starting value of ATT must be 0 (which means the agent cannot be in an attack state while moving). The other two parameters with similar properties as ATT are HEAL and MP (for this particular script).

The second possibility is "Value starts with an *arbitrary* value and does not change over the event." The example is the HP parameter. As mentioned above, the HP parameter relates to the health state of the agent and has nothing to do with movement per se. Hence, the agent can have any value of HP when a movement is initiated. This value is either set by a higher level process in initiating the movement or as a result of some processes, such as an Attack process that changes its HP value. And, unless the agent encounters other processes that change its HP (such as being attacked by an Enemy agent) in the process of movement, the movement process itself does not change the HP so it remains the same throughout the event. Specific value of HP is thus not a synchronic necessary precondition for the movement to begin.

The third possibility is "Value starts with a *specific* value but changes over the event," such as the MOV parameter. As mentioned above, the MOV value always starts at 0 at the beginning of the movement event and ends up with the value 1. Therefore, it is a synchronic necessary precondition for the movement to start but it does not retain the same value as the value at the beginning.

The fourth possibility is "Value starts with an *arbitrary* value and changes over the event," such as the TIME parameter. As mentioned above, a movement event can begin at any time and because time progresses, its value changes in the course of the movement event. The other two parameters that are similar to TIME in this regard are the X and Y location parameters, which is a consequence of the fact that a movement event can begin at any location and the location changes over the course of the event.

7.2 Counterfactual Scripts and Correlation Graphs of Parameters

Before delving into the discussion on learning and problem solving with respect to the StarCraft environment, there are some general processes in connection with scripts that have not been addressed before that we will discuss first in this section.

If one compares the parameters associated with each instance of an event across different instances, sometimes one observes that the values of the parameters change in a regular manner from instance to instance with respect to the values of some other parameters. We will use a simple example to illustrate the situation.

Figure 7.6a shows an object, represented by a circular shape, being acted on by a force, and moving an elemental distance and stopping – this is a typical situation in our real world when there is friction acting against the object involved, otherwise the object will keep on moving. Similar to before (e.g., Chap. 2, Fig. 2.15), we characterize the movement of the object with potential changes in five parameters – the change of its *absolute* location (ΔAL) and angle (ΔAA), the change of its *relative* distance (ΔRD) and angle (ΔRA) relative to some fixed points, and the change of its mass (ΔOM). We use the object's original position as a reference point to measure the relative distance, and the direction of the force as the reference for the relative angle of the consequential movement of the object. These are RD and RA respectively in Fig. 7.6a. For the absolute location, there are in principle two co-ordinate values, ALX and ALY, but we use AL to represent either ALX or ALY for the ease of discussion. The elemental movement predicate would be *Mov*(OBJ,

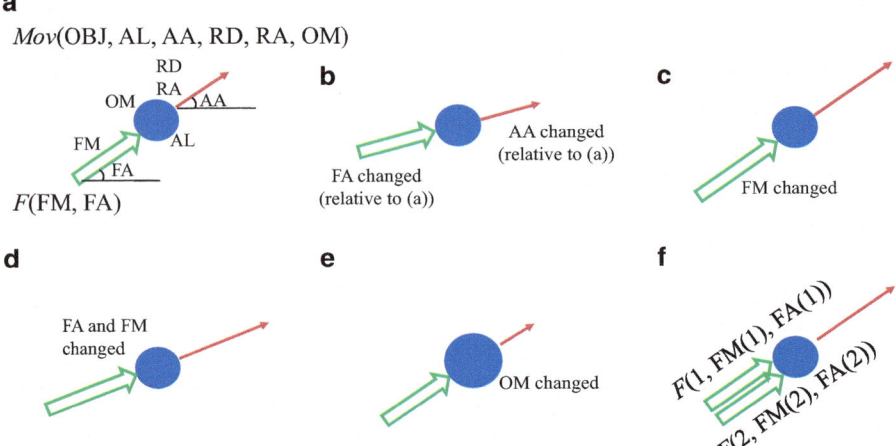

Fig. 7.6 Force and object movement to illustrate the idea of parameter correlation. OBJ is the *blue* circular object. (**a**) Parameters associated with force and movement. (**b**) Same force magnitude as in **a** but different force direction. (**c**) Same force direction as in **a** but different force magnitude. (**d**) Both force angle and magnitude are different than that in **a**. (**e**) Mass of OBJect changes relative to that in **a** but force direction and magnitude remain the same. (**f**) Two forces of the same direction and magnitude as in **a** act on the OBJect

AL, AA, RD, RA, OM) with the last five arguments representing the states of the object after the movement. The first argument, OBJ, refers to a specific object (like *Object*(X) in Chap. 6). As for the force, it has two parameters, its magnitude, FM, and absolute angle, FA. The force predicate would be F(FM, FA).

Now, suppose in an earlier learning process similar to the causal learning of force and movement that we discussed in Chap. 6, Sect. 6.4, a heuristic has been encoded that says that typically physical rules do not change with absolute location if there are no charged objects/environments involved. The system now observes what happens to the other parameters as forces F of different angles and magnitudes are applied to the object OBJ at possibly different absolute locations.

Now, suppose there is another instance, shown in Fig. 7.6b, that has been observed in which the starting values of all the other parameters (e.g., the magnitude of the force and the mass of the object, etc.) remain the same except that the absolute angle, FA, of the force is different. The system will then construct the FORCE-MOVEMENT-SCRIPT as shown in Fig. 7.7 in which there is a CFI (COUNTERFACTUAL INFORMATION) portion that captures the correlations between the changes of the parameters before and after the force application. In the script, the F(ORCE) and OBJ(ECT) are represented as two separate entities each with its own sub-script, and they interact through a causal connection (represented by the red arrow connecting F and *Mov*). For the *Mov* predicate, only the RD and RA parameters are shown in the interest of space.

In the CFI portion of the script in Fig. 7.7, graphs are shown relating the various parameters of the force F and the object's movement *Mov*. The starting values of

7.2 Counterfactual Scripts and Correlation Graphs of Parameters

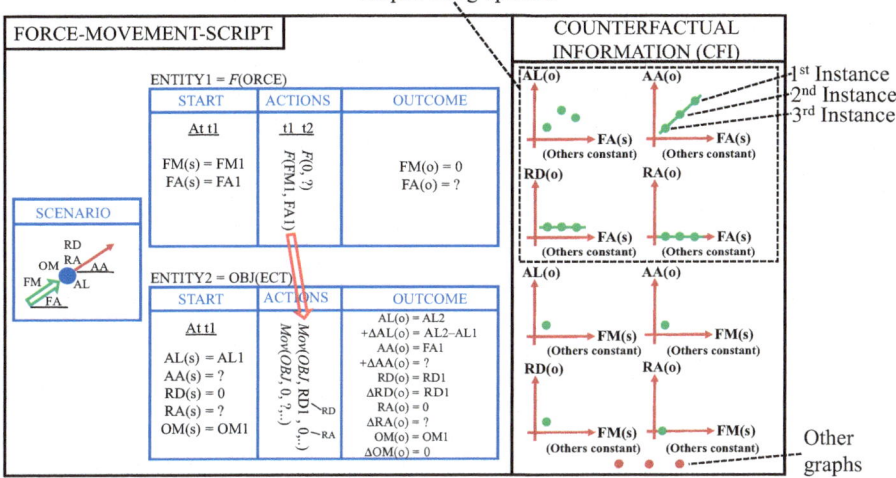

Fig. 7.7 FORCE-MOVEMENT-SCRIPT created as a result of experience with force and movement. In the interest of space, not all the parameters of *Mov* are shown. ParameterX(s) stands for the value of ParameterX at the START state and ParemeterX(o), at the OUTCOME state. *Green lines* in the COUNTERFACTUAL INFORMATION (CFI) portion of the script indicates generalization of the relationships between the parameters. See text for detailed explanation

the parameters associated with the object, such as AL (absolute location) and OM (object mass) can also affect the outcome of the other parameters but these are not shown in Fig. 7.7 and will be discussed later.

There are four graphs depicting the relationships between FA, the angle of the force F, and the outcome values of the four parameters AL, AA, RD, and RA of the object, given that the starting values of the other parameters such as FM and OM are constant (from instance to instance). (We omit the outcome values of the OM parameter for now in the interest of space and this will be discussed later.) We use FA(s) to represent the value of FA at the START of the script and FA(o) to represent the value of FA in the OUTCOME of the script, and similarly for the other parameters. The starting values of AA and RA are undefined – before F acts on the object, there is no movement so there is no "direction" associated with the object. These are indicated with a "?". Each green dot in the graphs represents the data point from one event (or one instance of the physical observation/event). For the other graphs in which the horizontal axis is FM(s), the data for the second instance/event are not indicated as currently the focus is on encoding the effects of the change of FA while keeping FM constant. At the very bottom of the CFI portion we show some dotted lines representing some other graphs not shown, and these would be the graphs encoding the change of OM (if any) with respect to changes in FM (s) and FA(s), the change of RD with respect to change of OM(s), etc.

In the first two graphs of AL(o) vs FA(s) and AA(o) vs FA(s), it can be seen that the value of AL(o) or AA(o) changes with respect to the change in the value of FA (s). In fact, the value of AA(o) not only changes with FA(s) but it is always equal to the value of FA(s) in both instances. In the third graph of RD(o) vs FA(s) it can be seen that the value of RD(o) does not change with the value of FA(s), and in the fourth graph of RA(o) vs FA(s) the value of RD(o) does not change with the value of FA(s) and is also always 0.

So far two instances of observation have been made – Fig. 7.6a, b. If one more instance of observation is made with yet a third value of FA(s), it is most likely that for the AL(o) vs FA(s) graph, the AL(o) values bear no regular correlation with FA(s) whereas for the AA(o) vs FA(s) graph, the value of AA(o) would continue to have the same value as FA(s) (Fig. 7.7, CFI portion). As for the other two graphs of RD(o) and RA(o) vs FA(s), they would continue to be "flat" – i.e., their values do not change with FA(s).

At this stage, using dual- or triple-instance generalization, one can say that AA(o) is always equal to FA(s), meaning that the direction of *MOV*ement is the same as the direction of *F*orce application, and RD(o) is the same no matter the value of FA, meaning that the amount of movement is the same no matter the direction of *F*orce application. As for the RA(o) vs FA(s) graph, it conveys the same information as the AA(o) vs FA(s) graph, namely the direction of movement is the same as the direction of force application (which means RA(o) is always 0). We use a green line coursing through the green dots to represent the constructed correlation and generalization between the two parameters involved. For values of the parameters involved that have not been observed/experienced, the noological system can use the lines to project, inductively, the expected behavior of the physical system.

If the physics of our reality is different from this, or that the object may be acted on by a force field such as an electric field as depicted in Fig. 6.6 of Chap. 6, and the amount of movement of the object given the same amount of force becomes dependent on the direction of force application, then the information will be encoded accordingly in these graphs in the CFI portion of the script.

Now, suppose there is another instance shown in Fig. 7.6c in which the absolute angle, FA, of the force, as well as the other parameters associated with the object are the same as that of Fig. 7.6a, but the magnitude of the force, FM, is different – with a larger value as shown in the form of a longer arrow. The consequence is that the movement of the object will be over a longer distance than in the previous situations – i.e. RD is larger. Now, after this event/instance and perhaps a few more similar ones are observed, more information can be added to the CFI portion of the FORCE-MOVEMENT-SCRIPT as shown in Fig. 7.8. Basically, after one or two more instances that allow for dual or triple-instance generalization, a proportional relationship is found between FM(s) and RD(o) – i.e., the larger the force, the larger the movement of the object proportionally, given that FA is constant. There is no regular correlation between FM(s) and AL(o), and AA(o) and RA(o) remain unchanged relative to the change of FM(s).

Now, suppose the second situation encountered is not like that of Fig. 7.6b, or Fig. 7.6c in which the parameters are the same as that of Fig. 7.6a except for *one* parameter that has been changed (absolute angle of the force in Fig. 7.6b and magnitude of the force in Fig. 7.6c), but instead both parameters of absolute angle and magnitude of the force (FA and FM) change as shown in Fig. 7.6d. In

7.2 Counterfactual Scripts and Correlation Graphs of Parameters

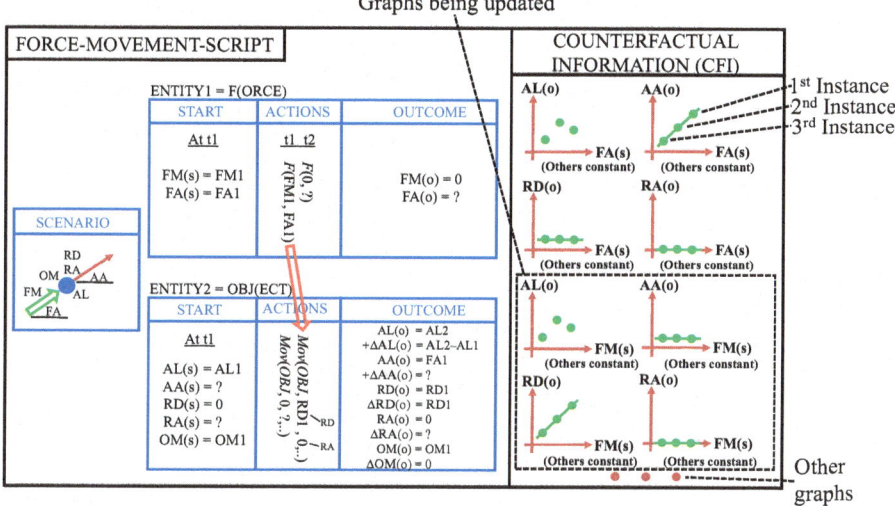

Fig. 7.8 After the situation of Fig. 7.6c, in which a force of a larger magnitude than Fig. 7.6a was applied, while other parameters remain constant, the CFI portion of the script of Fig. 7.7 is updated

this case, the system would not be able to form any generalization at this stage by comparing the instances of Fig. 7.6a, d. This instance will simply be stored for future situations in which there will be an appropriate instance to compare with. For example, if the situation of Fig. 7.6b is later encountered, which, when compared with the situation of Fig. 7.6d, differ only in one parameter's value, both of these situations can then be used to construct the RD(o) vs FM(s) graph.

Now we shall discuss two other parameters associated with the object, AL and OM, that we have omitted earlier. For AL, its starting value, AL(s), could in principle affect the relationship between FM(s) or FA(s) on the one hand, and the changes in the values of the object's parameters – AL, AA, RD, RA, and OM – on the other. As mentioned above, in Chap. 6, Sect. 6.4, a situation was depicted in which an electric field can affect the effect of a force on an object depending on the location of the object. If this is the case, a graph of, say, RD(o) vs AL(s) (all other starting parameter staying unchanged from instance to instance) can capture this relationship.

For OM, as with AL, there are two aspects to consider. One is whether OM can change, just like AL, before and after the force application. In non-relativistic physics, OM would remain constant no matter the object's speed. However, OM can change with speed in relativistic physics (Lorentz et al. 1923). In our situation, however, we assume that the object would travel for a short distance after the force acts on it and stop. In which case, OM(o) and OM(s) would be the same since the speed is 0 before or after the event, even if relativistic effects as we know them are taken into consideration in the process of the movement. This is of course not to rule out the possibility of an alternative physical law in which the gained mass stays, and

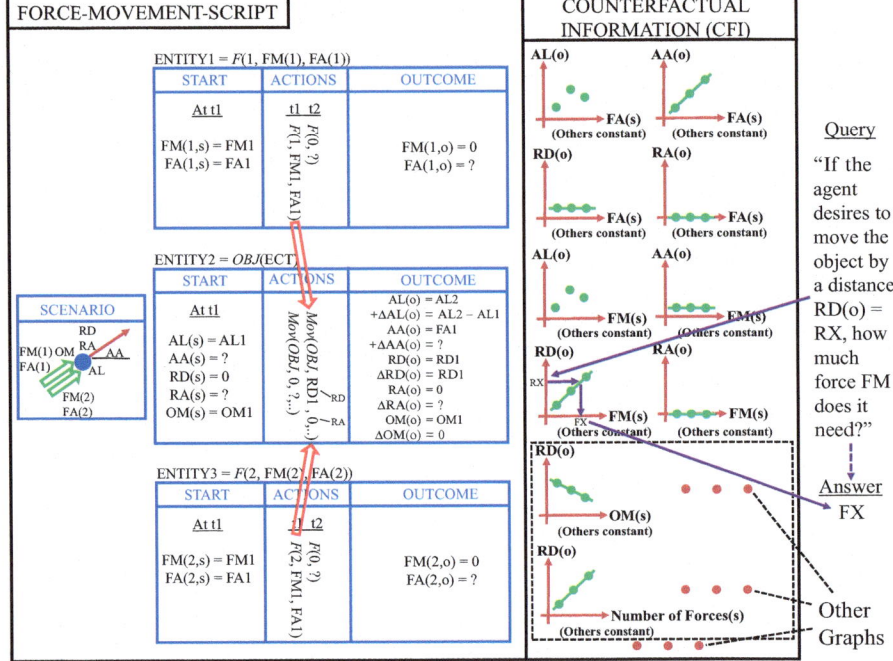

Fig. 7.9 Counterfactual information that captures change of displacement with respect to mass of object, as well as to the change of the number of forces applied to the object in different instances. A query for information from the CFI portion is shown. Each force is represented as a separate causal entity and there is one more parameter associated with the force – the force "ID", "1" and "2" associated with the two forces

therefore OM(o) > OM(s). If so, our learning algorithm will encode this in a graph just like it has done for the other parameters that change.

The other aspect is whether the starting value of OM, OM(s), will affect the other parameters. In the real world as we know it, in fact it will – given the same amount of applied force, the larger the mass the lower the speed that will be gained by the object acted on by the force, and if there is friction present, which in our case we assume that there is, the distance traveled by the object, RD, will be smaller, given all other parameters being the same. This situation is shown in Fig. 7.6e and will be captured as an inversely proportional relationship in a RD(o) vs OM(s) graph (other parameters being constant) and encoded in the CFI portion of the script accordingly. This is shown in Fig. 7.9 in the dotted box.

Yet another parameter that can be varied in the situation of Fig. 7.6 is the number of forces acting on the object. A two-force scenario is shown in Fig. 7.6f. If the additional forces have the same magnitude and angle of force application as that in earlier instances such as that in Fig. 7.6a, the amount of movement of the object, RD (o), will be proportionally increased. This is shown in the last graph of Fig. 7.9 at the bottom of the CFI portion in the dotted box. In Fig. 7.9 it is also shown that the additional forces are represented as additional causal entities acting on the object. There is an additional parameter associated with the force – the "ID" of the force, shown as "1" and "2" in the figure.

In a physical situation, the counterfactual graphs in Fig. 7.9 can be used to predict what happens when certain input parameters are set at certain values, even for situations other than those that have been observed or experienced before. This is generalization and "forward reasoning." The generalization is inductive, of course, and the system may be surprised to find a departure from the predicted value. If so, the system would register the departure in the graphs. Forward reasoning with Fig. 7.9 could also be useful in a mental simulation process in which certain actions could be part of a problem solving process, and their effects have to be simulated in order to arrive at a proposed solution.

More often, a noological system would need to carry out "backward reasoning" – given that it desires a certain effect(s), what would the action(s) have to be to achieve that. This information is also readily available in the counterfactual graphs. An example is shown in Fig. 7.9 – "If the agent desires to move the object by a distance RD(o) = RX, how much force does it need?" The answer can be read off the RD(o) vs FM(s) graph, and the answer would be, FX, say.

Figure 7.10 summarizes the above process of the construction and updating of the CFI portion of a script. The process begins with an observation of an instance of an event (such as the above force-movement event) with certain parameters (such as the location, the magnitude and angle of force applied, etc.). The Script Base is then queried to see if there is an earlier instance which differs from this current instance in the starting value of only one parameter, while the starting values of the other parameters are the same. The Script Base basically stores all the different scripts of various events that have been encountered earlier. There are two kinds of scripts. One kind is specific, which is basically a script that encodes an event that has been encountered the first time. One example would be the specific movement script depicted in Chap. 2, Fig. 2.17a. The other is general, which is constructed from more than one instance of the same event in which some parameters have been generalized based on the instances observed, such as the general movement script depicted in Chap. 2, Fig. 2.18. In some of the scripts, there could already be a CFI portion containing the records of some earlier encountered instances to which the current instance can be compared, such as the CFI portion of the FORCE-MOVE-MENT-SCRIPT of Fig. 7.9. The process then constructs or updates the counterfactual information accordingly.

There are two situations in which this query of the Script Base will not lead to useful generalization. One is, the event is observed the first time. Then, the new instance is simply encoded in the Script Base as a specific script for future comparisons. The other is, the current instance differs from some other earlier instances of the same event in more than one starting parameter values. An example would be Fig. 7.6d compared with Fig. 7.6a as discussed earlier. In this case, the instance is encoded as a specific instance which may be useful in future situations in which generalizations with yet other instances may be possible.

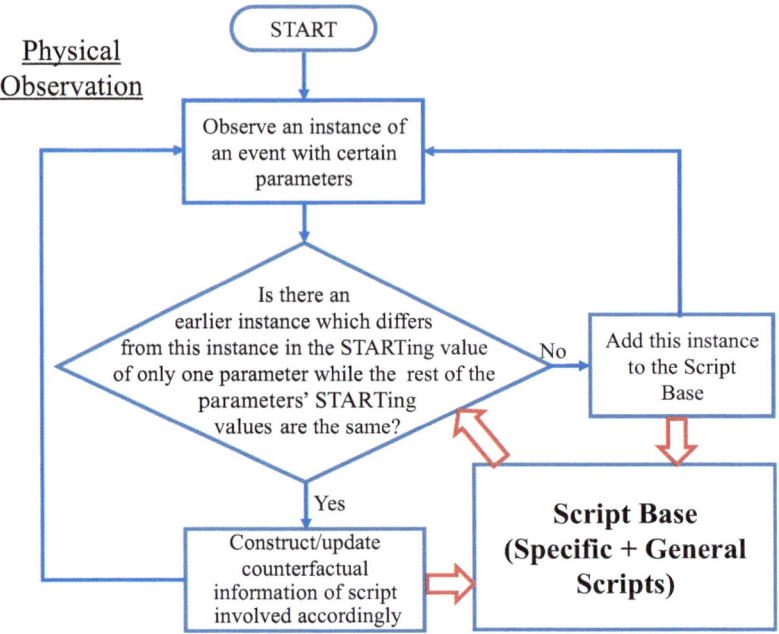

Fig. 7.10 Physical observation and the construction and updating of the CFI portion of a script

Often, a long time may elapse for the right instances of certain event to appear in the environment to provide the right parameter values to construct the counterfactual graphs. This could be detrimental to the noological system because there could be a need for it to use information derivable from the graphs for certain urgent problem solving processes that may improve its survivability. The noological system would therefore seek to derive further useful information from the environment by actively performing "physical experiments." Typically, in an experiment in a "scientific" setting, all parameters are "controlled" at the same value as before except the one that is of interest for the setting/scenario of concern. Therefore, the value of this parameter of interest would be varied while others are kept constant, and the effect(s) observed and encoded accordingly into the counterfactual graphs. The process is shown in Fig. 7.11. The same process takes place in human beings and other animals for the learning of day-to-day concepts through interaction with and observation of the environment.

Even though in this section we have used a relatively simple example of a force-movement event to illustrate how the CFI portion of the script can be constructed, the basic devices and processes are general and could be applied to a wide variety of situations. Below we will apply these to the StarCraft game situation in which there is a need to learn battle tactics and strategies to win the game.

7.3 Desperation and the Exhaustiveness of Observations and Experiments

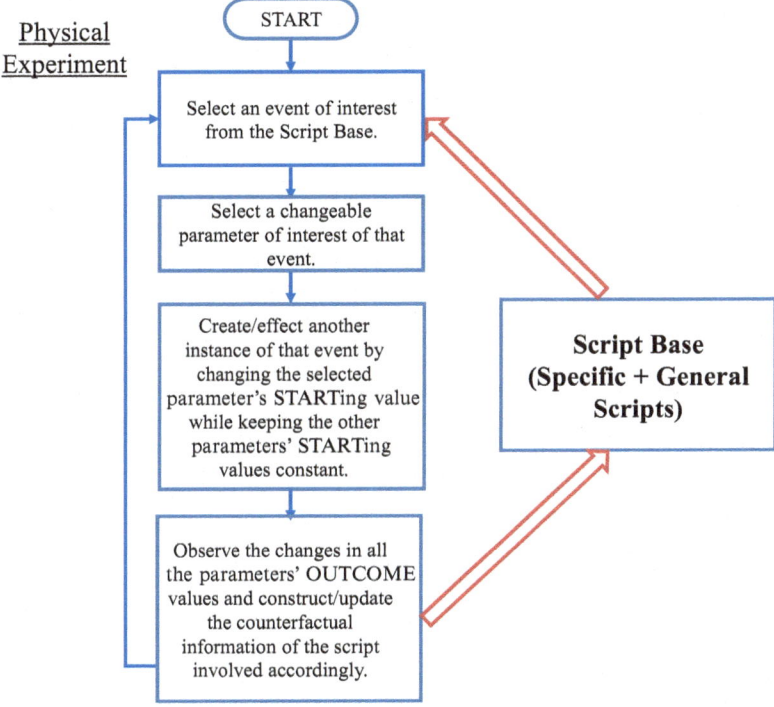

Fig. 7.11 Physical experiment conducted by a noological system to enrich the CFI portion of a script

7.3 Desperation and the Exhaustiveness of Observations and Experiments

A point to note about the Physical Observation and Physical Experiment processes, as discussed in the previous section and summarized in Figs. 7.10 and 7.11 respectively, is that there is no attempt at executing an exhaustive process to observe or experimentally search the entire space of all possible values of the various parameters to characterize the various causal rules relating the parameters involved. For example, consider the AA(o) vs FA(s) graph of the FORCE-MOVEMENT-SCRIPT of Fig. 7.7. The graph shows that the consequential direction of movement of an object involved, as a result of the application of a force on the object, AA(o), is the same as the direction of the force application FA(s), other parameters being constant (e.g., the mass, OM, of the object and the magnitude, FM, of the force involved are constant). This rule is derived from the two instances of observation (or experiment) of Fig. 7.6a, b in which the mass and the magnitude of the force involved are the same across the two instances. However, these "other parameters" stay at a *particular* value across the instances considered. What if these other parameters are constant but have some other values in other pairs of instances?

In principle, to have a complete picture of the rules relating the various parameters involved, an exhaustive observational or experimental process should be carried out in which these other parameters are observed or set at various different values, and then the relationship of the two parameters of concern are observed to see if there are locations in the parameter value space at which the relationship between the two parameters are different (say, from the "proportional and equal" relationship of the AA(o) vs FA(s) graph of the script in Fig. 7.7, for the case of the AA(o) and FA(s) parameters). However, as mentioned above, for observation this is often impractical as the relevant situations may not present themselves for a long time, and for experiment, this would involve a large amount of search effort.

Recall in Chap. 2, Sect. 2.5, we discussed the connection between the level of *desperation* of a noological system and its propensity to effect generalization on observed rules about the environment. Under the pressure to apply a rule to solve certain problem for the purpose of survival, generalization may be effected so that it is better to have some rules to try out than not. The situation is similar here. A noological system learns the rules between force and movement such as in the example of the previous section in order to apply them to some problems. If the system is desperate to apply the rules, then it will make do with whatever rules that have been established with specific "other parameter" values and generalize them and apply them accordingly, even without a high confidence on the generalized rules. If the system is not desperate, then it can afford to be more "scientific" and spend the time to carry out more exhaustive observations and experiments to more thoroughly explore the parameter value space to have a more complete and accurate picture of the rules involved. The situation is similar to that summarized in Fig. 2.13.

7.4 Rapid Learning and Problem Solving in StarCraft

In this section we apply the various learning and problem solving mechanisms in the foregoing discussion in this book to the learning of battle strategies in StarCraft. The same basic mechanisms are employed: effective causal learning as discussed in Chap. 2, small amount of exploration and training instances, construction of scripts for knowledge chunking for improving the efficiency of problem solving, affective competition as discussed in Chap. 3, Sect. 3.3, and learning of heuristics as discussed in Chap. 2, Sect. 2.6. A more involved example of heuristics learning will be encountered in this section.

There are two levels of causal learning discussed in this section. There is a somewhat "tactical" level at which the Self agent learns how to engage/attack an individual Enemy agent (we are assuming there is no prior built-in knowledge on how to do this). This is discussed in Sect. 7.4.1. Building on the tactical level, there is a more "strategic" level at which a number of agents coordinate among

7.4 Rapid Learning and Problem Solving in StarCraft 301

themselves to attack a group of Enemy agents, learning the optimal strategy from "physical experiments" – experiences created deliberately much like in the process of Fig. 7.11 to speed up learning so that immediate successful problem solving for survival in a battle situation can be effected. This is discussed in Sect. 7.4.3.

7.4.1 Causal Learning to Engage/Attack Individual Enemy Agents

Suppose the starting state of the battle situation is as depicted in Fig. 7.1. Suppose also earlier that the Self agent has learned the scripts of straight-line movements as discussed in Sect. 7.1.1. Now, we need to define a battle-related goal for the Self agent. This would be to Engage/Attack an individual Enemy agent. As mentioned above, we are assuming that there is no built-in knowledge on how to do this and it has to be learned. The StarCraft parameter ATT (Fig. 7.2) encodes the Attack State of the agents. When $ATT = 1$, the agents involved are engaged in an attack situation which consists of mutual exchange of fire between two agents. This is a "tactical" level of engagement.

In Fig. 7.12 we provide an overview of the backward chaining problem solving process to be used in the current situation. We begin with a top-level GOAL of Engaging/Attacking the Enemy agent, which corresponds to achieving $ATT = 1$. The basic process consists of first trying to see if the Script Base has any script or causal rule (a causal rule can be thought of as a script with only one step of action) that has $ATT = 1$ in its OUTCOME portion, and if not, the system would attempt to learn a relevant script. If the Script Base does have an appropriate script (the "first" script), then that is invoked and its preconditions (in the START portion) are examined to see if they are already satisfied. If some of them are not satisfied, then the Script Base is queried again to see if there are any scripts that can be used to satisfy them (i.e., with the appropriate states in their OUTCOME portions that will match up with the preconditions/START state of the first script). This process continues until all the preconditions are already satisfied by the current situation, just like in any backward chaining process.

The process depicted in Fig. 7.12 is no different than a typical backward chaining process except that in our framework, all the causal rules or scripts involved for each stage of the problem solving process must be learned from interacting with the environment. Once these rules are learned, they can be backward chained as shown in Fig. 7.12.

Below we assume that the Self agent begins with no script that has $ATT = 1$ in the OUTCOME portion. Therefore, the agent cannot activate any script to satisfy the GOAL right away. The Self agent then begins wandering in the environment in the hopes of discovering an appropriate rule and this is shown in Fig. 7.13. This is the exploration process in our learning framework.

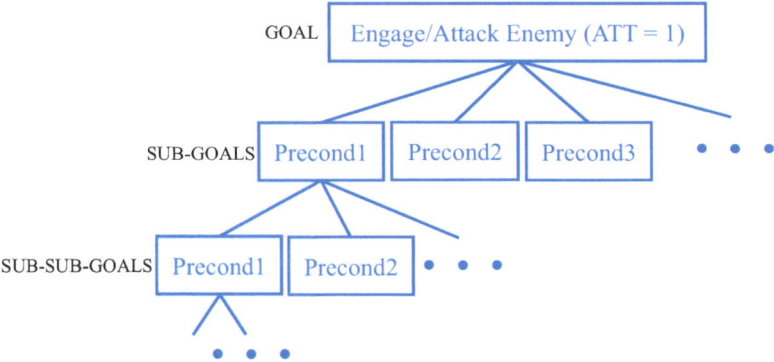

Fig. 7.12 Backward chaining process for achieving an Engage/Attack situation with an individual Enemy agent

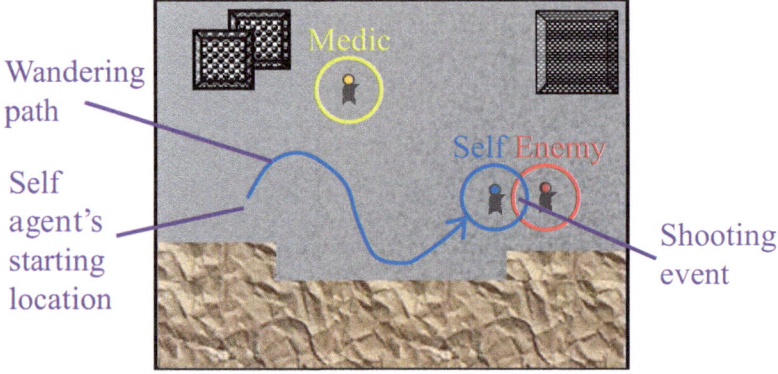

Fig. 7.13 The Self agent wanders in the environment to discover the causal rule for Engaging/Attacking the Enemy agent. During the Attack event, the agents shoot at each other, shown with animated graphical simulation of gun shots in the StarCraft game that is not shown here

In Fig. 7.13 it is shown that after some wandering around, the Self agent happens to come close to the Enemy agent. It turns out that there is a built-in rule in the StarCraft system which is when an agent from one side of the battle is at a certain distance from an agent from the other side (124 units), an Attack event (that involves both agents shooting at each other, shown with graphical simulation in StarCraft) will ensue, and if the agents do not then move away from each other, the shooting will continue until one of the agents "dies" as a result. In the process of shooting, the HP values of both agents decrease. The built-in rule is such that in order to trigger an Attack event, the distance between the agents involved must change from a particular value to another value, and a small number of parameters must have certain absolute values, while most of the other parameters' changes or absolute values do not matter. "Death" in the StarCraft sense is defined as $HP = 0$. At the moment of the beginning of the Attack event, the ATT value will also change

7.4 Rapid Learning and Problem Solving in StarCraft

from 0 to 1 and will change back to 0 after the shooting ends. These changes in ATT, HP, etc. will show up in the time log of all the parameters such as that shown in Fig. 7.2.

In some settings of the StarCraft environment, when two agents are engaged in a shooting event, the Enemy agent – the one controlled by StarCraft – may move away halfway through the event, or more Enemy agents may come and reinforce the original agent. For our simulations, we have arranged the settings so that the number of Enemy agents stays at a fixed number from the beginning of the shooting until at least one of the agents on either side dies.

Now, our Self agent begins with no knowledge of this built-in rule for engagement with the Enemy agent. In the wandering process when it reaches a certain distance from the Enemy agent, the StarCraft system activates the shooting/attack event and changes ATT from 0 to 1. These parameter changes are captured from the StarCraft time log and shown in Fig. 7.14. The causal learning mechanism detects this, and the change of distance from a certain value (128 units) just prior to the change of ATT to a value (124 units) at which ATT becomes 1 is identified to be a *tentative* necessary cause of the change in ATT (Chap. 2). The reason why the distance changes by an amount of 4 units from 128 to 124 is that in StarCraft, each time a move command is given to the agent, its relative distance to the other agent changes by 4 units. There is no one-unit change in the values of this parameter.

Figure 7.14 shows the above-mentioned change of the DIST (distance between the agent) parameter from a value of 128 to a value of 124 followed by a change of ATT from 0 to 1 in two blue boxes. There are also other parameters that change, such as TIME, AG, LAGL, X, and Y (highlighted in green boxes). Also, the values of the other parameters that do not change, such as HP and MP, have specific values at this moment of the initiation of the Attack event. Based on the rapid effective causal learning process as described in Chap. 2, at this moment those parameters that change will be encoded as diachronic preconditions and those that do not are encoded as synchronic preconditions. A diachronic precondition means that the *change* in the parameter involved is a necessary precondition for the consequence to take place and a synchronic precondition means that the *particular value* of the parameter involved is a necessary precondition for the consequence to take place. Therefore, after this first instance of an Attack event, these parameters are encoded accordingly as the preconditions for the change of ATT from 0 to 1. This is the initiation of the Attack event, which is the current top level GOAL for the Self agent (Fig. 7.12). In general, of course, the same set of preconditions can give rise to more than one consequence – i.e., other than ATT, there may be another parameter that changes as the same time as well. Therefore, at this stage, a causal rule with the above preconditions can be extracted as follows:

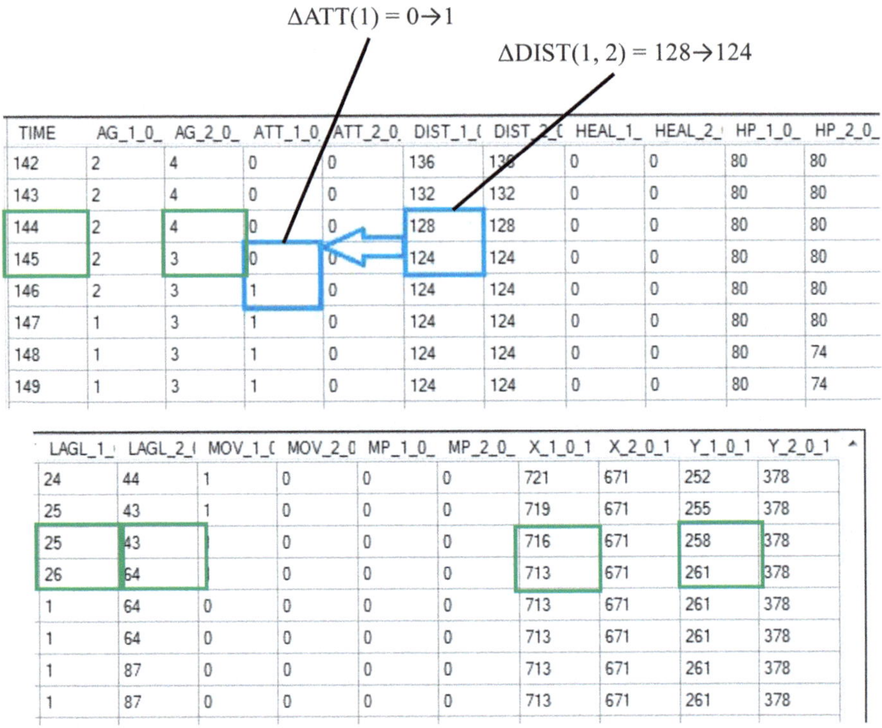

ΔATT(1) = 0→1

ΔDIST(1, 2) = 128→124

TIME	AG_1_0_	AG_2_0_	ATT_1_0	ATT_2_0	DIST_1_0	DIST_2_0	HEAL_1	HEAL_2	HP_1_0	HP_2_0
142	2	4	0	0	136	136	0	0	80	80
143	2	4	0	0	132	132	0	0	80	80
144	2	4	0	0	128	128	0	0	80	80
145	2	3	0	0	124	124	0	0	80	80
146	2	3	1	0	124	124	0	0	80	80
147	1	3	1	0	124	124	0	0	80	80
148	1	3	1	0	124	124	0	0	80	74
149	1	3	1	0	124	124	0	0	80	74

LAGL_1	LAGL_2	MOV_1_0	MOV_2_0	MP_1_0	MP_2_0	X_1_0_1	X_2_0_1	Y_1_0_1	Y_2_0_1
24	44	1	0	0	0	721	671	252	378
25	43	1	0	0	0	719	671	255	378
25	43	1	0	0	0	716	671	258	378
26	64	1	0	0	0	713	671	261	378
1	64	0	0	0	0	713	671	261	378
1	64	0	0	0	0	713	671	261	378
1	87	0	0	0	0	713	671	261	378
1	87	0	0	0	0	713	671	261	378

Fig. 7.14 StarCraft parameter changes at the beginning of an Attack event (©2014 IEEE. Reprinted, with permission, from Ho, S.-B. and Liausvia, F., "A Rapid Learning and Problem Solving Method," Proceedings of the IEEE Symposium on Computational Intelligence, Page 112, Fig. 2)

$$[\Delta\text{TIME} = 144 \rightarrow 145 \wedge \Delta\text{AG}(2) = 4 \rightarrow 3 \wedge \Delta\text{DIST}(1, 2) = 128 \rightarrow 124 \ldots$$
$$\wedge \text{HEAL}(1) = 0 \wedge \text{HEAL}(2) = 0 \wedge \text{HP}(1) = 80 \wedge \text{HP}(2) = 80 \ldots] \text{at time t}$$
$$\rightarrow [\Delta\text{ATT}(1) = 0 \rightarrow 1] \text{at time t} + 1$$

(7.1)

Recall that the basic algorithm of causal learning based on the discussion in Chap. 2, and as encapsulated in Appendix B, begins with identifying an event (a change of state/value of some parameter) and looks "backward" temporally to identify diachronic and synchronic preconditions/causes. This is basically how Rule (7.1) is created. There are of course other events as shown in Fig. 7.14 that will also give rise to other potential causal rules. However, after a "causal learning filtering" process, spurious causalities will be eliminated. This will become clearer after we discuss the process associated with Rule (7.1).

As soon as the ATT state changes to 1, there will be a series of shooting events that take place until one of the agents "dies" – i.e., its HP becomes 0. This sequence of shooting events is then captured in the form of an ATTACK-SCRIPT

7.4 Rapid Learning and Problem Solving in StarCraft

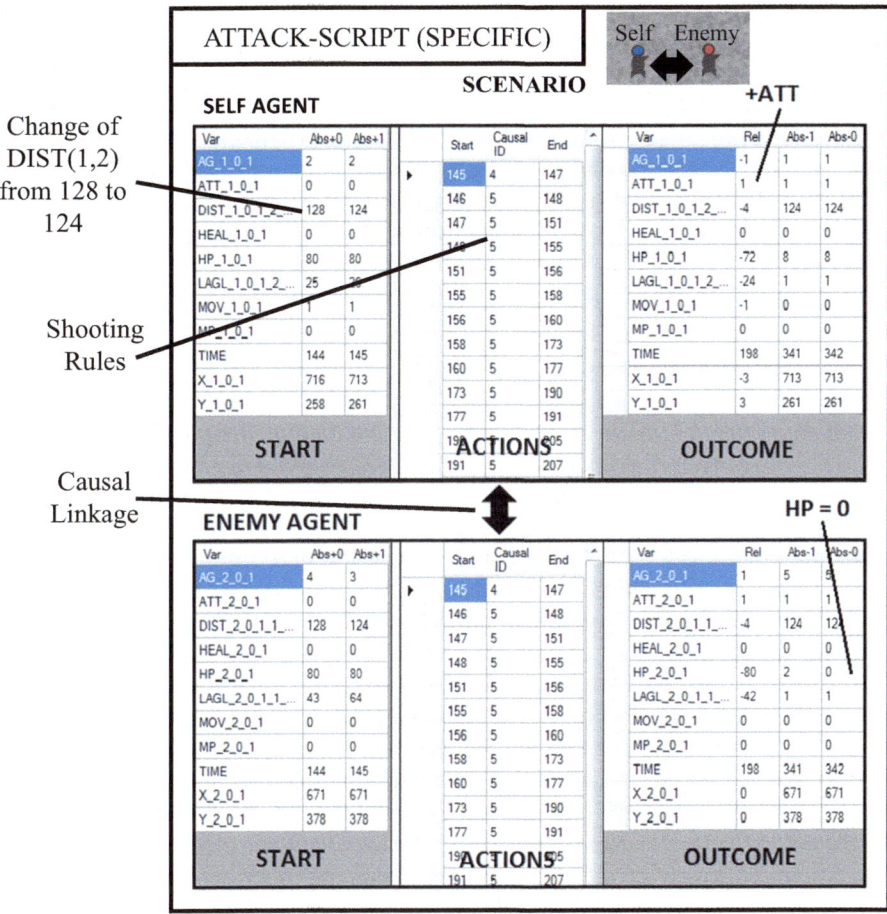

Fig. 7.15 A *specific* ATTACK-SCRIPT. Each entity, the Self or Enemy agent, has its own sub-script (©2014 IEEE. Reprinted, with permission, from Ho, S.-B. and Liausvia, F., "A Rapid Learning and Problem Solving Method," Proceedings of the IEEE Symposium on Computational Intelligence, Page 114, Fig. 4)

(SPECIFIC) as shown in Fig. 7.15, and the preconditions of the Attack event in Rule (7.1) become the START portion of the script.

In this script, as in the FORCE-MOVEMENT-SCRIPT of Fig. 7.9, etc., each entity's associated script is represented as a subscript. Therefore, there is a sub-script each for the Self and Enemy agents with their corresponding associated parameters. The sequence of shooting events is encoded in the ACTIONS portion of the script. As the series of shooting events (of one agent) are causally linked to the reduction in HP (of the other agent), established through causal reasoning in the same vein as that described in Chap. 2, the two scripts are shown to be connected with a Causal Linkage.

In the OUTCOME portion of the ATTACK-SCRIPT (SPECIFIC), one can see that the HP value of the Enemy agent becomes 0.[1] The entire ATTACK-SCRIPT is extracted in a similar vein as that of the STARCRAFT-MOVEMENT-SCRIPT as depicted in Figs. 7.3 and 7.4: $\Delta ATT(1) = 0 \rightarrow 1$ is being used as a starting boundary of the attack event and $\Delta ATT(1) = 1 \rightarrow 0$ is being used as the ending boundary, much like the MOV parameter in Fig. 7.4. (ATT value changing from 0 to 1 is also a desired built-in goal for battles as shown in Fig. 7.12). In general, identifying event boundaries is a complex issue which will require further investigations (Leong and Kwok 2011; Shipley and Zacks 2008).

Suppose now at this time the scenario is restarted with the Self and Enemy agents starting at different locations than in Fig. 7.13 – say, the locations as shown in Fig. 7.16. Suppose also now the Self agent is to satisfy the GOAL of Fig. 7.12 which is to Engage/Attack Enemy. The Script Base is queried to see if there is a script in which the ATT parameter gets changed from 0 to 1 as coded in the OUTCOME portion. The ATTACK-SCRIPT (SPECIFIC) of Fig. 7.15 is retrieved as it does have such a change in ATT (indicated as +ATT in the figure).

However, the ATTACK-SCRIPT (SPECIFIC) specifies very specific starting conditions, such as the TIME of the event must start at a certain time, and the Self and Enemy agents must be at specific X and Y values, etc., before the script is applicable. Even if we can apply the STARCRAFT-MOVEMENT-SCRIPT of Fig. 7.3 to change the location of the Self agent from its current starting location to this specific location, because we are not providing a means for the Self agent to *move* the Enemy agent to its corresponding specific location, it cannot apply the script to the situation. The specific value for the TIME parameter also presents a problem as the current situation definitely happens at a later time frame than that specified by the ATTACK-SCRIPT (SPECIFIC) because the event corresponding to that script happened earlier. So unless "winding back the clock" is allowed, this condition also cannot be met. The backward chaining process of Fig. 7.12 is hence stuck as these preconditions and possibly other preconditions cannot be satisfied right away, and there are no means to satisfy them. The ATTACK-SCRIPT (SPECIFIC) hence cannot be applied to general situations.

Now, as mentioned above, and as built into the mechanisms of StarCraft, for the Attack event to trigger, the only precondition that matters is the change of the distance between the agents involved from 128 units to 124 units, and the other parameters can change in other ways or have other synchronic values at the moment of the Attack event. This can be discovered in subsequent exploration processes in

---

[1] Suppose many instances of Attack events are triggered. What would happen in StarCraft is that the situation in which HP = 0 (death at the end) for the Enemy agent does not take place all the time. Sometimes, it is the Self agent whose HP value becomes zero. Nevertheless, this script, other than initiating an attack event (i.e. ATT → 1), also provides *one way* to kill the Enemy agent. Therefore, the script can also be called ATTACK-AND-KILL-SCRIPT and can be used by a Self agent accordingly. In Sect. 7.4.3.3 we show how the different HP OUTCOMES (i.e., sometimes the Enemy agent dies and sometimes the Self agent dies) can be organized in the counterfactual portion of a script.

7.4 Rapid Learning and Problem Solving in StarCraft 307

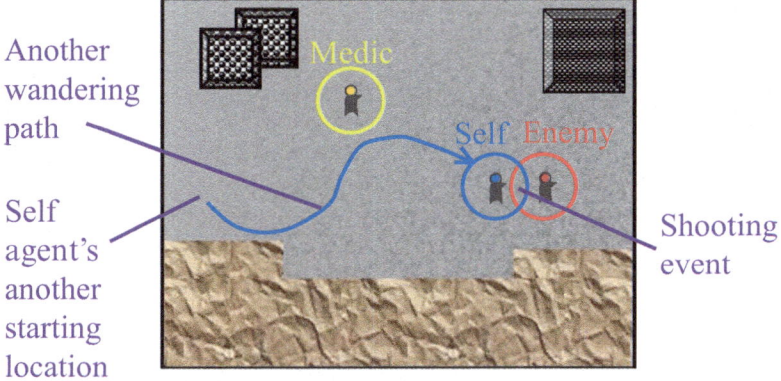

Fig. 7.16 Another episode of wandering of the Self agent, and the meeting of the Self and Enemy agents at a different location than in Fig. 7.13

which the Self agent explores along a different path at a different time with different settings for some of the other parameters, and arrives at a different triggering position for an Attack event as shown in Fig. 7.16 compared to that in Fig. 7.13. Through dual instance generalization as discussed in Chap. 2, many of the preconditions such as TIME, AG, X, and Y would be eliminated.

Suppose in a second instance of triggering an Attack event, some parameters are different than in the first instance but the Self agent's HP is still set at 80 (recall that HP for the Self agent was at the value 80 in Rule (7.1) and the script of Fig. 7.15). Then, "HP = 80" will not be eliminated yet and will still be labeled as a necessary precondition in the more general script created after this second Attack event has completed (i.e., when one of the Self and Enemy agents dies). Perhaps there is yet another instance in which HP is not set at 80 at the triggering of an Attack event, then HP = 80 would be removed as a necessary synchronic precondition for the general script created. (This could happen if the Self agent had participated in an earlier Attack event without dying, and its HP value was reduced to less than 80, or the Self agent could have simply begun with a different HP value set by some other control processes.)

Figure 7.17 shows that after a small number of instances, a general attack script, ATTACK-SCRIPT (GENERAL), can be derived. The causal learning process that we described above and in Chap. 2 functions like a "Causal Learning Filter" to filter away the unnecessary preconditions. Parameters that have been generalized and become "unnecessary" preconditions that can take on any values are colored in green, while those parameters that are "necessary" preconditions that must have specific values are colored in yellow.

As mentioned above, other than Attack, there are also other events that could potentially give rise to other causal rules (the discussion associated with Fig. 7.14 and Rule 7.1). For example, in Fig. 7.14, other than the two blue boxes containing events (changes of states) that are linked together, there can also be a number of other links, such as those between the events in the green boxes as well as those

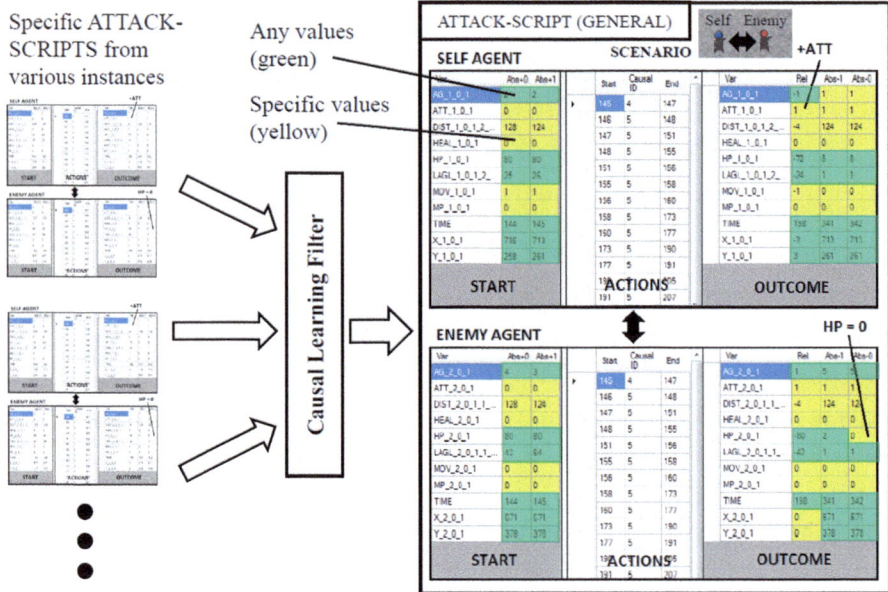

Fig. 7.17 A *general* ATTACK-SCRIPT is derived after a process of "filtering" through causal learning. Parameters that have been generalized and become "unnecessary" preconditions that can take on any values are colored in *green*, while those parameters that are "necessary" preconditions that must have specific values are colored in *yellow* (Right illustration above: ©2014 IEEE. Reprinted, with permission, from Ho, S.-B. and Liausvia, F., "A Rapid Learning and Problem Solving Method," Proceedings of the IEEE Symposium on Computational Intelligence for Human-like Intelligence, Page 114, Fig. 4)

between the events in the green and blue boxes. However, if these do not represent consistent and repeatable correlations, after a few instances of observation, the system would "filter" them away.

Figure 7.17 shows that there are two parameters with specific values that we have not highlighted hitherto that remain as necessary preconditions. They are the HEAL and MP parameters. These two parameters are actually not relevant in the Attack situation. They do not hamper the triggering of the Attack event because usually they would have the specified specific values of 0 at the onset of an Attack event anyway. The HEAL parameter is set to the value "1" when the agent is being "healed" by the Medic agent – i.e., having its HP point increased (this could happen because the agent may have suffered a decrease in its HP value because of an earlier Attack event, and now it is seeking the help of the Medic agent). Otherwise it remains at value 0. This Medic healing mechanism will be described in detail below. The MP parameter is only relevant when it is a parameter of the Medic. It records how much "energy reserve" the medic has to deliver its "healing" effort. For "fighting" agents such as the Self and Enemy agents, the value is always 0. In the subsequent discussion we will see that the Medic's MP actually gets reduced after it has helped to increase the HP of a fighting agent.

7.4 Rapid Learning and Problem Solving in StarCraft

Therefore, suppose at a certain point in time the Self or Enemy agent is being healed by a Medic agent, making HEAL $= 1$, it would have to disengage from the healing process before it can potentially be engaged in an Attack process (with HEAL $= 0$). Now that HEAL $= 0$, if the agent then attempts to engage in an Attack event using the script of Fig. 7.17, the Attack event will not be hampered by this necessary, though in a sense irrelevant, precondition.

In Fig. 7.17 the MOV parameter for the Self agent, highlighted in yellow which means it is a necessary precondition, is one because it has been moving toward the Enemy agent at the moment of engagement.

Note also that because the Enemy agent stays stationary through the Attack process, its location's X and Y values remain unchanged – this can be seen in the bottom right corner of the Enemy agent part of the script in which the Rel values (derived from subtracting the value in the Abs$+0$ column from the value in the Abs-0 column) of X_2_0_1 and Y_2_0_1 are both 0.

Now, after the learning of the general ATTACK-SCRIPT of Fig. 7.17, many of the parameter values have been generalized and the script can then be applicable to a general situation. Again, given a new scenario of some placement of Self and Enemy agents, suppose the ATTACK-SCRIPT (GENERAL) is selected because the high level GOAL of Engage/Attack Enemy (ATT $= 1$) is activated (the OUTCOME portion of the ATTACK-SCRIPT (GENERAL) shows that ATT has been changed from 0 to 1 so this is a potential script to be used). The next step is to check if the START conditions of this script could be satisfied as dictated by the backward chaining process of Fig. 7.12.

Suppose in the beginning of this new scenario, the Self and the Enemy agents are at different locations from all the other starting locations they were at in the earlier instances (such as those in Figs. 7.13 and 7.16) and have a different, say, HP value compared to those in the earlier instances (say, now HP $= 40$ units). Such a scenario is shown in Fig. 7.18. The general ATTACK-SCRIPT stipulates that some preconditions must be met, such as the values of HEAL, MP, etc. must be 0, and specifically the relative distance between the Self and Enemy agents, DIST(1, 2), must be 128 units changing to 124 units, but many of the other parameters can be of arbitrary values (such as HP). Now, as mentioned earlier, unless the Self agent is engaged in a "healing" process with a Medic agent, the HEAL and MP parameters are satisfied. As the Self agent is currently quite far from the Enemy agent (Fig. 7.18), the DIST(1, 2) $= 128$ is not satisfied. Still having the built-in goal of Engage/Attack Enemy as depicted in Fig. 7.12, the Self agent engages a backward chaining process to satisfy the START conditions of the ATTACK-SCRIPT (GENERAL).

Suppose the Self agent has learned the straight-line STARCRAFT-MOVEMENT-SCRIPT (GENERAL) of Fig. 7.3 earlier. Now, in the backward chaining process, it searches the Script Base for a script that can change its location. Since the stipulation of DIST(1, 2) $= 128$ in the ATTACK-SCRIPT (GENERAL) is in the form of relative distance which can correspond to many sets of absolute locations' X and Y values, we built in a procedure to generate an arbitrary set of X, Y values corresponding to DIST(1, 2) $= 128$ from the current X and Y values of the

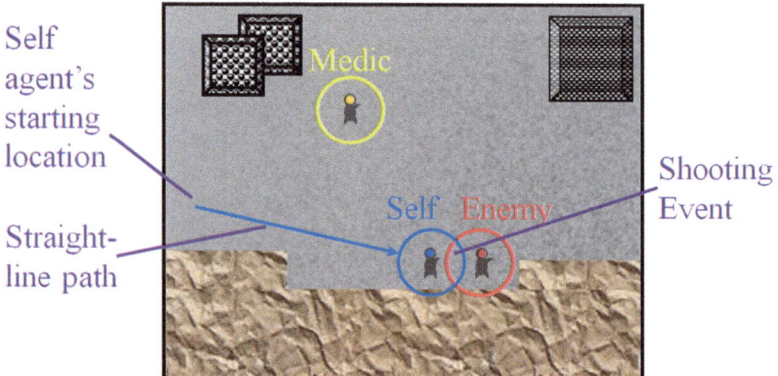

Fig. 7.18 Having learned the general ATTACK-SCRIPT, the Self agent can head toward an Enemy agent situated at an arbitrary location, from an arbitrary location, at an arbitrary time in a straight-line using the STARCRAFT-MOVEMENT-SCRIPT (GENERAL) to engage the Enemy agent in an Attack event

Enemy agent. (When the ATTACK-SCRIPT (GENERAL) is retrieved, its arbitrary Enemy agent's location's X and Y values are instantiated with the current known X and Y values of the Enemy agent's location.) Since the OUTCOME portion of the STARCRAFT-MOVEMENT-SCRIPT (GENERAL) shows that the script changes the absolute location – the X and Y values – of the agent involved (reflected in the Rel column of the script), the Self agent thus identifies this to be a usable script and retrieves it to generate a straight-line movement solution to bring the Self agent from the current starting location to a point that is at $DIST(1, 2) = 128$ from the Enemy Agent (an arbitrary one of the set of X, Y values computed above). And then, following that, it has to generate a movement from $DIST(1, 2) = 128$ to $DIST(1, 2) = 124$ to trigger the Attack event. Hence, with the two learned general scripts – STARCRAFT-MOVEMENT-SCRIPT (GENERAL) and ATTACK-SCRIPT (GENERAL) – the Self agent can solve the problem and achieve the GOAL (namely Engage/Attack Enemy) rapidly, with a purposeful and directed set of actions to move from a start location to meet the Enemy agent to trigger an Attack event, with no more need to wander around to achieve the same end. A similar detailed computational process would also apply to the process in Fig. 1.5 of Chap. 1.

The resultant straight-line path is shown in Fig. 7.18 and the corresponding combined MOVEMENT-TO-ATTACK-SCRIPT (GENERAL) is shown in Fig. 7.19. This is an example of a script construction process as a result of problem solving as depicted in Figs. 2.19 and 2.28 of Chap. 2. For two scripts to be combined, the ending values of the specific parameters of the preceding script must be the same or made to be the same as the starting values of the specific parameters of the succeeding script. In the case of Fig. 7.19, one can see that the ATT, HEAL, MOV, and MP values all match at the end of the STARCRAFT-MOVEMENT-SCRIPT (GENERAL) and at the beginning of the ATTACK-SCRIPT (GENERAL). For the value of DIST, the ATTACK-SCRIPT (GENERAL)

7.4 Rapid Learning and Problem Solving in StarCraft

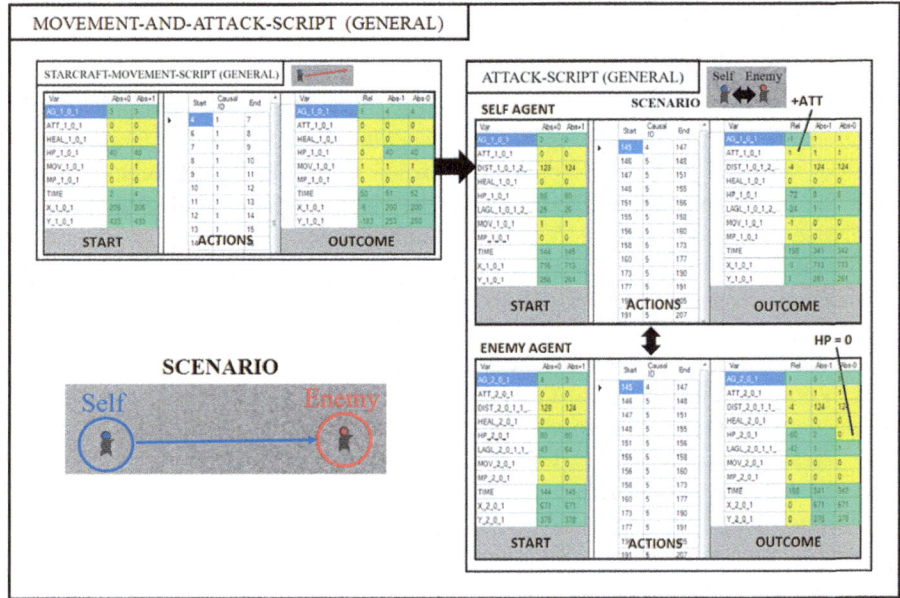

Fig. 7.19 Combining two learned chunks of knowledge as a result of a backward chaining process – one STARCRAFT-MOVEMENT-SCRIPT (GENERAL) and one ATTACT-SCRIPT (GENERAL) – to arrive at a solution: MOVEMENT-AND-ATTACK-SCRIPT (GENERAL), to attack an Enemy agent at an arbitrary location, from an arbitrary starting location of the Self agent, at an arbitrary time (©2014 IEEE. Reprinted, with permission, from Ho, S.-B. and Liausvia, F., "A Rapid Learning and Problem Solving Method," Proceedings of the IEEE Symposium on Computational Intelligence for Human-like Intelligence, Page 116, Fig. 7 (*left* illustration above) and Page 114, Fig. 4 (*right* illustration above))

specifies that it must be 128 units. As discussed above, we arbitrarily generate a set of specific X and Y values corresponding to this and then we set the ending X and Y values of the STARCRAFT-MOVEMENT-SCRIPT (GENERAL) to be these values. Thus, this location becomes a "GOAL" for the STARCRAFT-MOVEMENT-SCRIPT (GENERAL), resulting in the general STARCRAFT-MOVEMENT-SCRIPT being *instantiated* – it will stipulate a *specific* movement sequence from the current start location of the Self agent (instantiating the X and Y values of the START portion of the script) to this GOAL location, which will result in the triggering of the Attack event.

This combined MOVEMENT-AND-ATTACK-SCRIPT (GENERAL) allows the Self agent to move to attack an Enemy agent at an arbitrary time, given both are at arbitrary locations and with arbitrary values for many of their parameters.

This incremental knowledge chunking process discussed above, that resulted in the knowledge chunk as shown in the complex script of Fig. 7.19, is a more general process and creates a more general kind of chunked knowledge structure than that described for the spatial movement scenario of Chap. 3, Sect. 3.5. This would have applications to a wide variety of domains.

7.4.1.1 The Learning of the INCREASE-HP-SCRIPT

We now turn to detailing the mechanisms associated with the Medic agent that have been illustrated above in the various figures of the StarCraft environment. As mentioned earlier, the Medic agent can "heal" the fighting agent such as the Self agent or the Enemy agent by increasing their HP values. This is particularly useful after the agent has suffered some loss to its HP as a result of an Attack event, otherwise continued decrease in HP will result in Death – when HP is 0. There are Medic agents specifically available for either the Self or the Enemy sides. For the subsequent discussion, we are only concerned with the Medic agent for the Self side. This is circled with a yellow circle in Fig. 7.1, etc.

The StarCraft built-in rule for the Medic agent to be able to "charge-up" the HP of a fighting agent is similar to that for triggering an Attack event – the fighting agent involved must come to within a certain distance of the Medic agent. Figure 7.20 shows a situation in which the Self agent comes close enough to a Medic agent in the process of wandering and discovers the rule to get its HP charged-up.

Again, similar to the case of the learning of the ATTACK-SCRIPT (GENERAL), after a few instances of triggering a HP charge-up event, the Self agent learns a general INCREAST-HP-SCRIPT (GENERAL) as shown in Fig. 7.21. It can be seen that the relative distance between the Self and Medic agents, as captured in the DIST parameter, must change from 53 units to 50 units before the Increase-HP event will trigger.

In Fig. 7.21, it can be seen that the HP of the Self agent increases from its value in the START state to the OUTCOME state. The specific HP value in the START state is not a necessary precondition for the charging event – one can charge the agent starting from any HP value. The MP value of the Medic agent at the OUTCOM state is seen to have decreased from that in the START state.

Again, similar to the case of the Attack situation, given a goal to increase its HP, if the Self agent already has learned the STARCRAFT-MOVEMENT-SCRIPT (GENERAL) earlier and now it has just learned the INCREASE-HP-SCRIPT (GENERAL), it can combine these together in a problem solving process and generate a MOVEMENT-AND-INCREASE-HP-SCRIPT (GENERAL) as shown in Fig. 7.22. I.e., at any given moment and at any location, should it find the need to increase its HP, it will head straight toward a Medic agent to achieve that purpose without having to wander around to discover the function of the Medic agent fortuitously as in the earlier learning phase.

7.4.1.2 The Continuous Process of Problem Solving and Script Construction

As was discussed in Chap. 3, Sect. 3.5, our learning and problem solving paradigm is one of building ever complex scripts from a process of incremental chunking. Each time an incrementally more difficult problem is presented to the system, it

7.4 Rapid Learning and Problem Solving in StarCraft

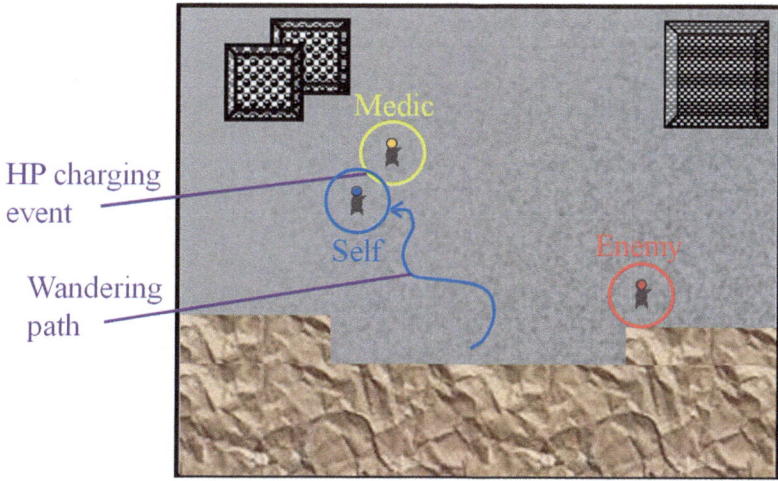

Fig. 7.20 A Self agent wanders around and accidentally discovers the rule to get its HP charged up by a Medic agent

solves the problem by using whatever means at its disposal – whether it be an available script in the Script Base, or some exploration in a blind search process to uncover new scripts – and then the derived solutions are stored as scripts (a chunked piece of knowledge) for future use. This way, ever increasingly complex problems can be tackled within a reasonable amount of time. The alternative process of problem solving (PS) and script construction looks like this:

PS → Chunked Knowledge(Scripts) → PS→Chunked Knowledge(Scripts)...

In our earlier discussion in Chap. 3, Sect. 3.5, we used a scenario of simple actions and movements to illustrate the idea of incremental chunking without applying the more general script construction processes to encode chunked knowledge. In the foregoing discussion in this chapter, what we have done was to combine the script construction processes described in Chap. 2 as well as the idea of incremental chunking in Chap. 3, apply them to the StarCraft game environment, and illustrate how this idea led to a process of rapid learning and problem solving.

7.4.2 Affective Competition and Control: Anxiousness Driven Processes

Having established the learning, script construction, and problem solving mechanisms in a basic StarCraft situation that includes an Attack event and an Increase-HP event, we now use the platform to illustrate affective competition and control. (Refer to Chap. 3, Sect. 3.3 for an earlier discussion on this). This is a "top level"

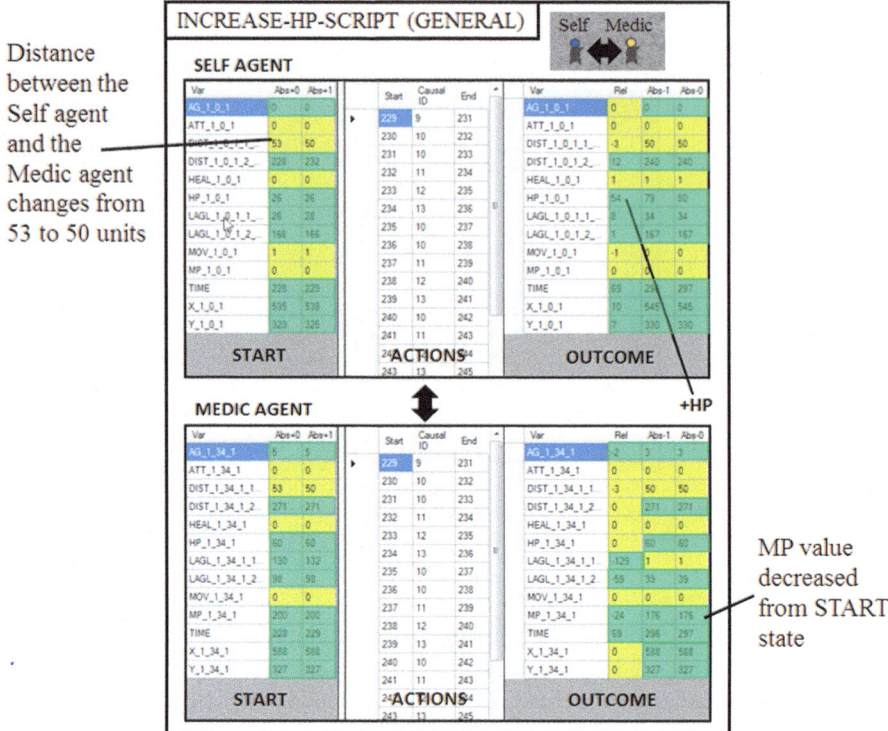

Fig. 7.21 A general INCREASE-HP-SCRIPT (©2014 IEEE. Reprinted, with permission, from Ho, S.-B. and Liausvia, F., "A Rapid Learning and Problem Solving Method," Proceedings of the IEEE Symposium on Computational Intelligence, Page 114, Fig. 5)

process in which multiple (in this case dual) sources of need priority compete for dominance and subsequently direct behaviors in such a way as to ensure the best overall satisfaction of the needs of the noological system involved.

The basic idea of the process is as follows. In the above scenario, the Self agent has two basic goals. We have highlighted and discussed in detail one of them – the Engage/Attack Enemy goal (Fig. 7.12) – and will now emphasize another one – the need of the Self agent to maintain its HP level at some reasonable value. The maintenance of the HP level serves one major purpose, which is to prevent the HP level from reducing to 0 thus leading to death. An indirect consequence of this is so that it could continue to have the "ammunition" to shoot at and kill the Enemy agent.

Typically, a noological system would have a "hungry" signal, which is when its internal energy reserve has gone below a certain level. In a typical natural noological system such as that of a human being, the internal energy and "health" are separate – i.e., if one is wounded, the health level goes down, but the internal energy could still stay at the same level for a while. In the StarCraft environment,

7.4 Rapid Learning and Problem Solving in StarCraft

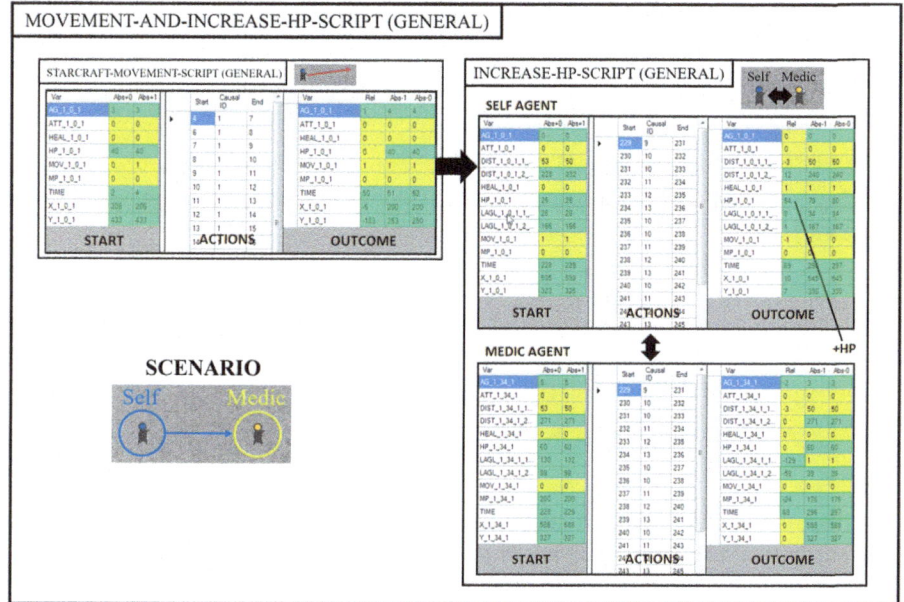

Fig. 7.22 A general MOVEMENT-AND-INCREASE-HP-SCRIPT (©2014 IEEE. Reprinted, with permission, from Ho, S.-B. and Liausvia, F., "A Rapid Learning and Problem Solving Method," Proceedings of the IEEE Symposium on Computational Intelligence for Human-like Intelligence, Page 116, Fig. 7 (*left* illustration above) and Page 114, Fig. 5 (*right* illustration above))

this situation is simplified as there is only one signal – the HP. In a natural noological system, when the internal energy level goes below a threshold, some priority is given for it to look for food to increase this level, and this priority increases as the level goes lower. In fact, the system enters a state of "anxiousness" – we term it "anxiousness of low internal energy" and a priority is established to address the reduction of the anxiousness.[2]

However, typically a noological system does not exist simply to serve one goal – say, that of maintaining its internal energy at a high level. Even the lowest form animals have at least one other important need which is to procreate to sustain the species. When there is more than one need, the needs will compete with each other for dominance. The more dominant need will be addressed first. If we reformulate this in terms of anxiousness, suppose each need gives rise to one kind of anxiousness – say "anxiousness of low internal energy level" and "anxiousness of lack of procreative activities" – the animal/agent/noological system will select the need

---

[2] As mentioned in footnote 3 of Chap. 3, *fear* and *anxiousness* are used synonymously here and the rest of the book. There are subtle differences which we will not particularly address in this book. E. g., typically it is "fear of the snake causes him to run" and "I am anxious that I cannot finish my homework by tonight." In these examples, it seems to be a matter of degree.

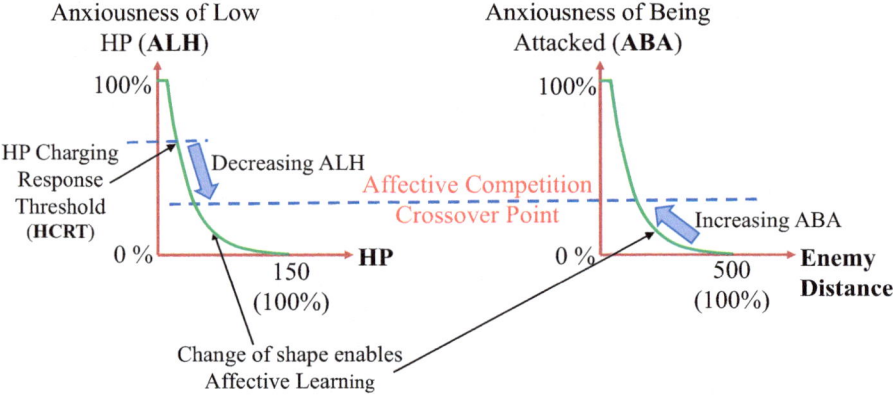

Fig. 7.23 Affective competition in StarCraft

that has the higher level of anxiousness to address first. Suppose a system's anxiousness of a low internal energy state has caused it to engage in food hunting and consumption, and suppose it does succeed in finding some food and reduce this anxiousness. Should it continue to look for more food to further reduce this anxiousness or should it switch to procreation? This will depend on whether it has already managed to reduce the low internal energy level anxiousness *below* that of the anxiousness level of lack of procreative activities. Figure 7.23 illustrates the idea of affective competition using the StarCraft situation.

On the left of Fig. 7.23 is a graph of Anxiousness of Low HP (ALH) vs the HP level. It shows that when the HP level is at 150, the ALH is at 0 (like a "full stomach" scenario). As the HP level reduces, the ALH level increases in an exponential manner very rapidly toward 100 % as the HP reduces toward 0, and it saturates at 100 % at some low level of HP.

In the case of StarCraft, there is no procreation scenario but there is one situation which generates affective competition. When the Self agent is engaged in an HP charging situation, it may be attacked by the Enemy agent. As the Enemy agent approaches it while it is charging its HP, it may want to disengage from the HP charging and engage the Enemy, otherwise it may be killed. Therefore, there is an Anxiousness of Being Attack (ABA) that can be defined as a function of, say, the distance of the Enemy agent from the Self agent. This is shown on the right side of Fig. 7.23 with a same exponential shape graph of ABA vs Enemy Distance. We have set Enemy Distance = 500 units as the point at which ABA becomes 0.

We have also defined a level of HP at which the Self agent becomes "really hungry" and it has to seek the Medic agent to charge up its HP – the "HP Charging Response Threshold (HCRT)." This is set at about 20 % HP level on the left graph of Fig. 7.23.

While the exact shape of the exponential graphs in Fig. 7.23 can be changed as a result of learning – e.g., the Self agent may learn that its level of ABA is too high or too low for a certain value of Enemy Distance and adjust that accordingly – we will

7.4 Rapid Learning and Problem Solving in StarCraft   317

assume for now that these shapes are constant in a particular situation of Self-Enemy interaction. (See Chap. 3, Sect. 3.3.1 for a short discussion on the learning issue associated with this.)

Now suppose the scenario begins with the Self agent being engaged in an Attack event with the Enemy agent such as that in Fig. 7.18. At some point, the HP level of the Self agent falls below a certain level such that its ALH increases above the HCRT. It then disengages from the Attack event with the Enemy agent and heads straight toward the Medic agent to charge its HP as shown in Fig. 7.24a. Now, while charging up its HP, its ALH level reduces according to the left graph of Fig. 7.23. As this is happening, suppose the Enemy agent moves toward it to attempt to initiate an Attack event as shown in Fig. 7.24b. As the Enemy agent nears, the Self agent's ABA increases according to the right graph of Fig. 7.23. There is a point at which the level of ABA exceeds the level of ALH – the Affective Competition Crossover Point. At that point, the Self agent will disengage from the HP charging event with the Medic and head straight toward the Enemy agent to engage it as shown in Fig. 7.24b. There is a goal change at this point from "Filling up HP" to "Attack/Engage the Enemy."

Continuing the process, if in this second Attack event, the Self agent's ALH increases above HCRT again, it will then again disengage from the Attack event and proceed to charge up its HP again and so on. Typically, if the Enemy agent is not provided with any Medic agent to charge up its HP, the Self agent will finally win the fight as it is able to sustain its HP at a relatively high level continually while depleting the Enemy agent's HP. This YouTube video shows a simulation of this process: https://www.youtube.com/watch?v=lo0woEt7L78_9 (or see video on https://noologyblog.wordpress.com: StarCraft Battle: Affective Competition).

One can see from the video that the entire process resembles a realistic situation in which an agent or animal is attempting to address more than one need present.

7.4.3 Causal Learning of Battle Strategies

Unlike reinforcement learning (Sutton and Barto 1998), our rapid effective causal learning platform allows the rapid learning of causal rules related to battle situations and the application of these to rapidly formulate effective battle strategies, much like what a human commander would do in a battle field. Also, the knowledge learned is represented in an explicit manner.

In this section we describe the representational and reasoning principles as well as computer simulations for a more complex battle scenario involving more than one agent on the sides of the Self and Enemy. The StarCraft Self agents build on the basic knowledge learned about engaging/attacking individual Enemy agents as discussed in Sect. 7.4.1 to organize/strategize more complex battle activities involving more than one agent on both sides.

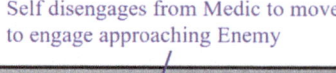
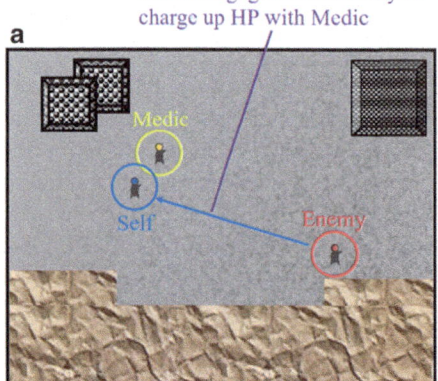
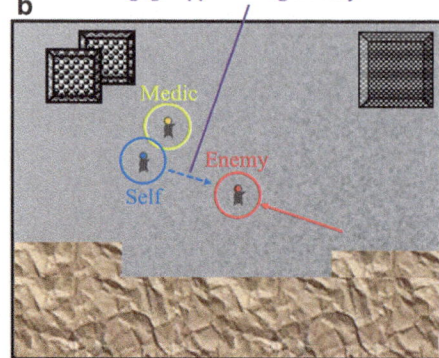

Fig. 7.24 Affective competition. (**a**) Initially, Self agent disengages from an Attack event with the Enemy agent to find a Medic to charge its HP because its ALH has increased above the HCRT. (**b**) As the Enemy approaches it while it is charging its HP, its ABA increases until an Affective Competition Crossover Point is reached, at which time it then disengages from the HP charging activity to engage the approaching Enemy agent

7.4.3.1 Commander Anxiousness

We employ a commander mode of operation in a StarCraft setting, i.e., a Commander agent (not explicitly shown in the StarCraft environment) directs a number of his fighting agents (Self agents) to engage the Enemy agents and learns about the likelihoods of success and failure of certain modes of engagement, thereby allowing her to formulate a battle strategy that allows her to win most of the time.

We begin by defining the goal of the Commander. The ultimate goal of a typical commander could be promotion, glory, etc. but in our current simulation we use "anxiousness to win the battle" to drive the Commander's behavior. The function to be maximized is:

$$\begin{aligned}\textbf{AV-HP-DIFF}&(\text{Averaged HP Difference}) = \\ &\text{Averaged sum of all Self agents' HPs minus} \\ &\text{averaged sum of all Enemy agents' HPs}\end{aligned} \quad (7.2)$$

Figure 7.25 shows the typical scenario in which there is a number of Self and Enemy agents deployed in a certain manner over the battle scene. The Anxiousness of Commander (ACDR) vs AV-HP-DIFF graph shows a similar pattern as the anxiousness graphs of Fig. 7.23 – when AV-HP-DIFF is 0, the commander anxiousness is very high, as it means on average the enemy is as strong as his side. The goal is to reduce the ACDR to 0, which means the total HP of the Self side is at the highest possible level relative to the Enemy side. In a starting scenario when both sides have agents that are charged up to their respective maximum HP, the ACDR will be high. The Commander hence attempts to retrieve scripts from the Script Base to see if there are any scripts that can help to reduce ACDR.

7.4 Rapid Learning and Problem Solving in StarCraft 319

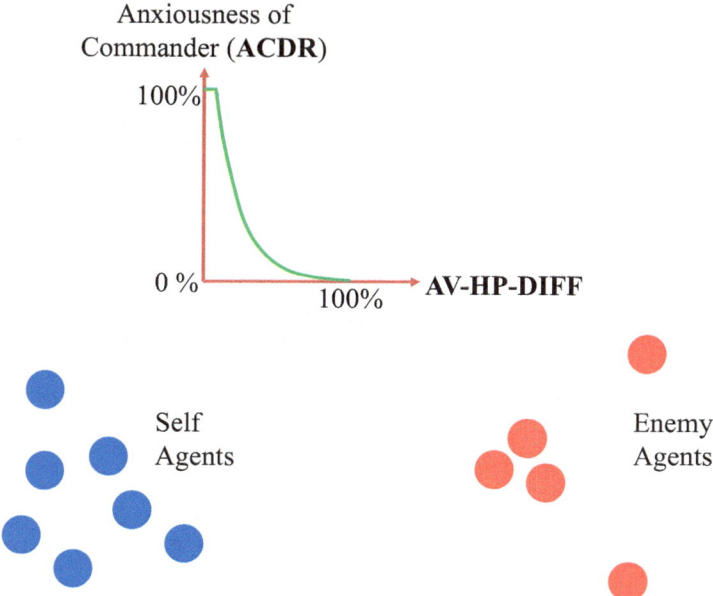

Fig. 7.25 The commander mode of operation – the Commander attempts to minimize an Anxiousness of Commander (ACDR) measure which is related in an inverse manner to the Averaged HP Difference (AV-HP-DIFF) between the Self and the Enemy agents

7.4.3.2 The Modes of Learning

Three modes of learning are employed in the learning processes – the observation mode, the physical experiment mode, and the mental experiment mode. We have discussed mental experiments quite extensively in the previous chapter (Chap. 6) in connection with the SMGO problem. Below, we summarize and discuss these modes of learning with respect to our current scenario. These modes can be further divided along the line of passive vs active learning as follows:

1. **Observation Mode (Passive Learning with Role Reversal)** – This is the mechanism that has been used in most of the foregoing discussions - the agent/Commander would observe the spatiotemporal activities in the world and capture and encode them in the form of event scripts. The usual generalization over location and other parameters is carried out and the event script is stored as a general script such as those discussed in Sect. 7.2. Figure 7.10 shows the process of conducting physical observation and constructing the CFI portion of the scripts involved. If there is more than one agent engaged in the same activity, they are captured together into a "super-script" such as those shown in Figs. 7.17 and 7.19. Sometimes, the observation is made on Enemy agents effecting certain actions on the Self agents (e.g., two Enemy agents attacking one Self agent, with certain consequences). This is then captured and role-

reversed into a separate script prescribing how the Self agents can also effect the same actions on the Enemy agents and expect the same consequences.
2. **Physical Experiment Mode (Active Learning)** – Often, as discussed in connection with Fig. 7.10, it may take a long time to observe events in the world that could inform the system of useful causal rules or scripts to be used for certain purposes. This is when the agent/Commander needs to actively carry out physical experiments (using a process similar to that described in Fig. 7.11) to effect certain actions on the environment and observe the consequences so that useful knowledge can be gained faster. Therefore, this is carried out when (i) the agent/Commander desires to obtain more knowledge about the environment in preparation for future uses, and earlier learning from observation alone was deemed insufficient; and (ii) the knowledge for certain causes and effects was not observed earlier, and there is a need for the knowledge for the current problem solving process. Physical experiments may be guided by scripts learned earlier. (E.g., the Script Base already contains a FORCE-MOVEMENT-SCRIPT, and now there is a need to know what happens when there is more force – Sect. 7.2.)
3. **Mental Experiment Mode (Active Learning)** – When the consequences of the individual sequences of actions (scripts) are known but the combined effects of these scripts are not known, the agent/Commander then carries out an internal simulation to determine these effects, and form generalizations accordingly. Also, the agent/Commander may generate simulations with different parameters to see how they affect yet other parameters to construct/refine the CFI portion of the script involved – e.g., in Chap. 3, Fig. 3.1, having known the basic elemental movement rules, generate different paths of movement to construct the ΔE vs Trajectory graph in the CFI portion of the script.

7.4.3.3 The MOVEMENT-AND-ATTACK Script with Counterfactual Information

Let us begin with the earlier discussed MOVEMENT-AND-ATTACK SCRIPT (GENERAL) in Fig. 7.19. In that script, the Self agent (and it could also be an Enemy agent if the role were reversed) moves toward the Enemy agent and an Attack event ensues, with the Self and Enemy agents locked causally in an exchange of shooting at each other. The script in Fig. 7.19 shows that finally the Enemy agent's HP goes to 0 and it dies.

Now, there could be many instances of MOVEMENT and ATTACK and each time the outcome could be different. In the StarCraft environment in general, and depending on some internal settings (capturing the probability of winning on either side), even if the starting conditions are the same, sometimes the Self agent would get killed instead. Certainly, if some of the starting conditions are different – e.g., the Self agent attacking from a different direction – the Self agent may end up dying. This is a kind of counterfactual information – "had it been in another instance, the outcome could have been..." In the same spirit as the capturing and encoding of counterfactual information in the earlier situations/scenarios

7.4 Rapid Learning and Problem Solving in StarCraft 321

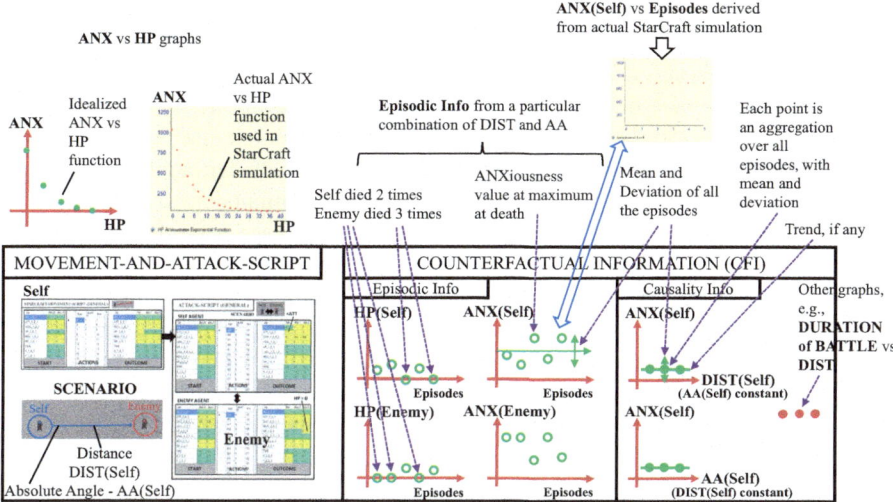

Fig. 7.26 MOVEMENT-AND-ATTACK-SCRIPT with counterfactual information (CFI) (See text for explanation)

(Figs. 7.7, 7.8, and 7.9, and other situations in earlier chapters), we capture and encode the counterfactual information for the MOVEMENT-AND-ATTACK-SCRIPT as shown in Fig. 7.26.

In Fig. 7.26 the basic MOVEMENT-AND-ATTACK portion of the script is enhanced with a CFI portion consisting of two sub-portions – Episodic Info and Causality Info. The Episodic Info portion records what happens to various parameters in the various episodes of "movement and attack." Typically, what is recorded are the values of these parameters at the OUTCOME of the event. As mentioned above, often, even if an event contains the same starting parameters (e.g., starting location and direction of attack, HP value, etc.) as an earlier event, the outcome could be different (e.g., which agent dies) due to probabilistic behavior of the system. The HP(Self) and ANX(Self) graphs in the script in the Episodic Info portion show the outcome of five episodes of similar Attack events with the same starting parameter settings of DIST and AA (see the SCENARIO portion of the script for the meaning of these parameters). ANX and HP are related through the graphs in the top left corner of the figure (the shape of the graph is similar to that in Fig. 7.23). We show both an idealized version of an ANX vs HP graph for ease of illustration as well as one that was actually used in some StarCraft simulations. The Causality Info portion is similar to the CFI portion of the FORCE-MOVEMENT-SCRIPT of Fig. 7.7 – basically it contains correlation information on parameters associated with the event involved: i.e., had a certain starting parameter been different (e.g., the direction of attack), how would some of the outcome parameters be different. The construction process of this CFI portion of the script is similar to that described for the FORCE-MOVEMENT script of Sect. 7.2 and that is encapsulated in Figs. 7.10 and 7.11.

Now, as discussed in the previous section, there are two ways this information can be collected. One is through observation of events happening in the environment in an Observation Mode of learning. So, if there is a movement and attack event that takes place due to some problem solving requirements, the parameter values are encoded accordingly (see Fig. 7.10 and associated discussion). Another mode of learning is the Physical Experiment Mode in which deliberate attempts are made by an "experiment initiator," in this case the Commander, to send an agent (in this case a Self agent) to generate an Attack event so see what the OUTCOME may be (see Fig. 7.11 and associated discussion). The Commander can do this many times with different starting parameters to see how they affect the outcome of the battle as measured by the HP and ANX parameters, and whether there are any patterns of correlation. (The Commander could also be interested in the outcome of other parameters, such as the degree of his agent being "battle hardened" or other measures as a result of each Attack event.) This information will be useful for the Commander to formulate optimal battle strategies as will be discussed in subsequent sections.

The process of physical experiment – the Physical Experiment Mode of learning – performed by the Commander here is similar to that described in Fig. 7.11, except that again because of probabilistic behavior of the system, the same experiment with the same starting parameter settings is usually conducted a number of times to see what the probability of certain outcome (say, the Enemy agent dying) is. This is captured in the Episodic Info portion of the script of Fig. 7.26.

In Fig. 7.26 we show that the HP(Self) parameter at the end of an Attack event is a parameter of concern, and the value of HP(Self) is collected over a small number of episodes, and in this case, five episodes. The statistics on HP(Enemy) is also collected – for this example shown, the Self agent died two times and the Enemy agent died three times. (The exact HP(Enemy) value at the end of a battle may not be known in a real battle unless the Enemy agent dies, which means HP(Enemy) must be 0.) There are other values in the OUTCOME portion of the MOVEMENT-AND-ATTACK-SCRIPT that could also be collected but we omit these for clarity.

Now, even though a Self agent can win in an Attack event, thus achieving HP (Enemy) = 0, and this would be advantageous to the Commander, there is another parameter that is also of concern. Suppose the Self agent's HP value at the end of a winning Attack event has been reduced to a low level. Even though the Self agent is "alive," it would be in a state of high anxiousness, and the Commander's ACDR (Fig. 7.25) would be high, which is undesirable. As mentioned above, there is a graph relating ANX and HP in the top left corner of Fig. 7.26. In the Episodic Info portion of the script, one can see the corresponding ANX(Self) and ANX(Enemy) episodic information derived from the HP episodic information. It shows that the ANX(Self) value is at its maximum in the two episodes in which the Self dies. Near the top right corner of Fig. 7.26 is a graph showing the actual ANX(Self) vs Episodes data derived from some StarCraft simulations (for some situations and settings, it may turn out that the Self agent always dies in a 1 vs 1 battle). It is also shown that the Mean and Deviation of the episodic data are computed from the raw

7.4 Rapid Learning and Problem Solving in StarCraft

episodic data and this information will be useful for the Causality Info portion of the script as will be discussed below.

Now, the episodic data in the Episodic Info portion of the script are obtained from a particular combination of parameters for the movement and attack scenario (as indicated in the top center portion of Fig. 7.26). In a *one* Self agent vs *one* Enemy agent SCENARIO as shown on the bottom left part of the script in Fig. 7.26, two parameters are involved. They are the DIST(Self) – distance of Self to Enemy agents – and the AA(Self) – the absolute angle of Self agent's direction of movement toward the Enemy agent for attack. The Episodic Info is thus collected with both of these parameters kept constant. (I.e., the Commander will send one Self agent to the Enemy agent, say, five times from exactly the same distance away and heading in exactly the same direction.) The purpose of this is so that these data can be aggregated for the Causality Info portion of the script to be described below.

In the Causality Info portion of the script, one can see that in the spirit of "controlled physical experiment" as shown in Fig. 7.11, given the available parameters in the SCENARIO (DIST(Self) and AA(Self)),[3] one parameter is varied while the other parameter is kept constant while physical experiments are performed (which, in the current SCENARIO, involves the Commander sending Self agents to be involved in Attack events with the Enemy agents to observe the consequences). Since in the current SCENARIO, there are just two parameters, DIST(Self) and AA(Self), firstly AA(Self) is kept constant (at some arbitrary value) and DIST(Self) is varied to see how it affects the ANX(Self) values over a number of episodes (in this case, 5). (I.e., the Commander sends one Self agent to attack one Enemy agent always from one direction but she varies the distance from which she sends the Self agent.) Then, DIST(Self) is kept constant (at some arbitrary value) and AA(Self) is varied to see its effects on ANX(Self). (I.e., the Self agent is sent from different directions but from exactly the same distance away to attack the Enemy agent.) The Mean and Deviation of the effects of these experiments (in this case, the effect on ANX(Self)) are plotted on the graphs in the Causality Info portion as shown. The results of the physical experiments as shown in the graphs in Fig. 7.26 are that the DIST(Self) and the AA(Self) parameters have no correlation with the level of ANX(Self) in general. This means no matter how far the Commander sends the Self agents from or in what direction she sends them from to attack the Enemy agent, the consequence on ANX(Self) is the same. There might be other graphs showing the effects of distance and angle of attack to other parameters which are not shown in Fig. 7.26 in the interest of space. Also, in the Causality Info portion, a green line linking all the data points shows the trend, if any, that can provide useful inductive information for the Commander.

---

[3] For now, we assume other parameters, such as the starting HP value of the Self and Enemy agents, are always fixed at a certain value. There are also other parameters such as the absolute locations of the agents involved. We assume there are heuristics learned earlier, such as that discussed in Sect. 6.4, Chap. 6, that supply the knowledge that absolute locations typically do not affect physical processes unless the entities involved are electrically or magnetically charged.

The script in Fig. 7.26 thus captures the basic useful information for one to one combats between Self and Enemy agents that will be useful for the Commander to formulate her battle strategies later.

7.4.3.4 Desperation and the Exhaustiveness of Physical Experiments

Note that the physical experiments described above are not exhaustive in the sense that the entire space of parameter values is not thoroughly searched. Each time a parameter is varied, say DIST(Self), the other (AA(Self)) is kept constant at a certain arbitrary value. However, there might be other values of the second parameter that may give rise to a different relationship between ANX(Self) and DIST(Self). Therefore, these experiments are just quick and dirty experiments the Commander conducts to quickly get some idea about the rules governing these parameters so that she can plan some reasonably optimal strategies quickly. The strategies that result are by no means optimal in an absolute sense. Later we will see that this quick and dirty method does allow the Commander to devise strategies that allow her to win the battles most of the time.

Recall in Sect. 2.5 of Chap. 2 and Sect. 7.3 of this chapter, we discussed the connection between the level of *desperation* of a noological system and its propensity to effect generalization on observed rules about the environment, as summarized in Fig. 2.13. Under the pressure of having to deal with a battle environment in which some reasonable strategies must be formulated to deal with the enemy quickly, an exhaustive search of the parameter value space is not an option. However, if the desperation level is low, there is then the luxury of executing a more "scientific" process of thoroughly exploring the entire space of the parameter values involved so that a complete picture can be discovered, providing some truly optimal solutions. We will assume that the Commander is faced with a highly or moderately desperate situation in the current discussion on the StarCraft battle environment.

7.4.3.5 Heuristics Learning and Encoding

In Sect. 7.4.3.3 we mentioned that in the physical experiments with one vs one battle scenarios between the Self and Enemy agents, it was learned that the ANX (Self) was not dependent on the value of the two parameters DIST(Self) and AA (Self) of the battle scenario. This information is represented in the graphs in the Causality Info portion of the script involved (Fig. 7.26). This information actually represents valuable heuristic information that should be explicitly encoded to benefit future problem solving and/or physical or mental experiment situations. Explicit symbolic encoding facilitates generalization processes which we shall illustrate in the subsequent discussions. Figure 7.27 shows the conversion of the graphical information into a symbolic form encoding the heuristics involved.

7.4 Rapid Learning and Problem Solving in StarCraft

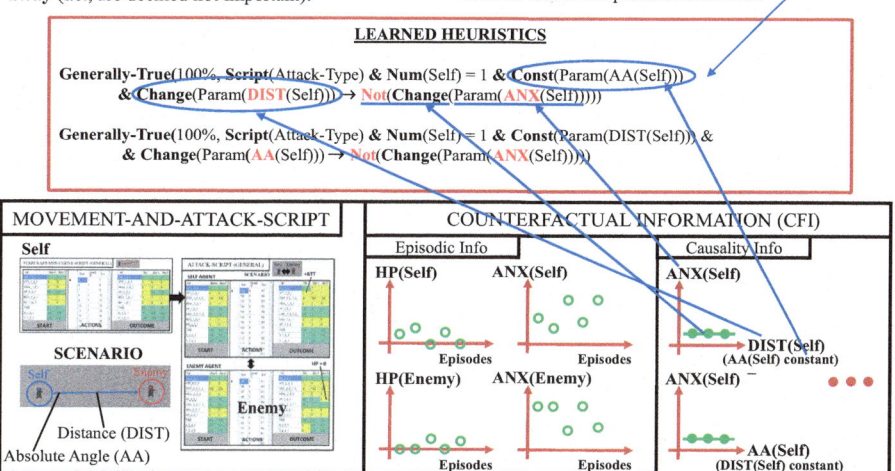

Assuming that the "importance" of other parameters such as the **Time** at which the event happens and the **Absolute Location** of the entire event have been learned earlier and these parameters have been **generalized away** (i.e., are deemed not important).

There are other parameters such as the starting **HP** of the Self agent that have not been generalized away, unlike **Time** and **Absolute Location**, and that may have a bearing on the outcome of the Attack event but we have omitted them for clarity. Otherwise, they will enter as one of the preconditions here:

Fig. 7.27 Symbolic encoding of the heuristics learned in connection with the script of Fig. 7.26 (See text for explanation)

In Fig. 7.27, we assume that there exists a general conversion process that converts the Causality Info into the LEARNED HEURISTICS in predicate logic form as shown. The first heuristic basically says that it is Generally-True (with 100 % confidence) that if the Script involved is of Attack-Type, the Num(ber) of Self agents involved is 1, the AA(Self) parameter is kept Constant, and the DIST (Self) parameter is Changed, it has no consequence on the Change in the ANX(Self) parameter (because the graph is flat).

As for the rest of the discussion involving heuristics, only one Enemy agent is involved, so we have omitted one condition – Num(Enemy) = 1 – in the interest of space in Fig. 7.27. We have also omitted other parameters such as the starting HP of the Self agent (in all the physical experiments, this is kept constant), as explained in the top right part of the figure. On the top left part of the figure, it is mentioned that it is assumed that parameters such as Time and Absolute Location have been found to be always "not important" and have been generalized away (see Sect. 6.4, Chap. 6 for a discussion of a similar process).

The second heuristic in Fig. 7.27 concerns the Change of the AA(Self) parameter, which also results in Not(Change) in the ANX(Self) parameter.

7.4.3.6 Heuristics Learning as a Causal Learning Process

In the Causality Info portion of the scripts in Figs. 7.26 and 7.27 or the CFI portion of the scripts in the various figures in Sect. 7.2, there are a number of graphs relating two variables that are based on data collected over a number of instances. In Fig. 7.27 we also show how heuristics expressed in symbolic logical form can be extracted from these graphs. If one examines the heuristic rules extracted in Fig. 7.27, one can see that they look very similar to the causal rules that we have been formulating in the foregoing discussions in various chapters – there are synchronic and diachronic preconditions defined in connection with the parameters involved and they "cause" certain effects in terms of the changes of some parameter values (diachronic effects). Whereas in many of the causal rules discussed in Chap. 6, say, in which the diachronic effect is a *Move* predicate which involves the change in an object's location, which is a parameter of the object, here in the case of Fig. 7.27 it is the "movement"/change of the ANX(Self) parameter that is the diachronic effect. Therefore, we can conceptualize the heuristic rules as causal rules, and they are learned in a similar vein as the learning of those causal rules in the foregoing discussions, e.g., the learning of the movement/non-movement rules in Chap. 6.

Figure 7.28 shows how the learning or extraction of the heuristics in Fig. 7.27 can be interpreted as a causal learning process as discussed in Chap. 2 and elsewhere. Instead of considering the changes of some parameters across *time*, here the changes of parameters are characterized across *instances* of experiences/events. Those parameters that do not change across instances are the necessary "synchronic enabling conditions" (see Fig. 2.8). The change that is deliberately made on a parameter across instances (like the application of a "force" – in this case, Change(Param(AA(Self)))) is the "diachronic cause," and the change(s) in the parameter(s) as a result is the "diachronic effect." In the case of Fig. 7.28, the effect is such that the parameter involved – ANX(Self) – does not change.

The bottom of Fig. 7.28 shows the corresponding heuristic rule of the causal picture above which looks similar to a typical causal rule in the foregoing discussions.

7.4.3.7 Continued Commander Physical Experiments with More Self Agents

As mentioned in Sect. 7.4.3.3, to construct the CFI portion of the script quickly so that it can be used to formulate optimal battle strategy, the Commander executes physical experiments (Fig. 7.11) by sending the Self agent deliberately to attack the Enemy agent a number of times. The script of Fig. 7.26 was derived from a 1 Self agent vs 1 Enemy agent Attack event. The Commander would also like to know what happens if more than one Self agent is sent to attack one Enemy agent. (Now the number of Self and Enemy agents are also parameters whose values are to be

7.4 Rapid Learning and Problem Solving in StarCraft

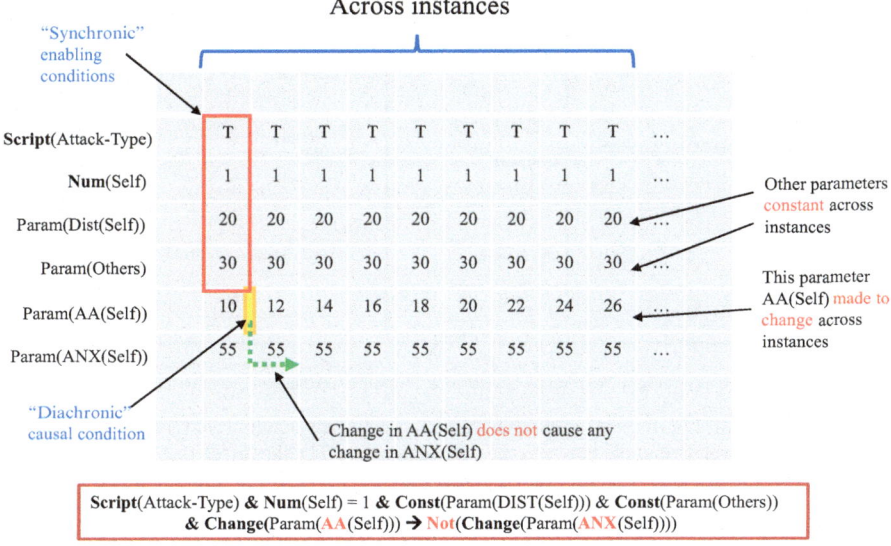

Fig. 7.28 The causal learning interpretation of heuristics learning

varied in the process of Fig. 7.11.) Therefore, she sends more than 1 Self agent to attack 1 Enemy agent over a number of instances to collect statistics to construct the CFI. Figure 7.29 shows the script capturing the results of five trials of physical experiments with 2 Self agents vs 1 Enemy agent.

As can be seen in the SCENARIO portion of the script, with 2 Self agents, Self1 and Self2, there are two additional parameters compared to the case of 1 Self vs 1 Enemy agent scenario of Fig. 7.26. These parameters are RA, the relative angle between Self1 and Self2 agents, and TD, the difference between the starting time at which the two Self agents begin heading toward the Enemy agent to initiate the Attack event. These parameters are supplied by the sensory systems when they become available because of the scenario. The earlier parameters, DIST and AA, also take on slightly different definitions here. DIST(Self1+Self2) means the averaged distance of both agents and AA(Self1+Self2) means the averaged absolute angle along which the Self agents head toward the Enemy agent. In all the physical experiments, both Self agents begin their approach toward the Enemy agent from the same distance. Therefore, DIST(Self1+Self2) = DIST(Self1) = DIST(Self2). The value of DIST(Self1+Self2) is of course varied over different instances.

In this new scenario, the script is structured accordingly, with Self1 and Self2 agents executing a movement script each and then meeting the Enemy agent and triggering an Attack event.

Again, similar physical experiments as in the one Self vs one Enemy agent scenario above are carried out, except that for the current scenario, more parameters are involved – DIST, AA, RA, and TD. The experiments consist of varying one parameter while keeping the rest constant (at some arbitrary values), and for each

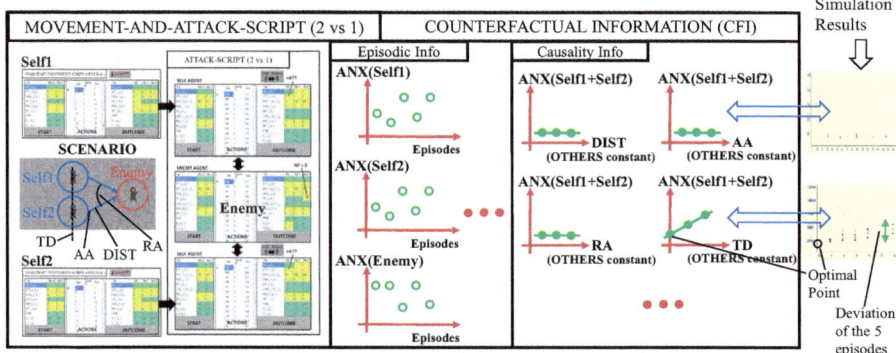

Fig. 7.29 Script of 2 Self agents (Self1 and Self2) vs 1 Enemy agent. Two more parameters are present – RA, the relative angle between the Self agents, and TD, the difference between the time at which the two Self agents begin heading toward the Enemy agent to initiate the Attack event. These parameters are supplied by the sensory systems when they become available because of the scenario. Both Self agents begin their approach toward the Enemy agent from the same DIST

parameter being varied, the same experiment is performed a number of times, typically five times. Again, as noted at the end of Sect. 7.4.3.4, the experiments are not exhaustive in terms of covering the entire parameter value space, but they are quick and dirty attempts to derive some useful rules for the Commander to plan reasonable winning strategies.

In the Causality Info portion of the script, instead of plotting the anxiousness of each Self agent vs the various parameters separately, the combined anxiousness of both agents, ANX(Self1+Self2), vs the various parameters are plotted.[4] This is because the Commander's concern is the *averaged* sum of all the Self agents' HP, as encapsulated in Eq. (7.2), which is related to the sum of all the Self agents' anxiousness. Some results from simulation experiments are shown on the right side of Fig. 7.29. The deviation of the results from the five episodes of experiments for the parameter concerned is indicated as a double green arrow.

As can be seen in the Causality Info portion of the script, the change of all the parameters except one does not result in the change of ANX(Self1+Self2). The change of TD, while other parameters are kept constant, changes ANX(Self1 +Self2). There is an optimal point at which ANX(Self1+Self2) is minimal, which is when $TD = 0$. The change of the other parameters, DIST, AA, and RA does not change ANX(Self1+Self2). The reason why the optimal point is at $TD = 0$ is that

---

[4] A question may be raised here as to whether selecting to compute ANX(Self1+Self2) is a built-in process or something the system can figure out by itself. Since earlier in Sect. 7.4.3.1 we have defined minimizing ACDR (Anxiousness of Commander) to be a goal and ACDR is related to the summation of the anxiousness of individual Self agents, there is a way to select to computer ANX (Self1+Self2) automatically through a high level reasoning process. However, we will not delve into this in this book and will leave this for future investigations.

7.4 Rapid Learning and Problem Solving in StarCraft

this corresponds to when both Self agents start moving at the same time and arriving at the Attack event at the same time (the speed of movement for both agents is the same). This results in both Self agents inflicting the maximum amount of "damage" on the Enemy agent as they both shoot at the Enemy at the same time. If $TD \neq 0$, one Self agent would begin shooting first, and the Enemy would at some point of time have to deal with only one Self agent. This causes the Self agent's HP to be reduced more. Thus, the Commander finds this out through physical experiments. The optimal point can be queried in the same manner as that illustrated in Fig. 3.2 when the Commander is formulating an optimal battle plan, which will be described in subsequent sections.

In the same vein as the process of heuristics extraction as illustrated in Fig. 7.27, Fig. 7.30 illustrates the extraction/learning of heuristics represented in predicate logic form for the current scenario. What is highlighted in blue arrows is the process of converting the TD effect on ANX(Self1+Self2) into a symbolic heuristic form.

7.4.3.8 Further Heuristic Generalization

Figure 7.31 shows how the heuristics learned from the 1 Self agent vs 1 Enemy agent scenario shown in Fig. 7.27 and those learned from the 2 Self agents vs 1 Enemy agent scenario shown in Fig. 7.30 can be further processed to create more general heuristics. Note that the number of Self agents involved in a battle is also a parameter – Num – that has a value to be varied. In the figure, "Self++" means adding all the Self agents involved together, irrespective of how many there are. The meanings of parameters such as DIST(Self++) = DIST(Self1+Self2+...) are as defined above in the previous section for the 2 Self agents vs 1 Enemy agent case.

Figure 7.31 shows that the fact that the changes of the two parameters, DIST and AA, do not change the total ANX value of the Self agents involved is "generally" true for *any number* of Self agents involved in the battle. This is useful information as it implies that for further physical experiments involving even more Self agents, there is no need to conduct experiments that vary DIST and AA – DIST and AA are *not important* in terms of being able to affect ANX. This can drastically reduce the search effort involved.

Suppose the Commander carries out another series of physical experiments (as directed by a process like that in Fig. 7.11) involving 3 Self agents vs 1 Enemy agents, varying only the parameters RA and TD, as DIST and AA have been established in the earlier experiments to be not important. Now, with 3 Self agents, Self1, Self2, and Self3, there are 3 RA (relative angle) values to consider. More experiments on varying these RAs have to be carried out but because there is no need to conduct experiments for DIST and AA, the overall number of experiments is capped. This demonstrates the power of the use of heuristics.

The results of the experiments will reveal that the change of RAs between the Self agents does not affect the value of ANX(Self1+Self2+Self3). Thus, a further generalization can be made that the value of RA is not important in further experiments. Moreover, TD(Self++) is found to be important, and has to have a

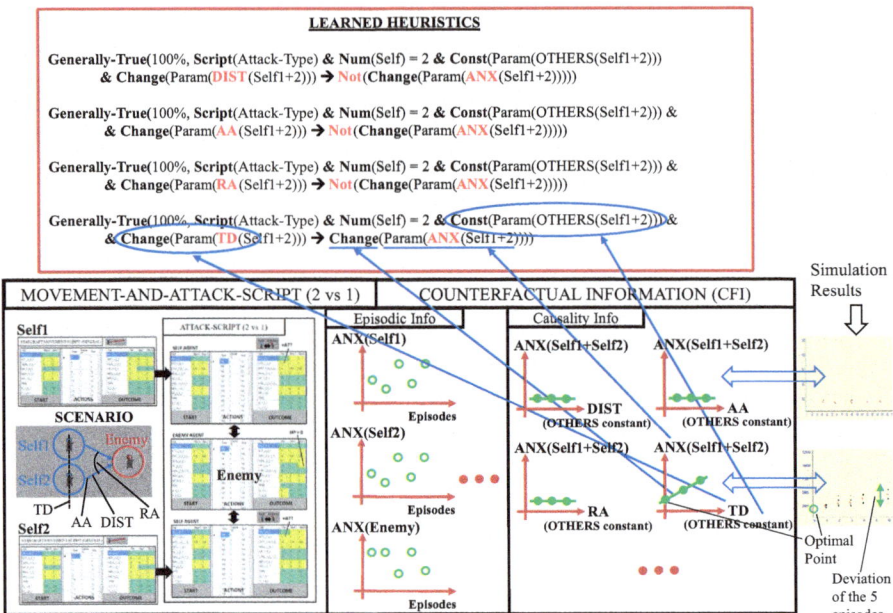

Fig. 7.30 Symbolic encoding of heuristics learned in connection with Fig. 7.29 through a similar process as that described for Fig. 7.27

certain value, 0, in order to minimize ANX(Self++). This means that in subsequent experiments when more Self agents are considered, there is no need to vary DIST, AA, and RA, but the starting time of all the Self agents sent to attack the Enemy agent must be the same (i.e., TD(Self++) = 0).

This situation is shown in Fig. 7.32 together with the picture of where the process of heuristic generalization and "filtering" takes place in the overall forward search process of Fig. 2.28. The figure also shows the modules involved for the discussion on heuristic generalization. This can be contrasted with the highlighted portion of the picture in Fig. 3.13 of Chap. 3 which did not involve processes on heuristics learning. Now, we have demonstrated using a different domain than that considered in Chap. 2, Sect. 2.6 the mechanisms of heuristic rules learning and encoding, and the function of Heuristics Filter in drastically reducing problem solving search space.

7.4.3.9 An Improved Battle Difficulty Measure

In Sect. 7.4.3.1, we defined the goal of the Commander as attempting to maximize the difference between the HP levels of the Self and Enemy agents, as encapsulated in Eq. (7.2). However, in the various scripts discussed above, we used the averaged

7.4 Rapid Learning and Problem Solving in StarCraft

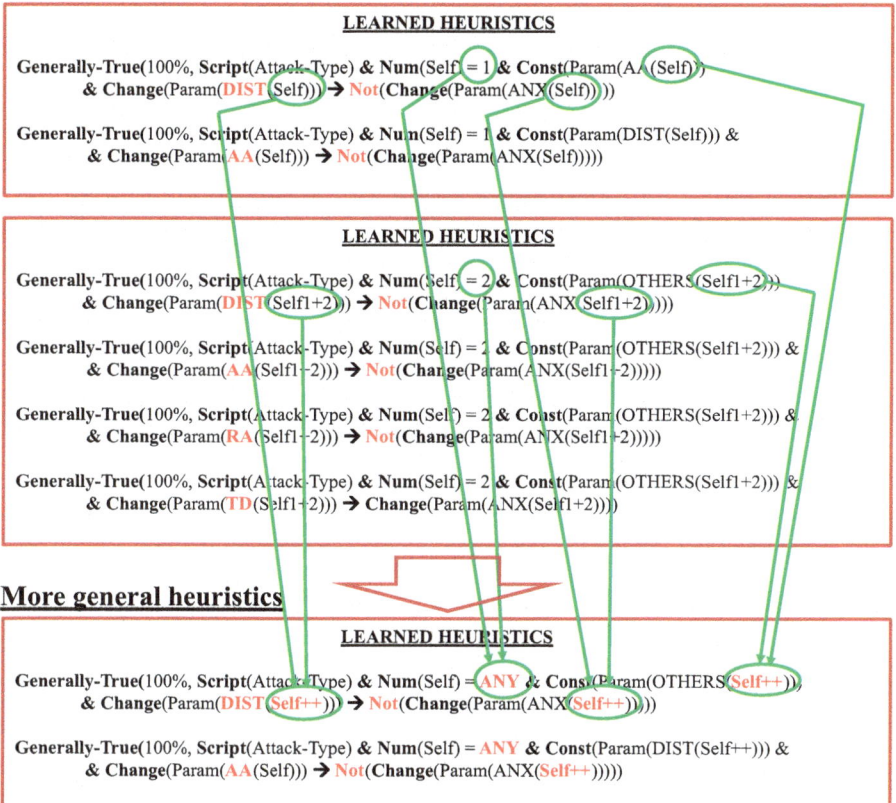

Fig. 7.31 Further generalizations on the heuristics from Figs. 7.27 and 7.30. "Self++" means averaging all the Self agents' parameter values together, irrespective of how many there are. The final generalizations imply that DIST and AA are generally *not important* in affecting ANX

anxiousness of all the Self agents portion as a measure of the success of the Attack event. We did not take into consideration the Enemy agent's HP or anxiousness as this information is often unavailable. We are betting on the fact that in general, if the result of the battle is such that the anxiousness on the Self side is the least, the situation will be advantageous to the Commander on the Self side.

In this section, we propose a slightly better measure of success of a battle. Because each time the Commander sends a Self agent to fight in a battle, there is "effort" required, reflecting a kind of resource commitment, hence this effort should be part of the measure. Otherwise, it will always be advantageous to send more Self agents as that will guarantee a kill on the Enemy agent. And then, the longer the engagement in an Attack event with an Enemy agent, the worse the situation, as other Enemy agents may arrive for the reinforcement of the current Enemy agent. There is a trade-off here, if the Commander sends too few agents to conserve the effort, the time taken to finish the battle, even if at the end it is a winning situation for the Commander, may be longer.

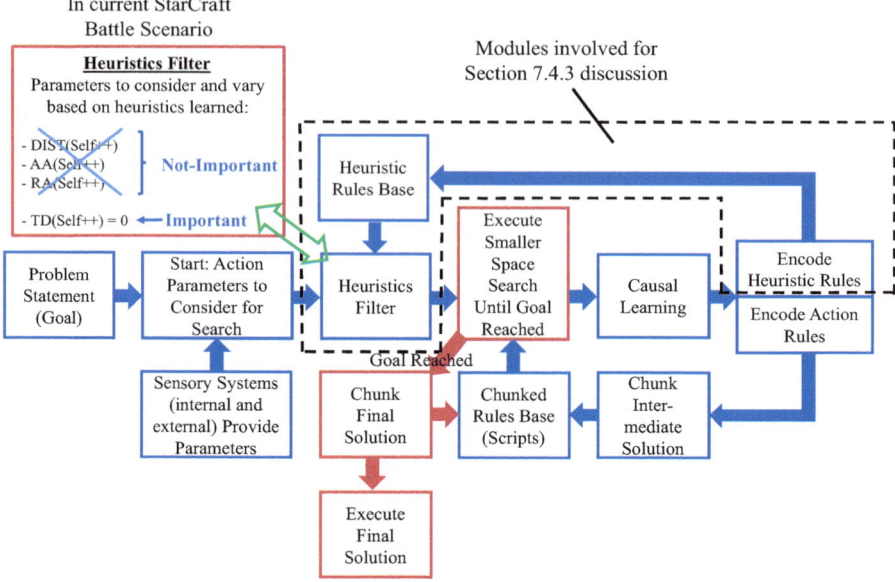

Fig. 7.32 Heuristic generalization after a 3 Self agents vs 1 Enemy agent physical experiments and its role in reducing the search space of further experiments. From now on, the only parameter that is Important that should be changed in further experiments is TD (See text for explanation)

Therefore, we define a Battle Difficulty (BATT-DIFFTY) measure for a battle based on three parameters as follows[5]:

$$\text{BATT} - \text{DIFFTY} = \textbf{Effort} \ (30 \text{ units per Self agent sent}) + \\ \textbf{Total ANXiousness} \ (\text{total of all Self agents' anxiousness}) + \\ \textbf{Duration of Battle} \ (\text{until one side wins by killing all on the other side})$$

(7.3)

Our tests revealed that the resultant optimal solution found (e.g., the optimal number of Self agents to send to Attack a number of Enemy agents) is not sensitive

---

[5] Again, a same question like that discussed in footnote 4 can be raised here – how can this function be learned? Or derived from some other built-in knowledge? The Effort portion of this function relates to energy expansion, and this could come from the basic knowledge that anything that has to do with energy expansion has to be taken into consideration in any task, as in the case of spatial movement discussed in Chap. 3, Sect. 3.1. As for Duration of Battle, this can come from earlier experiences in which too much time spent on an Attack event resulted in the enemy being able to reinforce its agents, and that resulted in the Self side being more likely to lose in the battle. The complete reasoning and learning process for this is left as a challenge for future investigations, but it is certainly doable within our framework or a logical extension of it.

7.4.3.10 Higher Order Counterfactual Information

So far, based on all the physical experiments with 1, 2, and 3 Self agents vs 1 Enemy agent, the Commander has learned the general heuristics that the various parameters associated with the Self agents (DIST, AA, and RA) are not important except that TD must be 0 (all agents start toward the Enemy at the same time). The only parameter left to further experiment with would be the number of Self agents sent.

Using the new BATT-DIFFTY measure of Eq. (7.3), and using the simulation data from the earlier experiments with 1, 2, and 3 Self agents, it can be seen that there is a general trend of reduction of BATT-DIFFTY value as the number of Self agent increases. As the number of Self agents sent to battle increases further, there may be a point at which BATT-DIFFTY increases instead. Therefore, the Commander continues to experiment with 4, 5, 6, ... Self agents and the collected data are organized in the CFI portion of a MOVEMENT-AND-ATTACK-SCRIPT (N vs 1) as shown in Fig. 7.33.[6]

In Fig. 7.33 it can be seen in the BATT-DIFFTY vs Number of Self agents graph that there is an optimal number of Self agents sent to the battle as the graph's trend is "downward" first as the number of Self agents sent increases and then the graph goes "upward."[7] The graph shows the Mean and Deviation of the various episodes of experiments. The BATT-DIFFTY value of every episode is also shown. The Mean value is used to identify the optimal number of Self agents and it is 3. This means on average if 3 Self agents are sent to a battle, the BATT-DIFFTY is minimal.[8] Of course, there may be some episodes in which the number of Self

---

[6] This entire process of observing a trend of BATT-DIFFTY changing as a result of the change of the number of Self agents involved in a battle and then triggering the continued experiments by varying the number of Self agents is certainly learnable within our causal learning framework, but it would be necessary to build some representations at a meta-level – a level at which the physical experimental process itself is explicitly represented in some scripts and/or logical forms. As mentioned at the end of Chap. 6, Sect. 6.6, "thinking and reasoning" processes are also a kind of "problem solving script." In our current simulations, we use a build-in procedure that changes this "number of agents" parameter in the spirit of the algorithm in Fig. 7.11.

[7] A question can be asked about the construction of the CFI portion and the attendant BATT-DIFFTY vs Number of Self Agents graph. How does the system automatically know to construct this graph? Again, there has to be a meta-level explicit representation of the physical experimental process and the script construction process – the system has an internal knowledge that it is varying the number of agents in these experiments, and it knows that it is targeting to compute BATT-DIFFTY as a *goal* of these experiments. Therefore, there would be a meta-level script representing this internal knowledge explicitly that is subject to some meta-level learning process.

[8] This first round of simulation results was derived from a certain setting in the StarCraft environment and the settings were such that there was a slight advantage on the enemy side. Therefore, to guarantee winning over the one Enemy agent, more Self agents are needed. Later in Sect. 7.4.3.12, we change this setting and a different picture emerges.

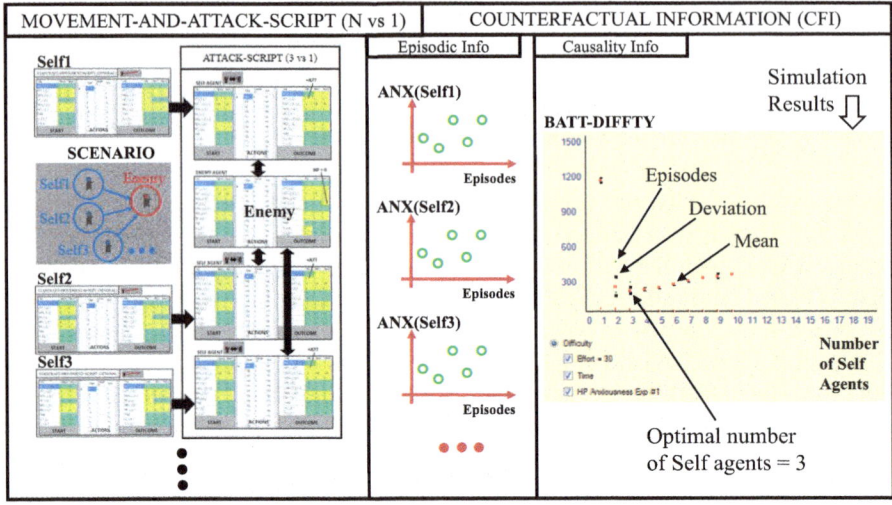

Fig. 7.33 Script derived from N Self agents vs 1 Enemy agent experiments (combining scripts of 1 vs 1, 2 vs 1, 3 vs 1, etc. experiments). The Causality Info portion provides the optimal number of Self agents vs 1 Enemy agent in terms of achieving a minimal value of BATT-DIFFTY on average. The deviation and mean values for BATT-DIFFTY are shown as *blue* and *red dots* respectively. The mean value is the basis for extracting the optimal number of Self agents

agents sent is 2 and yet a lower BATT-DIFFTY is achieved. But what we are after here is the average outcome.

Having collected the counterfactual information regarding N vs 1 agents battle scenario, the next series of experiments that the Commander would perform would be to change the number of Enemy agents to see if certain generalizations can be obtained, because battle situations are not always restricted to N Self agents vs only 1 Enemy agents.[9]

Unlike the situation in which the number of Self agents is varied, in which the number is under the control of the Commander, to change the number of Enemy agents, the Commander would have to rely on opportunistic situations to conduct the experiments. So, if there is a congregation of, say, two or three Enemy agents somewhere in the battlefield, the Commander would send certain number of Self

---

[9] Knowing what parameters in the environment are relevant for experimentation is a perennially interesting issue. Firstly, the number of Enemy agent observed is a parameter provided by the visual system. Suppose earlier, the Commander had come across only situations that had only 1 Enemy. She would form a generalization that the Enemy agents always come singly. The moment there is a situation in which there is more than 1 Enemy agent, she can immediately over-generalize and form the impression that the Enemy agents can come in any number. This could provide an impetus for her to begin the multi-Enemy agent simulations. This is environment-inspired experimentation. Alternatively, we can have an "imaginative" Commander who would simply take any parameter and carry out a "what-if-this-has-a-different-value" process. This requires resource commitment, of course, and the Commander must have some indication that exploration along that direction may be worthwhile.

7.4 Rapid Learning and Problem Solving in StarCraft 335

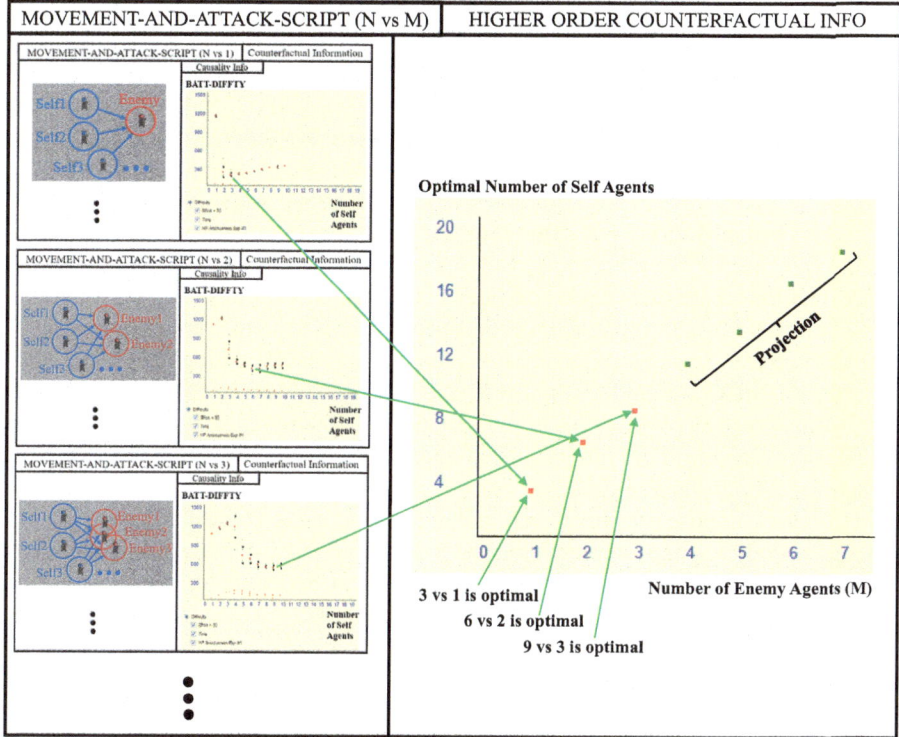

Fig. 7.34 Script containing HIGER ORDER COUNTERFACTUAL INFORMATION (HIGHER ORDER CFI) that relates Optimal Number of Self Agents to the Number of Enemy Agents. There is a trend on the data collected from engaging a small number of Enemy agents that can be projected to a larger number of Enemy agents

agents to engage them in an Attack event, and the outcome is then observed. After a certain number of episodes, the collected information can be organized as shown in Fig. 7.34.

What is shown in Fig. 7.34 is a MOVEMENT-AND-ATTACK-SCRIPT (N vs M) with HIGHER ORDER COUNTERFACTUAL INFORMATION (HIGHER ORDER CFI). The graph plotted is Optimal Number of Self Agents vs Number of Enemy Agents (M).[10] The purpose of creating this graph is to see if there is a general trend. The experiments are performed for 1, 2, and 3 Enemy agents (and sending 1, 2, 3, and more Self agents to see what the optimal number of Self agents is for each of these numbers of Enemy agents). It can be seen that there is a straight-line trend for $M = 1$, 2, and 3: the optimal numbers of Self agents are 3, 6, and 9 respectively. Therefore, a generalization could be made through a straight-line projection to a larger number of Enemy agents. (If there is no straight-line, a simple

---

[10] Again, the automatic construction of this graph would be based on a similar process as that discussed in footnote 6 concerning the construction of the graph in Fig. 7.33.

generalization cannot be made.) This way, no matter how many Enemy agents are the target of attack, the Commander knows how many of Self agents to send that would achieve an optimal value for BATT-DIFFTY (Eq. 7.3). Of course, whether she would actually send the number of Self agents would depend on whether that number of Self agents is available and whether there are other priorities.

7.4.3.11 Formulation of Battle Strategies

Armed with the knowledge, as encapsulated in the script of Fig. 7.34, of the number of Self agents to send to battle with the Enemy agents to obtain an optimal BATT-DIFFTY measure, the Commander can plan optimal battle strategies.

Suppose the Commander is not resource constrained and has sufficient number of Self agents to deploy, then for any group of M Enemy agents, the Commander simply sends an optimal number of Self agents based on the graph in the HIGHER ORDER CFI portion of Fig. 7.34.

However, suppose now the Commander is resource constrained and has a limited number of Self agents to deploy, then she has to devise an optimal strategy based on the knowledge of Fig. 7.34. Figure 7.35 shows an example in which there are 5 Enemy agents in total and they are deployed in groups of 1, 2, and 2 agents, and the Commander also has a total of 5 Self agents to deploy for the battle.

The Commander first *mentally simulates* a straightforward division of the Self agents into three groups of 1, 2, and 2 agents and send them to attack the 1, 2, and 2 Enemy agents groups respectively and simultaneously. The mental simulation based on the knowledge in the scripts of Figs. 7.33 and 7.34 results in the outcome that all 5 Self agents would be killed, given the current setting of the StarCraft environment. (In Fig. 7.33 it can be seen that if 1 Self agent is sent to attack 1 Enemy agent, the Self agent always dies. This situation is similar for 2 Self agents vs 2 Enemy agents, as can be seen in one of the three graphs on the left side of the script in Fig. 7.34.)

The Commander next tries a "divide and conquer" strategy – sending her Self agents to tackle one group of Enemy agents at a time. However, she needs to formulate a right order of execution – which group of Enemy agents to attack first and how many Self agents to send for each attack.

The script of Fig. 7.34 provides the Commander the knowledge to execute a mental simulation to formulate an optimal strategy. The script of Fig. 7.34, other than providing the information on the optimal number of Self agents for each Attack event, also provides the expected BATT-DIFFTY value for that particular optimal situation (see the left side of the script containing the three graphs). The Commander performs an exhaustive search of all possible combinations of sending the Self agents to tackle the groups of Enemy agents sequentially and selects the one that has the lowest total BATT-DIFFTY, but the search space is not large as she only needs to consider the optimal number of Self agents to be sent for each sequential step of each combination. In addition, she is constrained by the small number of Self agents she has at her disposal.

Fig. 7.35 5 Self agents to attack groups of 1, 2, and 2 Enemy agents

Suppose the Commander first considers tackling one of the groups of 2 Enemy agents first. Figure 7.34 dictates that the optimal number of Self agents to send is 6. However, since only 5 Self agents are available, the Commander can only send 5 Self agents, and would expect a less than optimal outcome, which is all that she can hope for. Thus sending 5 Self agents is sending "as optimally as allowable number of Self agents" for this situation.

Therefore, the exhaustive search process proceeds as follows. Consider all combinations of attacking the 1, 2, and 2 Enemy agents groups sequentially (i.e., attacking the groups in the order of 1, 2, and 2, or 2, 1, and 2, etc.), and for each combination, send "as optimally as allowable number of Self agents" at each step of the sequential attack process, based on the script of Fig. 7.34 and the constraint of how many Self agents are available. Since the Commander can send only "as optimally as allowable number of Self agents" at each step, the outcome may not be optimal in that some of the Self agents may die (e.g., if 4 Self agents were to attack 2 Enemy agents, both Enemy agents always die but sometimes 1 or 2 of the Self agents may also die), and the mental simulation would use the typical outcome as encountered in the earlier experimental episodes to predict the likely number of Self agents remaining. These remaining Self agents would then be deployed in mental simulation for the next step of the sequential battle process. The mental simulation tallies up the total BATT-DIFFTY for each of the combinations of attack sequence. The combination corresponding to the lowest BATT-DIFFTY is selected as the optimal battle strategy.

Figure 7.36 shows the execution of the optimal battle strategy in a StarCraft simulation of the battle situation of Figs. 7.35 and 7.36a shows that 3 Self agents were sent to attack 1 Enemy agent. Figure 7.36b shows that the 1 Enemy agent died and none of the Self agent died. In Fig. 7.36c, all the 5 Self agents then proceeded to attack 1 of the 2 Enemy agents groups. (Sending 5 Self agents is not the optimal number to attack 2 Enemy agents but as mentioned above, this is the "as optimally as allowable number of Self agents.") Figure 7.36d shows the outcome – both the Enemy agents died but 1 of the Self agents also died. Next, in Fig. 7.36e, the remaining 4 Self agents proceeded to attack the last 2 Enemy agents group. Figure 7.36f shows the outcome – both the Enemy agents died and 2 of the Self agents also died. The battle outcome is a win for the Commander with 2 Self agents still alive while all Enemy agents are killed. The video of the process is here: https://

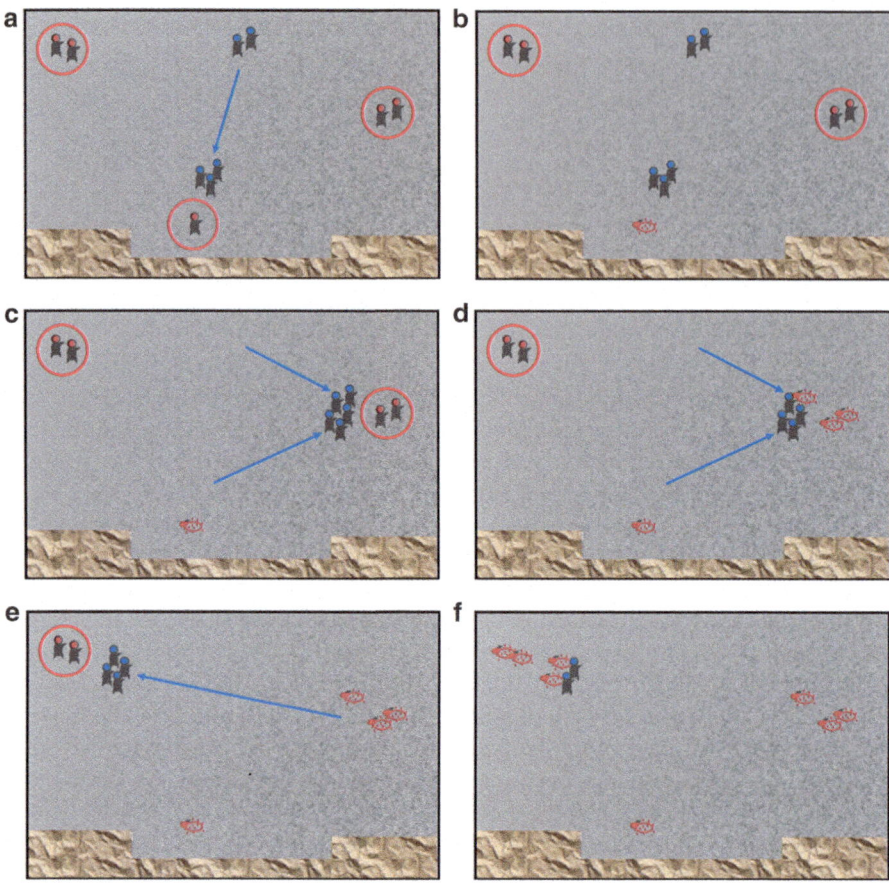

Fig. 7.36 StarCraft simulations: 5 Self agents vs 5 Enemy agents separated into 3 groups of 1, 2, and 2 agents. *Red circles* mark the Enemy agents, the rest are Self agents. (**a**) Commander sends 3 Self agents (optimal number) to attack 1 Enemy agent. (**b**) Enemy agent dies. No death on Self's side. (**c**) Commander then sends all 5 agents to attack a group of 2 Enemy agents. (This is as optimally as allowable number of Self agents to be sent.) (**d**) Both Enemy agents are killed while 1 Self agent is killed. (**e**) Commander then sends all 4 remaining Self agents to attack the last 2 Enemy agents. (**f**) This results in all Enemy agents being killed and 2 Self agents being killed. Overall the battle is won, with 2 of 5 Self agents surviving and all Enemy agents killed. Video: https://www.youtube.com/watch?v=v7yjY5UZuQo (or see video on https://noologyblog.wordpress.com: Battle involving 5 Self Agents vs 1, 2, 2, Enemy Agents)

www.youtube.com/watch?v=v7yjY5UZuQo (or see video on https://noologyblog.wordpress.com: Battle involving 5 Self Agents vs 1, 2, 2, Enemy Agents).

Note that the outcome of the "physical" execution may differ a bit from the outcome of the mental simulation because during each Attack event, the number of deaths of the agents on each side is probabilistic and the mental simulation is based on the most typical outcome experienced in the episodes of Attack events encountered in the learning phase.

7.4 Rapid Learning and Problem Solving in StarCraft

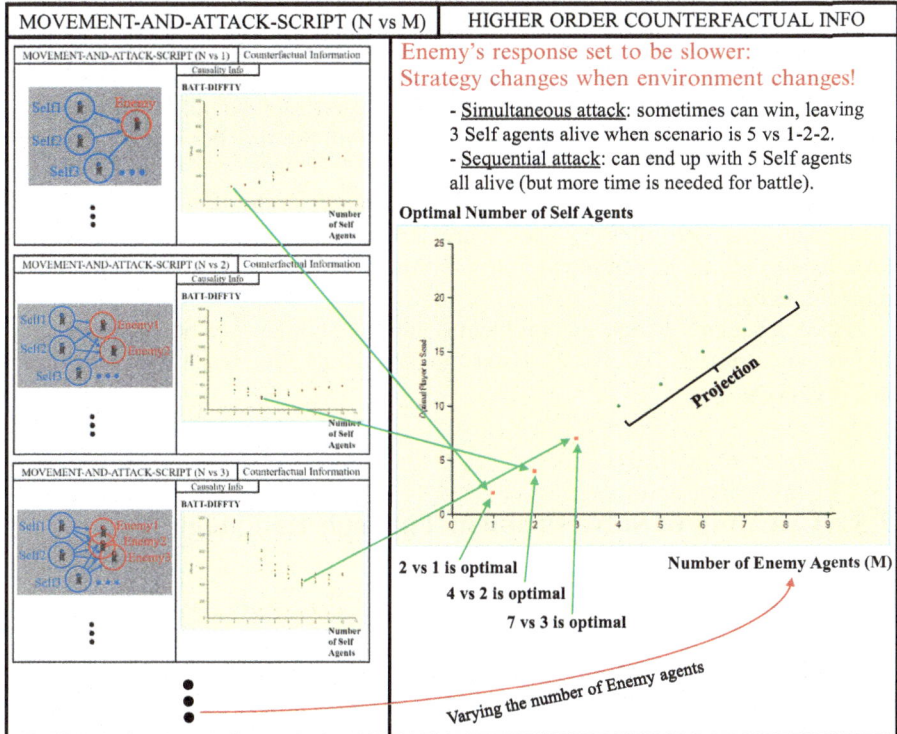

Fig. 7.37 The resultant learned MOVEMENT-AND-ATTACK-SCRIPT (N vs M) after the Enemy agents' response has been adjusted to be slower

7.4.3.12 Change of Battle Strategy as the Environment Changes

As we had mentioned in Sect. 7.4.3.3, there are settings in StarCraft that can be adjusted to change the probabilistic outcome of an Attack event – i.e., whether the Self agent or the Enemy agent is more likely to be killed. In Fig. 7.37 we show the resultant MOVEMENT-AND-ATTACK-SCRIPT (N vs M) after setting the response of the Enemy agent to be slower than in the foregoing discussion, and after going through the entire learning process as discussed in this Sect. 7.4.3. The Self agent is now more likely to kill the Enemy agent and the optimal number of Self agents for a particular number of Enemy agent is reduced. For example, for the case of 1 Enemy agent, now the optimal number of Self agents is 2, reduced from 3 before (Fig. 7.34).

The change in battle strategy for the battle scenario of Fig. 7.35 ensues (executing again the entire battle strategy formulation process discussed in the previous section). Whereas before, simultaneous attacks of the 5 Self agents (i.e., sending the same number of Self agents to attack each group of the Enemy agents) always resulted in the death of all Self agents, now the Self's side sometimes wins, leaving 3 Self agents alive while all the Enemy agents are killed. Sequential attacks, such as attacking the 1, 2, and 2 Enemy agents groups in that order, can end up with 5 Self agents all alive while all Enemy agents are killed, even though more time is needed for the battle. If a situation calls for simultaneous attacks to save time (i.e., wipe out the Enemy agents faster), the battle strategy chosen now could be simultaneous attacks, as it does win sometimes, and may be worth the risk in executing in some battle situations.

Unlike in reinforcement learning, in our learning paradigm, when the environment changes, new rules of battles are learned and new battle strategies are formulated very rapidly, allowing the Commander to respond rapidly to the environmental change.

Battle strategies which are problem solutions should also be represented as scripts and stored for future quick deployment. We leave this to future investigations.

7.5 Learning to Solve Problem Through Language

In the same vein as was discussed in Sect. 5.4, with explicit and grounded representations of the various concepts used in the discussion in this chapter, another way to provide a system with a problem solution is through the experience and knowledge of other systems communicated in natural language form. Suppose a Self agent has learned the STARCRAFT-MOVEMENT-SCRIPT of Fig. 7.3 and the ATTACK-SCRIPT (GENERAL) of Fig. 7.17 but it has not learned yet how to put these together. The Self agent is at the stage just prior to the discussion centered around Fig. 7.18, which is starting from the GOAL of Engage/Attack the Enemy agent situated at an arbitrary location with the Self agent starting from an arbitrary location. The subsequent problem solving process discussed in connection with Fig. 7.18, through a backward chaining process, resulted in a chunked script of Fig. 7.19 – a MOVEMENT-AND-ATTACK-SCRIPT (GENERAL). Now suppose at this point, instead of proceeding with the problem solving process, the Self agent gets an advice from another Self agent or some other entities in a natural language form as follows:

To attack an enemy, you must move near it first. (7.4)

Figure 7.38 shows how the advice of Rule (7.4) can be put together into a problem solution. Assuming that the sentential mapping to internal script representation is known, i.e., the system knows the grammar of the language used to receive her advice from, we show the mapping of the individual constituents of the sentence – the "words" – to the grounded representations of the corresponding concepts.

When someone provides an advice that says you "must" do things in a certain way, that means in his/her knowledge base, there is only one known way to do it. This is the meaning of the word "must." "Near" connotes a region of space that is not far from the object involved. "First" provides the temporal order to join the scripts together – in this case, the MOVEMENT-SCRIPT first. The resultant script

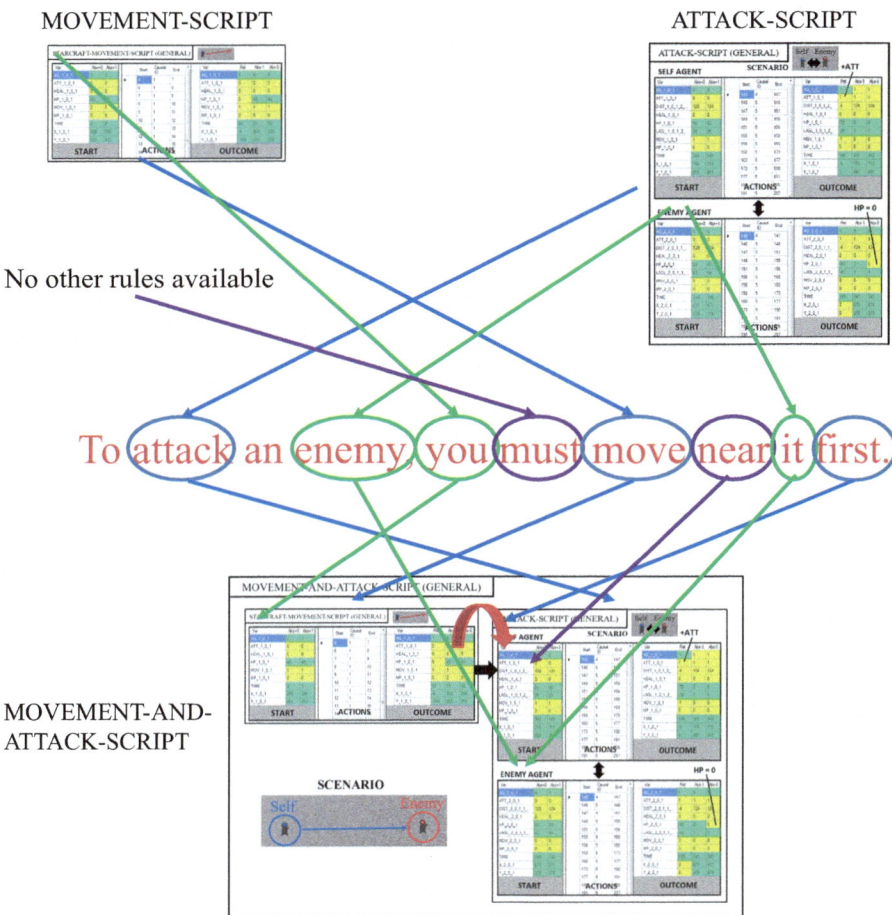

Fig. 7.38 Assembling a problem solution from natural language instruction (©2014 IEEE. Reprinted, with permission, from Ho, S.-B. and Liausvia, F., "A Rapid Learning and Problem Solving Method," Proceedings of the IEEE Symposium on Computational Intelligence for Human-like Intelligence, Page 117, Fig. 7 (MOVEMENT-SCRIPT above) and Page 114, Fig. 4 (ATTACK-SCRIPT above))

is a ready-to-use problem solving script that is learned through language rapidly. This is yet another illustration of "knowing is knowing how to act."

7.6 Summary

In this chapter we showed how all the noological learning, representational, and processing devices discussed in the preceding chapters, from rapid causal learning to learning of script and counterfactual script, knowledge chunking, learning of heuristics, and affective competition could be put together to rapidly learn and formulate battle tactics and strategies in a StarCraft game environment. Many of the complexities of the StarCraft environment were ignored and the focus was on a few key activities such as movement, attack, and fill-up HP, but we managed to demonstrate how a "deep thinking and quick learning" process (Sect. 6.6, Chap. 6) can be applied to address the problem in a "deep" way (e.g., as emphasized in Fig. 1.17, Chap. 1) in terms of addressing the issues all the way from rapid learning of concepts/scripts to problem solving.

Hence the noological processes discussed in this chapter for addressing the StarCraft issue cover the "depth" aspects of noological processing as depicted in Figs. 1.2 and 1.17 in Chap. 1: from perceptual processes to semantic and episodic memory processes, motivational and affective processes, conceptual processes, goal formation processes, action planning processing, and above all, learning processes that permeate all the various aspects of processing. The processes in this chapter basically implement the overall architecture for a noological system as depicted in Fig. 1.7 in Chap. 1 as well as the detailed processes as depicted in Figs. 2.28 and 3.23 in Chaps. 2 and 3 respectively.

This chapter also raised a number of interesting learning issues that have not been addressed. These are the issues noted in footnotes 4, 5, 6, 7, 9, and 10. To address these requires the representation of the internal mental processes themselves in script form. This will be an interesting challenge for future investigations.

References

Ho, S.-B., & Liausvia, F. (2014). Rapid learning and problem solving. In *Proceedings of the IEEE symposium series on computational intelligence for human-like intelligence*, Orlando, Florida (pp. 110–117). Piscataway: IEEE Press.

Leong, H. W., & Kwok, K. (2011). Towards robust agent behaviors in model and simulation: Situation filling in with commonsense knowledge. In *20th Conference on Behavior Representation in Modeling & Simulation*, Sundance, Utah (pp. 41–48). BRIMS Society.

Lorentz, H. A., Einstein, A., Minkowski, H., & Weyl, H. (1923) *The principle of relativity: A collection of original memoirs on the special and general relativity*. London: Constable.

Shipley, T. F., & Zacks, J. M. (2008). *Understanding events: From perception to action*. Oxford: Oxford University Press.

StarCraft II. (2015). http://us.battle.net/sc2/en/. Blazzard Entertainment, Inc.

Sutton, R. S., & Barto, A. G. (1998). *Reinforcement learning: An introduction*. Cambridge, MA: MIT Press.

Wender, S., & Watson, I. (2012). Applying reinforcement learning to small scale combat in the real-time strategy game StarCraft: Broodwar. In *Proceedings of the IEEE Conference on Computational Intelligence and Games (CIG)*, Granada (pp. 402-408). Piscataway: IEEE Press.

Chapter 8
A Grand Challenge for Noology and Computational Intelligence

Abstract In this chapter, a carefully designed micro-environment, called the Shield-and-Shelter (SAS) micro-environment, is described as a benchmark for adaptive autonomous intelligent agents (AAIA). This benchmark specifies what an AAIA needs to be able to do to be considered fully adaptive and autonomous. The various processes from perception to detailed action execution, including motivational and affective processes, are engaged by the SAS micro-environment. Compared to other benchmarks often used in AI, this is a benchmark that is more relevant for assessing a noological system's adaptive and intelligent abilities, which include the system's motivational and affective processes.

Keywords Micro-environment • Micro-world • Shield • Shelter • Perceptual process • Motivational process • Affective process • Conceptual process • Goal formation process • Learning process • Affective computing

In the previous chapter we applied the various rapid learning and problem solving methods developed in the foregoing chapters to the StarCraft game environment, addressing in the process the various noological processing aspects as depicted in Figs. 1.2 and 1.17 of Chap. 1. In this chapter, we propose a Shield-and-Shelter (SAS) micro-environment as a benchmark for adaptive autonomous intelligent agents (AAIA). This benchmark presents a challenge to the application of the various noological processing devices discussed in the foregoing chapters to address the issues involved. Like the StarCraft environment, the SAS micro-environment engages the various aspects of noological processing as depicted in Figs. 1.2 and 1.17 which we deemed are necessary for a full and adequate characterization of a noological system. However, the SAS micro-environment is simpler in some ways than the StarCraft environment but more complex in some other ways. In the next chapter, we present a solution to address a part of the SAS micro-environment. To address the SAS micro-environment fully, representational and processing devices may be required that are extensions to those that we have discussed so far in this book, and we present this as a grand challenge for future investigations.

As the SAS micro-environment has been fully defined in Ho (2013a), we extract from Sections II, III, IV, and VI of that paper (with publisher's permission) and present them in the following discussion with minor modifications of the text and figures.

8.1 The Shield-and-Shelter (SAS) Micro-environment[1]

8.1.1 Basic Considerations

As mentioned in Chap. 1 and the above, a micro-environment that is useful for the computational study of intelligent processes must have the right level of complexity to elicit the various critical aspects to be addressed as depicted in Figs. 1.2 and 1.17. Among other things, "time" is an important factor that has to be emphasized and addressed. The environment, whether external or internal, is not static – activities take place to support or challenge the survival of the agent. A meaningful environment would also contain entities functioning in various roles and interacting with other entities. These interactions would elicit actions taken by the intelligent agent to satisfy its internal needs. Therefore, the following minimal features should be present in the environment:

- Entities (intelligent agents and other objects that can potentially interact with each other).
- Temporality (to allow activities to take place).
- Activities and interactions between entities (to elicit various functionalities).
- Agent's internal needs (this is the internal micro-environment that produces goals and motivations).
- Agent's actions (these allow the agent to act in ways to satisfy its internal needs)
- Agent's perceptual system (this assists the agent in seeking solutions to its problems).

With these minimal features, the intelligent actions taken by the intelligent agent, whether they are external actions taken to satisfy certain goals or internal conceptual processes (mental actions) for the subsequent elaborations of external actions, would be *embodied, situated, and grounded*. Figure 8.1 depicts a minimal micro-environment that has a simple Agent (that has the ability to move about), a Projectile that can threaten the Agent (by hurtling toward the Agent), and "Walls" that can function as Shields that the Agent can potentially use to protect itself. The Agent and Projectile can potentially move about on their own and the Shield can be

---

[1] This section is derived, with minor modifications, from the following source: ©2013 IEEE. Reprinted, with permission, from Ho, S.-B., "A Grand Challenge for Computational Intelligence: A Micro-Environment Benchmark for Adaptive Autonomous Intelligent Agents," Proceedings of the IEEE Symposium Series on Computational Intelligence – Intelligent Agents, Pages 45–48, Section II.

8.1 The Shield-and-Shelter (SAS) Micro-environment

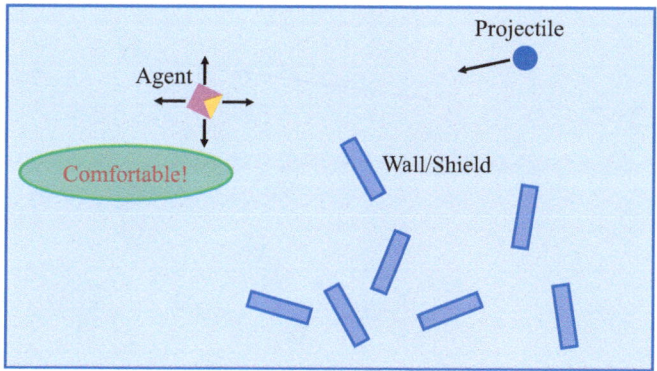

Fig. 8.1 The SAS micro-environment (external and internal) (©2013 IEEE. Reprinted, with permission, from Ho, S.-B., "A Grand Challenge for Computational Intelligence: A Micro-Environment Benchmark for Adaptive Autonomous Intelligent Agents," Proceedings of the IEEE Symposium Series on Computational Intelligence – Intelligent Agents, Page 46, Fig. 2)

moved by the Agent. The Agent has an internal state and currently it is in a state of Comfort as it is not under any threat.

8.1.2 Activities in the Micro-environment

Figure 8.2 depicts how an internal Pain signal can arise in the Agent – when the Projectile appears and hurtles toward it, and it is hit by the Projectile. The Pain signal defines a minimally simple pre-specified avoidance need in the *internal* micro-environment.

Figure 8.3 depicts how a complex affective state can arise from this simple micro-environment. Having experienced Pain, the Agent, on seeing the mere appearance of the Projectile, even when the Projectile is stationary, would enter a state of Anxiousness[2] because of its prior experience that informs it about what this situation may imply for the future – a possible future Pain experience. (Compare with Fig. 8.1 in which the Agent is in a state of Comfort despite the Projectile's presence because it has not experienced the Pain arising from the Projectile yet). The state of Anxiousness will prime the Agent to be ready for potential harm from the Projectile and search for solutions to avoid the harm. One possible solution that the Agent can find is to move aside and avoid the Projectile when it hurtles toward it again, and this is illustrated in Fig. 8.4. The Agent can prime itself for this

---

[2] As mentioned in footnote 3 of Chap. 3, *fear* and *anxiousness* are used synonymously here and the rest of the book. There are subtle differences which we will not particularly address here. E.g., typically it is "fear of the snake causes him to run" and "I am anxious that I cannot finish my homework by tonight." In these examples, it seems to be a matter of degree.

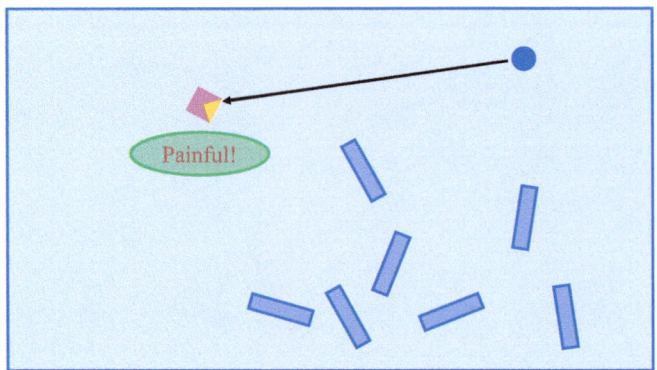

Fig. 8.2 Pain signal as a built-in simple avoidance need in the SAS *internal* environment (©2013 IEEE. Reprinted, with permission, from Ho, S.-B., "A Grand Challenge for Computational Intelligence: A Micro-Environment Benchmark for Adaptive Autonomous Intelligent Agents," Proceedings of the IEEE Symposium Series on Computational Intelligence – Intelligent Agents, Page 46, Fig. 3)

avoidance action so that harm can be avoided. When the Agent is successful in avoiding the projectile, it enters a state of Relief.[3]

However, if the Agent has prior observations on certain physical phenomena in the micro-environment that it can take advantage of to protect itself from the Projectile's harm, it can plan ahead for a different solution to a Relieved state. For example, Fig. 8.5 illustrates an earlier occasion in which the Projectile was deflected by a "Shield" (the blue Wall) when it hit the Shield. (Here we make some simplifying assumptions on physics – assuming that the Shield that is hit by the Projectile does not move).

If such knowledge of physical interactions has earlier been acquired by the Agent, then on the appearance of the Projectile which leads to the Agent entering the state of Anxiousness (such as that depicted in Fig. 8.3), the Agent may then be able to plan and act ahead to relieve itself of its anxiousness – way before the Projectile starts to move. This is illustrated in Fig. 8.6 in which the Agent seeks "protection" from the Projectile in advance by hiding behind the Shield observed earlier that could deflect the Projectile.

Here we would like to emphasize that the Agent's learning of the use of the Shield (the blue Wall) to protect itself (Fig. 8.6) should be acquired in a rapid and general way. I.e., the traditional method of reinforcement learning alone (e.g., Sutton and Barto 1998) will not suffice as it requires many repetitions of the situations and it also does not allow generalization of the potential Shield function to the other blue Walls (see related discussion in Appendix A). To be able to do this, the Agent must be able to learn and identify *causes* of events (Friston 2010; Passingham and Wise 2012; Pearl 2009; Sloman 2005). Similarly, in the case of

---

[3] In here and subsequent scenarios, we assume that the Projectile disappears after either hitting or not hitting the Agent and reappears somewhere else.

8.1 The Shield-and-Shelter (SAS) Micro-environment

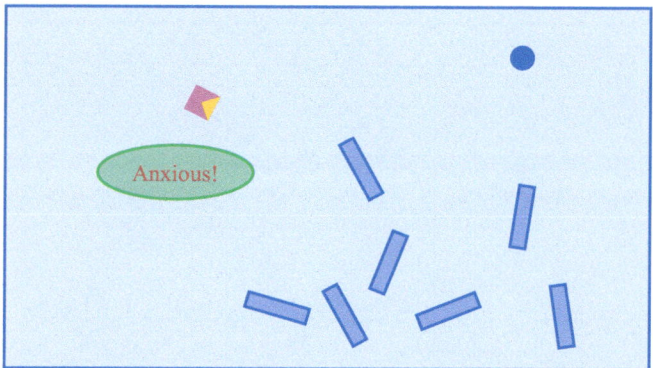

Fig. 8.3 Having experienced pain, anxiousness can arise when the projectile is present (©2013 IEEE. Reprinted, with permission, from Ho, S.-B., "A Grand Challenge for Computational Intelligence: A Micro-Environment Benchmark for Adaptive Autonomous Intelligent Agents," Proceedings of the IEEE Symposium Series on Computational Intelligence – Intelligent Agents, Page 46, Fig. 4)

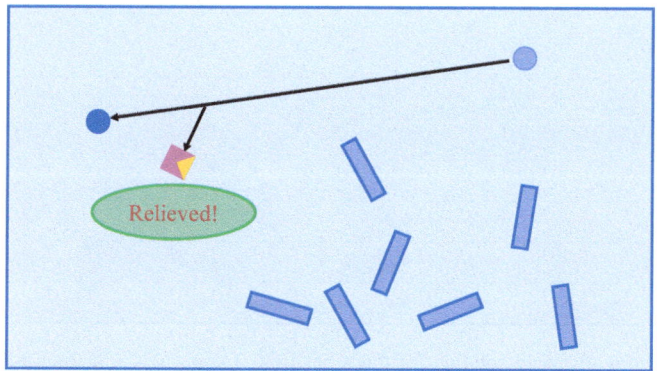

Fig. 8.4 A possible relief for the agent from the projectile – moving aside when the projectile comes hurtling toward it (©2013 IEEE. Reprinted, with permission, from Ho, S.-B., "A Grand Challenge for Computational Intelligence: A Micro-Environment Benchmark for Adaptive Autonomous Intelligent Agents," Proceedings of the IEEE Symposium Series on Computational Intelligence – Intelligent Agents, Page 46, Fig. 5)

the other "Pain avoidance" solution in Fig. 8.4 in which the Agent moves aside to avoid the Projectile, being able to identify the *causes* (which include the Projectile and its *trajectory*) of the earlier Pain event would allow it to concoct the solution rapidly. Ho (2013b) describes a method for uncovering simple causations using counterfactual analysis – e.g., "had the force not been there, the object would not have moved, hence the force is probably the cause of the movement" – and this requires an episodic memory for the storage and retrieval of past events for comparisons across time. Ho (2014) describes a temporal correlation-based method

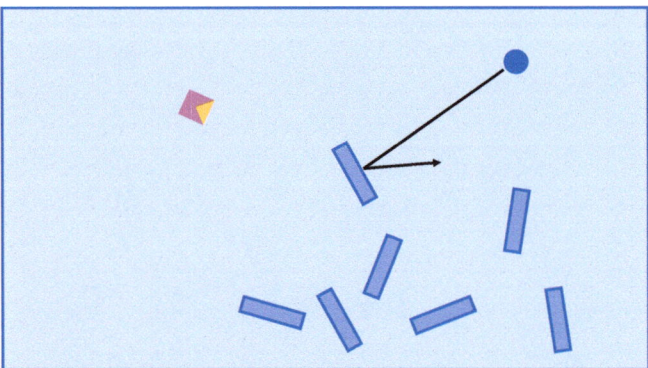

Fig. 8.5 A shield deflecting a projectile (©2013 IEEE. Reprinted, with permission, from Ho, S.-B., "A Grand Challenge for Computational Intelligence: A Micro-Environment Benchmark for Adaptive Autonomous Intelligent Agents," Proceedings of the IEEE Symposium Series on Computational Intelligence – Intelligent Agents, Page 46, Fig. 6)

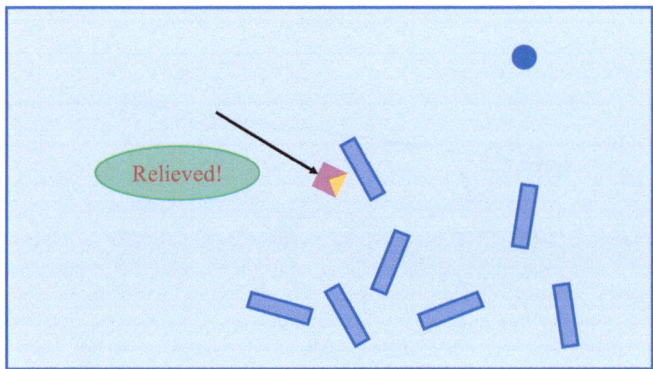

Fig. 8.6 A more advanced solution for the relief from anxiousness – finding a shield (©2013 IEEE. Reprinted, with permission, from Ho, S.-B., "A Grand Challenge for Computational Intelligence: A Micro-Environment Benchmark for Adaptive Autonomous Intelligent Agents," Proceedings of the IEEE Symposium Series on Computational Intelligence – Intelligent Agents, Page 47, Fig. 7)

for rapid effective learning of causality which has been discussed in detail in Chap. 2.

The micro-environment and the potential activities (both external to the Agent or internal to it – e.g., its affective states) that can arise in the Agent as depicted in Figs. 8.1, 8.2, 8.3, 8.4, 8.5, and 8.6 address some very fundamental and critical issues that are common to all noological processes. In this micro-environment, we have defined and built-in a simple value signal – a Pain signal. Yet, with the concept of time and hence future potential activities, the Agent can experience a higher level affective state – Anxiousness (of Pain) – and the exact situation in which it may arise is not built into the Agent and is instead contingent upon what the Agent may

8.1 The Shield-and-Shelter (SAS) Micro-environment

experience in the "real" world. The need for Relief from Anxiousness and Pain hence drive the Agent to solve problems and seek certain solutions. Relief thus defines an *affective goal* of the Agent and this gives rise to the motivational aspects of intelligent processes.

In Fig. 8.5, the function and hence the concept of Shield is merely physical – that of being able to deflect the trajectory of the Projectile. In Fig. 8.6, the function and hence concept of Shield acquired by the Agent is *embodied, situated, and grounded*. The concept of Shield now incorporates the movement, activities, and problem solving process of the Agent itself in its attempt to find a solution to reach a final affective state (of being Relieved) from an initial affective state (of being Anxious), in addition to the physical functions of the Shield (which is that of being able to deflect the Projectile). Thus in one fell swoop, this micro-environment addresses all the ten aspects of intelligent processes illustrated in Figs. 1.2 and 1.17 of Chap. 1:

- **Basic perceptual processes** operating on the external environment represented in a raw, analogical and pictorial form are engaged.
- **Higher level perceptual processes** are engaged including the processing of the motion of the entities in the environment, hence giving rise to concepts of activities.
- **Attentional processes** are engaged to allow the Agent to focus on the relevant stimuli for conceptualizations and actions – e.g., attending to and identifying the Projectile as a possible source of Pain, and attending to the relevant elements in the physical interactions between entities to characterize concepts such as Shield.
- **Memory processes** such as episodic memory are engaged to remember past episodes of experience that have relevance for identifying *causes* for future actions and conceptualizations (e.g., the activity of the Projectile that causes Pain – Fig. 8.2 – or the interaction between the Projectile and the Shield that uncovers a causal agent that can function to deflect a Projectile – Fig. 8.5). Semantic memory processes are engaged in the formation of concepts such as Shield and Shelter.
- **Affective processes** are engaged – from Pain to Anxiousness, Relief, etc.
- **Conceptual processes** that are fully *embodied, situated*, and *grounded* are engaged, as illustrated by the fully embodied, situated, and grounded concepts of Shield and Shelter that arose in the Agent as it seeks solutions and carries out actions to satisfy its internal needs, driven by its motivational and affective processes.
- **Goal Formation processes** are engaged – the goal of achieving certain affective states drives the formation of subgoals and the attendant plans and actions to be taken to attain the desired affective states.
- **Actions** are planned to effect the desired goal state.
- **Detailed actions** are executed to effect the desired goal state.
- **Learning Processes** – the Agent only has a minimal built-in goal (in this case, "avoid Pain") and a built-in response to Pain (when the Projectile hits it). All other knowledge and sub-goals necessary for problem solving are learned by

observing and interacting with the environment. Even the knowledge that the blue Walls can be moved in a certain way (so as to enable the construction of a "Shelter") as would be described later in Figs. 8.10 and 8.11 is gained through learning (e.g., through the Agent pushing the objects and observing the resultant motions, much like the learning processes discussed in Sect. 3.5, Chap. 3, and in particular, the discussion in connection with Fig. 3.12).

The Agent in the above micro-environment is autonomous in the sense that only very minimal, basic, and general goals are built-in (in this case, "avoid Pain") and any further goals or subgoals (e.g., hiding behind a Shield) necessary for the ultimate achievement of the built-in goal is discovered by the agent in the process with no human intervention.

8.1.3 Further Activities and Concepts

Can further functional concepts other than Shield arise in this micro-environment? Suppose after the Agent has found Relief from the Projectile as depicted in the situation in Fig. 8.6, it is then threatened again by the appearance of the Projectile in another position as shown in Fig. 8.7, which causes it to have to find another solution to escape from this threat, as shown in Fig. 8.8. And if, after this temporary Relief from the Projectile, the Agent is further repeatedly threatened by the Projectile and has to repeatedly enter the state of Anxiousness, it may then enter a state of being "Stressed" as depicted in Fig. 8.9.

To alleviate Stress, the Agent may attempt to find a more permanent solution – that of actively constructing a "Shelter" to protect itself on a more permanent basis as shown in Fig. 8.10. Unlike the concept of Shield in which the Agent merely makes use of a particular physical object – the Wall – by positioning itself strategically with respect to it and the Projectile, the concept of Shelter involves not only solving for a solution that involves just the Agent's own movement but also an active part on the Agent to push the pieces of Shields in order to construct the Shelter. There is a lot more embodiment and situatedness involved here. This Shelter represents a permanent solution to its Anxiousness problem and it arrives at an affective state of Comfort which is something more permanent than Relieved. (Here we have made simplifying assumptions about physics – unlike in the real world, the Projectile is not able to dislodge the individual Shields if it comes into contact with them so that the Relief and Comfort are permanent – even though the Shields can be moved by the Agent as mentioned above). Figure 8.11 shows how the Agent moves the Shields in the process of constructing a Shelter.[4]

---

[4] Section 3.5 of Chap. 3 shows the complete process of learning to handle physical objects such as Walls to construct a Shelter – discussed in the context of incremental chunking for problem solving.

8.2 The Generality of the SAS Micro-environment 351

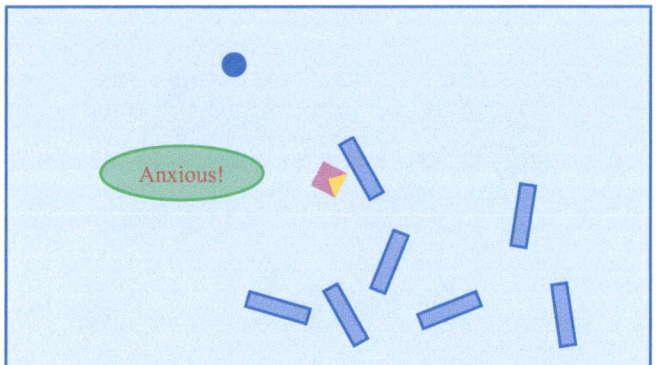

Fig. 8.7 Further appearances of the projectile in a "threatening" position (©2013 IEEE. Reprinted, with permission, from Ho, S.-B., "A Grand Challenge for Computational Intelligence: A Micro-Environment Benchmark for Adaptive Autonomous Intelligent Agents," Proceedings of the IEEE Symposium Series on Computational Intelligence – Intelligent Agents, Page 48, Fig. 8)

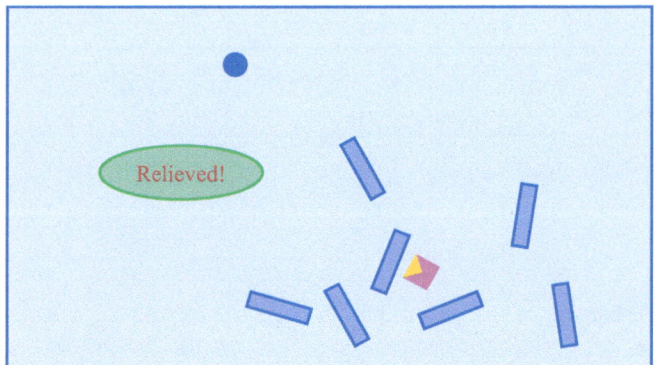

Fig. 8.8 Agent finds another shield to be relieved (©2013 IEEE. Reprinted, with permission, from Ho, S.-B., "A Grand Challenge for Computational Intelligence: A Micro-Environment Benchmark for Adaptive Autonomous Intelligent Agents," Proceedings of the IEEE Symposium Series on Computational Intelligence – Intelligent Agents, Page 48, Fig. 9)

8.2 The Generality of the SAS Micro-environment[5]

In the previous section we described a basic SAS micro-environment to address the various levels of intelligent processes as depicted in Figs. 1.2 and 1.17 of Chap. 1. Any attempt at designing an intelligent agent to capture the processes necessary for the agent to behavior in an intelligent manner as described in Sect. 8.1 – i.e.,

---

[5] This section is derived, with minor modifications, from the following source: ©2013 IEEE. Reprinted, with permission, from Ho, S.-B., "A Grand Challenge for Computational Intelligence: A Micro-Environment Benchmark for Adaptive Autonomous Intelligent Agents," Proceedings of

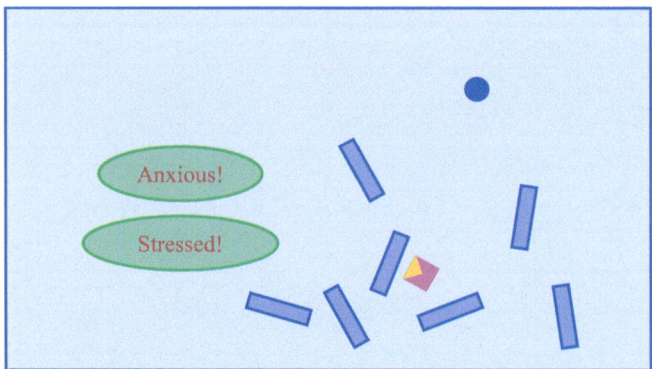

Fig. 8.9 Another appearance of the projectile at a different threatening location. Agent enters a state of stressed after repeated experiences of Anxiousness (©2013 IEEE. Reprinted, with permission, from Ho, S.-B., "A Grand Challenge for Computational Intelligence: A Micro-Environment Benchmark for Adaptive Autonomous Intelligent Agents," Proceedings of the IEEE Symposium Series on Computational Intelligence – Intelligent Agents, Page 48, Fig. 10)

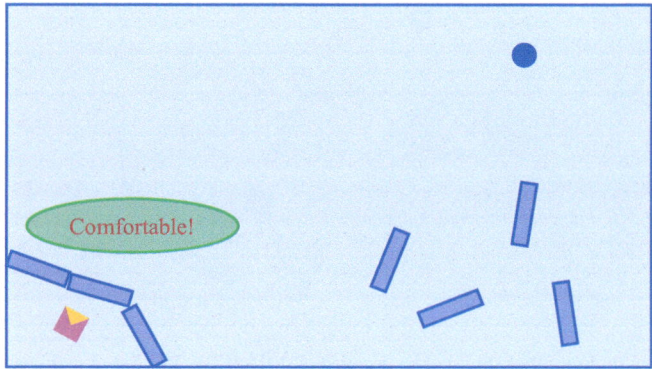

Fig. 8.10 A permanent solution to avoid the state of stressed – the construction of a shelter to arrive at a state of comfort (©2013 IEEE. Reprinted, with permission, from Ho, S.-B., "A Grand Challenge for Computational Intelligence: A Micro-Environment Benchmark for Adaptive Autonomous Intelligent Agents," Proceedings of the IEEE Symposium Series on Computational Intelligence – Intelligent Agents, Page 48, Fig. 11)

seeking solutions to satisfy its internal needs, aided by its experiences with the environment and the concepts acquired in its process of interacting with the environment – must be general and extendable should the environment change.

For example, one way the environment may be different is depicted in Fig. 8.12, in which the blue Walls are now not able to function as Shields because they allow the Projectile to penetrate them, and instead, there is another group of red Walls

---

the IEEE Symposium Series on Computational Intelligence – Intelligent Agents, Pages 48–49, Section III.

8.2 The Generality of the SAS Micro-environment

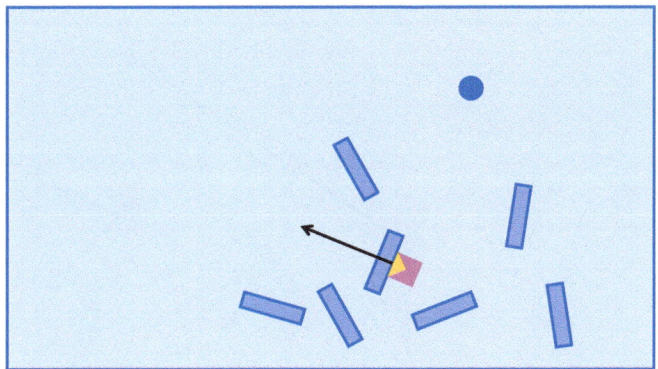

Fig. 8.11 Agent pushing a shield/wall to construct a shelter (©2013 IEEE. Reprinted, with permission, from Ho, S.-B., "A Grand Challenge for Computational Intelligence: A Micro-Environment Benchmark for Adaptive Autonomous Intelligent Agents," Proceedings of the IEEE Symposium Series on Computational Intelligence – Intelligent Agents, Page 48, Fig. 12)

than can function as Shields. The Agent will then learn rapidly (with only a few training instances) and select the correct kind of objects accordingly for its Shield and Shelter functions. Yet another example would be if we have built-in not only the internal signal Pain but also the Hunger signal – i.e., we have expanded the internal micro-environment – the Agent would then occasionally need to leave the Comfort of its Shelter to look for Food, taking the risk that in the process it might be threatened by the Projectile, as depicted in Fig. 8.13. The Agent would then have to strike an optimized balance between the need to avoid Pain and the need to alleviate Hunger.[6]

Therefore the behavior of the agent must be adaptive with respect to various possible extensions of the external and internal environments. The intelligent agent must have minimal built-in mechanisms to allow it to adapt to changes in the environment. The agent must not have pre-conceived notions of how entities interact with each other in the environment, and instead it would learn the regularities by observing the interactions in the environment and use the learned knowledge to solve problems to satisfy its internal needs. The agent could have some built-in sensory capabilities, such as the ability to perceive the location and orientation of the various entities in the environment and the ability to detect their motion. The internal signals such as Pain and Hunger would be pre-specified. In between these built-in mechanisms, the intelligent agent must learn from the environment to satisfy its internal needs. How might a general intelligent agent be constructed to meet this requirement? This is a grand challenge posed for noology and computational intelligence.

---

[6] This is addressed to some extent by affective competition as discussed in Sect. 3.3 of Chap. 3 and Sect. 7.4.2 of Chap. 7.

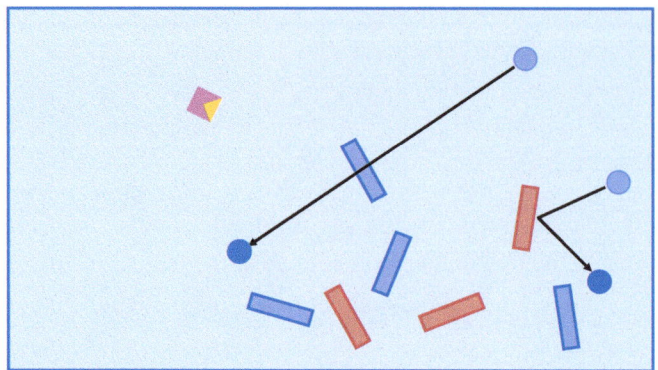

Fig. 8.12 Alternative physics – blue walls are not able to deflect the projectile. Red walls are able to do so (©2013 IEEE. Reprinted, with permission, from Ho, S.-B., "A Grand Challenge for Computational Intelligence: A Micro-Environment Benchmark for Adaptive Autonomous Intelligent Agents," Proceedings of the IEEE Symposium Series on Computational Intelligence – Intelligent Agents, Page 48, Fig. 13)

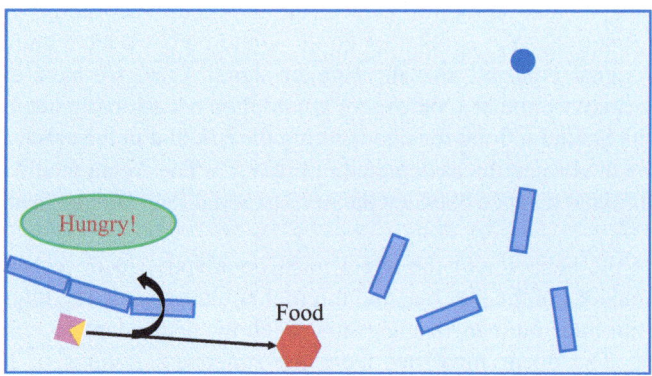

Fig. 8.13 Agent leaving the shelter to look for food in an expanded internal environment with the Hunger signal (©2013 IEEE. Reprinted, with permission, from Ho, S.-B., "A Grand Challenge for Computational Intelligence: A Micro-Environment Benchmark for Adaptive Autonomous Intelligent Agents," Proceedings of the IEEE Symposium Series on Computational Intelligence – Intelligent Agents, Page 49, Fig. 14)

8.3 The Specifications of the SAS Micro-environment Benchmark[7]

Before proceeding to discuss the solutions for addressing part of the SAS micro-environment benchmark in the next chapter, in this section we state the full specifications of the SAS micro-environment benchmark. The benchmark assumes

---

[7] This section is derived, with minor modifications, from the following source: ©2013 IEEE. Reprinted, with permission, from Ho, S.-B., "A Grand Challenge for Computational Intelligence: A Micro-Environment Benchmark for Adaptive Autonomous Intelligent Agents," Proceedings of the IEEE Symposium Series on Computational Intelligence – Intelligent Agents, Page 49, Section IV.

8.3 The Specifications of the SAS Micro-environment Benchmark

that the intelligent agent is fed a relatively processed perceptual input – that there is no need to do low level vision to uncover the location, orientation, and shape of the various objects from pixel-level perceptual input. The specifications are as follows:

- The intelligent mechanisms controlling the Agent in the micro-environment must direct the Agent to exhibit the behaviors shown in Figs. 8.1, 8.2, 8.3, 8.4, 8.5, 8.6, 8.7, 8.8, 8.9, 8.10, and 8.11 given the various interactions between the entities in the environment.
- For exhibiting the behaviors shown in Figs. 8.1, 8.2, 8.3, 8.4, 8.5, 8.6, 8.7, 8.8, 8.9, 8.10, and 8.11, the only built-in internal value signal is Pain (and it is triggered by the Projectile hitting the Agent) and a built-in motivational goal is "Avoid Pain."
- For exhibiting the behaviors shown in Figs. 8.1, 8.2, 8.3, 8.4, 8.5, 8.6, 8.7, 8.8, 8.9, 8.10, and 8.11, the only built-in perceptual processing abilities of the Agent are (i) the ability to know the location, orientation, and shape of the various objects (i.e., visual ability is assumed) and (ii) the ability to detect elemental (pixel by pixel level) motions of the various objects.
- For exhibiting the behaviors shown in Figs. 8.1, 8.2, 8.3, 8.4, 8.5, 8.6, 8.7, 8.8, 8.9, 8.10, and 8.11, the Agent must not have any pre-conceived notions of the physical behaviors of the objects – e.g., no pre-conceived notion of "trajectories" and the outcomes of physical interactions between the objects such as the Shield object being able to deflect the Projectile, etc. – and no pre-conceived concepts such as Shield and Shelter. It also has no built-in knowledge of how its own actions can move the objects in the environment in certain manners (Fig. 8.11) and this must be learned from its interactions with the various objects.[8]
- For exhibiting the behaviors shown in Figs. 8.5 and 8.6 – i.e., from the Agent's observation of an event showing an object being able to deflect the Projectile to the application of the concept as a Shield to protect itself – the Agent must be able to learn the concept of Shield rapidly after a single or a small number of instances of observation, and the concept thus encoded should include generalizations to other similar "Shield" objects so that it can subsequently use the other similar objects as Shields in the situations of, say, Figs. 8.7, 8.8, 8.9, and 8.10. To achieve this, the Agent must be able to learn concepts by identifying *causes* (Friston 2010; Ho 2013b, 2014; Pearl 2009; Passingham and Wise 2012; Sloman 2005) rather than able to carry out only pure reinforcement learning (Sutton and Barto 1998), which is slow and normally requires a large number of situation repetitions. The same requirement is expected for the behavior shown in Fig. 8.4 – moving aside to avoid the Projectile.
- The Agent must be able to behave accordingly should the internal and/or external environment change like in Figs. 8.12 and 8.13 with no necessity to make changes to its basic noological processing mechanisms incorporating the various aspects of processing as depicted in Figs. 1.2 and 1.17 of Chap. 1.

---

[8] Much like the learning process discussed in Sect. 3.5, Chap. 3.

8.4 Conclusion[9]

In this chapter, we described a SAS micro-environment benchmark that possesses a minimal set of features that engages the various aspects of noological processes from perception to action, thus ensuring that the critical issues of how the various processes are characterized and how they interact with each other can be addressed (Fig. 1.2, Chap. 1). Because the SAS micro-environment engages a small width of noological processing (Fig. 1.17, Chap. 1), the problem is deemed to be more tractable. The main purpose of the micro-environment benchmark is to pose a challenge to noology and computational intelligence to construct a noological system to address the issues involved, and a satisfactory system would be one that incorporates general mechanisms that are scalable as the environment (both internal and external) becomes more complex.

Note that many emotional states can only be elicited when the environment involves more than one agent. Examples are anger, grief, admiration, envy, gloat, contempt, etc. Therefore, to investigate these, the basic SAS micro-environment would have to be extended to include more than one agent interacting with each other.

Problem

Before proceeding to read the next chapter for the solution to the SAS micro-environment problem using the general approach propounded in this book, the reader should pause for a thought here on how the other existing methods in AI/computational intelligence – e.g., A* search, reinforcement learning, traditional AI problem solving, neural networks, deep learning, etc. – can or cannot be brought to bear on this problem.

References

Friston, K. (2010). The free-energy principle: A unified brain theory? *Nature Reviews Neuroscience, 11*, 127–138.

Ho, S.-B. (2013a). A grand challenge for computational intelligence: A micro-environment benchmark for adaptive autonomous agents. In *Proceedings of the IEEE symposium series on computational intelligence – Intelligent agents*, Singapore (pp. 44–53). Piscataway: IEEE Press.

---

[9] Part of this section is derived, with minor modifications, from the following source: ©2013 IEEE. Reprinted, with permission, from Ho, S.-B., "A Grand Challenge for Computational Intelligence: A Micro-Environment Benchmark for Adaptive Autonomous Intelligent Agents," Proceedings of the IEEE Symposium Series on Computational Intelligence – Intelligent Agents, Page 45–48, Section VI.

Ho, S.-B. (2013b). Operational representation – A unifying representation for activity learning and problem solving. In *AAAI 2013 fall symposium technical reports-FS-13-02*, Arlington, Virginia (pp. 34–40). Palo Alto: AAAI.

Ho, S.-B. (2014). On effective causal learning. In *Proceedings of the 7th international conference on artificial general intelligence,* Quebec City, Canada (pp. 43–52). Berlin: Springer-Verlag.

Passingham, R. E., & Wise, S. P. (2012). *The neurobiology of the prefrontal cortex*. Oxford: Oxford University Press.

Pearl, J. (2009). *Causality*. Cambridge: Cambridge University Press.

Sloman, S. (2005). *Causal models: How people think about the world and its alternatives*. Oxford: Oxford University Press.

Sutton, R. S., & Barto, A. G. (1998). *Reinforcement learning: An introduction*. Cambridge, MA: MIT Press.

Chapter 9
Affect Driven Noological Processes

Abstract In this chapter, a partial solution to the grand challenge posed in the previous chapter is presented. Again, all the noological processing devices that were developed earlier are brought to bear on the problem, notably those from Chaps. 6 and 7. This further affirms that the devices developed in this book have sufficient generality and are applicable to a wide range of scenarios. It is found, in the process of addressing the SAS micro-environment benchmark, that a type of personality trait, neuroticism, has to be introduced to characterize the agent involved and this personality characterization influences the outcome of the problem solving process.

Keywords Micro-environment • Micro-world • Script • Causal learning • Causal reasoning • Affective computing • Anxiousness • Personality • Neuroticism

In Chap. 8 (and Ho 2013) we posed a grand challenge to noology and computational intelligence – we proposed to use a micro-environment, termed the SAS (Shield-and-Shelter) micro-environment, as a benchmark to address a number of issues covering a wide range of noological phenomena, from perception to memory, attention, problem solving, action planning, and action. We articulated the principles in the introductory chapter, Chap. 1, that there are a number of fundamental issues underlying the phenomenon of intelligence that must be addressed together, and with a micro-environment such as the SAS micro-environment, these issues become manageable. At the same time, the SAS micro-environment is designed in such a way that all major issues in noological processing are addressed simultaneously with respect to the underlying driving forces of the noological system. This allows us to gain a full understanding of the computational and representational issues involved and how they can be put together in the operations of a complete noological system. And then, it is expected that these mechanisms are scalable to address the full real-world environment.

In the preceding chapters leading to the current chapter, a number of fundamental noological mechanisms has been established. These include effective rapid causal learning, affect-driven processes, affective competition, learning of extended sequence of actions/behaviors (to serve certain needs), learning of meta-level heuristic rules, etc. These mechanisms will now be brought to bear on the SAS micro-environment problem. We will divide the SAS micro-environment problem into various stages and discuss them in separate sections. These stages include learning that certain activities in the environment can cause pain – an internal state to be avoided, learning that certain actions can avoid the pain consequence, learning about certain physical properties of the objects in the environment that can give rise to certain interactions, and these interactions can be exploited to solve the problem of pain avoidance, etc. We provide detailed computational mechanisms to address part of the SAS micro-environment challenge while providing only a rough outlined solution to the rest of it, relegating the detailed computational solutions to future investigations.

9.1 Learning and Encoding Knowledge on Pain-Causing Activities

The relatively simple narrative of the SAS micro-environment (Chap. 8 and Ho 2013) begins with an Agent initially in a state of Comfort (no Pain, and Pain being the only negative reinforcement signal to be avoided), and then a Projectile appears at a certain location and hurtles toward it and hits it. At that moment, a Pain signal is experienced (this is a built-in response). The situation could be probabilistic in that after each appearance of the Projectile, it does nothing and disappears, thus not leading to any pain experienced by the Agent. Or, after each appearance of the Projectile, it hurtles in a direction that does not lead to hitting the Agent, and the Agent again does not experience any Pain. Each of these situations can be captured as a script and their probabilities of being realized recorded. In order to focus our effort on addressing the various representational and computational issues involved, we consider only the basic stipulation of the SAS micro-environment that the Projectile always hurtles toward the Agent and hits it or at least attempts to hit it. The probabilistic situations could be added on separately later.

Figure 9.1 shows the sequence of events from the appearance of the projectile to the experience of pain by the Agent in the SAS micro-environment. Figure 9.1 is basically the same as Fig. 8.2 but we repeat it here for the ease of reference.

The top part of Fig. 9.2 re-represents the interaction between the Projectile and the Agent along a spatiotemporal axis from left to right and the bottom part of the figure is the PROJECTILE-CAUSING-PAIN-SCRIPT captured and learned from the event. There are many ways to capture a sequence of activities into an event or a script but here a simple rule is used – the boundary of the "projectile causing pain" event is demarcated by a change in activities. The *Materialization* of the Projectile is the first activity within the time frame of interest and it is captured as the first activity of the event/script. When the Projectile hits (*Contacts*) the Agent, a *Pain* sensation is registered in the Agent and the Projectile stops. This is the end of the event/script.

To be consistent with the methods of script construction in the foregoing discussion, the Agent and the Projectile should both have their own sub-scripts

9.1 Learning and Encoding Knowledge on Pain-Causing Activities

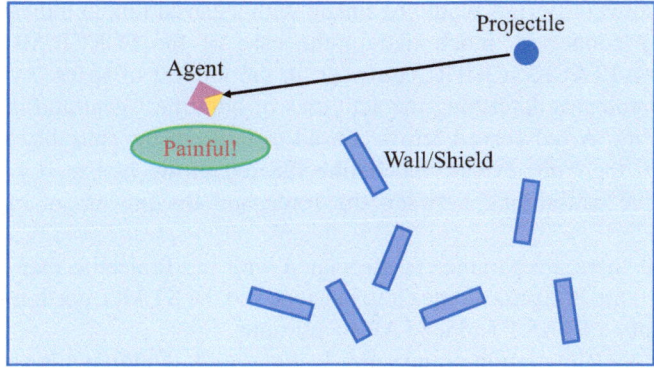

Fig. 9.1 Agent experiences pain as a result of the projectile hurtling from afar and hitting it (Same as Fig. 8.2. ©2013 IEEE. Reprinted, with permission, from Ho, S.-B., "A Grand Challenge for Computational Intelligence: A Micro-Environment Benchmark for Adaptive Autonomous Intelligent Agents," Proceedings of the IEEE Symposium Series on Computational Intelligence – Intelligent Agents, Page 46, Fig. 3.)

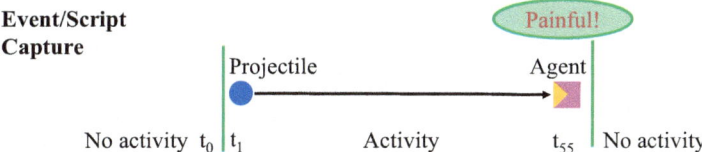

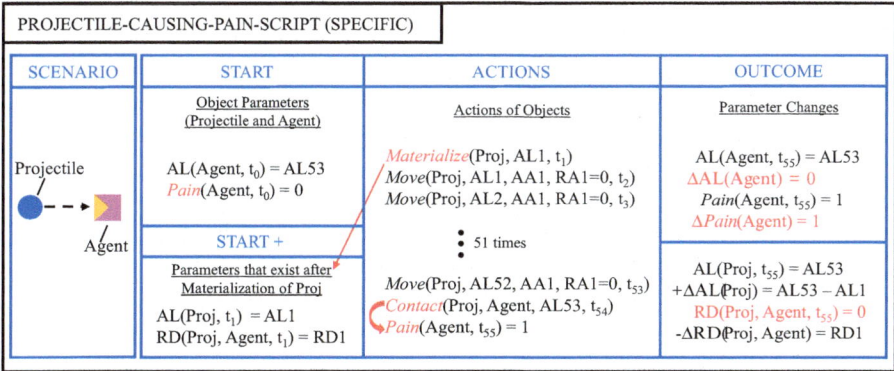

Fig. 9.2 A specific PROJECTILE-CAUSING-PAIN-SCRIPT. The *red curved arrow* shows the causality between the *Contact* event and the *Pain* event. The START+ portion contains parameters that come into existence after the *Materialization* of the Projectile. "Proj" is a short-form of "Projectile"

much like in the case of the *F*(ORCE) and *OBJ*(ECT) in the FORCE-MOVEMENT-SCRIPT of Fig. 7.7 of Chap. 7 or the Self and Enemy agents in the ATTACK-SCRIPT of Fig. 7.15 of the same chapter as they are separate entities.

And then these sub-scripts would be linked with a causal link to indicate how they are causally connected, much like in the case of the FORCE-MOVEMENT-SCRIPT or ATTACK-SCRIPT. However, in the interest of space, we have combined the predicates describing the activities of both the Agent and the Projectile into one script. A red curved arrow is used to indicate the causality between the *Contact* and the *Pain* events, much like the red arrow in Fig. 7.7 of Chap. 7 indicating the causal link between the force and the movement of the object involved.

Note that there are parameters associated with the Projectile that do not exist initially until the Projectile comes into existence so the START portion of the script is divided into a START and a START+ portions.

The SCENARIO portion of the script includes only the Projectile and the Agent and excludes the other objects – the Walls – in the environment. This is achieved through the four general heuristics – Rules 6.16, 6.17, 6.18, 6.19 – that have been learned earlier and discussed in Chap. 6. These heuristics basically say that if a physical object is not charged it will not influence the physical activities of other objects that are not charged. Assuming that the Agent in the current situation has the pre-knowledge that the Walls are not charged objects, based on the visual characteristics (these could be like the BLUE Walls in Chap. 6, Fig. 6.7), and that the Projectile and Agent itself are also not charged, the Walls are then initially excluded from the SCENARIO portion of the script of Fig. 9.2. I.e., the Walls do not influence the movement activity of the Projectile toward the Agent.

This is a *specific* script derived from one particular instance in that all of the parameters involved have to have specific values. For example, the event begins at time t_0 and ends at time t_{55}, and the Projectile *Materializes* at location AL1 at time t_1. The Agent is at location AL53 at time t_1 and is till located at AL53 at time t_{55}. The first *Move* activity of the Projectile is from the location AL1 at the specific time t_1, and it is in the absolute direction AA1 and relative direction RA1 (these directions have the same meanings as those defined in Fig. 2.15 of Chap. 2). Suppose in a second instance the Projectile appears again at a different time and at a slightly different distance from the Agent and hurtles toward the Agent at a slightly different absolute angle AA, then a *general* PROJECTILE-CAUSING-PAIN-SCRIPT can be learned and encoded as shown in Fig. 9.3.

In Fig. 9.3, the parameters that have been generalized are preceded by a blue "*". This is because these parameters have different values in the second instance, and based on dual instance generalization as discussed in connection with Fig. 2.13 as well as elsewhere in Chap. 2, these parameters are deemed "unimportant" for the execution of the script. In Fig. 9.3, we mark those parameters and their corresponding values that are "important" in red. These values are constant across instances and "survive" the process of dual instance generalization. So, what the general script of Fig. 9.3 basically says is that a Projectile can materialize anywhere and at any time and head toward the Agent (RA = 0), which could also be situated anywhere, and the Agent will receive a Pain signal at some time later.

Armed with this script, the Agent can use it to anticipate what may happen in the future. Suppose now the earlier episode is over and the Projectile disappears. Now,

9.1 Learning and Encoding Knowledge on Pain-Causing Activities 363

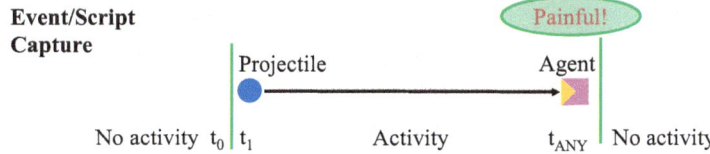

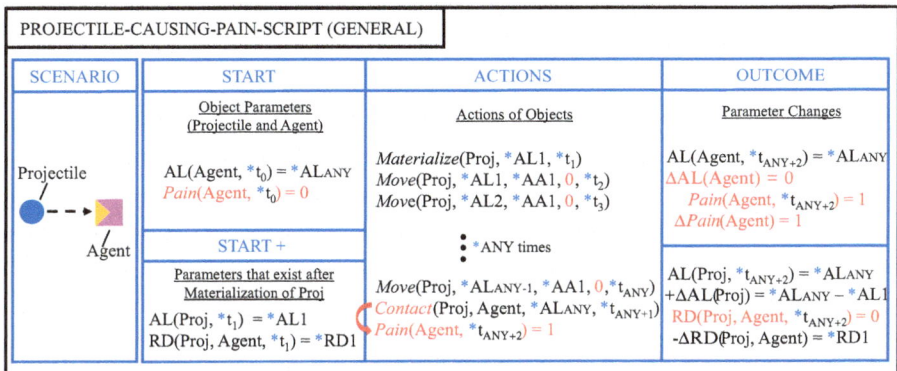

Fig. 9.3 A *general* PROJECTILE-CAUSING-PAIN-SCRIPT. Parameters that have been generalized are preceded with a *blue* "*". Parameters that have stayed constant over the instances are highlighted in *red*

suppose a Projectile appears again at a location close to the first one as shown in Fig. 9.4 (similar to Fig. 8.3 but reproduced here for the ease of reference). Before the Projectile has the time to carry out any action, this activity of *Materialize* (Proj, some_location, some_time) is matched to the various scripts in the Agent's Script Base and the PROJECTILE-CAUSING-PAIN-SCRIPT is retrieved. The script predicts that at some future time there will be a Pain event in connection with the Projectile's projected activities.[1] Therefore, even before the Projectile begins to move, the Agent has the ability to predict that it will move based on the mental simulation which is based on a known script, and move in a certain manner as dictated by the script that results in some Pain to the Agent some time in the future. At this time, the Agent enters a state called "Being Anxious" – the anticipation of a future negative, undesired event. In this case, the undesired event is Pain, which is built-in to be something undesirable. The ability to enter a state of anxiousness in

---

[1] This can be inferred from just reading off the OUTCOME portion of the script or a mental simulation can be carried out through the use of the ACTIONS portion of the script. If the script is matched to the environment with high certainly, there is no need to carry out the mental simulation. However, similar to the situation as discussed in connection with Figs. 6.9 and 6.10 in Chap. 6, sometimes only a closest matched script is retrieved and there may be things in the environment that may affect the expected outcome of the script – a good example would be an obstacle between the Agent and the Projectile much like in the SMGO problem of Chap. 6. In that case, a simulation will reveal that the Projectile cannot reach the Agent so no future Pain event would ensue.

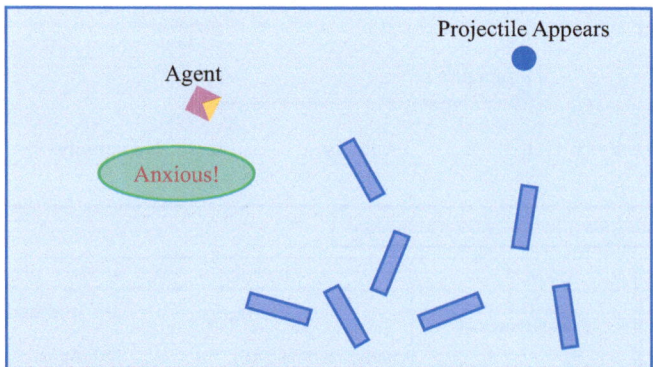

Fig. 9.4 Having had the earlier experience of the projectile causing pain in the agent, now the mere appearance of the projectile even before any movement on the part of the projectile would immediately cause a state of anxiousness in the Agent (Similar to Fig. 8.3. ©2013 IEEE. Reprinted, with permission, from Ho, S.-B., "A Grand Challenge for Computational Intelligence: A Micro-Environment Benchmark for Adaptive Autonomous Intelligent Agents," Proceedings of the IEEE Symposium Series on Computational Intelligence – Intelligent Agents, Page 46, Fig. 4)

anticipation of any future negative event is also a built-in feature of any noological system. The function of anxiousness is to direct the Agent away from other currently engaged activities of lower priorities and attempt to address the issues related to a possible future negative consequence that is more damaging to the Agent. Basically there is a process of anxiousness and needs competition as discussed in Sect. 3.3 of Chap. 3 and Sect. 7.4.2 of Chap. 7.

9.2 Anxiousness Driven Noological Processes

In general, the two major aspects of the situation that affect anxiousness are:

- Can the Agent find a solution and in time?
- If a solution exists, can the Agent execute it successfully and in time?

Therefore, anxiousness would increase as time approaches the impending "doom," and increase exponentially, much like that depicted in Fig. 3.8 of Chap. 3. Given the PROJECTILE-CAUSING-PAIN-SCRIPT, which encodes the speed of the Projectile implicitly in its specifications, it is possible for the Agent to anticipate the amount of time needed to end up in the Pain situation from the moment the Projectile appears. Therefore, it is able to anticipate how much anxiousness it should have as time progresses. In the same spirit as the discussion associated with Fig. 3.8 of Chap. 3 on anxiousness driven actions, in the current situation an Anxiousness of Projectile-Causing-Pain (APP) vs Time to Projectile Hitting Agent (THA) graph is created as shown in Fig. 9.5 immediately upon the Agent's "realization" that it should be anxious about being in the state of Pain

9.2 Anxiousness Driven Noological Processes

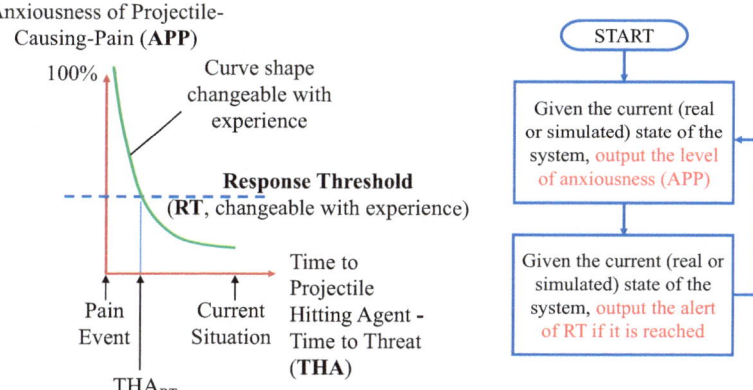

Fig. 9.5 Anxiousness of Projectile-Causing-Pain (*APP*) vs Time to Projectile Hitting Agent (*THA*)

some time in the future. This anxiousness graph, once set up, would present a competition to other affective priorities much like that discussed in connection with the StarCraft scenario in Sect. 3.3 of Chap. 3 or Sect. 7.4.2 of Chap. 7. Hence the level of anxiousness at each moment in time would have a big influence on the Agent's choices of activities and its decision to pick a certain course of action. The Agent's choice of action would also be connected to its degree of desperation (Fig. 2.13, Chap. 2).

As with the graph in Fig. 3.8, there is a response threshold (RT) associated with the APP graph that will trigger the Agent into action to avoid the negative consequence that is associated with the anxiousness involved. The time corresponding to this action is THA_{RT}. This RT and the shape of the graph are changeable as a result of experience and learning as discussed in Sect. 3.3 of Chap. 3. There is a process that continually monitors the current value of THA and outputs the value of APP for the information of the Agent as shown on the right side of Fig. 9.5. When the RT value is reached, it is also output for the Agent to take the necessary actions. The graph can be used in the real world situation or in a mental simulation process when the value of THA is known and the corresponding APP value can be derived.

Suppose currently the Agent is not engaged in other particularly important tasks, or it may just be observing the environment to attempt to learn some new knowledge, based on the affective competition priority as discussed in Sect. 3.3, Chap. 3, the avoidance of future Pain to the agent is definitely of high priority. Therefore, in the next section, we discuss how the agent would deal with this situation.

Figure 9.6 summarizes the process above. As discussed in footnote 1 above, there may or may not be a need to carry out the full detail of the step of "Mental Simulation based on script." The Agent can actually just read directly from the

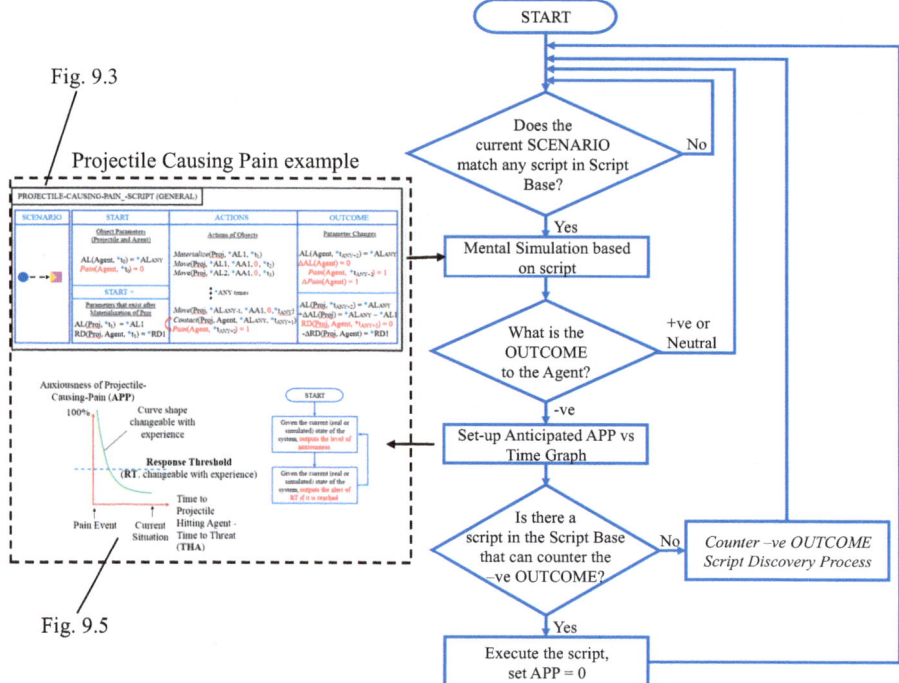

Fig. 9.6 Mental simulation and discovery of −ve, +ve or neutral future consequences of a current scenario and the handling process. A part of the process is similar to a part of Fig. 6.11 of Chap. 6 – here, instead of *Counter-Thwarting Script Discovery Process*, it is *Counter −ve OUTCOME Script Discovery Process*

OUTCOME of the script, once the script has been matched, to determine whether there is a negative OUTCOME (in the case of the PROJECTILE-CAUSING-PAIN-SCRIPT, the negative OUTCOME of $Pain(\text{Agent}, *t_{ANY+2}) = 1$ can be read directly from the script of Fig. 9.3). However, unless the SCENARIO portion of the scrip is matched to the real world situation with a high certainty, it is good to carry out the simulation in full as there could be unexpected interactions that may arise in the course of carrying out the ACTIONS of the script.

As mentioned in footnote 1, an example would be that there is a physical obstacle between the Projectile and the Agent but the Agent does not have a script that informs on what is expected when there is an obstacle. Consider the situation discussed in connection with Fig. 6.9b in Chap. 6, in which the "best-match" script is used to provide a solution to the Agent to guide its actions before the Agent has any experience with the *Wall* that would present an impediment to the movement to *Object*(X), and the situation ended up that there was a surprise impediment. Similarly, there could be surprises in the current scenario. In the case of Fig. 6.9b, the surprise is not discovered until the solution's physical execution time. However, the Agent could also have experienced and learned the concept of physical impediment elsewhere separately from the experience of Projectile-

Causing-Pain, and can actually activate this knowledge in the course of mental simulation and discover that the Projectile will not reach the Agent (in the case that there indeed is an obstacle between the Projectile and the Agent), thus discovering the surprise "mentally" in the course of mental simulation. When that happens, the will be no anxiousness on the part of the Agent arising from being hit by the Projectile.

A note should be made about +ve or neutral OUTCOME in Fig. 9.6. An opposite situation to that of the Projectile-Causing-Pain experience example could be something like a "Hurtling-Food-Causing-Decrease-in-Hunger" – i.e., a piece of food hurtles toward the Agent and "hitting" it, and assuming a simplified scenario, causing an alleviation of its hunger upon contact. That would be a positive situation that gives rise to the emotion of Hope rather than Anxiousness and there is no action needed to try and thwart the movement of the hurtling food. Similarly, a neutral outcome could be something that does not lead to Pain or anything positive, such as a piece of sponge that hurtles toward a person and hitting a person, unless it is a signal of something further, positive or negative, to come.

9.3 Solutions to Avoid Future Negative Outcome

Similar to the process depicted in Fig. 6.11 in which when certain expected script's actions were thwarted in the process of executing the script, the system first check the Script Base to see if there is an existing script that can help avoid the expected Pain in the current situation/SCENARIO (Fig. 9.6). To do this, the current SCENARIO together with a desired Not-Pain OUTCOME are used to query the Script Base to see if a method/script already exists to achieve this end. If so, the script is retrieved and executed and APP is set to 0 (no anxiousness). Otherwise, causal reasoning and exploration (by activating available elemental actions, etc.) are carried out to concoct a plan to avoid the impending Pain in a *Counter −ve OUTCOME Script Discovery Process*. We shall assume that the first option of an earlier-learned counter −ve OUTCOME script in the Script Base is not available and the second option is activated. (If the second option is successful – i.e., a Pain avoidance plan can be discovered and successfully executed – then the method will be stored as a script for future use.)

The following sections describe the second option of *Counter −ve OUTCOME Script Discovery Process*. This process is identical to the *Counter Thwarting Script Discovery Process* of Fig. 6.21, Chap. 6, except that the *"Counter Thwarting"* portion is replaced by *"Counter −ve OUTCOME."* Figure 9.7 shows the *Counter −ve OUTCOME Script Discovery Process*. The green path in the figure shows the process to be discussed below.

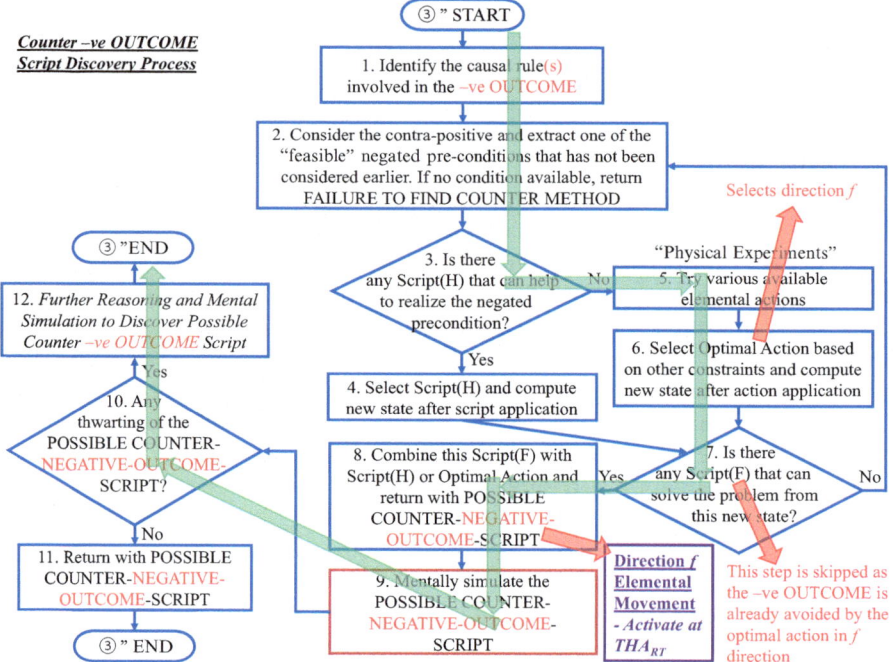

Fig. 9.7 The *Counter –ve OUTCOME Script Discovery Process* which is identical to the process of Fig. 6.21. The places where the two processes are different are highlighted in *red*. The *green arrows* show the flow of the process to be discussed below. The *purple box* indicates a POSSIBLE-COUNTER-NEGATIVE-OUTCOME-SCRIPT. (See text for explanation)

9.3.1 Causal Reasoning to Identify the Cause of Negative Outcome

This section describes Steps 1, 2, and 3 of Fig. 9.7 as applied to the current Projectile-Causing-Pain problem. The PROJECTILE-CAUSING-PAIN-SCRIPT learned in Fig. 9.3 basically stipulates that whenever a Projectile appears somewhere, it will always hurtle toward the Agent and cause Pain to the Agent. Now, the script encodes a series of actions that finally leads to the Pain event. In the situation described in Sect. 6.5.2 in Chap. 6 and encapsulated in Rule 6.21, there was a set of conjunctive *synchronic* preconditions, each of which could be disjunctively negated (Eq. 6.23) to effect a negated diachronic consequence (Eq. 6.22). As a dual to the situation in Sect. 6.5.2, if a *chain* of actions, which represents a set of conjunctive *diachronic* preconditions, leads to a final consequential activity (a final diachronic consequence), then interrupting *any one* of the actions in the chain would interrupt the final consequential activity. The actions in the chain form a "conjunction over time" leading to a consequence – in this case, $Pain(\text{Agent}, *t_{ANY+2})$. Hence, if the desired outcome is $Not(Pain(\text{Agent}, *t_{ANY+2}))$, then taking the contra-positive will

9.3 Solutions to Avoid Future Negative Outcome

lead to negating *all* the actions in the chain *disjunctively*. In general, the situation is as follows:

$$\text{Action-1}|t_1 \wedge \text{Action-2}|t_2 \wedge \text{Action-3}|t_3 \wedge \ldots \ldots \rightarrow \text{Consequence}|t_x \quad (9.1)$$

$$\neg \text{Consequence}|t_x \rightarrow \neg \text{Action-1}|t_1 \vee \neg \text{Action-2}|t_2 \vee \neg \text{Action-3}|t_{3\ldots\ldots} \quad (9.2)$$

Therefore, since as soon as the Projectile appears or materializes, the sequence of actions stipulated by the ACTIONS portion of the PROJECTILE-CAUSING-PAIN-SCRIPT of Fig. 9.3 will occur, leading to $Pain(\text{Agent}, *t_{ANY+2}) = 1$, and hence to achieve $Not(Pain(\text{Agent}, *t_{ANY+2}))$ (or $Pain(\text{Agent}, *t_{ANY+2}) = 0$), the Agent can take actions or solve problems to achieve one of the following:

$$Not(Materialize(\text{Proj}, *AL1, *t_1)) \vee$$
$$Not(Move(\text{Proj}, *AL1, *AA1, 0, *t_2)) \vee$$
$$Not(Move(\text{Proj}, *AL2, *AA1, 0, *t_3)) \vee$$
$$\vdots \quad (9.3)$$
$$Not(Move(\text{Proj}, *AL_{ANY-1}, *AA1, 0, *t_{ANY})) \vee$$
$$Not(Contact(\text{Proj}, \text{Agent}, *AL_{ANY}, *t_{ANY+1}))$$

The first condition involves the prevention of the *Materialization* of the Projectile. If the Agent has this ability, the problem is solved right away. Suppose the Agent does not have this ability, the list above stipulates that the Agent can also create a $Not(Move(\text{Proj}, *, *, 0, *))$ situation at any one of the locations along the trajectory of the Projectile hurtling toward it. Now, the original $Move(\text{Proj}, *, *, 0, *)$ stipulates that the Projectile is moving *directly toward* the Agent (RA = 0). Therefore, $Not(Move(\text{Proj}, *, *, 0, *)$ stipulates that as long as the Projectile is made not to move in the direction directly toward the Agent, it will not arrive to *Contact* the Agent and cause *Pain* to the Agent further down the road.

One of the ways to do this would be to exert a force from a distance (from where the Agent is currently located) to thwart or deflect the movement of the Projectile toward the Agent. Another way would be to hurtle something from the current location of the Agent to intercept and deflect the Projectile anywhere along the trajectory. Yet another way would be to destroy/dematerialize the Projectile somewhere along its trajectory so that its "movement toward the Agent" becomes a non-issue.

Suppose all the above options are not available – the Agent does not have the ability of exerting a force from a distance, hurtling another object, or dematerializing the Projectile. The Agent is then left with the last one – $Not(Contact(\text{Proj}, \text{Agent}, *AL_{ANY}, *t_{ANY+1}))$ – which stipulates achieving a not-contact situation when the Projectile arrives at the location of the Agent. As the Agent is able to move itself, achieving the $Not(Contact(\ldots))$ condition may be possible. We are assuming that the method to exploit this option immediately – a script that has been learned earlier that can be activated to immediately achieve the $Not(Contact(...))$ condition – is not readily available (thus failing the testing step of Step 3 and bypassing Step 4 of Fig. 9.7). In the following sections we will discuss two ways a script can be discovered to achieve this. So far, we have covered Steps 1, 2, and 3 of Fig. 9.7 and in the subsequent sections we will begin with Step 5.

9.3.2 Identifying a Method to Remove the Cause of Negative Outcome

Recall that the phenomenon observed and the script captured (Fig. 9.3) as described above for the Projectile-Causing-Pain event is such that the Projectile stops after it contacts the Agent. (In the real world, it is of course possible to have the Projectile keep moving forward, after hitting the Agent, and maybe pushing the Agent along. Both that situation and the current one would result in similar problem solving processes). Figure 9.8 depicts the scenario at the moment when the Projectile stops at the end of the mental simulation using the script of Fig. 9.3 (with a concomitant generation of a Pain signal in the Agent). The Projectile at the predicted/expected end position is represented as a light blue circle at the end of the mental simulation based on the script in Fig. 9.3.

In Fig. 9.8, the sizes of the Projectile and Agent are enlarged and furthermore, the Agent, considered to be an "extended rigid object," is divided into a collection of circularly shaped elements of finite size and number. In the spirit of the discussion in Chap. 4, an extended rigid object is made up of a collection of elemental objects that are rigidly linked to each other, much like the "attached" state of two elemental objects discussed in connection with Fig. 4.24a of Chap. 4. The property of a rigid object is such that all the elemental parts (objects) move in unison much like the elemental objects that are attached together in Fig. 4.25b. In the current situation and general situations, it does not matter which point on the Agent is being contacted as far as the reasoning process leading to the appropriate action is concerned. We have selected a point along one side of the "rectangular" Agent for this contact event – in earlier discussion, we identified that side as the "face" of the Agent. It could have been one "corner" of the Agent.

The elemental part of concern on the Agent is the part (the elemental circular object) that contacts the ANTI-GOAL. Let us denote this *Relevant-Part*(Agent). The system first considers what the desired action is on that part and then the entire Agent, being an extended rigid object, would move in unison with that part. In Fig. 6.19, Chap. 6, there is a script – the EMAWAG-SCRIPT – that can result in a $Not(Contact(...))$ (or $\neg Contact(...)$) situation between two objects (as encoded in the OUTCOME portion of the script). Even though the shape and color of the *Wall* in the EMAWAG-SCRIPT are different from the shape and color of the Projectile in the current situation, we assume that this is the best and a reasonable match

9.3 Solutions to Avoid Future Negative Outcome

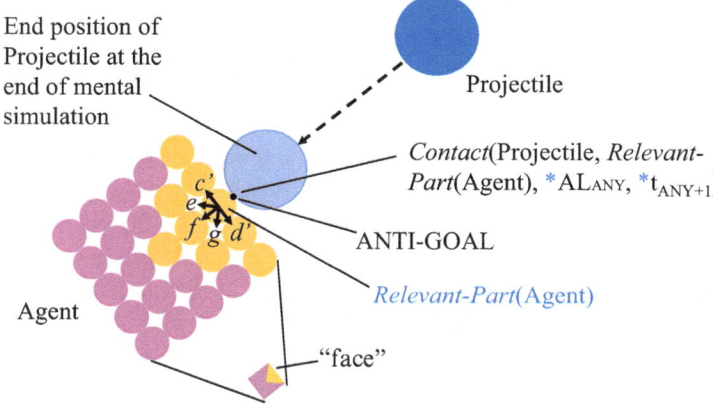

Fig. 9.8 Projectile hurtles toward the agent and stops when it makes contact with the agent – at the *light blue circle* location – as a result of mental simulation based on the PROJECTILE-CAUSING-PAIN-SCRIPT of Fig. 9.3. The agent is enlarged into an "extended rigid object" consisting of a number of elemental parts/objects. The point at which the agent contacts the projectile is the ANTI-GOAL. *Relevant-Part* (Agent) is the elemental part of the agent contacting the projectile at the projectile's end position

provided by the Script Base and this script is retrieved to guide the action of the Agent in the current situation.[2]

Therefore, we assume that the *Relevant-Part*(Agent) and Projectile of Fig. 9.8 correspond to and are bound to *Object*(X) and *Wall* in Fig. 6.19 respectively – we assume the two arguments in the ¬Contact predicate in the EMAWAG-SCRIPT are interchangeable as a piece of built-in knowledge (i.e., ¬*Contact*(*Object*(X), *Wall* (AL)) = ¬*Contact*(*Wall*(AL), *Object*(X)). As a result, the EMAWAG-SCRIPT is activated and applied accordingly.

The EMAWAG-SCRIPT prescribes a number of "directions of escape" in its CFI portion (Fig. 6.19). The Agent can choose *any* of these directions to effect an elemental movement to achieve the ¬*Contact* consequence. If the Agent applies the *maximum* distance heuristic as discussed in Chap. 2, Sect. 2.6.4, then it would choose to move elementally in direction *f*, which allows it to "escape most quickly" away from the ANTI-GOAL – the contact point between the Agent and the

---

[2] There are two possibilities here. One possibility is that the match is not exact, given the EMAWAG-SCRIPT as it is represented in Fig. 6.19, and a moderate desperation in solving the current problem would result in the system accepting a certain non-perfect degree of match and the EMAWAG-SCRIPT is tried out. Another possibility is that the EMAWAG-SCRIPT could have earlier been generalized in terms of the shape and color of the *Wall* and the *Object*(X), and so the match between the current scenario and the SCENARIO in the EMAWAG-SCRIPT is exact. In fact, if the match is not exact, after the current process in which this best match is used to solve the current problem, and if the actions succeed in achieving the *Not-Contact* condition, then the EMAWAG-SCRIPT of Fig. 6.19 can be generalized immediately to dictate arbitrary shape and color for the two entities involved in the SCENARIO.

projected last position of the Projectile (Step 6, Fig. 9.7). And, as the Agent is a rigid body, the movement of all the elemental parts on the Agent are always in unison, therefore the entire extended "body" of the Agent represented by all the elemental parts in Fig. 9.7 move elementally in the same direction f as that of the *Relevant-Part*(Agent).

Therefore, if the EMAWAG-SCRIPT of Fig. 6.19 is available, the Agent can bypass Step 5 of Fig. 9.7 and proceed to Step 6 to select direction f as discussed above. Otherwise, the Agent can experiment physically by trying out different directions of movements (Step 5) and learn and encode something like the EMAWAG-SCRIPT as discussed in Chap. 6 and then select f as the direction to achieve the *Not (Contact(...))* condition. Step 7 of Fig. 9.7 is also bypassed as by selecting to move elementally in the direction f, the –ve OUTCOME (the ANTI-GOAL) is already avoided (for the earlier case of the SMGO problem in Chap. 6, after moving elementally from a thwarted position, there is still the need to head toward the GOAL, therefore for that case there is a need to identify "Script(F)" in this step).

9.3.2.1 Timing of Action Execution to Counter Negative Outcome

Next, we consider the issues of the timing of the execution of the action to achieve the ¬*Contact* consequence to counter the negative outcome of Pain. Earlier, according to Eq. (6.23) of Chap. 6 or Eq. (9.3) of this chapter, we have reasoned that to prevent a certain effect from taking place, we need to negate *any one* of its necessary preconditions. However, having picked one of the desired actions (such as one of the "escape" actions above based on the EMAWAG-SCRIPT for the current example), we need to consider *when* a good time to initiate the action is. Unlike the case in Eq. (6.23) in which the disjunctive preconditions are *synchronic*, in the current situation, as encapsulated in Eq. (9.3), the disjunctive preconditions that can be realized to achieve the *Not(Pain*(Agent, *$*t_{ANY+2}$)) are *diachronic* – there is a time associated with each of them and they follow one another in a chain. We state the following built-in rule:

If a condition is necessary to cause an effect and the condition is expected
to appear at time T, then to prevent the effect from happening,
the condition must be prevented BEFORE or AT time T. (9.4)

This rule is applicable to both the conjunctive diachronic preconditions of Rule Eq. (9.3) as well as that of the conjunctive synchronic preconditions of Eq. (6.23) if time is associated with each of the synchronic preconditions. Translated to the current situation, since the *Contact* event between the Projectile and the Agent is a necessary condition for the *Pain* consequence, so if it is anticipated to take place at time T, then it must be prevented *before* time T. Therefore, in the current scenario, as discussed in connection with Eq. (9.3) above, suppose we select to achieve one of the *Not(Move(...))* conditions, say, *Not(Move*(Proj, * AL2, * AA1, 0, *t_3)) (the third step), and if a method has been found to achieve that (such as perhaps using a force to deflect the movement of the Projectile, as discussed above), then the method must be implemented before *t_3 (remember a "*" means it can be *any* value but in this case it would be three elemental time frames away from the beginning of the script). This is to prevent *Move*(Proj, * AL2, * AA1, 0, *t_3) from happening. In the current consideration in which we have picked instead the *Not (Contact(...))* option to prevent the *Pain* event, the action to achieve the *Not (Contact(...))* situation must be activated one or more than one elemental time frame *before* the Projectile arrives at its projected "final" position, which is the "End position of Projectile at the end of mental simulation" as indicated in Fig. 9.8 when the *Contact* event takes place.

This still represents a wide range of time instances at which the Agent can take the escape action as discussed in association with Fig. 9.8. Typically, a noological system does not wait until the very last time instance to take actions to avoid negative expected consequences. As mentioned above in association with Fig. 9.5, a state of anxiousness exists in a noological system when a future negative outcome is expected and it increases as the time of the negative outcome approaches. There is a response threshold, RT (learnable/changeable with experience), that corresponds to the level of anxiousness reached at which a noological system would act to reduce/remove that anxiousness. There is a corresponding time at which the RT is reached, which is THA_{RT}. We assume that the Agent will wait for as long as possible to activate the EMAWAG-SCRIPT of Fig. 6.19 to achieve the ¬*Contact* (...) situation according to the maximum distance heuristic – by moving in the direction *f*. This is so that in case there is some change in the situation that renders the future Pain event unrealizable, the Agent would then be better off devoting its effort to attend to other activities rather than acting to avoid this Pain event prematurely.[3] The Agent should also not act later than THA_{RT} as that is presumably a threshold that has been learned earlier to be an ideal lower-bound time to respond to anxiousness (see discussion in Sect. 3.3.1 of Chap. 3). Therefore, we assume that the Agent would act at exactly THA_{RT} to effect the actions to achieve ¬*Contact* (...), whether in mental simulation or in executing the actions involved in the real world.

In Step 8 of Fig. 9.7 we indicate the creation of a POSSIBLE COUNTER-NEGATIVE-OUTCOME-SCRIPT in a purple box. The script basically stipulates an elemental movement in direction *f*, activated at THA_{RT}. The consideration of activation time for an action did not arise in the process of the solution for the SMGO problem in Chap. 6 even though the overall process of the *Counter −ve OUTCOME Script Discovery* is identical to that in Chap. 6, Fig. 6.21. This is because in the current problem the Projectile, unlike the *Wall* of the SMGO problem, is a moving "ANTI-GOAL" and also there is a Pain state that gives rise

---

[3] This strategy could also be a result of some learning process but here we assume it is built-in.

to a time course of anxiousness in connection with it, unlike in the case of the SMGO problem.

9.3.2.2 Attempted Script to Counter Negative Outcome

Having devised a POSSIBLE COUNTER-NEGATIVE-OUTCOME-SCRIPT to avoid Pain in Step 8 (purple box) of Fig. 9.7, we arrive at Step 9. As mentioned before in Chap. 6 and above, any proposed script is only a possible solution as when it is either mentally or physically simulated, there may be unforeseen effects that render it unworkable. Therefore, we now carry out a mental simulation to see if the script can indeed work.

Figure 9.9 shows the results of the mental simulation. Whereas in the case of the SMGO problem in Chap. 6, the simulation involved only the script of *Object* (X) whose movement was thwarted by the *Wall* (Fig. 6.21), the simulation of the scenario here involves the script of the Projectile as the Projectile moves, unlike the *Wall* which is stationary. (The correspondence is *Object*(X) in Fig. 6.21 = Agent here, and *Wall* in Fig. 6.21 = Projectile here.)

Now, what is shown in Fig. 9.9 is that the Agent had effected the script to counter the −ve OUTCOME by moving elementally in direction f at time THA_{RT}, expecting the Projectile to stop at the earlier end position in Fig. 9.8 so that ¬*Contact*(...) is achieved, but the script of Fig. 9.3 (the PROJECTILE-CAUS-ING-PAIN-SCRIPT (GENERAL)) stipulates that the Projectile would only stop when it contacts the Agent, wherever it may be. This gives rise to an anticipated *Pain*(...) and hence the POSSIBLE COUNTER-NEGATIVE-OUTCOME-SCRIPT is thwarted in Step 10 of Fig. 9.7 and the process flows to Step 12 – *Further Reasoning and Mental Simulation to Discover Possible Counter −ve OUTCOME script*. This activates a process similar to that of Fig. 6.25 in Chap. 6 as shown in Fig. 9.10 which attempts to find yet another possible solution by composing a longer sequence of actions much like in the case of the SMGO problem (the solution of which is shown in Figs. 6.30 and 6.31) or like in the case of Fig. 6.28.

The process of Fig. 9.10 actually returns NO SOLUTION to the current problem because of the limiting condition at the end of Step 4. However, in the next section we consider some situations under which a solution is possible.

9.3.2.3 Conditions Under Which a Solution Is Possible

In our current scenario in which only one possible optimal elemental action was found in the process of Fig. 9.7 (that of moving elementally in direction f), Step 2 of Fig. 9.10 is bypassed and the process flows to Step 3. The script of effecting an elemental movement in the direction f at time THA_{RT} (the "purple box script" of Fig. 9.7) is then chosen to be the script to be composed into a long sequence of actions in Steps 3 and 4. For each application of this elemental movement script for

9.3 Solutions to Avoid Future Negative Outcome

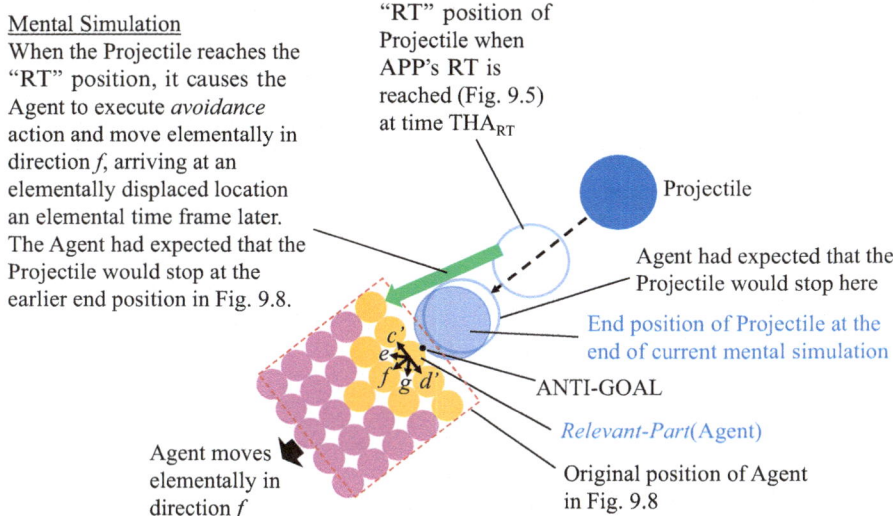

Mental Simulation
When the Projectile reaches the "RT" position, it causes the Agent to execute *avoidance* action and move elementally in direction f, arriving at an elementally displaced location an elemental time frame later. The Agent had expected that the Projectile would stop at the earlier end position in Fig. 9.8.

"RT" position of Projectile when APP's RT is reached (Fig. 9.5) at time THA_{RT}

Projectile

Agent had expected that the Projectile would stop here

End position of Projectile at the end of current mental simulation

ANTI-GOAL

Relevant-Part(Agent)

Agent moves elementally in direction f

Original position of Agent in Fig. 9.8

Fig. 9.9 Mental simulation of Step 9 of Fig. 9.7. The agent moves elementally in direction f and had expected the projectile to stop at the end position of Fig. 9.8 but the projectile goes on to contact the agent as stipulated by the script in Fig. 9.3

the Agent to attempt to escape from the Projectile, the mental simulation of Fig. 9.9 is carried out (Step (i) in Step 4 of Fig. 9.10).[4] Each time, because the Agent has moved an elemental distance further away from its original location, the time at which RT (Fig. 9.5) is reached is a little later, hence there would be a sequence of increasing later times $THA_{RT} + \Delta$, $THA_{RT} + 2\Delta$, ... at which the direction f elemental movement is effected. And each time, the test for whether "the −ve OUTCOME is removed after each mental simulation of Fig. 9.9" in Step (i) of Step 4 of Fig. 9.10 would fail and the "Repeat purple box script in Fig. 9.7" process is effected. The resultant attempted escape script is a long sequence of elemental movements in the direction f as encoded in the script AGENT-AVOID-PAIN-SCRIPT as shown in Fig. 9.11.

Now, as mentioned above, at the end of Step 4 there is a limiting condition which terminates the process at some point otherwise an infinitely long process may result. For the case of the SMGO problem of Chap. 6, there could be a very long or infinitely long *Wall* and hence the process must be terminated and a NO SOLUTION returned. We are assuming that there is an energy limit to the *Agent/Object* (X) in the case of the SMGO problem so that a very long or an infinitely long

---

[4] If the full process of the script of Fig. 9.3 is repeatedly used in the simulation of Fig. 9.9, it would be computationally expensive. However, as mentioned earlier, only the OUTCOME portion of the script needs to be checked for the consequence/result of the script's application, therefore full simulation is not needed. Also, the Agent can generalize from the results of a few instances of similar simulation results to conclude that the repeated simulation of Fig. 9.9 would give rise to the same results.

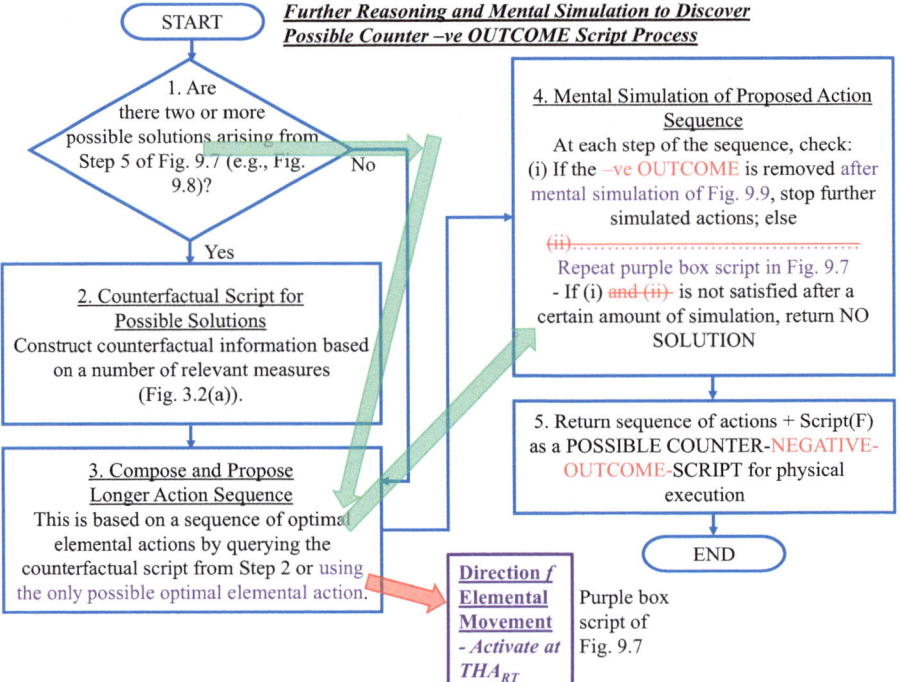

Fig. 9.10 The *Further Reasoning and Mental Simulation to Discover Possible Counter −ve OUTCOME Script Process*, similar to that of Fig. 6.25. The *red words* and cancellations highlight the differences between this process here and that in Fig. 6.25. The *purple words* represent processes unique to the current scenario's implementations but are still basically the same processes as in Fig. 6.25. The *green* path shows the process flow in the current scenario

sequence of actions is not a viable solution. Similarly, in the current scenario, the continued application of the direction f elemental movement that allows the Agent to keep avoiding the projected "final" location of the Projectile, which is where the Projectile contacts the Agent and causes Pain, is not viable if the Agent's energy is finite and it has to stop moving at some point in time. However, the solution is viable in an ideal situation in which the Agent does have an infinite amount of energy and the last part of Step 4 can be ignored and a solution is returned in a form of infinitely repeating application of an elemental movement in the direction f as shown in the AGENT-AVOID-PAIN-SCRIPT in Fig. 9.11. This is basically "keep running away from the Projectile in direction f starting from time THA_{RT}."

Another situation in which a solution is possible is if the Projectile loses momentum and stops after some time and the Agent could "outrun" it even if the Agent may later finally run out of energy itself.

In both cases, the AGENT-AVOID-PAIN-SCRIPT could be composed and proposed as a possible viable solution to the situation, and the Agent can simply execute it in the real world as a best bet to avoid Pain in the given situation.

9.3 Solutions to Avoid Future Negative Outcome

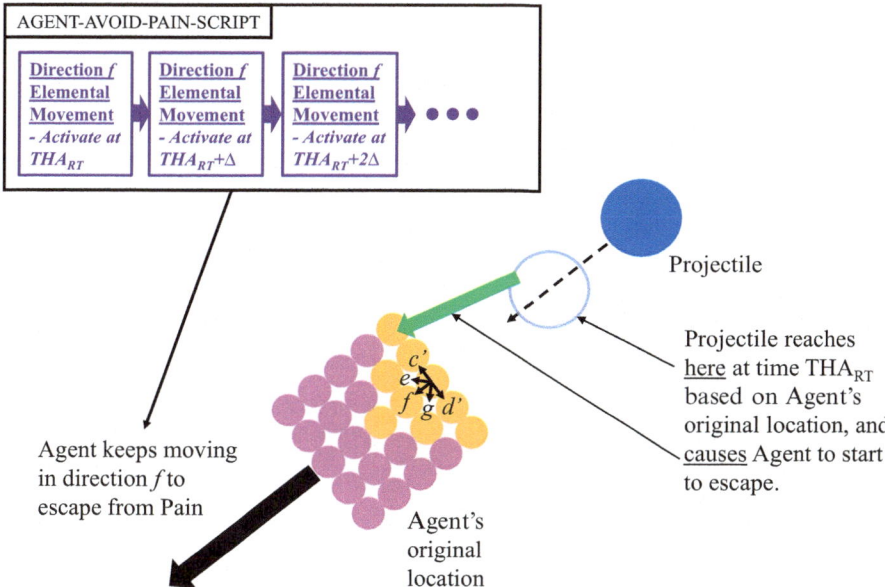

Fig. 9.11 A possibly viable AGENT-AVOID-PAIN-SCRIPT

9.3.3 A Second Method to Remove the Cause of Negative Outcome

The AGENT-AVOID-PAIN-SCRIPT of the previous section is basically a "keep running directly away from the Projectile" method which is energy consuming and will not lead to any realistic solution as the Agent may tire and stop and get hit, unless the Projectile loses momentum and stops first, or unless the situation is ideal and the Agent does indeed have an infinite amount of energy. A better energy conserving method is shown in Fig. 9.12 which is the same as Fig. 8.4 of Chap. 8 – the Agent simply "steps aside" to let the Projectile pass through, requiring a small amount of energy expenditure. In this section we discuss the reasoning processes involved in devising such a solution.

9.3.3.1 Analogous Situations Inspired Solution

There are many possible reasoning scenarios that could arrive at a solution such as that of Fig. 9.12. However, assuming that the Agent has some prior experience with other scenarios of problem solving that are analogous to the current situation, those scenarios could be exploited to derive a solution relatively quickly.

Recall that in Chap. 6, in the SMGO problem in which the Agent considered various options to counter its thwarted attempt to move toward the GOAL, one of the options, as discussed in connection with Eq. (6.23), was to move the *Wall*

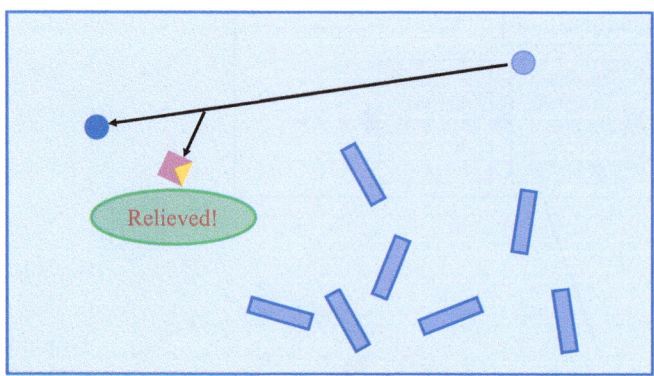

Fig. 9.12 A possible relief from the projectile for the agent – moving aside when the projectile comes hurtling toward it (From Fig. 8.4. ©2013 IEEE. Reprinted, with permission, from Ho, S.-B., "A Grand Challenge for Computational Intelligence: A Micro-Environment Benchmark for Adaptive Autonomous Intelligent Agents," Proceedings of the IEEE Symposium Series on Computational Intelligence – Intelligent Agents, Page 46, Fig. 5)

involved in some manner to remove the movement impediment (refer to e.g., Fig. 6.12 for the *Wall* and other elements in the scenario). This was discussed in connection with Option (4) following Eq. (6.23) and we mentioned then that we would relegate the discussion to this chapter as the solution based on the movement of the *Wall* for the SMGO has some similarity with that in the current scenario.

In Fig. 9.13 we reproduce part of Fig. 6.29a from Chap. 6 with additional items relevant to the current discussion to illustrate the idea. Recall that Fig. 6.21a illustrated a sequence of movement in the *d* direction of *Object*(X) that resulted in a *Not*(*Contact*(…)…) situation between *Object*(X) and the *Wall*. Had the *Wall* been moved in the *opposite* direction as shown with a dark red arrow in Fig. 9.13, the same *Not*(*Contact*(…)…) situation could also be achieved for *Object*(X) while it stays put at its Current Starting Position without having to move in direction *d*, and consequently, *Object*(X) can head straight toward the GOAL and the problem is solved. What kind of knowledge would the Agent need to possess to facilitate a reasoning process such as this?

In Fig. 9.13 we introduce the concept of an "elemental location" – W – which is like a virtual object. This object is like a "relative location" to part of the *Wall* and for W, it is relative to a corner of the *Wall*. When *Object*(X) arrives at location *y* in the mental simulation, it can be said to be *Next-to*(*Object*(X), *Rel-Loc*(W)). W is like an object attached to the *Wall* – *Attached*(*Wall*, *Rel-Loc*(W)) – and would move with the *Wall*. In general, there are many such locations attached to the *Wall* all around the *Wall* but we just show one that is relevant to the current discussion for the sake of clarity.

Figure 9.14 shows how the *Wall* is moved in the opposite direction toward the Current Starting Position of *Object*(X), by the *same amount* of movement as the earlier movement of *Object*(X) from its Current Starting Position to position *y*, to achieve the same effect/goal of *Next-to*(*Object*(X), *Rel-Loc*(W)). For the SMGO

9.3 Solutions to Avoid Future Negative Outcome

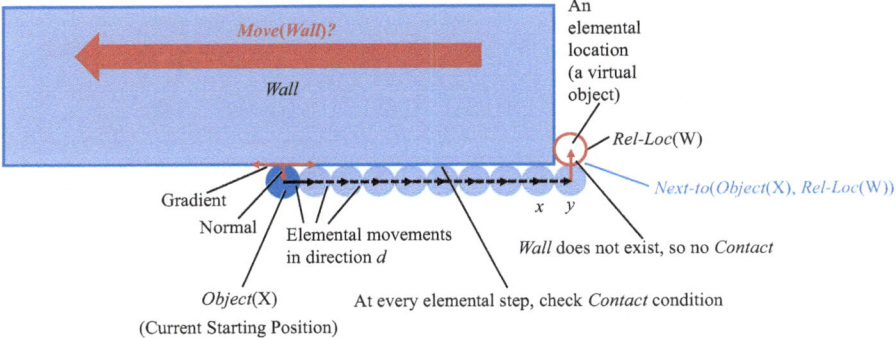

Fig. 9.13 The possible opposite movement of the *Wall* to achieve the same effect of *Not(Contact* (...)...) achieved by the movement of the Agent, shown as a *dark red arrow*, for the SMGO problem of Chap. 6. *Rel-Loc*(W) is like a virtual object attached to the *Wall* and when *Object* (X) arrives at position *y*, *Next-to(Object*(X), *Rel-Loc*(W)) is true. (Part of Fig. 6.29a is reproduced here with the *dark red arrow* added)

problem, it would be solved at this point as now *Object*(X) can move toward the GOAL unimpeded, as the *Wall* has been "move cleared of the way."

The movement of the *Wall* begins with the reasoned movement of *Rel-Loc*(W). As mentioned above, one way to achieve *Next-to(Object*(X), *Rel-Loc*(W)), hence the removal of the ANTI-GOAL, is for *Object*(X) to move in a series of elemental direction *d* movements. The other way would be to move *Rel-Loc*(W) in the opposite direction by the same amount to achieve the same removal of the ANTI-GOAL. This process is derived from a piece of knowledge that says that in spatial movement, if object A moves in a certain straight-line to meet object B, object B can move likewise in the opposite direction to meet object A. This encodes a kind of knowledge on "relative movement." How can such a piece of knowledge be learned? This would be discussed shortly in connection with Fig. 9.15 below.

In any case, once it is reasoned that a possible solution is to move *Rel-Loc*(W) in the opposite direction by the same amount as *Object*(X)'s movement to position *y* to achieve the removal of the ANTI-GOAL, because the *Wall* is attached to *Rel-Loc*(W), all the elemental parts on it move in the same manner as *Rel-Loc*(W) (movement of extended object – Fig. 6.25b, Chap. 4). Hence the *Wall* moves in the opposite direction and by the same amount as shown in Fig. 9.14. The green arrow in Fig. 9.14 shows the connection between the movement of *Rel-Loc*(W) and the *Wall*.

Next we consider the acquisition of the idea of relative movement as mentioned above. In the top part of Fig. 9.15 we consider the scenario consisting of two objects, *Obj*(A) and *Obj*(B), spaced some distance apart. The GOAL is *Next-to(Obj* (A), *Obj*(B), any location and time). This is similar to the scenario of Fig. 9.3 in which the Projectile hurtles toward the Agent except that there is no Pain signal generated here. One of the possible solutions that a problem solving process would

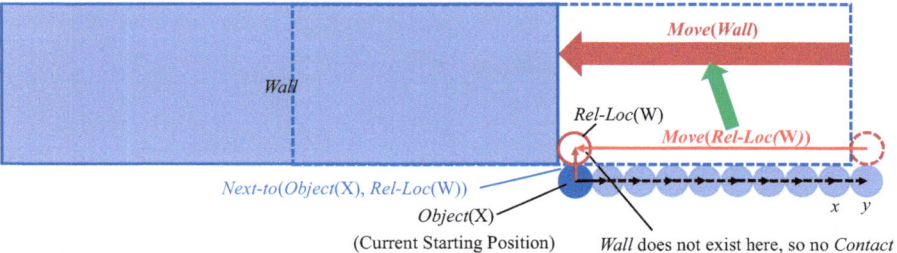

Fig. 9.14 The *Wall* moving in the opposite direction and by the same amount as the movement of *Object*(X) in Fig. 9.13 to achieve the same goal/effect of *Next-to*(*Object*(X), *Rel-Loc*(W)). For the SMGO problem, it would be solved at this point as *Object*(X) can now move toward the GOAL unimpeded. The *green arrow* shows that it is the movement of *Rel-Loc*(W) that results in the movement of the *Wall* as *Rel-Loc*(W) is attached to the *Wall* and together they move as an extended object (Fig. 4.25b, Chap. 4)

GOAL:
*Next-to(Obj(A), Obj(B),
any location and time)*

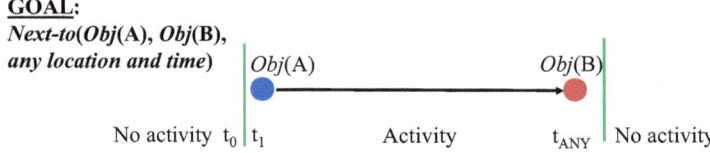

OBJECT(A)-MOVES-TO-CONTACT-OBJECT(B)-SCRIPT (GENERAL)			
SCENARIO	START	ACTIONS	OUTCOME
●--→● *Obj*(A) *Obj*(B)	Object Parameters (*Obj*(A) and *Obj*(B)) AL(*Obj*(A),*t_0) = *AL1 AL(*Obj*(B),*t_0) = *AL$_{ANY}$ RD(*Obj*(A),*Obj*(B),*t_0) = *RD1 *Next-to*(*Obj*(A),*Obj*(B), *AL$_{ANY}$,*t_0) = 0	Actions of Objects *Move*(*Obj*(A),*AL1,*AA1,0,*t_2) *Move*(*Obj*(A),*AL2,*AA1,0,*t_3) ⋮ *ANY times *Move*(*Obj*(A),*AL$_{ANY-1}$,*AA1,0,*t_{ANY}) *Next-to*(*Obj*(A),*Obj*(B),*AL$_{ANY}$,*t_{ANY+1}) =1	Parameter Changes AL(*Obj*(A),*t_{ANY+2}) = *AL$_{ANY}$ − δ +ΔAL(*Obj*(A)) = *AL$_{ANY}$ − δ + *AL1 AL(*Obj*(B),*t_{ANY+2}) = *AL$_{ANY}$ ΔAL(*Obj*(B)) = 0 RD(*Obj*(A),*Obj*(B),*t_{ANY+2}) = 0 −ΔRD(*Obj*(A),*Obj*(B)) = *RD1 Δ*Next-to* (*Obj*(A),*Obj*(B), *AL$_{ANY}$,*t_0) = 1

Fig. 9.15 A script for *Obj*(A) to move *Next-to Obj*(B)

generate is shown at the bottom part of Fig. 9.15 in the form of the OBJECT(A)-MOVES-TO-CONTACT-OBJECT(B)-SCRIPT – *Obj*(A) moves next to *Obj*(B).

As can be seen, the general script of Fig. 9.15 specifies any starting locations for *Obj*(A) (*AL1) and *Obj*(B) (*AL$_{ANY}$) and the ending locations for *Obj*(A) is right next to *Obj*(B), *AL$_{ANY}$ − δ, while *Obj*(B) stays at *AL$_{ANY}$. The *Next-to* (*Obj*(A), *Obj*(B), *AL$_{ANY}$, *t_0) condition changes from 0 to 1 at the end of the movement.

Now, given the same GOAL statement at the top part of Fig. 9.15 – Next-to(*Obj* (A), *Obj*(B), *any location and time*) – for the same scenario of *Obj*(A) and *Obj*

9.3 Solutions to Avoid Future Negative Outcome

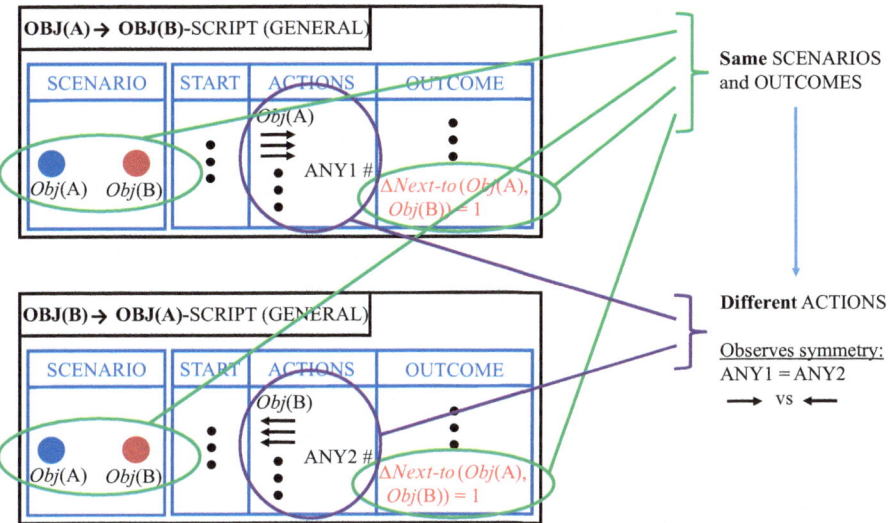

Fig. 9.16 Two scripts with the same SCENARIOS – two objects, *Obj*(A) and *Obj*(B) spaced some distance apart – and the same GOALS/OUTCOMES: the objects being next to each other. The first one is the same as that of Fig. 9.15 and the second one is moving *Obj*(B) toward *Obj*(A). The two alternative solutions are different but symmetrical

(B) spaced some distance apart, suppose the option is available in which *Obj*(B) could be moved as well as *Obj*(A), then another possible solution could also be generated in a problem solving process which is that of moving *Obj*(B) toward *Obj*(A) instead. Figure 9.16 shows both the scripts of Fig. 9.15 as well as this "*Obj*(B) moves to *Obj*(A) script" for comparison.

Hence, by mapping the corresponding entities in Fig. 9.14 to those in Fig. 9.16, i.e., *Object*(X) = *Obj*(A) and *Rel-Loc*(W) = *Obj*(B), the paired solutions of Fig. 9.16 provide the alternative solution of moving *Rel-Loc*(W) by an amount equal to and in a direction opposite to that of moving *Object*(X) to achieve the same effect of *Next-to*(*Object*(X), *Rel-Loc*(W)), thus moving the *Wall* in the same manner as a result.

Then, by mapping the *Wall* and *Object*(X) of Fig. 9.14 to the Agent and Projectile of Fig. 9.8 respectively, a solution for the Agent of Fig. 9.8 to achieve a *Not*(*Contact*(...)) situation with the Projectile (the removal of the ANTI-GOAL) is to implement exactly the same solution for the *Wall* to achieve a *Not*(*Contact*(...)) situation with *Object*(X) – that of moving the Agent until it is "cleared off" the Projectile's final projected position in Fig. 9.8 in the same manner that the Wall is moved until it is "cleared off" *Object*(X)'s Current Starting Position. The solution is shown in Fig. 9.17. In a similar vein as the earlier solution of Fig. 9.11, the Agent begins the "moving aside" actions at the point at which the Projectile reaches the "RT" position.

In the above discussion, we ignored the size of the Projectile and assumed that the Projectile is an elemental object, and because each time the Agent would move

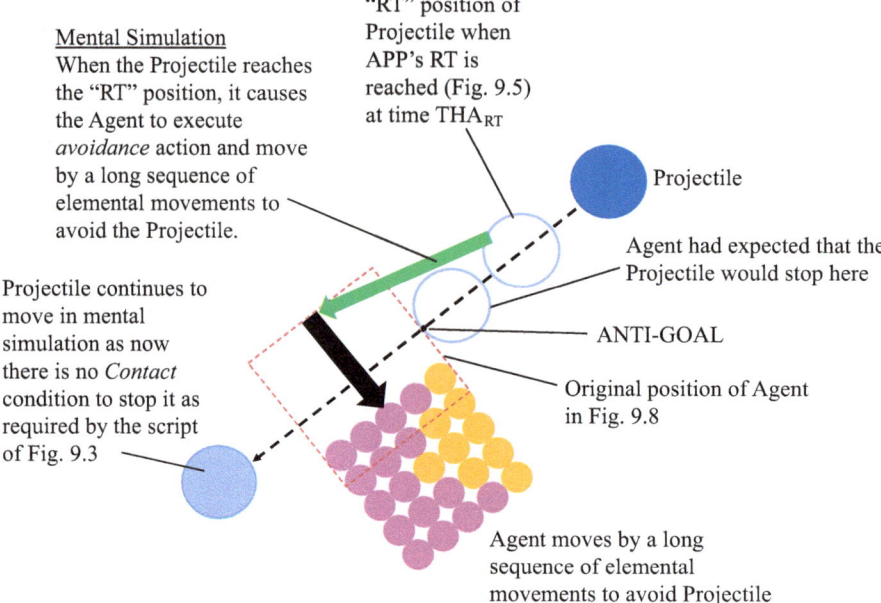

Fig. 9.17 A solution in which the agent "moves aside" to avoid the projectile

some distance similar to the movement of the *Wall* in Fig. 9.14, it would "cleared off" the Projectile after the last "movement aside." However, if the Projectile is sizeable and not an elemental object, then the Agent may not have moved enough in a first try to clear off the Projectile, and in a mental simulation in which the Projectile is moved toward the Agent to see if this first attempted solution works (the process of Fig. 9.7), another ANTI-GOAL of the Projectile contacting the Agent would be created. Similar to the repeated application of the same action attempts to achieve the final removal of ANTI-GOAL of Figs. 6.28 and 6.29, the Agent would now repeat the same attempt to "move aside" (in mental simulation) until it is cleared off the Projectile's final position (as shown in Fig. 9.17). This sequence of actions is then returned as a possible solution to be executed in the "real world."

There are other possible ways to move the Agent aside to stay cleared off the path of the Projectile. Instead of moving in a path perpendicular to that of the Projectile as in the case of Fig. 9.17, the Agent could also move in other directions to avoid the Projectile (such as the one shown in Fig. 9.12 which is not perpendicular to the path of the Projectile). However, the solution of Fig. 9.17 could be reasoned out rapidly through the process described above and it also happens to be the most efficient solution in terms of energy usage – by moving in a direction perpendicular to the Projectile's path, a minimum amount of energy is expanded to stay clear off the Projectile's path.

There is also another kind of possible movement of the *Wall* in Fig. 9.14 to allow it to move clear off *Object*(X)'s intended path toward the GOAL – the *Wall* could be rotated/swung about some points so that none of its parts is in the way of *Object* (X). The Agent of our current problem can also do likewise to avoid being hit by the Projectile. We leave the detail of the reasoning process involved for future investigations.

9.4 Further Projectile Avoidance Situations and Methods

The "stepping aside" solution of Sect. 9.3.3 is a permanent solution and the Agent is forever Relieved if newly appearing Projectiles would still head toward the same earlier location of the Agent – the Agent has to step aside once and for all. This method would also be the only method if the Agent has no other physical options such as those discussed in connection with Eq. (9.3) at its disposal. However, if newly appearing Projectiles always home in onto the Agent no matter how the Agent steps aside to avoid the earlier Projectile, such as in the situation of Fig. 9.18, the "stepping aside" method will require the Agent to "keep stepping aside." Certainly this method is still better than the "keep running away from the Projectile" method of Fig. 9.11 discussed in Sect. 9.3.2.3, as each stepping aside action does not require much energy, even though each time the Agent would still need to enter the state of Anxiousness when a new Projectile appears. However, suppose a "shielding" function is discovered such as that depicted in Fig. 8.5 (reproduced here in Fig. 9.19) – a *Wall* that can deflect the Projectile – then a more permanent solution is possible with a once and for all energy expenditure and Anxiousness experience. The Agent would "hide" behind an appropriate *Wall* and the Projectile would be deflected by it – such as that shown in Fig. 8.6 (reproduced here as Fig. 9.20). The solution would provide permanent Relief with no further Anxiousness if the Projectile always appears at the same location even though it can hurtle itself toward the Agent in different directions such as shown in Fig. 9.18.

The scenario of Fig. 9.19 is very similar to that of the SMGO problem discussed in Chap. 6 (e.g., Fig. 6.9b) in which an *Object*(X) attempts to reach a GOAL on the other side of a *Wall* and is thwarted in its attempt because of the impediment presented by the *Wall* involved. In the situation of Fig. 9.19, the Projectile would have reached some location (GOAL) beyond the Wall had it not been impeded by the Wall. Of course in the SMGO situation *Object*(X) comes to a halt when it reaches the *Wall* but in Fig. 9.19 we depict that the Projectile gets deflected by the Wall. We shall simplify the discussion by assuming that the Projectile actually stops when it hits the Wall – the effect of being thwarted from attempting to reach the other side of the Wall is the same.

Recall that in the discussion in connection with Fig. 6.9b, the *Agent* involved, after having observed that *Object*(X) could not penetrate two points on the *Wall* in the direction of the GOAL, made a generalization that all BLUE parts of the *Wall* are impenetrable and formulated Rule 6.21 to encode this knowledge. A similar

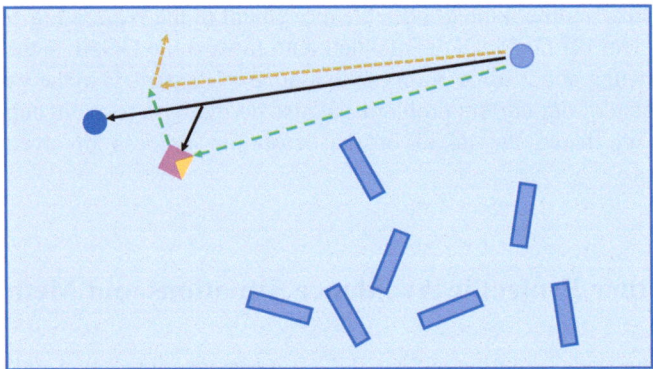

Fig. 9.18 Newly appearing projectiles keep homing in onto the agent no matter where the agent moves to (assuming projectiles always appear at the same location). First, the projectile hurtles along the *black path*, and next, after the agent has stepped aside, the *green dotted path*, and so on (©2013 IEEE. Reprinted, with permission, from Ho, S.-B., "A Grand Challenge for Computational Intelligence: A Micro-Environment Benchmark for Adaptive Autonomous Intelligent Agents," Proceedings of the IEEE Symposium Series on Computational Intelligence – Intelligent Agents, Page 46, Fig. 5)

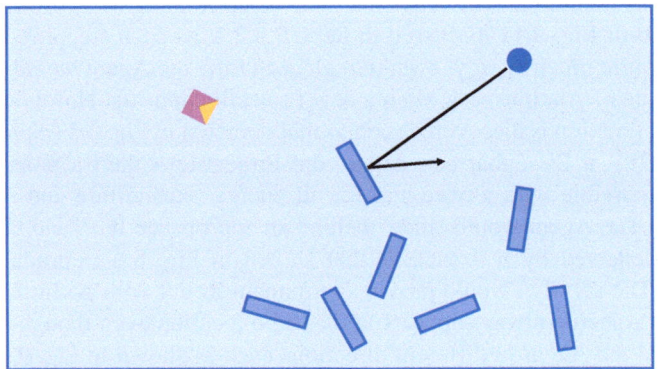

Fig. 9.19 A shield deflecting a projectile (Same as Fig. 8.5. ©2013 IEEE. Reprinted, with permission, from Ho, S.-B., "A Grand Challenge for Computational Intelligence: A Micro-Environment Benchmark for Adaptive Autonomous Intelligent Agents," Proceedings of the IEEE Symposium Series on Computational Intelligence – Intelligent Agents, Page 46, Fig. 6)

kind of generalization can take place here with regards to the observation of Fig. 9.19. Suppose two instance of the Wall deflecting/halting the Projectile has been observed, each instance consisting of the Projectile hitting a different part of the Wall. The Agent then generalizes to obtain an equation similar to that of Rule 6.21.

Now, in the SMGO problem, the *Agent* found solutions to counter the thwarting of the movement of *Object*(X) but in the case here, we are assuming that the Projectile always hits the Wall and stops or hits the Wall and deflects and stops.

9.4 Further Projectile Avoidance Situations and Methods

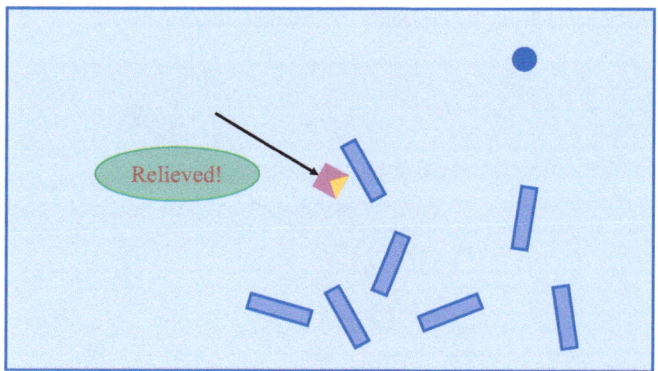

Fig. 9.20 A more permanent solution for the relief of anxiousness due to the situation of Fig. 9.18 (Same as Fig. 8.6. ©2013 IEEE. Reprinted, with permission, from Ho, S.-B., "A Grand Challenge for Computational Intelligence: A Micro-Environment Benchmark for Adaptive Autonomous Intelligent Agents," Proceedings of the IEEE Symposium Series on Computational Intelligence – Intelligent Agents, Page 47, Fig. 7)

There is no ability or motivation on the Projectile's part to generate a complex trajectory to go around the Wall to overcome the impediment presented by the Wall.[5]

In the following we outline a process that allows the Agent to reason out how to use the Wall to protect itself but will leave computational details to future investigations. Consider Fig. 9.21 in which the Agent now, using the analogy with the SMGO scenario, conceives of the effect of the Wall as removing many potential GOALS that the Projectile could have reached, each of which is represented by a red cross over a circle containing the alphabet "G." For each of the possible trajectories of the Projectile, there is a whole line of "G's" being obviated by the Wall. The Agent then considers, in mental simulations (as the rules governing the Projectile and the Wall are all known), all other possible trajectories of the Projectile that would be impeded by the Wall (with appropriate generalizations so that not every possible trajectory needs to be mentally simulated) and generates an area "behind" the Wall as shown in Fig. 9.21 within which the Projectile cannot reach from its current location.

Once the area of no accessibility to the Projectile has been identified, the Agent calculates the shortest trajectory to one of the locations within the area, shown as "A" in Fig. 9.21. If the Agent moves to that location, it would have expanded a minimal amount of energy to seek protection from the Projectile.

However, it is often the case that when one seeks protection behind a Wall from "projectiles" – such as bullets, one tends to move closer to the Wall such as along trajectory B in Fig. 9.21. This is because in a realistic real-world situation in which one is facing a shooter that runs about with a gun, it is likely that the starting

---

[5] Suppose Projectiles are like bullets, there may be new technologies in the future that allow the bullets' trajectories to be curved or programmed in any way the shooter wishes.

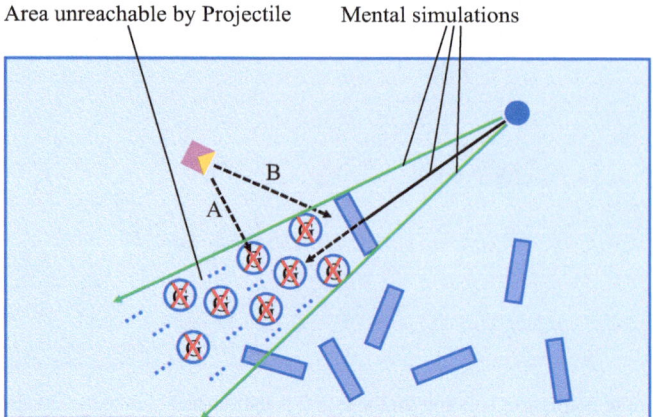

Fig. 9.21 Agent reasons out an area "behind" the Wall that is not reachable by the Projectile. Trajectory A is the shortest distance from the agent's current location to a "protected" location. Trajectory B, however, is more often used to seek protection

locations of the projectiles/bullets are not fixed. If that is the case, hiding at the end point of trajectory A is not safe as a slight lateral shift of the starting location of the projectile will lead to the Wall not being able to deflect the projectile hurtling toward the Agent. However, if the Agent moves nearer the Wall along trajectory B at the expense of energy and hides right behind the Wall, and probably also positioning itself near the "center" of the corresponding side of the Wall, it will have a better chance of not being hit by a projectile with an uncertain starting location. A mental simulation by jittering the starting location of the projectile will allow this "safe location behind the Wall" to be reasoned out. Therefore, there are other factors that lead to trajectory B being chosen that are not specified in the original SAS microenvironment as defined in Chap. 8.

9.4.1 Structure Construction and Neuroticism Driven Processes

If the uncertainty of the Projectile's starting location is greatly increased and the situation ends up something like that depicted in Figs. 8.7, 8.8, and 8.9 in which the Agent has to keep moving over quite long distances to seek new Shields/Walls, and in the meantime repeatedly experiencing highly Anxious states, the Agent would then attempt to construct a structure that provides a longer term solution. The difference between this and the "hiding behind shields" situation is that now the Agent has to actively move the Walls to construct the structure, while before it simply relied on existing structures such as the Walls as laid out in the original configuration. This process was shown in Figs. 8.10 and 8.11.

Given this experience with the Projectile's possible starting locations in Figs. 8.2, 8.7, and 8.9, there can be a range of different solutions to the problem that has to do with the Agent's personality disposition. We define a personality measure of Neuroticism that is proportional to the degree of the Agent's negative outlook on experienced negative outcomes. (Neuroticism is one of the big five personality traits identified by psychologists – e.g., Sagiv et al. 2002). So, if the Agent had experienced the Projectile coming from only a certain range of directions toward it, but then it assumes that the next time the Projectile may come from a wider range of other directions, then its Neuroticism is high. Otherwise it is low.

Figure 9.22 shows the three starting locations for the Projectile from Figs. 8.2, 8.7, and 8.9 and together they form a possible range of starting locations. If the Agent is not particularly neurotic, it would simple expand a minimal amount of energy and construct the "extended shield" as shown in the figure. This is adequate protection provided that the starting locations of the Projectiles keep more or less within the range experienced so far.

There are two situations in which the Agent may construct a more extensive structure to shield itself from the Projectile. One is, if the Projectiles start appearing in yet other locations and hurtle toward it from yet other directions outside the range shown in Fig. 9.22, it may then need to extend the shield to cover those directions even if it is not particularly neurotic. The other is, the Agent is simply neurotic and expects worse future possibilities of negative outcomes without experiencing them directly. It then generalizes the possible starting locations of the Projectiles to a wider range. One possibility, as shown in Fig. 9.23, is that it assumes that the Projectile could come from *all* possible directions, indicated by the area covered by the yellow and green dashed arrows. It then constructs a Shelter – an extended and enclosed structure of a number of Shields – to protect itself accordingly.

There is of course a price to pay for being neurotic, which is that more energy expenditure is needed to implement the solution to prepare for the worst possible outcomes. Whether an Agent would actually take actions to address its neuroticism would depend on the outcome of the tussle between the two opposing constraints.

9.5 Summary and Future Investigations

As can be seen from the discussion in this chapter, we have only described the detailed computational mechanisms to address part of the various issues put forth in the SAS micro-environment of Chap. 8. Sections 9.1, 9.2, and 9.3 provided detailed computational mechanisms to address the SAS micro-environment up till Fig. 8.4. Section 9.4 provided some degree of computational account for using a Shield (the Wall) to protect the Agent, as specified in Figs. 8.5 and 8.6. As for the processes leading to the construction of the Shelter from Figs. 8.7, 8.8, 8.9, 8.10, and 8.11, as discussed in Sect. 9.4.1 above, only a very rough outline was provided. More complex computational devices are needed to address these issues in future investigations.

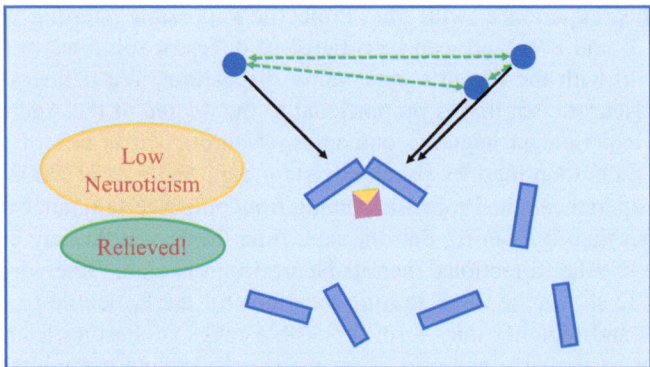

Fig. 9.22 An "extended shield" constructed to protect the agent from the possible starting locations of the projectiles so far experienced, with low neuroticism (a personality trait). The *green dashed arrows* show the range of the Projectile starting locations experienced so far (these locations are taken from Figs. 8.2, 8.7, and 8.9)

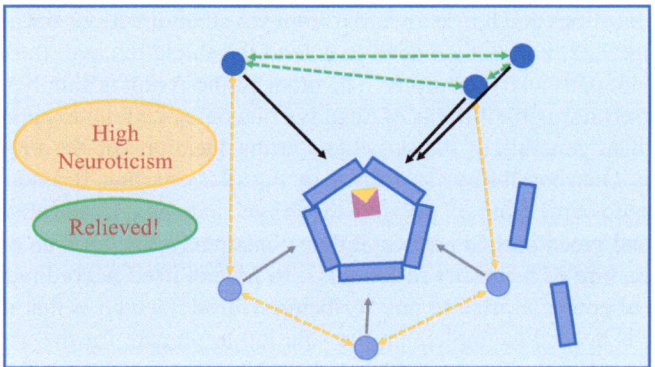

Fig. 9.23 Agent is highly neurotic and assumes that the projectiles could come from all possible directions – the *yellow dashed arrows* show the extended range of possible projectile starting locations compared to that in Fig. 9.22. It then constructs a shelter – an extended and enclosed structure made up of a number of shields – to protect itself

The solutions provided to address the SAS micro-environment up till Fig. 8.4 – i.e., up till using the "stepping aside method" to avoid the Projectile – as discussed in Sects. 9.1, 9.2, and 9.3 use all of the methods and principles developed in various preceding chapters of the book. This affirms that the methods and principles developed earlier are general and can be applied to a wide range of domains.

One major achievement of this chapter is to show how an affective state such as Anxiousness could drive the noological processes toward providing a solution to address the problems at hand. This elucidates the *function* of Anxiousness in computational terms and serves as a model for addressing other kinds of emotions.

In Sect. 9.4.1, another affective disposition, Neuroticism, was also discussed in terms of its functional role in driving the Agent to arrive at a certain solution for the problem involved, though no computational details were provided. We present this as a challenge for noology and computational intelligence.

Lastly, the paradigm propounded in this book in general on explicit and grounded representations of concepts allows the learning of problem solutions through language. For example, consider if one were to provide a piece of advice to the Agent through a natural language sentence such as "To avoid being hit by the Projectile, you need to move aside when it comes hurtling toward you." The Agent can map the individual words to the grounded and explicit conceptual representations that it has of the various concepts involved and construct an internal representation for the advice and execute it accordingly. We did not flesh out the detailed computational mechanisms for this in this chapter, but with our paradigm, as was pointed out in a number of other places throughout this book – such as Sects. 5.4 and 7.5 – learning to solve problem through language is possible.

Problems

1. Solve the rest of the SAS micro-environment as defined in Chap. 8
2. At the end of Sect. 9.3 we mentioned an alternative solution to that of Fig. 9.17 in which the Agent "moves aside" for the Projectile to pass by, and that is to instead *swing* itself around some pivoting points to likewise avoid the Projectile. Devise a solution such as this using similar devices as we have discussed.
3. Recently there has been attempts at using reinforcement learning to learn to play a number of "Atari games" such as Space Invaders (deepmind.com). The shooting of "bullets" between the player and the "space invaders," and the player having to learn to move the object under her control to avoid the bullets has some similarity to the "projectile and agent" problem tackled in this chapter. Formulate an effective causal learning framework along the line of this book to learn to play games such as Space Invaders as well as other similar games.

References

Ho, S.-B. (2013). A grand challenge for computational intelligence – A micro-environment benchmark for adaptive autonomous agents. In *Proceedings of the IEEE symposium series on computational intelligence on intelligent agents* (pp. 44–53) Singapore. Piscataway: IEEE Press.

Sagiv, L., Schwartz, S. H., & Knafo, A. (2002). The big five personality factors and personal values. *Personality and Social Psychology Bulletin, 28*, 789–801.

Chapter 10
Summary and Beyond

Abstract In this chapter, other than summarizing the achievement of this book, a number of important issues are discussed. These include the issue of scaling – how can the principles and methodologies developed in this book be scaled up to handle the real-world environment, both internal and external to the agent involved? Another critically important issue is the representation and learning of mental procedures. This will imbue the agent with vastly improved adaptive abilities. Other issues such as the constraints imposed by personality, culture and social situations on behavior are briefly discussed. Lastly, comparisons are made between the paradigm developed in this book and the various methodologies of traditional AI.

Keywords Scaling • Learnable mental process • Personality • Culture • Social situation • Connectionism • Symbolic processing • Causal learning • Reinforcement learning • Deep thinking • Quick learning • Evolution

In this book we set out to construct a new paradigm for the characterization and understanding of intelligent systems. We believe the time is ripe for restructuring some of the existing concepts in AI specifically and cognitive science in general, as well as adding some new concepts and methods to found a new discipline to deal with the issues of intelligent systems and attempt to provide a "theory" of intelligence.

Firstly, we deemed that a new name is necessary for this new discipline, and we have called it "noology" – the study of intelligent systems – which are in turned called "noological systems." "Intelligent" systems comprise cognitive, affective, motivational, and motor processing of all kinds operating in concert to produce intelligent behavior and hence noological systems (as opposed to cognitive systems, or cognitive-affective systems, etc.) is a better way of referring to them.

Secondly, a computational approach is adopted for the characterization of various noological processes and various computational and representational constructs used in traditional AI, such as predicate logic, scripts, search, and heuristics, are used here but they are deployed differently in the new paradigm. For example, the basic concept of forward and backward search is deemed necessary for basic noological functions but extensive search is not a noologically realistic process. Instead, a small amount of forward search combined with causal learning and learning of heuristics is deemed a noologically realistic process, and this was thoroughly investigated in Chaps. 2, 3, 6, 7, and 9. In a problem solving process, the amount of backward search is drastically reduced through the earlier learned knowledge that is chunked in *scripts*. Heuristics emerged as part of the causal learning processes (Chaps. 2, 6, and 7) and these greatly reduce the complexity of search processes (whereas in traditional AI, heuristics are hand-coded). Scripts, a kind of structure for organizing knowledge devised by Schank and Abelson (1977), have been enhanced with causal learning to allow them to be acquired from experience with the external world (Chaps. 2, 3, 6, 7, and 9). Predicate logical statements are used to represent knowledge about the external and internal worlds but there are two aspects of their deployment that are different from that in traditional AI. One is, the predicates used are well-defined grounded concepts and not any arbitrarily defined concepts, and higher level predicates/concepts can be built up from grounded predicates/concepts (Chap. 4). The other is, predicate logical statements are packaged in scripts for problem solving through search, instead of just being used directly in logical inference (Nilsson 1982; Russell and Norvig 2010).

Thirdly, a noological system is characterized fundamentally as a system that has some built-in internal goals/needs and the principal processing backbone of a noological system is the problem solving processes that attempt to satisfy those goals/needs. Other functions such as perception and conceptualization are service functions that assist in the problem solving processes in terms of providing information for the system to learn useful causal rules and apply them to particular situations.

Fourthly, emphasis is placed on the computational role of various affective and personality processes in the functioning of the entire noological system (Chaps. 8 and 9), whereas in traditional AI the issues on affective and personality processes are shunned (e.g., Russell and Norvig 2010), and in other disciplines of cognitive science such as psychology and neuroscience, no computational characterization of affective and personality processes is forthcoming.

Fifthly, a novel learning paradigm that is noologically realistic – causal learning, which is rapid and requires only a small number of training instances – is formulated (Chap. 2) and applied to various noological processes throughout the book.

Arising from these considerations and characterizations, we can concretely define *intelligence* as the ability of a system to identify causality and effect generalizations on the attendant causal rules for effective problem solving with respect to some internal goals of the system, in order to maximize the adaptability and survivability of the system involved.[1]

Also, arising from the expositions in the book, we identify a *script* as a fundamental unit of intelligent behavior. (Script is discussed in Chaps. 2, 3, 6, 7, and 9.)

---

[1] Contrast this with the often-used method of defining intelligence in AI, which is the Turing test. We believe our characterization is better as the Turing test is ill-defined (e.g., there are open-ended questions on the test such as the kinds of questions that should be asked for the test and how long should the test be conducted before any definitive conclusion can be made). Moreover, our characterization also prescribes a method for the construction of an intelligent system.

Each script consists of a START state, a sequence of one or more than one step of ACTIONS, an OUTCOME portion, and a COUNTERFACTUAL INFORMATION portion. Scripts are learned in a number of learning scenarios. Given a goal, scripts can be used to solve problems rapidly. Another kind of fundamental units of noological processes is the atomic operational representations of grounded concepts as discussed in Chap. 4.

Furthermore, in contrast to the "deep learning" paradigm currently subscribed by a large number of researchers in AI (e.g., LeCun et al. 2015), what we have demonstrated instead is a "deep thinking, quick learning" paradigm (discussed in Sect. 6.6, Chap. 6) that allows rapid learning of various concepts that is more akin to the learning processes found in natural intelligence systems (such as human beings). The paradigm also provides the mechanisms for the explicit representation of the knowledge contents learned that facilitates symbolic communication of the knowledge involved between noological systems, that in turn facilitates further rapid learning.

Therefore, in summary, a principled and systematic consideration of noological processes led to the following basic principles for characterizing and constructing noological systems[2]:

- A noological system is characterized as primarily consisting of a processing backbone that executes problem solving to achieve a set of built-in primary goals which must be explicitly defined and represented. The primary goals or needs constitute the bio-noo boundary (Chap. 1).
- Motivational and affective processes lie at the core of noological processing and must be adequately computationally characterized. (Sects. 1.4 and 3.3, Chaps. 8 and 9).
- Rapid effective causal learning provides the core learning mechanism for various critical noological processes (Chap. 2).
- The perceptual and conceptual processes perform a service function to the problem solving processes – they generalize and organize knowledge learned (using causal learning) from the noological system's observation of and interaction with the environment to assist with problem solving (Chaps. 1 and 6).
- Learning of scripts (consisting of start state, action steps, outcome/goal, and counterfactual information) from direct experience with the environment enables knowledge chunking and rapid problem solving. This is part of the perceptual/conceptual processes. Scripts are noologically efficacious fundamental units of intelligence that can be composed to create further noologically efficacious units of intelligence that improve problem solving efficiency, in the same vein that atoms are composed into molecules that can perform more complex functions. (Sect. 2.6.2, Chaps. 6, 7, and 8).

---

[2] To contrast the principles laid out here with the usual emphases placed on these issues in AI and the cognitive sciences, see the discussion at the beginning of Chapter 1 in connection with the statements of these principles.

- Learning of heuristics further accelerates problem solving. Similarly, this derives from the perceptual/conceptual processes. (Sects. 2.6.1 and 6.4, and Chap. 7, specifically Sects. 7.4.3.5, 7.4.3.6, and 7.4.3.8).
- All knowledge and concepts represented within the noological system must be semantically grounded – this lies at the heart of providing the mechanisms for a machine to "really understand" the meaning of the knowledge and concepts that it employs in various thinking and reasoning tasks. There exists a set of ground level atomic concepts that function as fundamental units for the characterization of arbitrarily complex activities in reality. (Chaps. 1 and 4 in general, and specifically Sects. 4.5, 5.4 and 7.5).

As has been stated in the beginning of the book (Chap. 1), the emphasis of our investigations is to explore and establish fundamental principles of noology by addressing the issues associated with a thin and deep micro-environment. We have successfully done that by applying the principles above to relatively simple but general problems such as the SMG problem (Chap. 2), the SMGO problem (Chap. 6), the problems presented by the StarCraft game environment (Chap. 7), and parts of the SAS micro-environment benchmark (Chap. 9). Though more work is needed to apply these principles to more complex environments such as those found in the real world, we believe the principles that we have laid down are fundamentally sufficient to deal with these.

In the following we will look beyond the discussions in this book and outline how the basic principles may be applied to more complex external and internal environments. We will also discuss one issue raised at the end of Sect. 6.6, Chap. 6 regarding the learning of the internal mental processes, such as the process of causal learning itself discussed in Chap. 2, the various internal "deep thinking" processes that were discussed in connection with the SMGO problem in Chap. 6, etc. "Learning how to think" is certainly important for the adaptability of a noological system.

10.1 Scaling Up to the Complex External Environment

In this section we will not delve into the details but will provide a sketch of how the principles of noology established in this book can be applied to a complex real world problem. The problem we wish to present for discussion is non-other than the acquisition of the Restaurant Script presented in Schank and Abelson (1977) (first discussed in Chap. 2, Sect. 2.6.2). The script is a prime example of a real-world script. And as emphasized above and illustrated throughout this book, scripts are important fundamental processing units that support and generate intelligent behavior, and hence being able to learn these scripts in general, and scripts such as the Restaurant Script in particular, from observation of and experience with the world is crucial for a noological system.

10.1 Scaling Up to the Complex External Environment

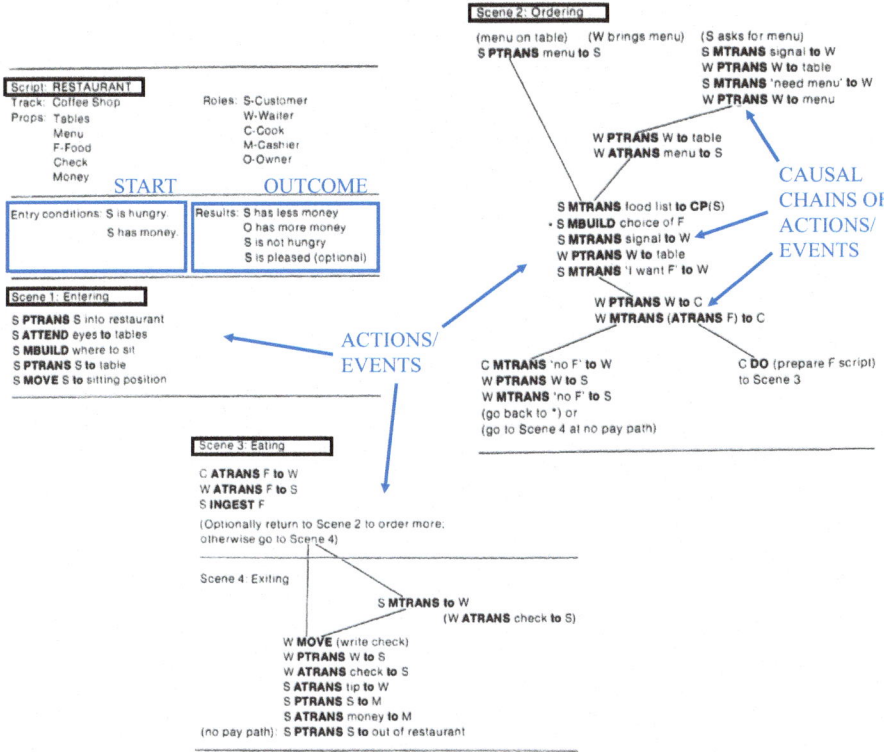

Fig. 10.1 Schank and Abelson's (1977) Restaurant Script (Republished with permission of Taylor and Francis Group LLC Books, from Scripts Plans Goals and Understanding, Roger Schank and Robert Abelson, 1977; permission conveyed through Copyright Clearance Center, Inc.). *PTRANS* Physical Transfer, *MTRANS* Mental Transfer, *ATRANS* Abstract Transfer – e.g., transfer of possession; *MBUILD* Mentally Building representation of sorts. The correspondences between some of its structures and the START, ACTIONS, and OUTCOME portions of our script's constructs are shown

Figure 10.1 shows a complete version of the Restaurant Script from Schank and Abelson (1977) (Fig. 2.17b of Chap. 2 shows only part of the same script). As pointed out in connection with Fig. 2.17b, the Entry conditions portion of the script corresponds to the START portion in our script framework (e.g., see Fig. 2.17a of Chap. 2), the Results portion of the script corresponds to the OUTCOME portion in our script framework, and the actions/events in the Scenes (Scenes 1, 2, and 3 in Fig. 10.1a) correspond to the ACTIONS portion in our script framework. These are shown in blue rectangles and words.

Scenes 1, 2, and 3 in Fig. 10.1 consist basically of causal chains of events – for each event, its preceding event must take place before it could take place. In a typical restaurant scenario, Scene 1, Entering, consists of only one possible sequence of events. Scene 2, Ordering, however, could start with S – the Customer – picking up the menu herself (PTRANSing or Physically Transfering the menu to herself) or asking for the menu from W (Waiter). The rest of the script is quite self-explanatory.

In Schank and Abelson (1977) it was mentioned briefly that the Restaurant Script could be learned by a child "by being dragged through the experience enough times" (page 222 of Schank and Abelson 1977). However, due to the limitation of technology at that time, there was no concrete implemented computational systems that demonstrated the acquisition of scripts such as this through direct experience (i.e., a "robot child" acquiring the Restaurant Script in this manner). In order to do that, firstly there must be relatively sophisticated computer vision technologies that allow the entire spatiotemporal experience to be captured and characterized – what objects were seen, encountered, interacted with, what actions and events took place, etc. Secondly, there must be some learning methods that allow the various causal chains of events to be learned and encoded rapidly to construct the script involved. These technologies were not available then, but are available now.

In Fig. 10.2 we illustrate the spatiotemporal picture of a typical trip to a restaurant. The event of visiting a restaurant consists of sub-events of ordering food, eating food, paying and leaving, etc. The sub-events are basically spatiotemporal patterns of activities generated by the actors – in this case, the Customers and the Waiter. In Fig. 10.2 we use a dashed green hexagon at a location to represent the amount of time spent at the location for the actor involved. The event begins with the Customers walking from the door area to the chairs and sitting down, while the Waiter waits for them to decide what they would like to order (assuming that the menus are already at the table). When signaled, the Waiter then walks to a location next to one of the customers to take the order. After that, the Waiter walks to the Kitchen Area to pick up the food. He then serves the food to the customers by bringing it to their table at different locations, and then he returns to his original location. The Customers then eat the food. When the customers have finished eating, they signal the Waiter again to pay for the food (this trajectory is the same as when the Waiter comes over to take the order). After that they walk to the door and leave the restaurant.

Using the current state of the art of computer vision technologies (e.g., Forsyth and Ponce 2011; Shapiro and Stockman 2001; Szeliski 2010, etc.), the trajectories of these actors and other objects involved can be captured. Here we assume that a noological system endowed with the power of computer vision is positioned somewhere in the restaurant to observe all the activities to learn and construct the Restaurant Script, or the noological system could be one of the Customers (a robot). Activities such as gesturing to the Waiter and eating the food can also be captured and characterized with current state of the art computer vision techniques (e.g., Wang and Mori 2009; Yuan et al. 2011, etc.). These spatiotemporal patterns observed are then organized into meaningful subunits much like those operational representations of concepts discussed in Chap. 4.

Rapid effective causal learning such as those described in Chap. 2 is used to build a causal chain of sub-events that characterize the Restaurant Script. For example, there is a temporal correlation between the Customers' signaling to the Waiter and the Waiter initiating actions to approach the table. Therefore, the

10.1 Scaling Up to the Complex External Environment

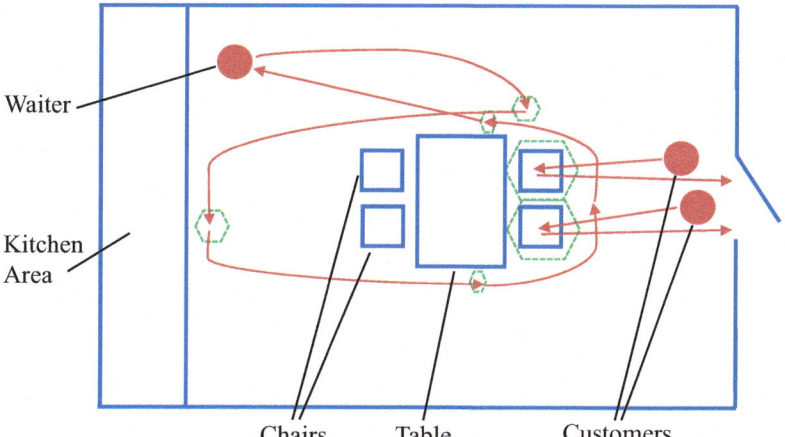

Fig. 10.2 An eat-at-restaurant event. *Dashed green* hexagons represent time spent at the corresponding locations. Customers' starting points are near the door and the Waiter's starting point is near the kitchen. (See text for explanation)

signaling to the Waiter is the potential cause of him approaching the table to take the orders, much like how the causal rules are created as described in Chap. 2. The details are discussed in Chap. 2.

Scripts can be acquired under two situations (Fig. 2.19, Chap. 2). The noological system can observe the events that is happening in the real world (itself possibly being a participant – such as being one of the Customers) and build the scripts based on that experience. A script can also be created in the course of problem solving – e.g., when searching for a sequence of actions that alleviates one's hunger. Incremental chunking can participate in greatly reducing the computational time needed for the construction of the scripts in both situations as described below.

In the first situation in which the noological system observes the spatiotemporal patterns representing the events, say, in a restaurant, and attempts to build a script out of them, the segmentation of the stream of spatiotemporal patterns taking place in the restaurant into the various steps of ordering food, eating food, paying for the food, etc. can be difficult without some prior knowledge. However, the chunks of knowledge that have been learned earlier can facilitate the construction of the current Restaurant Script. For example, a simpler "Eating Script" might have been learned in a different context – e.g., eating at home. This sub-script involves using the hand to pick up the food and transferring it to the mouth, and the process continues until the food is completely (or largely) consumed. So, having observed this eating process through a computer vision module in a separate occasion and stored it as an Eating Script, when a similar pattern of spatiotemporal patterns is now observed in the restaurant, the noological system automatically characterizes it as an eating process, and the eating process is then carved out of the continuous stream of spatiotemporal pattern observed in the restaurant. Likewise, the ordering process, the paying process, etc. can be carved out based on prior known scripts and then chained into a meaningful causal sequence of sub-events to build the current Restaurant Script.

Continuing a similar process, a Restaurant Script can be embedded in a larger Shopping Script which includes other events such as Buying Shoes Script and Buying Clothes Script which together achieve the purpose of acquiring certain desired collection of goods.

This process can be thought of as "learning about the world from the ground up" – simpler and more fundamental knowledge and concepts about the world are first learned in an "infant/child" stage and then they are used to build complex knowledge. That way, the higher level, complex knowledge is *grounded* in the fundamental knowledge about the world. At the same time, this incremental learning process facilitates problem solving.

In the second situation, the noological system would begin with a goal of, say, acquiring a certain desired collection of goods. If the system has earlier learned the individual scripts detailing the methods of acquiring the various types of goods (e.g., Buying Shoes Script, Borrowing Clothes Script, Making Toys Script, etc.), it can then solve the problem by assembling the individual scripts together. The problem solving/learning process is rapid as the sub-problems have already been encoded and solved before. This is the spirit behind the incremental chunking process described in Chap. 3, Sect. 3.5. Thus, our framework provides the mechanisms for learning ever increasingly complex concepts rapidly, and the framework is extendable to learning concepts in the complex real-world environment.

In recent years there has been some research efforts as regards learning script-like knowledge structure from video or textual data derived from the Internet (Chambers and Jurafsky 2008; Manshadi et al. 2008; Regneri et al. 2010; Tu et al. 2014). This is a good indication that researchers are channeling their efforts into this area of research. However, in these efforts, the "script" learned does not contain the complete structure of script as articulated in this book, which includes the specifications of scenario/start state, actions, outcome, and counterfactual information. The reason why these components of a script must be present before the script structure is complete is that the script then acquires what we term "noological efficacy" so that it can participate effectively in accelerating problem solving processes (stated in the fifth principle of noology in Chap. 1 and at the beginning of this chapter, and discussed also in Sect. 3.10.2 of Chap. 3). The original script structure as investigated by Schank and Abelson (1977) also contains 3 of these components (start state, actions, and outcome) and are hence noologically efficacious. In addition, our scripts are enhanced with counterfactual information. The efforts of Schank and Abelson and the others focus primarily on using scripts for question-answering, whereas our effort focuses on problem solving, thus our script structure contains more elements. In the first principle of noology as stated in Chap. 1 and the beginning of this chapter, problem solving with respect to primary goals constitutes the processing backbone of a noological system; therefore, it is of primary importance. Future research can certainly benefit from these earlier efforts in script learning as well as our more complete formulation of script structure and the learning methods described in this book.

In terms of the locus of script processing in the human brain, as mentioned in Sect. 3.10.2 in Chap. 3, a number of neuroscientists have identified it to be in the prefrontal cortex (Barbey et al. 2008; Krueger and Grafman 2008; Wood and Grafman 2003; Wood et al. 2005). They believe the human prefrontal cortex stores Structured Event Complexes (SECs), and these SECs are representations composed of higher-order *goal-oriented sequence of events* that are involved in the planning and monitoring of complex behavior. From a computational perspective, the SECs are non-other than scripts in our noological processing framework.

10.2 Scaling Up to the Complex Internal Environment

As articulated in Chap. 1 and discussed throughout the book, notably in Chaps. 3, 7, 8, and 9, a noological system's problem solving and action priorities are driven by its internal needs and affective states. We have discussed in detail in a number of places the simple needs of alleviating hunger (Sects. 1.2, 3.6, 7.4.1.1, and 8.2) and avoidance of pain (Sect. 8.1, and Chap. 9) and the computational processes associated with them. We have also discussed the roles of affective states such as desperation (Chap. 2, Sect. 2.5, in influencing generalization of causal rules) and anxiousness/fear (Chaps. 8 and 9, in driving the system to solve an imminent problem, after which the system arrives at the emotion state of relief) and detailed the computational processes associated with them. We also discussed a personality trait, neuroticism, in connection with a noological system's propensity to solve a problem in a certain manner (Chap. 9, Sect. 9.4.1). However, the full-blown internal environment associated with a typical natural noological system such as a human being is quite complex. There is a large number of needs such as those shown in Fig. 1.9 and Table 1.1 in Chap. 1. There is also a large number of possible emotional states associated with a typical natural noological system (Plutchik 2002; Reeve 2009).

Let us begin with the lowest level of the Maslow hierarchy (Maslow 1954), the Physiological level, as shown in Fig. 10.3 which is the same as Fig. 1.9a and is reproduced here for ease of reference. The basic Physiological needs are food, water, sex, temperature comfort, breathing, sleep, and excretion. How do we consider these item in our computational framework?

Firstly, we group food, water, sex, and temperature comfort into one group and breathing, sleep, and excretion into another. The first group consists of needs that are satisfied by external causal entities. The discovery of the causal entities involved for all these items can be similar to that of the discovery of food to alleviate hunger discussed in Sects. 1.2, 3.6, 7.4.1.1, and 8.2. This can be achieved by either accidental discovery, an instinct in knowing that entities with certain sensory qualities (shape, color, texture, smell, sound, etc.) can satisfy certain needs, or someone else supplying the entities to the noological system (e.g., at infantile stage, one is given food or water by the caregiver). The last two methods would vastly reduce the blind search time required by accidental discovery. In any of these

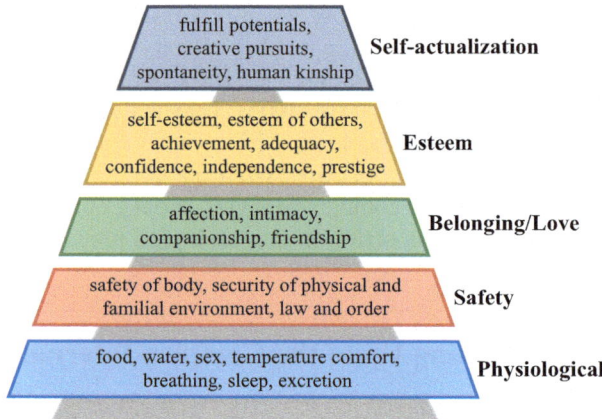

Fig. 10.3 Maslow hierarchy of needs (Maslow 1954, same as Fig. 1.9a)

methods, the moment the noological system experiences a change in its corresponding internal state given an interaction with a certain external causal entity, it will learn the causal rule associated with it just like in the case for food as discussed (e.g., the moment the caregiver supplies water to an infant's mouth, the infant experiences thirst quenching, or the moment a fan is used to blow wind at the infant, the infant experiences a body-temperature cooling effect, etc., and the infant then learns the corresponding causal rules accordingly). Having learned the causal rules, if the needs arise later, the noological system would know how to act to satisfy the needs (e.g., if it is thirsty, it will seek water; if it is hot, it will seek cooling devices or places). This process also works similarly for the learning of a number of specific taste rewards and punishers as shown in Table 1.1 of Chap. 1 (tastes of saltiness, sweetness, bitterness, etc.) or those specific rewards and punishers in the olfactory, somatosensory, visual, and auditory domains – e.g., one sees flowers the first time and experiences a reward of "visual happiness," and then one creates a corresponding causal rule accordingly.

As for the second group of needs consisting of breathing, sleep, and excretion, there are no external causal entities that cause these to be satisfied. The physiology of an animal or human would automatically cause these to be satisfied. However, there could be external agent(s) that may *prevent* these from being satisfied. E.g., the noological system learns that by pinching its nose and closing its mouth or by being in an enclosed space, it cannot breathe, or by being in a noisy and bright place, it cannot sleep, etc. The same causal learning process can take place but now it is learning about "*cause-not-happen*."

Therefore, for all these basic needs, the noological system can basically deploy causal learning to learn the attendant causal rules to satisfy these needs (or to avoid hindering the built-in automatic needs-satisfying physiological mechanisms). However, what is needed is for the builder of the noological system (or for nature, evolution) to build in the various needs accordingly to expand the space of needs for the system. This would scale up the internal environment. This is the bio-noo boundary as discussed in Sect. 1.4.

10.2 Scaling Up to the Complex Internal Environment

Interestingly, for artificial noological systems such as robots, the basic needs are few such as shown in Fig. 1.9b – they could just be battery power and/or fuel.

Next, we proceed one level above the basic, Physiological, level of the Maslow hierarchy and consider the level of Safety. Take the first item: safety of the body. This has a close parallel to the situation encountered in the SAS micro-environment as discussed in Chaps. 8 and 9 – a Projectile hurtling toward and harming an Agent. The problem is basically to learn about the causal entities that can cause harm to one's body and learn ways to avoid them. As can be seen in the partial solution to the SAS micro-environment problem in Chap. 9, the learning process is quite complex but we have shown how that can be handled within our paradigm. In the case of the Projectile hitting the Agent situation, firstly a script about the process is learned. Then, a process of problem solving identifies the cause(s) involved in the Pain experience of the Agent and the solution(s) to avoid the Pain. To scale up to the complex environment, naturally much more complex scripts and problem solving processes are involved. The learning of scripts that deal with the complex external environment has been discussed in the previous section. It is expected that the core problem solving processes involved are similar to the processes discussed in Chap. 9 with perhaps some further enhancements. In the real world, there is a multitude of situations in which harm to the body may happen – e.g., being hit by a car, being hit by falling objects, being shot or stab by others, skidding and falling onto the ground, falling off from a high level, etc. Learning about these possibilities and working out solutions to avoid them would be one of the higher priorities of a noological system. But first the noological system must be built-in with the need to protect its body, and the properties of the environment that lead to possible harm to the body and the corresponding ways to avoid them would then be learned through observation of and interaction with the environment.

Going down the list of the Safety level, there are security of physical and familial environment, law and order, etc. Again, these needs must be built-in, and some extensions to the various representational and computational mechanisms discussed in the current book will be needed to handle these. A similar process is required for the various needs at the higher levels of Belonging/Love, Esteem, and Self-actualization. Again, for a robot, these levels also tend to be a lot simpler, such as that shown in Fig. 1.9b. Suffice it to emphasize here that further investigations are needed and the various noological processing mechanisms formulated in this book would be a good starting point for the endeavor.

Whether the full complex internal environment such as that laid out in the Maslow hierarchy of Fig. 10.3 is to be considered and its corresponding computational mechanisms to be worked out depends on what kind of purpose one has in mind. If the purpose is to build a relatively simple autonomous robot, something akin to the autonomous systems' needs of Fig. 1.9b plus some enhancements may be sufficient. It is unlikely that one would build an artificial being that requires the full range of internal motivations and needs like that in the Maslow hierarchy, or for that matter the full range of rewards and punishers of Rolls as laid out in Table 1.1, as many of these items arise from the biological underpinning of the noological system involved. However, if one needs to simulate human beings in high fidelity for the purpose of understanding human behavior, then the full range of the motivations and needs in both the Maslow hierarchy and the list of Roll's rewards and punishers, as well as perhaps other additional items not included in these two lists would be needed. Therefore, even if an artificial being itself does not require the full range of these motivations and needs to drive its own behavior, for it to better understand and interact with human beings, it must also have the built-in ability to simulate these motivations and needs that drive the behaviors of human beings.

In connection with the scaling up to handle more complex internal environment, there is the issue of needs and affective competition that we discussed in Sects. 3.3, 3.7, and 7.4.2. In these sections we considered primarily only competition between two affective priorities. When there are multiple priorities, and also a hierarchy of priorities such as the Maslow hierarchy, the situation will become complex. Associated with this would be the needs and affective priority learning – how the priorities may change as a result of experience. These are challenges for future investigations.

Likewise, future investigations could expand on the computational and functional paradigm we developed for the emotional state of anxiousness/fear to other emotions such as hope (opposite of fear – an anticipation of positive outcome), distress, relief, joy, sadness, anger, surprise, etc. (Plutchik 2002; Reeve 2009).

10.3 Explicit Representations of Learnable Mental Processes

At the end of Sect. 6.6, Chap. 6, we mentioned that the thinking and reasoning processes of Figs. 6.21 and 6.25 are also a kind of (problem solving) script which is amenable to learning. This applies also to the various other "mental procedures" operating on various kinds of representations throughout the book. For example, we formulated the rapid causal learning algorithm of Chap. 2 and applied it to the SMGO problem in Chap. 6, the StarCraft game environment in Chap. 7, and a number of other places. The algorithm is a procedure that operates on certain observed phenomena and their corresponding internal representations. There are also issues of mental procedures noted in footnotes 4, 5, 6, 7, 9, and 10 of Chap. 7. For these and the other mental procedures to be learnable and not built-in, a requirement would be that these procedures themselves can be explicitly represented, much like, say, the MOVEMENT-SCRIPT of Fig. 2.18 representing the "procedure" of movement.

In terms of type of knowledge representation, mental procedures or processes in the form of scripts are a kind of "declarative procedures." In traditional AI, typically two kinds of knowledge representations are distinguished. One is declarative representations such as predicate logic representation in which the knowledge involved is "declared" in a logical statement, and there is an inference process/procedure that operates on it. The declared knowledge can be easily changed

10.3 Explicit Representations of Learnable Mental Processes

through the modification of the logical statements, such as in the processes discussed in Chap. 6, but the inference procedure – the "mental process" – is typically built-in and not amenable to learning. The other is procedural representations in which knowledge is encoded as a series of steps of computer program actions, an early example of which is Winograd's procedural representations for language processing (Winograd 1973). These computer action steps which are mental processes are also rigidly built in as computer codes that cannot be modified easily except through the human programmer's intelligent intervention. Declarative procedures in the form of scripts which encode mental processes, however, are amenable to machine modification and hence learning.

We will consider as an example the process of observing some temporal correlations of events resulting in the encoding of a causal rule as articulated in Chap. 2. The basic idea of the rapid causal learning method of Chap. 2 says that "*if I (the Agent) observe that one event is followed closely by another event, I encode a rule that says that the first event causes the second event.*" We will refer to this as the *basic causal learning process*. There are other finer points in connection with this causal learning process (e.g., synchronic causal conditions, dual instance generalization, etc.) and they have been discussed in Chap. 2, but suppose we use the simple idea encapsulated in the sentence (in italics) above as a basis for the creation of a causal rule. How might we represent such an internal mental process explicitly in a declarative form?

A kind of representational framework called "conceptual dependency (CD) theory" devised earlier by Schank (1973) can provide just such a mechanism to represent the above mental process of causal rule creation. Firstly, let us introduce the basic ideas of CD. Figure 10.4a shows the representation of the sentence "Mark was angry because Jane kicked him" using CD. The core sentence "Jane kicked Mark" is represented with a double arrow and a single arrow linking the three entities, "Jane," "kick," and "Mark" at the top part of the representation. "o" stands for "object." The down arrow represents a causal relationship between Jane kicking Mark and Mark acquiring the state of being angry – the bottom part of the representation basically shows Mark changing from some other states to the state of being angry (in Schank (1973), an "up" arrow instead is used to represent the causal relationship).

Figure 10.4b shows the representation of the sentence "Mark ate the meat with a fork." The basic "Mark ate the meat" portion is represented by linking "Mark," "INGEST," and "meat" with a double arrow and a single arrow as shown, just like in Fig. 10.4a. Therefore, Mark's action is an INGEST action and the object of the ingestion is meat. Further to the right of Fig. 10.4b, there is an elaboration of the instrument used to transfer the meat into Mark's mouth. "I" is the INSTRUMENT case. "TRANS" is "physical transfer." The elaboration on the right basically represents the process of using a fork to transfer meat to Mark's mouth – the fork ATTACHes to the meat and is an object ("o") of the transfer process. The meat is transferred to a mouth POSSessed-BY Mark, which is Mark's mouth.

Figure 10.4c shows the CD representation of the concepts of "learn," "remember," and "forget." The idea of mental transfer – MTRANS – is introduce here. As

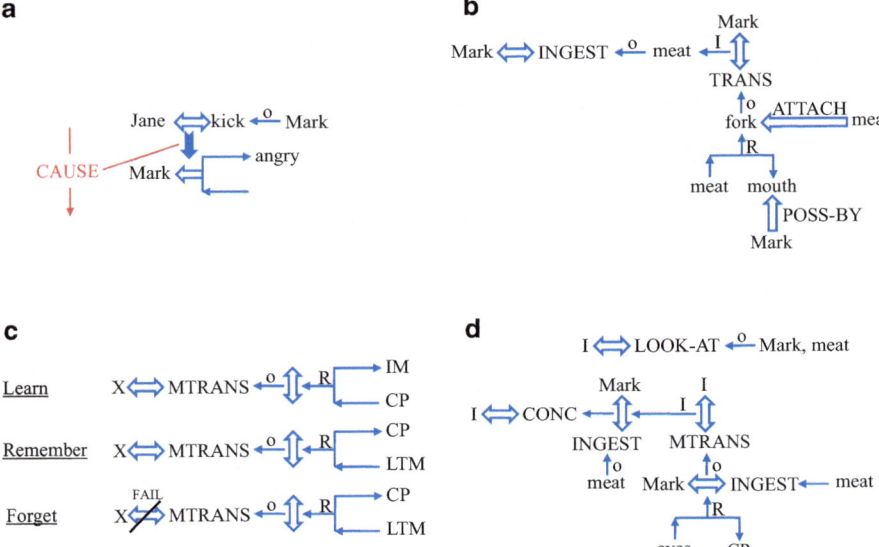

Fig. 10.4 (a) Conceptual dependency (CD) representation of "Mark was angry because Jane hit him." (b) "Mark ate meat with a fork." (c) "Learn," "remember," and "forget." (d) "I saw Mark eating meat." (All based on Schank 1973. See text for explanation)

opposed to TRANS in Fig. 10.4a, which is physically transferring some objects in the real world, MTRANS is *mentally* transferring some objects in the internal process of an agent. CP stands for conscious processor, IM stands for immediate memory, and LTM stands for long term memory. So, to "learn" is to mentally transfer something from the CP to the IM. To "remember" is to transfer something from the LTM to the CP. To forget is to fail to transfer something from the LTM to the CP.

Putting the above representational devices together, Fig. 10.4d shows the CD representation of the sentence "I saw Mark eating meat." "Mark INGEST meat" is a physical event that happens in the physical world. CONC stands for "CONCeptualize." Therefore "I" CONC the fact that "Mark INGEST meat." How do I do it? The I(nstrument) of my CONCeptualization is a Mental TRANSfer (MTRANS) of the observed physical event "Mark INGEST meat" from the *eyes* to the CP. At the same time, a separate fact exists which is "I am LOOKing AT Mark and the meat" (i.e., to *see* something I must be *looking* at that same something).

Using the CD representational devices above, we can now represent a more concrete version of the above basic causal learning process such as *"If I (the Agent) observe that the appearance of OBJ1 is followed closely by the appearance of OBJ2, I encode a rule that says that the appearance of OBJ1 causes the appearance of OBJ2."* Figure 10.5 shows how the above sentence and hence the process of basic causal learning described by it is represented using CD representation.

The top part of Fig. 10.5 encodes the process "I observe the appearance of OBJ1 at T1," in a similar vein to the encoding of "I saw Mark eating meat" of

10.3 Explicit Representations of Learnable Mental Processes

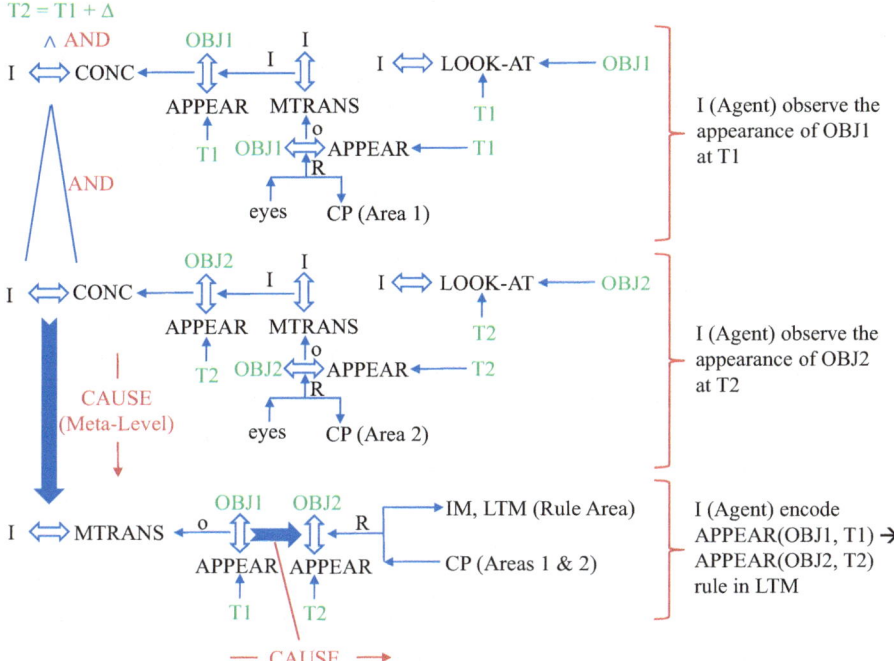

Fig. 10.5 Encoding of the basic causal learning process using CD representations. The represented process/rule is "If I (the Agent) observe that the appearance of OBJ1 is followed closely by the appearance of OBJ2, I encode a rule that says that the appearance of OBJ1 causes the appearance of OBJ2"

Fig. 10.4d. The middle part of Fig. 10.5 encodes the process "I observe the appearance of OBJ2 at T2." The condition that these two events must happen closely to each other is encoded as $T2 = T1 + \Delta$. The above three conditions must happen before a causal rule is encoded. This is shown as two conjunctions "AND" linking the three conditions together in the figure. When these conditions are satisfied, it CAUSEs the encoding of the causal rule, shown at the bottom part of Fig. 10.5.

When the appearance of OBJ1 takes place at T1, the mental representation of APPEAR(OBJ1, T1) (in the CD form) is transferred from the eyes to the CP (Area 1). A similar process takes place for the appearance of OBJ2, whose corresponding mental representation is stored in CP (Area 2). The creation of the causal rule APPEAR(OBJ1, T1) → APPEAR(OBJ2, T2) involves retrieving the corresponding representations in CP (Areas 1 and 2) and storing it in the Rule Area of IM or LTM in a more permanent form of storage.

In Fig. 10.5 we distinguish the CAUSE relation at two levels. One is the CAUSE relation between the appearances of OBJ1 and OBJ2. This is at the level of the causal knowledge captured. There is also a CAUSE which we mark "Meta-Level" in the figure that is at the level of the causal learning process itself that results in the causal knowledge capture. Hence, at both levels similar CD constructs are used to represent the processes/rules involved (the causal learning process itself, as stated in the italicized sentence above and encoded in the CD representation of Fig. 10.5, is also like a causal rule in turn). This implies that if processes/rules in the form of CD representations are learnable, they are learnable both for the basic level causal knowledge (such as APPEAR(OBJ1, T1) $\rightarrow$ APPEAR(OBJ2, T2)) as well as the causal learning process itself, or for that matter, for any other mental processes.

Therefore, even though we have only outlined a rough idea here, we have demonstrated that in principle mental processes are amenable to learning. For the case of the causal learning process, of course it would be advantageous to have a basic version built-in to a noological system so that it can immediately be used for survival purposes. However, being able to learn these processes means that the system can refine on its built-in mental processes, including its causal learning processes, to better adapt to the environment. For human beings at least, as we know, we are often able to transcend our built-in mental processes. Take the case of causality. We can imagine and reason about acausality. Therefore, we know when to apply causal reasoning and when not to for the purpose of achieving some goals. This provides us with yet another level of adaptability to better our chances of survival.

Interestingly, lurking among the CD representations that we have shown in Figs. 10.4 and 10.5 are the conceptual structures that provide the mechanisms for a machine to be "self-aware" or "conscious." Consciousness is a complex issue that we will not delve into here. Suffice it to note that self-awareness definitely allows a noological system to be more adaptive: for example, if the system *knows/understands/is aware* that it is certain actions of it that give rise to some consequences, desirable or undesirable, it certainly would be better able to improve on its future behavior. One can conceivably represent events such as "I remember that X happened" or "I remember that I remember that X happened" using the CD structure of Fig. 10.4c. There could be many levels of "remember" that require a recursive structure to capture the event involved and CD can easily handle that. Accordingly, one can also use similar structures to represent "I know that it was my X action that caused the Y consequence," and "I know that I know that it was my X action that caused the Y consequence," etc. Fig. 10.5, to some extent, captures the system's *knowing* its own causal reasoning process, and hence the process can be modified or improved on accordingly.

10.4 Perception, Semantic Networks, and Symbolic Inference

There are three main topics in traditional AI that are not covered in this book. They are perception, semantic networks, and symbolic inference. In this section we will describe the connection of these topics to the discussions in this book.

10.4 Perception, Semantic Networks, and Symbolic Inference

There is a vast effort in endowing machines with perceptual abilities, notably visual abilities in the area of computer vision (e.g., Forsyth and Ponce 2011; Shapiro and Stockman 2001; Szeliski 2010, and numerous other references in this area). In Chap. 1 we have emphasized that the role of vision or for that matter the other perceptual domains such as audition, olfaction, the tactile sense, and the taste sense is to provide a service function (at the perceptual level) for the main noological process of problem solving – they make problem solving much more efficient. In Chap. 6 we detailed how the visual sense can greatly assist problem solving in a directed and efficient manner. Since there has already been a large and somewhat successful effort in computationally characterizing vision in the various computer vision efforts, we do not replicate the discussion here but simply note that it is certainly indispensable for a noological system to have the various senses to assist it in problem solving processes. Once the visual processes are able to detect objects, actions, and map out the 2.5D sketch (Fig. 6.1, Chap. 6), the noological processes we discussed in this book can take off from there. Also, there is a possibility that the rapid causal learning mechanisms discussed in Chap. 2 can assist in the vision problem itself. As was discussed in the Neuroscience Review of Sect. 3.8 of Chap. 3, the general function of the cortex is "cause recovery." That certainly includes the visual cortex and the other sensory cortices. Therefore, further research is needed to see if the rapid causal learning mechanism can have any role to play in this.

Semantic network is a knowledge structure that organizes knowledge in the form of a graph consisting of nodes representing various concepts and links representing the relationships between these concepts. Typical kinds of links are *category* links – e.g., a bird is "a kind of" animal – and *part-of* links – e.g., a seat is a part of a chair. In Chap. 4 we have used a number of *part-of* conceptual structures. There could also be links that describe the various components of an event – e.g., for an eating event, there could be typical location, start-time, end-time, agents, and actions involved, etc. (containing information much like the Restaurant Script discussed in Sects. 2.6.2 and 10.1). Similar to the role of vision, the role of semantic networks is also a service function (at the conceptual level) to problem solving as discussed in Chap. 1 (e.g., Fig. 1.7). For example, if a chair is a kind of furniture, then it can probably be found in a furniture shop. If a bird is a kind of animal, then if most animals are edible, perhaps a bird is also edible. This knowledge organization can certainly help to short-cut problem solving processes. As for event type concepts, we deem that it is better organized in the form of scripts as has been discussed in detail in this book so that they can be learned and used effectively for problem solving. Scripts, like semantic networks, also performs a conceptual service function.

Therefore, symbolic inference, probabilistic or otherwise, if applied to a semantic network, such as in the examples above, can assist in the problem solving process. This has been investigated in various AI research effort (e.g., Russell and Norvig 2010 and a host of other AI literature) so we will not replicate it here. Similarly, quite a lot of research has been carried out in AI in connection with symbolic inference as applied to predicate logical representation of knowledge that can assist problem solving. Analogical reasoning can likewise assist in problem solving and we discussed an example in Sect. 9.3.3.1 in Chap. 9.

We do want to emphasize that the critical issue for knowledge representation is semantic grounding as has been discussed in Chaps. 1 and 4 and various other places. Whether the form of representation is in semantic networks or predicate logic, or for that matter scripts, the concepts involved must be grounded so that the noological system "really understands" the concepts involved and can deploy them effectively in problem solving processes. Perception facilitates the external symbolic grounding process as discussed in Chaps. 1 and 4, and the bio-noo boundary provides the internal ground as discussed in Sect. 1.4 of Chap. 1, Sect. 4.3.8 of Chap. 4, and a number of places throughout the book.

10.5 Personality, Culture, and Social Situations

There are three other important kinds of constraints on intelligent behavior that have not been discussed or discussed at length in this book. They are personality, culture, and social situations.

We have touched on personality issue in the solution to the SAS microenvironment in Sect. 9.4.1 of Chap. 9. Personality provides a layer of internal constraint on behavior. E.g., an introvert may have a strong internal need to acquire certain information from other persons but his personality restrains it (extraversion/introversion is one of big five personality dimensions identified – e.g., Sagiv et al. 2002). Culture provides an external layer of constraint on behavior. For example, one may have a desire to sing aloud in a public place, but depending on the culture one is embedded in, it may have to be suppressed.

Another external constraint on behavior is the social situation in which an agent finds itself. It has been found that emotional disclosure on social network is associated with the social network's properties in terms of size and density (Lin 2014). Certainly, in any social situation, whether physical or virtual, the type and number of other intelligent agents present would greatly constrain the behavior of the agent involved.

Future effort should be devoted to a computational understanding of these factors in shaping the overall behavior of a noological system/agent.

10.6 Comparisons with Methods in AI

In this book we have mentioned at a number of places that many of the existing methods in AI such as search and reinforcement learning are not adequate in addressing noological problems, as demonstrated in the SMGO problem (Chap. 6), the StarCraft problem (Chap. 7), and the SAS micro-environment problem (Chaps. 8 and 9). However, because our paradigm is also computational in nature, it serves to make some quick comparisons between our methods and the AI methods.

10.6.1 Connectionism vs Symbolic Processing

In this section we raise some issues with regards to "connectionism" vs "symbolic processing," two main paradigms in AI that have been employed in constructing AI systems. Even though this is a relatively old issue, and the debates on this have abated and some of the issues seem to be moot in light of new developments, we believe the new computational understanding of noological system expounded in this book can throw light on the issue, thus setting a direction for future research.

We restrict the version of connectionism to be discussed in this section to the "parallel distributed processing" variety expounded by Rumelhart et al. (1986a, b). There is also a version of connectionism that uses "localized representations" that is more skin to symbolic processing (represented by Fahlman 1979 and subsequent trend of research).

In the mid 1980s, a "connectionist" movement in AI began (or was re-ignited, as there were also earlier attempts at connectionism), triggered mainly by Rumelhart et al.'s book (1986a, b). The new paradigm was a response to the prevailing paradigm in AI at that time, which was the "symbolic processing" paradigm, which seemed to have brought no progress in AI.

In the early days of AI, it was thought that symbolic reasoning (such as that based on predicate logic) underlie intelligent behavior in human beings, and AI's goal was to replicate this in machines. Symbol processing also comes naturally to computers, so its implementation in computers is straightforward. However, symbolic processing suffered from one major problem, which was that often complex knowledge representational structures based on symbolic representations, such as the Restaurant Script of Schank and Abelson (1977) that we have discussed in Sect. 10.1 and a variety of places in this book, were hand-coded for the purpose of reasoning and not learned. This was not only impractical, but was also not in line with what natural intelligent systems do, which rely on learning to acquire most of its knowledge. Adaptability through learning is the hallmark of intelligence, which is expected in intelligent systems, artificial or natural.

However, up till 1986, there was no known general methods for learning. The original single-layered Perceptron (Rosenblatt 1962), which originated in early neural network research, was found not to be able to learn certain classes of problem (Minsky and Papert 1969). The announcement of the backpropagation algorithm (Rumelhart et al. 1986a which was actually discovered earlier by Werbos 1974), applied to a multi-layered neural network, was a game changer. In principle, the backpropagation neural network can learn any arbitrary input-output mapping. Thus, it was thought that this general learning mechanism can be employed to build general intelligent machines that are adaptive to any arbitrary environment. Moreover, there is another advantage of using neural network to encode knowledge, which is a concept need not be represented in one "node" (say, a computer memory location), but can instead be distributed over many nodes, in a form of "distributed representation" (Hinton et al. 1986), in contrast to the way symbols are represented in computers, which is "localized." This has two advantages. Firstly, the knowledge represented is resistant to damages of some of the neural network elements. Secondly, it facilitates automatic generalizations (Hinton et al. 1986).

However, after more than two decades of "connectionist movement," in terms of its contribution to building truly generally intelligent machines, connectionism fared no better than symbolic AI. Even though further development of neural network did lead to successes in the perceptual domain of intelligent functions – e.g., the application of deep learning to speech processing, object recognition, etc. (e.g., Deng and Yu 2014), in some other fields, symbolic processing fared better, in the form of Deep Blue (Hsu 2002 – a chess playing program that could beat world chess champions) and Watson (Ferricci et al. 2010 – a question-answering machine that can beat humans at the game of Jeopardy which requires encyclopedic knowledge), etc. Of course these are specialized, and not general, intelligent systems.

In this book we have demonstrated quite clearly that the main issues underlying the characterization of general intelligent systems are not the issues related only to conceptual representation (see the basic principles of noological systems stated at the beginning of this chapter), and for conceptual representation, the issue is also not one about the type of representation – whether it is connectionist or symbolic, or some other forms of representations – but instead it is whether the concept represented is properly grounded. For example, in Chap. 4, we have illustrated the grounded representation of the concept of Move in both a pictorial form and a predicate logic form. Both forms are grounded and provide the "meaning" of the concept involved to the noological system using the concept for reasoning and problem solving. That provides the system with the ability to "really understand" the concepts it operates with.

Moreover, as mentioned in the summary in the beginning of this chapter as well as in Chap. 1 (e.g., the discussion in connection with Fig. 1.7), there is another ground – the internal ground, consisting of the inner needs and motivations of a noological system – that has to be specified to fully characterize the system. Thus, the issues of connectionism vs symbol processing are moot.

With regards to the failure of the early symbolic processing AI systems in their abilities to learn complex symbolic representations, we have demonstrated in this book that by applying causal learning, the noological systems can acquire and encode complex scripts and heuristics that can be used for reasoning and problem solving. This, again, is a demonstration that the issues of connectionism vs symbol processing are moot. Hence a good direction to characterize general noological system should be toward refining the various principles outlined in this book.

We end this section with some comments on an observation made by Marvin Minsky:

> However, progress has been slow in other areas, for example, in the field of understanding natural language. This is because our computers have no access to the meanings of most ordinary words and phrases. To see the problem, consider a word like "string" or "rope." No

10.6 Comparisons with Methods in AI   411

computer today has any way to understand what those things mean. For example, you can pull something with a string, but you cannot push anything. You can tie a package with string, or fly a kite, but you cannot eat a string or make it into a balloon. (Minsky 1992)

To allow computers to really understand what a string really means, you need grounded representation of the concept of string, and this can be captured and characterized through the visual perceptual apparatus of a noological system. This was discussed in Chap. 4 and a few other places and specifically, examples of grounded representations for some sentences were discussed in Sect. 4.5 of Chap. 4, Sect. 5.4 of Chap. 5 and Sect. 7.5 of Chap. 7. Problem 1 of Chap. 4 implies that it is possible to represent grounded concept of elastic objects. Similarly, grounded concept of string can likewise be represented within our framework. In this book, semantic grounding is enshrined as one of the basic principles of noology.

10.6.2 Effective Causal Learning vs Reinforcement and Other Learning Mechanisms

In a number of places throughout the book, we have compared the currently often-used method of reinforcement learning (Sutton and Barto 1998) with our methods in general and effective causal learning in particular. Without going into the detailed comparisons between the methods, we ask some overarching questions regarding reinforcement learning: Could the method be used to tackle the SMGO problem of Chap. 6, the StarCraft problem of Chap. 7, and the SAS micro-environment problem of Chap. 8 in a noologically realistic manner? We believe our causal learning method of Chap. 2 may require further refinements, but we have shown how it can be used to tackle these problems in a noologically realistic manner – not only the problem solving processes are quick, they also resemble how natural noological systems would go about tackling them. On the other hand, reinforcement learning is not able to generate noologically realistic solutions to these problems and within noologically realistic time frames. Appendix A also provides a simple example to illustrate what a noologically realistic approach to the problem of learning requires that pure reinforcement learning is inadequate in addressing.

One interesting point to raise is, in reinforcement learning, "rules" are also learned in the process. These rules basically inform the system, given a certain state of the world, what the next best action is. How are these rules different from the causal rules that are learned through causal learning?

The major fundamental difference between these two kinds of learning in this regard is that the rules learned in reinforcement learning are "state-transition rules" – given a state of the world, what is the best action to take. (This is also illustrated in Appendix A.) The fundamental problem is that state-transition rules do not contain very powerful generalizations about the world such as the causal rules of our paradigm. Often, the entire "state of the world" is identified, and the ideal transition/action that was learned over many training episodes is picked. It is almost like "the entire world in this state" is the cause of the transition, not individual relevant entities within that world, such as a particular force applied to an object (thus causing it to move), a particular person's particular foot stepping on something (thus causing it to break), etc. As such, there is no real understanding of what the causal entities are, and hence there are no powerful general causal rules being learned and encoded. Reinforcement learning typically searches through a large space for optimal sequences of actions, and requires many episodes of learning in which positive and negative reinforcement signals are received. And as mentioned earlier in the introductory section of Chap. 2, animals wandering into a city and finding ways to survive in it do not have the luxury of many episodes of reinforcement learning – one negative reinforcement signal, such as being hit by a vehicle while crossing the road, will render the animal inoperative (Greenspan 2013). But, in reality, these animals do learn causalities quickly and survive.

There has been some effort in enhancing the basic reinforcement paradigm with a generalization component so that the state-transition rules are generalized (Tan et al. 2008). In some applications, it was shown that the speeds of reinforcement learning involved were improved. However, a large amount of learning was still needed because the generalization process does not identify causality and hence extract and learn the attendant causal rules that are even more powerful generalizations.

There are at least two other kinds of learning mechanisms used in traditional AI and the cognitive sciences, and they are the unsupervised and supervised learning mechanisms. The rapid effective causal learning process is basically an unsupervised learning process. As discussed in Sects. 3.8 and 3.9 of Chap. 3, this process is probably the process that takes place in the cerebral cortex which is thought to carry out unsupervised causal learning. There is another aspect of the unsupervised learning process within our framework of noological computations. It concerns the learning of the spatiotemporal patterns present in the world to capture knowledge in the form of scripts as discussed through the book. As discussed in Sect. 4.4 of Chap. 4, whereas efforts in pattern recognition such as that represented by Uhr and Vossler (1981) and Fukushima (1980) focus on extracting *spatial* patterns, the unsupervised learning for knowledge chunking within our framework basically learns *spatiotemporal* patterns through an unsupervised learning process capitalizing on causal learning.

As for supervised learning, we postulated in Sect. 3.9 of Chap. 3 that the storage of the rules or chunk rules in the causal learning process probably involves the cerebellum which is believed to be performing supervised learning. There is another aspect of supervised learning and that is associated with learning through language – i.e., a "teacher" provides the knowledge to the "student." This has been discussed in connection with semantic grounding in Sects. 4.5, 5.4, and 7.5. There is yet another aspect of supervised learning that learns spatial or spatiotemporal patterns. This will be discussed in the next section in connection with deep learning.

10.6.3 Deep Learning vs Deep Thinking and Quick Learning

At the end of Chap. 6, in Sect. 6.6, we contrasted the current popular method of deep learning (LeCun et al. 2015) with our method of "deep thinking and quick learning." Deep learning has been applied quite successfully to a number of tasks such as speech processing, natural language processing, and object recognition (e.g., Deng and Yu 2014). One interesting aspect of deep learning is when the method is applied to, say, object recognition, the emergent responses of "neuronal elements" in the intermediate layers of the multi-layer learning architecture automatically correspond to intermediate structures of the objects involved, and these are automatically organized in a hierarchical fashion – e.g., at the lower levels of the deep learning net, feature detectors emerge that correspond to features shared among many kinds of objects such as lines, curve segments, and corners, while at higher levels of the net, feature detectors emerge that detect the higher level parts of the objects involved such as the eyes and nose of a face – if the net has been trained to recognize faces (Fukushima 1980; LeCun et al. 2015). This maps very well onto neuroscience data in which it has been demonstrated that the various levels of processing in the visual cortex consist of feature detectors that detect increasingly higher level subparts of an object as one ascends in the cortical processing hierarchy (Mountcastle 1998; Hubel and Wiesel 1962, 1965).

In the various applications of deep learning so far (e.g., Deng and Yu 2014), the emphasis is on recognition and classification. This correspond primarily to the sensory processing/perception aspect of a noological system (Fig. 1.7). In the current book, we have not discussed any computational mechanisms in connection with perception. Deep learning may well fill this gap. Therefore, in this sense the coverage of our deep thinking and quick learning paradigm is complementary to the deep learning paradigm.

However, the currently available deep learning mechanisms are relatively slow. Despite the fact that the learning of the hierarchical feature detector aspect corresponds well to what is expected of a noologically realistic system, the speed of learning and the large amount of training examples aspect is not noologically realistic (e.g., Le et al. 2012). If this aspect of deep learning can be improved, it would serve as a noologistically realistic computational model. Future investigations could perhaps attempt to combine the causal learning method of Chap. 2 with the current deep learning methods to realize this.

10.7 A Note on Biology

As mentioned in the discussion in connection with Fig. 1.1 in Chap. 1, the paradigm developed in this book is potentially applicable to the understanding of cellular systems. As noted by some biological researchers (Albrecht-Buehler 2013; Ford 2009; Hameroff 1987), cells are intelligent systems in themselves, even though this is currently not a mainstream view. As suggested by Hameroff (1987), a cytoskeletal component, the microtubules, may be able to perform computation, and hence they constitute the "nervous systems" of the cell. The reason why something like this is necessary is that a single cell alone, such as the amoeba or the white blood cell, is able to exhibit behavior that is usually thought only possible in biological entities that have a nervous system. These single cells exhibit the behavior of goal-directed locomotion – e.g., the amoeba and the white blood cell could "actuate" different parts of their cell membrane, making them behave like limbs, and propel themselves in various directions, and usually with a goal in "mind," which is to pursue/home in to food (typically bacteria) and consume them. However, to understand what is taking place exactly in the cytoskeleton as well as the nucleus of a cell is no mean task: despite an incredible amount of research (in terms of millions of papers published) on the cell that took place since Hameroff's (1987) proposal of the possibility of the role of cytoskeleton in computation, biologists are still struggling to understand the functions of a bewildering array of components in the cytoskeleton and the nucleus (see, for example, Alberts et al. 2015, and a host of research papers in this field). The exact functions of and interactions between most of these components still elude biologists.

Therefore, currently and in the near future, mapping the noological paradigm articulated in this book to biological systems is still some way off, even though the noological angle could definitely inspire and help shape certain directions of research to further understand these systems.

However, we would like to make a note about an aspect of biological systems that could immediately benefit from the insights gained in our formulation of noological principles. Biological systems, whether they are single cellular or multi-cellular systems, are not static in their structures and designs. Over long time frames, evolutionary forces shape and modify these organisms, transforming them based on some goals, and these goals are typically identified to be "survival advantages." (Darwin 1859, and the subsequent evolutionary biology that he inspired). Therefore, evolution itself is an intelligent process, or at least the outcome of evolution appears to be the results of intelligent processes. Now, there is an interesting twist here. The current mainstream view of evolution is the Darwinian view (one other alternative being the Lamarckian view (Lamarck 1830)), which is mutation plus natural selection, which is a random, "unintelligent" search process even though the outcome is seemingly the results of intelligent actions. This scenario is also present in AI systems, in which blind, unintelligent search in problem solving can produce results that appear to be the work of intelligent cogitation and actions (Russell and Norvig 2010). The major drawback of blind, random search is the impossibly long time needed for complex problems. Now, as expounded in this book, a causal learning augmented search process can greatly speed up problem solving. Might evolution not have benefited from such a process? As articulated in our causal learning paradigm, the problem solving process initially begins with a modicum amount of random search (Chap. 2). Then, causal rules are quickly learned and encoded based on what is learned in this initial search process, and these guide further problem solving processes in an "intelligent" manner, resulting in these processes being vastly accelerated. This would be a compromised view between the extreme creationist view of the emergence of intelligent organisms requiring a super-being, and the purely random search of Darwinian evolution which requires an inordinate amount of time for any intelligent organisms to emerge – much like the scenario of the infinite monkey theorem in which it has been proven that getting a monkey to type randomly at a typewriter and given an infinite amount of time, the entire works of William Shakespeare could be created (Hoffmann and Hofmann 2001).

As has been investigated for some time, there is a methodology in AI called genetic algorithm which simulates the process of "random mutation plus selection" of biological evolution that has been applied to solve various problems in computation (Holland 1975; Banzhaf et al. 1998). Recently, there were some attempts to speed up genetic algorithms by adding a supervised learning component (e.g., Wang and Chang 2011). This supervised learning portion is of course "intelligently added" by the human researchers. In the absence of a super-being, evolution of course could not have proceeded in exactly the same manner. Most likely, the process is initially more akin to a random search. The "monkey" does not have to proceed from nothing to William Shakespeare in one giant step. It could, perhaps, go to school and be educated first. Much like our causal learning augmented search process, this random search process allows the system to encode certain useful causal rules, and then these rules are applied intelligently to obviate subsequent random search. To enable causal learning (in the same vein as building in the supervised learning mechanisms to improve genetic algorithms as mentioned above), some learning mechanisms must be present first, much like what we have described in this book. Presumably, these mechanisms could have arisen also through some random process over an initially relatively long period of time, then subsequently intelligent evolutionary learning would proceed at a much accelerated rate. Might evolution have proceeded along such a path? Evolution is certainly a "learning and problem solving" process, and in this view, an intelligent one too. There is still so much that is unknown about cellular mechanisms and there has been so many surprising discoveries such as discoveries that run counter to the original "central dogma of molecular biology" (Alberts et al. 2015) that this idea cannot be rejected at the outset and should be entertained. This view does not require a super-being and at the same time it does not require impossibly long random search times for evolution to achieve what it has achieved over roughly 3 billion years on Earth. Recently discovered Lamarckian-like epigenetic processes could further contribute to this process of intelligent evolution (Carey 2013). Someday, perhaps, biology could be subsumed under noology.

References

Alberts, B., Johnson, A., Lewis, J., Morgan, D., Raff, M., Roberts, K., & Walter, P. (2015). *Molecular biology of the cell* (6th ed.). New York: Garland Science, Taylor & Francis Group, LLC.

Albrecht-Buehler, G. (2013). *Cell intelligence*. http://www.basic.northwestern.edu/g-buehler/FRAME.HTM

Banzhaf, W., Nordin, P., Keller, R., & Francone, F. D. (1998). *Genetic programming: An introduction* (1st ed.). San Francisco: Morgan Kaufmann.

Barbey, A. K., Krueger, F., & Grafman, J. (2008). Structured event complexes in the medial prefrontal cortex support counterfactual representations for future planning. *Philosophical Transactions of the Royal Society Series B, 364*, 1291–1300.

Carey, N. (2013). *The epigenetics revolution: How modern biology is rewriting our understanding of genetics, disease, and inheritance.* New York: Columbia University Press.

Chambers, N., & Jurafsky, D. (2008). Unsupervised learning of narrative event chains. In *Proceedings of the annual meeting of the Association for Computational Linguistics: Human language technologies,* Columbus, Ohio (pp. 789–797). Madison: Omni Press.

Darwin, C. R. (1859). *On the origin of species by means of natural selection, or the preservation of favoured races in the struggle for life.* London: John Murray.

Deng, L., & Yu, D. (2014). *Deep learning methods and applications.* Hanover: Now Publishers.

Fahlman, S. E. (1979). *NETL, a system for representing and using real-world knowledge.* Cambridge, MA: MIT Press.

Ferricci, D., Brown, E., Chu-Carroll, J., Fan, J., Gondek, D., Kalyanpur, A. A., Lally, A., Murdock, W., Nyberg, E., Prager, J., Schlaefer, N., & Welty, C. (2010). Building Watson: An overview of the DeepQA project. *AI Magazine, 31*(3), 59–79.

Ford, B. J. (2009). On intelligence in cells: The case for whole cell biology. *Interdisciplinary Science Reviews, 34*(4), 350–365.

Forsyth, D. A., & Ponce, J. (2011). *Computer vision: A modern approach* (2nd ed.). Englewood Cliffs: Prentice Hall.

Fukushima, K. (1980). Neocognitron: A self-organizing neural network model for a mechanisms of pattern recognition unaffected by shift in position. *Biological Cybernetics, 36*, 193–202.

Greenspan, J. (2013). Coyotes in the crosswalks? Fuggedaboutit! *Scientific American, 309*(4), 17. New York: Scientific American.

Hameroff, S. R. (1987). *Ultimate computing: Biomolecular consciousness and nanotechnology.* Amsterdam: Elsevier Science Publishers B.V.

Hinton, G. E., McClelland, J. L., & Rumelhart, D. E. (1986). Distributed representations. In D. E. Rumelhart, J. L. McClelland, & PDP Research Group (Eds.), *Parallel distributed processing: Explorations in the microstructure of cognition* (Vol. 1). Cambridge, MA: MIT Press.

Hoffmann, U., & Hofmann, J. (2001). *Monkeys, typewriters and networks.* Berlin: Wissenschaftszentrum Berlin für Sozialforschung gGmbH (WZB).

Holland, J. H. (1975). *Adaptation in natural and artificial systems.* Ann Arbor: University of Michigan Press.

Hsu, F.-H. (2002). *Behind deep blue: Building the computer that defeated the world chess champion.* Princeton: Princeton University Press.

Hubel, D. H., & Wiesel, T. N. (1962). Receptive fields, binocular interaction and functional architecture in the cat's visual cortex. *Journal of Physiology, 160*, 106–154.

Hubel, D. H., & Wiesel, T. N. (1965). Receptive fields and functional architecture in two non-striate visual areas (18 and 19) of the cat. *Journal of Neurophysiology, 28*, 229–289.

Krueger, F., & Grafman, J. (2008). The human prefrontal cortex stores structured event complexes. In T. F. Shapley & J. M. Zacks (Eds.), *Understanding events: From perception to action.* Oxford: Oxford University Press.

Lamarck, J. B. (1830). *Philosophie Zoologique.* Paris: Germer Baillière.

Le, Q. V., Ranzato, M. A., Monga, R., Devin, M., Chen, K., Corrado, G. S., Dean, J., & Ng, A. Y. (2012). Building high-level features using large scale unsupervised learning. In *Proceedings of the 29th international conference on machine learning,* Edinburgh, Scotland, UK (pp. 81–88). Madison: Omnipress.

LeCun, Y., Bengio, Y., & Hinton, G. E. (2015). Deep learning. *Nature, 521*, 436–444.

Lin, H. (2014). *Sharing the positive or the negative? Understanding the context, motivation and consequence of emotional disclosure on facebook.* Ph.D. thesis, Nanyang Technological University, Singapore.

Manshadi, M., Swanson, R., Gordon, A. S. (2008). Learning a probabilistic model of event sequences from internet weblog stories. In *Proceedings of the 21st FLAIRS conference*, Coconut Grove, Florida (pp. 159–164). Menlo Park: AAAI Press.

Maslow, A. H. (1954). *Motivation and personality*. New York: Harper & Row.

Minsky, M. (1992). Future of AI technology. *Toshiba Review, 47*(7). http://web.media.mit.edu/~minsky/papers/CausalDiversity.txt.

Minsky, M., & Papert, S. (1969). *Perceptrons: An introduction to computational geometry*. Cambridge, MA: MIT Press.

Mountcastle, V. B. (1998). *Perceptual neuroscience: The cerebral cortex*. Cambridge, MA: Harvard University Press.

Nilsson, N. J. (1982). *Principles of artificial intelligence*. Los Altos: Morgan Kaufmann.

Plutchik, R. (2002). *Emotion and life*. Washington, DC: American Psychological Association.

Reeve, J. (2009). *Understanding motivation and emotion*. Hoboken: Wiley.

Regneri, M., Koller, A., & Pinkal, M. (2010). Learning script knowledge with Web experiments. In *Proceedings of the 48th annual meeting of the Association for Computational Linguistics*, Uppsala, Sweden (pp. 979–988). Stroudsburg: Association for Computational Linguistics.

Rosenblatt, F. (1962). *Principles of neurodynamics: Perceptrons and the theory of brain mechanisms*. New York: Spartan.

Rumelhart, D. E., Hinton, G. E., & Williams, R. J. (1986a). Learning internal representations by error propagation. In D. E. Rumelhart, J. L. McClelland, & the PDP Research Group (Eds.), *Parallel distributed processing: Explorations in the microstructure of cognition* (Vol. 1). Cambridge, MA: MIT Press.

Rumelhart, D. E., McClelland, J. L., & PDP Research Group. (1986b). *Parallel distributed processing: Exploration in the microstructure of cognition* (Vol. 1 & 2). Cambridge, MA: MIT Press.

Russell, S., & Norvig, P. (2010). *Artificial intelligence: A modern approach*. Upper Saddle River: Prentice Hall.

Sagiv, L., Schwartz, S. H., & Knafo, A. (2002). The big five personality factors and personal values. *Personality and Social Psychology Bulletin, 28*, 789–801.

Schank, R. C. (1973). Identification of conceptualization underlying natural language. In R. C. Schank & K. M. Colby (Eds.), *Computer models of thought and language*. San Francisco: W. H. Freeman and Company.

Schank, R., & Abelson, R. (1977). *Scripts, plans, goals and understanding*. Hillsdale: Lawrence Erlbaum Associates.

Shapiro, L. G., & Stockman, G. C. (2001). *Computer vision*. Upper Saddle River: Prentice Hall.

Sutton, R. S., & Barto, A. G. (1998). *Reinforcement learning: An introduction*. Cambridge, MA: MIT Press.

Szeliski, R. (2010). *Computer vision: Algorithms and applications*. Berlin: Springer.

Tan, A.-H., Lu, N., & Xiao, D. (2008). Integrating temporal difference methods and self-organizing neural networks for reinforcement learning with delayed evaluative feedback. *IEEE Transactions on Neural Networks, 19*(2), 230–244.

Tu, K., Meng, M., Lee, M. W., Choe, T. E., & Zhu, S.-C. (2014). Joint video and text parsing for understanding events and answering queries. *IEEE MultiMedia, 21*(2), 42–70.

Uhr, L., & Vossler, C. (1981). A pattern-recognition program that generates, evaluates, and adjusts its own operators. In E. A. Feigenbaum & J. Feldman (Eds.), *Computers and thought*. Malabar: Robert E. Krieger Publishing Company.

Wang, Z., & Chang, C. S. (2011). Supervisory evolutionary optimization strategy for adaptive maintenance schedules. In *Proceedings of the IEEE symposium on industrial electronics*, Gdansk, Poland (pp. 1137–1142). Piscataway: IEEE Press.

Wang, Y., & Mori, G. (2009). Human action recognition by semi-latent topic models. *IEEE Transactions on Pattern Analysis and Machines Intelligence, 31*(10), 1762–1774.

Werbos, P. J. (1974). *Beyond regression: New tools for prediction and analysis in the behavioral sciences*. Ph.D. thesis, Harvard University.

Winograd, T. (1973). A procedural model of language understanding. In R. C. Schank & K. M. Colby (Eds.), *Computer models of thought and language*. San Francisco: W. H. Freeman and Company.

Wood, J. N., & Grafman, J. (2003). Human prefrontal cortex: Processing and representational perspectives. *Nature Reviews Neuroscience, 4*, 139–147.

Wood, J. N., Tierney, M., Bidwell, L. A., & Grafman, J. (2005). Neural correlates of script event knowledge: A neuropsychological study following prefrontal injury. *Cortex, 41*(6), 796–804.

Yuan, J., Liu, Z., & Wu, Y. (2011). Discriminative video pattern search for efficient action detection. *IEEE Transactions on Pattern Analysis and Machines Intelligence, 33*(9), 1728–1743.

Appendices

Appendix A: Causal vs Reinforcement Learning

In this appendix we employ a simple example similar to that used in Sect. 1.2 of Chap. 1 to contrast the differences between pure reinforcement learning (Sutton and Barto 1998) and the rapid effective causal learning mechanisms motivated in the discussion in Sect. 1.2 and that is explained in detail in Chap. 2. In the paradigm described in this book, effective causal learning is a critical learning mechanism subserving all levels of noological processing.

Figure A.1 shows a simple "nano-world" consisting of 11 squares. There is an Agent and a piece of Food at some locations. The Agent has a choice of moving either to the right (R) or left (L) starting from any square. Below it is shown a typical search process produced by reinforcement learning. The circles represent the "states of the world," and in this case it would consist of the locations of the Agent and the Food. We also stipulate here that when the Agent is "touching" the Food, i.e., it is one square next to the Food, it is rewarded (much like in Fig. 1.5a). Each time a reward signal is generated, there is some algorithm (e.g., Q-learning, Sutton and Barto 1998) that will strengthen the weight associated with the action that results in the reward, and that signal is also propagated backward toward the starting state so that the Agent learns the entire sequence of correct actions leading to the reward (in this case, two consecutive rightward movements). The algorithm typically requires many cycles of weight updating as each time the weight associated with the action in the "reward direction" is only modified slightly.

The basic problem with pure reinforcement learning is, there is no generalization involved in the learning process. After having learned the correct sequence of actions to reach the Food on the right side of the nano-world, suppose in a new situation, the Food appears on the left side instead such as that shown in Fig. A.2a. The entire process has to be repeated to find the right sequence of actions to the Food on the left (now consisting of two leftward movements), even though it would seem commonsensical that if the Food is the cause of the reward and if it now

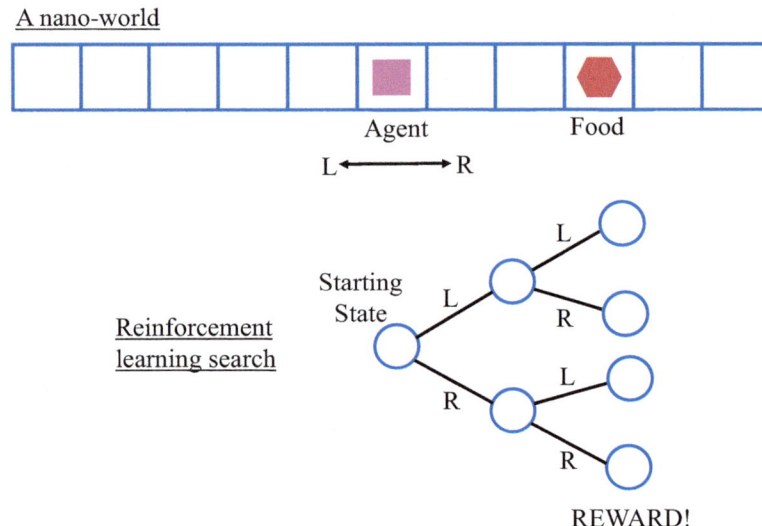

Fig. A.1 A nano-world with an Agent and a piece of Food, and the attendant reinforcement learning process

appears on the left side, one needs to just move in the left direction accordingly to claim the reward.

Another situation in which a seemingly simple generalization process would allow the transfer of learning from the situation in Fig. A.1 immediately to bear on the problem is shown in Fig. A.2b. In Fig. A.2b, the Food is shown farther way from the Agent on the right side. Again, an entire, and now more extensive, reinforcement search process has to be carried out to find the correct action sequence to reach the Food. If the nano-world were to just expand further with the Food placed at a yet farther location from the Agent, the process would quickly become combinatorial and unmanageable.

In a noologically realistic scenario, as discussed in Sect. 1.2 of Chap. 1, the Agent would be equipped with an Eye to identify the shapes of potential causal agents that it interacts with, as shown in Fig. A.3, and it learns causality and generalizes over parameters that are irrelevant. Much like that discussed in connection with Figs. 1.4 and 1.5 in Chap. 1, as soon as the first relatively simple scenario of learning takes place as in Fig. A.1, in which the Food is relatively nearby, the Agent would learn that the touching of the Food is the cause of its internal energy increase, and that the Food has a certain discernable shape. The Agent also learns about the nature of movement – that if one needs to reach some place, the most energy conserving way is to keep moving in the same direction toward it (through causal learning, as discussed in Sect. 3.1, Chap. 3). Some generalizations over some parameters are needed, as it might at first think that the Food must be situated at a certain location for it to be efficacious in supplying

Fig. A.2 (**a**) Food has been moved from the right to the left side. (**b**) Food is placed farther away from the Agent on the right side

energy, such as discussed in Sect. 1.2. After that, whether the Food is on the left side or very far away, it would just head straight toward it, exhibiting intelligence and common sense.

The fundamental problem with pure reinforcement learning is that it simply learns a sequence of actions blindly. These actions lead to certain rewards, but it does not learn the *causes* of the rewards. In noological systems, it should be causes that are learned, and that provides the maximum power of generalization. In this book, notably in Chap. 2, we present a general causal learning paradigm that is applicable to general situations.

Appendix B: Rapid Effective Causal Learning Algorithm

In Chap. 2 we described a process for the identification of diachronic and synchronic causes to achieve rapid unsupervised causal learning. The learning algorithm is given as follows:

Definitions of Terms

(i) An Event is defined as a change of state of an (O)bject. An (E)vent k at (T)ime m means the Object is in an old state at Time m-1 and is now in a new state at Time m.
(ii) E(k)T(m) = An Event k that happens at Time m.
(iii) [E(k)T(m)I(n)E(p)] = Event k happens at Time m followed by Event p after (I)nterval n. (A temporal correlational/causal rule)
(iv) EO(x)T(m) = Event or Object x present at Time m.
(v) [{EO(x)T(r)...}E(k)T(m)I(n)E(p)] : {EO(x)T(r)...} are synchronic causes for [E(k)T(m)I(n)E(p)]
(vi) EOA = a synchronic AND cause – a necessary cause.

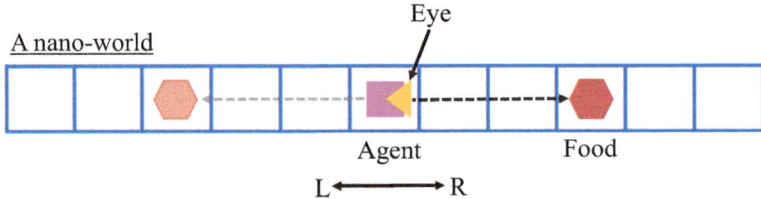

Fig. A.3 The Agent is equipped with sensory organs and it carries out causal reasoning to behave intelligently and commonsensically

 (vii) EOO = a synchronic OR cause.
(viii) EOTO = a (t)entative synchronic OR cause.

TRL = Tentative_Rule_List = nil
PRL = Possible_Rule_List = nil
CRL = Confirmed_Rule_List = nil

  Begin temporal observation (observation across time)
    For each time T(q),
      If there is an event E(p) at T(q), consider all events E(k) at T(m)
      where T(m) < T(q),
        Add all [{EOA(x)T(r)...}E(k)T(m)I(n)E(p)]
          where m <= r <= q, to TRL;
        If PRL = nil
          then PRL = TRL;
        else
          for each [{EOA(x)T(r)...}E(k)T(m)I(n)E(p)] on TRL
            if there exists [{EOA(x1)T(r1)...}E(k1)T(m1)I(n1)E(p1) on PRL
            such that E(k) = E(k1), T(m) <> T(m1), I(n) = I(n1), E(p) = E(p1)
              then Compose {Precondition} as follows:
                Retain all EOA(x) = EOA(x1) in {Precondition} as EOA(x);
                Convert all EOA(y) <> EOA(y1) to EOTO(y) and EOTO(y1)
                  in {Precondition};
                Put [{EOA(x)...EOTO(y)...EOTO(y1)...} E(k)T(m)I(n)E(p)]
                  on CRL;
                Remove [{EOA(x)T(r)...}E(k)T(m)I(n)E(p)] from TRL;
                Remove [{EOA(x1)T(r1)...}E(k1)T(m1)I(n1)E(p1) from PRL;
            if there exists [{EOA(x2)T(r2)...}E(k2)T(m2)I(n2)E(p2) on CRL
            such that E(k) = E(k2), T(m) <> T(m2), I(n) = I(n2), E(p) = E(p2)
              then Compose {Precondition} as follows:
                Retain all EOA(x) = EOA(x2) in {Precondition}as EOA(x);
                Convert all EOA(y) <> EOA(y2) to EOTO(y) and EOTO(y2)
                  in {Precondition};
                Put [{EOA(x)...EOTO(y)...EOTO(y2)...}E(k)T(m)I(n)E(p)]

on CRL;
Remove [{EOA(x)T(r)...}E(k)T(m)I(n)E(p)] from TRL;
PRL = PRL + TRL;
End (for)
End (If)
For each [{EOA(x)...EOTO(y)...EOTO(y1)}E(k)T(m)I(n)E(p)] on CRL;
Check if there exists an earlier {EOA(x)...}E(k) that does not give rise to an E(p) after I(n);
If so, convert all EOTO(y) to EOO(y) in the precondition of that rule as follows:
[{EOA(x)...EOO(y)...EOO(y1)...} E(k)T(m)I(n)E(p)];
End (For each time T(q))

End (Begin)

Index

A

A*
  algorithm, 61, 63, 112, 115, 238–240, 252, 281
  search, 63, 66, 67, 80, 90, 238, 281, 356
AAIA. *See* Adaptive autonomous intelligent agents (AAIAs)
AAIS. *See* Adaptive autonomous intelligent systems (AAISs)
ABA. *See* Anxiousness of Being Attacked (ABA)
Accelerated problem solving, 4, 5, 37, 394
Action at a distance, 74, 179, 182
Action planning, 6, 284, 359
Adaptive autonomous intelligent agents (AAIAs), 2, 7, 34, 130, 136, 137, 343, 361, 364, 378, 384, 385
Adaptive autonomous intelligent systems (AAISs), 2–4, 6, 8
Adaptive system, 3, 5, 83, 208, 406
Adult phase problems, 117
AEMCW-SCRIPT. *See* Allowable–elemental–movement–at–contact–with–Wall–script (AEMCW-SCRIPT)
Affect, 3, 4, 6, 89, 91, 103, 179, 191, 236, 293, 295, 296, 320, 322, 329, 364
Affect driven noological processes, 359–389
Affective competition, 100, 103, 105, 122, 284, 300, 313–318, 353, 359, 365, 402
Affective learning, 25–26, 103–104, 141
Affective processes, 4, 6, 8, 36, 60, 128, 129, 284, 342, 349, 393
Affective state, 3, 13, 25, 201, 345, 348–350, 388, 402
Affordance, 13, 14, 23, 24
Affordance chain, 23–25
After, concept of, 185
Agent Moving problem, 113, 114, 118
AGI. *See* Artificial General Intelligence (AGI)
AI. *See* Artificial Intelligence (AI)
ALH. *See* Anxiousness of Low HP (ALH)
Allowable–elemental–movement–at–contact–with–Wall–script (AEMCW-SCRIPT), 250–252, 261–266, 276
Amoeba, 414
Analogical reasoning, 408
Analogical representation, 71, 349
Anger, 79, 119, 356, 402
Anti-goal, 104–107
Anxiousness, 13–15, 25, 32, 34, 101, 102, 104, 119, 128, 172, 187, 313–317, 322, 328, 331, 345–350, 352, 364–367, 373, 374, 383, 385, 388, 402
  competition, 99–104
Anxiousness of Being Attacked (ABA), 102, 103, 316–318
Anxiousness of Low HP (ALH), 101–103, 316–318
Anxiousness of Projectile-Causing-Pain (APP), 364, 365, 367
Appear, concept of, 159
Arrive, concept of, 185–187, 217
Artificial General Intelligence (AGI), 54, 56, 64, 120, 122, 137
Artificial Intelligence (AI), 1–5, 7, 8, 18, 26, 29, 34–36, 41, 42, 61, 62, 79, 81, 86, 90, 95, 118, 147, 148, 213, 215–218, 238, 252, 280, 281, 283, 285, 356, 391–393
Atomic movement operators, 167–171
Atomic operational representation, 153, 156, 158–160, 163, 164, 171–172, 393

Atomic operators, 156, 157, 159, 160, 162, 163, 165–167, 171, 172, 177, 183, 184, 213
Attach, concept of, 180–182
Attach-link, 180, 181
ATTACK-SCRIPT, 304–310, 312, 340, 341
Attention, 6, 18, 55, 125, 359
Attentional process, 15, 129, 349
Autonomous system, 22, 25, 401
Avoidance of anti-goal, 79–80

B

Backward chaining, 11, 13, 24, 73, 74, 96, 119, 120, 214, 242, 243, 256, 274, 301, 302, 306, 309, 311, 340
Backward reasoning, 16, 73, 192, 201–205, 209, 297
Balans chair, 29
Barriers, 211–213, 271
Basal ganglia, 19, 125–138
Basal ganglia loop, 127–129
Basal-ganglionic-cortical loops, 125–128
Basic noological principles, 4, 36–37, 393
Battle Difficulty measure, 330–333
Battle strategies, 300–313, 336–340
Bayesian causal inference, 42–44
Biology, 1, 2, 413
Bio-noo boundary, 4, 20–24, 36, 90, 91, 151, 172, 393, 400, 408
Blind search, 9, 42, 62, 63, 108, 239, 313, 399
Blinking, concept of, 160
Broom, 147
Built-in need, 16, 17, 21, 90, 172

C

Caudate, 126
Causal knowledge, 4, 217, 222
Causal learning, 18
  augmented problem solving, 61–82, 90
  augmented search, 42, 63, 64, 67, 71, 80–82, 238, 414, 415
  of battle strategies, 301–313
  filter, 307
Causal role of sensory information, 221–282
Causal rule, 6, 11, 13, 15, 19, 46, 48, 52, 53, 55–57, 59, 82, 98–100, 107–109, 111, 112, 119, 120, 137, 141, 145, 153, 162, 187, 191–217, 226–230, 234, 235, 237, 255, 287, 289, 299, 301–304, 307, 317, 320, 326, 392, 403, 405
Cause of negative outcome, 368–383

Cause recovery, 41, 42, 46, 47, 138, 139
Cell as intelligent machines, 1, 413, 414
Cell as machines, 1, 413, 414
Cell intelligence, 1, 413, 414
Cerebellar-cortical loops, 129
Cerebellum, 19, 125, 129–137
Cerebral cortex, 18, 82, 125, 126, 128–135, 137–139
CFI. *See* Counterfactual information (CFI)
Changing goal, 104–107
Chunked Final Solution, 85
Chunked Intermediate Solution, 85
Chunked rule, 35, 109, 111–115, 117, 118, 137, 141, 207, 208
Chunked Rule Base, 85, 109, 112, 115, 119, 123, 137
Chunking. *See* Knowledge chunking
Cognition, 3, 22
Cognitive hierarchy construction, 163
Cognitive linguistics, 28, 146, 185–187, 217
Cognitive system, 3, 26, 140, 154, 164, 391
Comfort, state of, 150, 151, 345, 350, 352, 360
Commander anxiousness, 318
Complex external environment, 394
Complex internal environment, 401, 402
Computational power of recurrent neural networks, 131–134
Computer simulation, 103, 111, 113, 114, 116, 121, 122, 317
  of incremental chunking, 111–118
Conception, 16, 29–32, 146, 164, 173
Conception as service function, 29
Conceptual dependency (CD), 403–405
Conceptual grounding, 145–188
Conceptualization, 16, 18, 128, 146, 154, 174, 349, 392
Conceptual process, 4–6, 15, 16, 18, 36, 123, 284, 342, 344, 349, 394
Conjunctive diachronic causal condition, 59
Conjunctive diachronic condition, 58–60
Connectionism, 409
Consciousness, 172, 406
Continuous-Trajectory, concept of, 165, 166
Cookie cutter, 183, 184
Correlation graphs, 291–298
Corticostriatal loops, 125, 126
Counterfactual information (CFI), 5, 6, 36, 68, 76, 78, 85, 92–97, 249–251, 258, 269, 270, 292–299, 319–324, 326, 327, 333–336, 371, 393
Counterfactual script
  construction, 77, 85, 94, 269, 284
  query, 78, 95

Index   427

Counter Thwarting Script Discovery Process, 367
Counter-thwarting script, 244, 252, 253, 255, 259–262, 266, 267, 272, 366
Counter−ve OUTCOME script, 366–368, 373, 374, 376
Crawling robot problem, 86, 87
Culture, 57, 408
Cytoskeleton, 414

D

Darwinian evolution, 415
Darwinian view, 414
Declarative procedures, 403
Declarative representation
Deep Blue, 410
Deep learning, 281, 356, 393, 413
Deep thinking, 280–281, 342, 393, 413
Dematerialization, concept of, 176, 194
Depart, concept of, 159, 161
Desperation, 3, 59–61, 66, 71, 229, 232, 237, 281, 299–300, 324, 365, 399
Detach, concept of, 180, 198
Diachronic causal condition, 52, 53, 55
Diachronic causal factor, 52
Disappear, concept of, 159, 160
Distributed processing modules (DPM), 129–131, 135, 136
Dual instance generalization, 59, 66, 71, 228–231, 233, 255, 264, 269, 307, 362, 403

E

Eagerness, 60
ED. *See* Enemy Distance (ED)
Effective causality, 44–48, 50–52
Effective causal learning, 10–12, 16, 41–86, 99, 226, 284, 300, 411
Effort optimization, 90–97
Elastic object, 196–198
Electrically charged object, 182, 235, 237
Elemental action, 70, 75–79, 108, 111, 112, 246, 253, 266, 268, 270–274, 367, 374
ELEMENTAL-MOVEMENT-TO-AVOID-WALL-ANTI- GOAL-SCRIPT (EMAWAG-SCRIPT), 258, 370–372
Elemental representation, 155, 163
Emotion, 3, 4, 34, 60, 158, 367, 388, 399
Enemy Distance (ED), 102–104, 316

Energy
  depletion, 9, 97–99, 107, 120
  goal, 120–122
  need, 12–15
Engage/Attack enemy, 284, 300–314, 340
Epigenetic process, 415
Episodic memory, 15, 18, 157, 342, 347, 349
Esteem, 401
Evolution, 21
Evolutionary learning, 415
Evolution as intelligent problem solving, 20, 21, 90, 414
Exhaustiveness of experiment, 299–300, 324
Exhaustiveness of observation, 299–300
EX-IN causal rule, 119–122, 202
Existential atomic operator, 158–160
Experiential memory, 157, 159, 163, 165
Explicit representation of mental process, 402
Explicit temporal representation, 145, 163, 217
External ground, 16, 23

F

Fan, 147, 213
Fear, 101, 345, 402
Fluid, 153, 154, 162, 164, 187, 211–213
Food, 8, 9, 11–16, 20, 21, 23, 25, 32, 33, 36, 59, 82, 90, 91, 96, 97, 104, 119–123, 315, 316, 353, 354, 367, 396, 397, 399, 414
Force, concept of, 192–196, 291–297
FORCE-MOVEMENT-SCRIPT, 292–294, 297, 305, 361
Forward reasoning, 200, 204, 297
Forward search, 61, 83, 84, 86, 89, 107, 123, 330, 391
Forward search framework. *See* Noological forward search framework
Function, 3, 5, 6, 13, 15, 19, 20, 22, 29, 31, 34, 37, 60, 83, 90, 91, 122, 124–132, 134, 136–139, 141, 147, 184, 214, 236, 312, 316, 318, 330, 344, 349, 352, 353, 364, 388, 391, 392, 394
Functional definition, 29, 30, 32
Functional Reasoner, 30
Functional reasoning, 30, 31
Function of the prefrontal cortex, 140–141

G

General brain architecture, 125, 134–136
General causal rule, 13, 15, 53, 66, 123, 237, 239, 412
Globus pallidus, 126

Goal competition, 25–26
Goal formation, 6, 284, 349
Go, concept of, 26, 148
Gravity, 51, 52, 56, 57, 213, 229, 230, 235
Grounding, 6, 82
Grounding concept, 147
Ground level knowledge representation, 146–151
Ground level symbolic representations, 28, 150

H
HCRT. *See* HP Charging Response Threshold (HCRT)
Health Point (HP), 99–104, 106, 285, 288–291, 302–309, 312–322, 325, 328–331, 342
Heuristics
 filter, 84, 330
 generalization, 234–237, 329–330, 332
 learning, 35, 300, 324–330 (*see also* Learning of heuristics)
 rule base, 85, 123
Hierarchical Bayesian inference, 131
Higher order counterfactual information, 333–336
Houk distributed processing modules (DPM), 129–131
Houk module, 137
HP. *See* Health Point (HP)
HP Charging Response Threshold (HCRT), 101–104, 316, 317
Human needs, 22
Hunger
 alleviation, 24, 225
 strike, 22, 90, 91

I
IBM Watson System, 148
Immobile object, 176–179, 196–198, 201, 204, 209, 210, 246
INCREASE-HP-SCRIPT, 312, 314
Incremental chunking, 68, 108, 111–118, 205–208, 210, 218, 241, 312, 313, 350, 398
Incremental knowledge chunking, 89, 107–118, 122–124, 141, 311
Inductive competition for rule generalization, 230–234
Inductive heuristic generalization, 234–237
Inelastic object, 178, 179, 196–198, 208
Initial Infant Learning Phase, 111
Intelligence, definition, 6, 392

Intelligent evolution, 415
Intelligent evolutionary learning, 415
Internal ground(ing), 6, 16, 20–24, 26–29, 90, 151, 408
Internal need, 8–10, 12, 14, 16, 18, 20, 23, 24, 31, 35, 91, 92, 94, 99, 344, 349, 352, 353, 399

J
Jeopardy, game of, 148, 410

K
Knowledge chunking, 5, 14, 16, 35, 36, 83, 86, 107, 112, 117, 118, 284, 300, 341, 393

L
Lamarckian evolution, 414
Lamarckian-like epigenetic process, 415
Lamarckian view, 414
Lamprey, 132, 133
Language, 4, 125, 129, 131, 148, 186, 187, 213–215, 217, 281, 284, 341, 389, 413
Learning of heuristics, 5, 14, 16, 37, 61–83, 86, 218, 284, 300, 329, 341, 391, 394. *See also* Heuristics learning
Learning of scripts, 5, 36, 68, 83, 284, 341, 393
Learning through language, 284, 340, 412
Learning to learn, 4, 5
Learning to solve problem through language, 187, 213–215, 284, 340–341
Lightning, 50, 51
Locus of memory, 139–140
Loose-Appear, concept of, 158, 159
Loose-Disappear, concept of, 158, 159

M
Make-Appear, concept of, 159
Markov chain, 216
Maslow hierarchy, 21, 22, 25, 123, 399, 401, 402
Materialization, concept of, 175, 192
Maximum, concept of, 171, 172
Maximum distance heuristic, 79–80, 258, 259, 371
Memory, 20, 75, 91, 104, 134, 140, 145, 157, 224, 359, 404, 410
Memory process, 6, 128, 349
Mental process, 158, 277, 342, 394, 402, 403, 406

Index 429

Mental simulation, 200, 201, 260, 261, 266–269, 274–280, 297, 336–338, 363, 365–367, 370, 371, 373–376, 378, 382, 385, 386
Meta-level cause, 405
Meta-level heuristic rule, 236, 237, 359
Meta-level inductive heuristic generalization, 234–237
Meta-level information, 85
Meta-level script, 333
Micro-environment, 33–36, 107, 108, 130, 136, 344, 356, 359, 361, 364, 378, 384, 385, 394. *See also* Micro-world
Microtubule, 414
Microtubule computation, 414
Micro-world, 34, 35. *See also* Micro-environment
Minimum, concept of, 171, 172
Minimum distance heuristic, 65, 66, 76, 79–81, 85, 90
Mobile object, 176, 196, 197, 201, 202, 204, 205, 208, 209
Modes of Learning, 319–320
Momentum Cancellation, 195, 196
Momentum, concept of, 175
Momentum Continuation, 194, 195, 210–212
Monkey-and-bananas problem, 218
Most smile heuristic, 81
Motivation(al)
    competition, 25–26, 89, 123
    learning, 119–122, 136, 137
    process, 4, 6–8, 15, 36, 129, 137, 284, 342, 349, 393
Move, concept of, 161
Movement, concept of, 160–162
Movement-related atomic operators, 160–162
MOVEMENT-SCRIPT, 69, 70, 72, 73, 76, 85, 92–97, 105, 121, 241, 242, 252–256, 260, 268, 269, 272, 277, 286–288, 340, 341
MOVEMENT-TO-ATTACK-SCRIPT, 310

N
Naïve physics, 216
Nano-world, 8, 9, 12
Need, 16, 17, 21, 172
    competition, 97, 99–104
    satisfaction, 89–97
Negative outcome, cause of, 368–383
Neuroscience, 1, 5, 7, 18, 89, 124–141, 215
Neuroticism, 386–388
Non super-visual situation, 279–280

Noological efficacy, 5, 75, 141, 393, 398
Noological forward search framework, 83–86, 89, 109
Noologically efficacious unit, 5, 6, 37, 140, 393
Noologically realistic solutions, 118
Noological manipulatability and recomposability, 164
Noological principles, 4, 36–37, 92, 283, 393
Noological process, 4, 5, 18, 32, 136, 138, 139, 152–155, 162, 168, 170, 225, 252, 342, 348, 356, 359–389, 391–393
    architecture, 8–19, 260
    framework, 42, 83, 122–124, 140, 399
Noological realism, 391, 392, 411, 413
Noological system, 3–8, 15–19, 22, 29, 31, 33, 34, 36, 37, 42, 44, 46, 48, 49, 59, 61, 68, 75, 79–83, 90, 92, 99, 103, 118, 123, 124, 128, 132, 133, 138, 145, 147, 150–153, 155, 157, 158, 163, 167, 172, 173, 186, 187, 196, 208, 213, 215, 217, 221, 222, 224–226, 239, 242, 246, 248, 252, 279–281, 294, 297, 299, 300, 314, 315, 324, 342, 343, 356, 359, 364, 373, 391–394, 396–401, 407–410
    basic noological principles, 392, 393
Noology, 2, 217, 283, 343–356, 359, 389, 391, 394, 398, 415
Novel Instances of Concepts, 162

O
Obstruction, concept of, 178
Open-ended script, 274
Operational representation, 145–188, 191–217
Opportunistic situation, 48–51, 334

P
Pain, 42, 82, 84, 119, 128, 150, 151, 155, 158, 171, 172, 187, 202, 345–350, 353, 355, 360–365, 367–370, 372–374, 376, 379, 399, 401
Pain-causing activities, 360–364
Penetration, concept of, 178
Perception, 7, 16, 18, 19, 29–33, 35, 125, 136, 146, 147, 154, 185, 224, 356, 359, 392, 406
Perception as service function, 123, 138
Perceptual process, 6, 15, 138, 284, 349, 355
Persist, concept of, 158–160
Personality, 387, 388, 392, 408
Philosophy, 1

Phylogenetic changes of the cerebral cortex, 135
Physical experiment, 248, 257, 299, 301, 319, 320, 322–329, 332, 333
Physical observation, 293, 298, 299, 319
Physical Reasoner, 30
Physical World Array, 165, 166, 198, 200
Physics, 2, 30, 112, 117, 118, 122, 174, 178, 217, 249, 256, 294, 295, 346, 350, 354
Physiological need, 21, 399
Pictorial representation, 27, 92, 149, 217
Place, concept of, 185, 186
Predicate logic, 3, 28, 156, 188, 216, 325, 329, 391, 409
Prefrontal cortex, 18, 19, 83, 129, 131, 134, 137
Primary goal, 4, 8, 12, 14–16, 20–24, 32, 33, 36, 398
  explicit definition and representation, 4, 15, 393
Proactive installation, 58
Problem solving, accelerated, 5, 37, 107, 394, 398
Problem solving method, 115, 281, 284, 286, 287, 290, 304, 305, 308, 311, 314, 315, 341, 343
Procedural representation, 403
PROJECTILE-CAUSING-PAIN-SCRIPT, 360, 361, 363, 364, 368
Propulsion, concept of, 174–176
Proximal observation, 47
Psychology, 1, 5–8, 18, 213
Pull, concept of, 180–182, 198
Pull Test, 180, 181, 198, 199
Punishers, 400
Purchasing Need, 122, 123
Push, concept of, 180, 198
Putamen, 126

Q
Qualitative physics, 211, 216
Quick learning, 280–281, 342, 393, 413

R
Rapid causal learning, 43–44, 51–59, 86, 284, 341, 359, 403, 407
Rapid cause recovery, 48–51
Rapid effective causal learning, 3, 5, 6, 34–36, 42, 47, 56, 61, 66, 67, 281, 303, 317, 393
Rapid problem solving, 5, 6, 215, 393

Reasoning and problem solving, 145, 153, 162, 163, 167, 187, 188, 191, 192, 198–201, 214, 216, 410
Recognition-Generation Bidirectionality, 157, 162, 164, 186
Recovery from script thwarting, 244–280
Recurrent neural networks, 131–133, 216
  computational power of (*see* Computational power of recurrent neural network)
Recursive functions, 131
Reflection, concept of, 176–179, 196–198
Reinforcement learning, 5, 12, 19, 41, 42, 86, 128, 130–132, 134, 137, 138, 238, 281, 284, 317, 346, 355, 356, 408, 411, 412
Reinforcers, 22, 23
Relevant parameters for script, 70
Relief, state of, 346, 402
Representation of interactions, 174–182
Restaurant Script, 68, 69, 94–97, 99, 394–396, 398, 407, 409
Retroactive restoration, 52–54, 56, 57, 230
Robotic arm movement problem, 141
Rule confidence, 49, 55

S
Sadness, 119, 172, 402
Safety, 22, 401
SAS micro-environment, 128, 131, 343, 345, 351–356, 359, 360, 387, 388, 394, 401, 408
Scalability, 35, 86
Scalar parameter, 171, 172
Scaling up to the complex external environment, 394
Scaling up to the complex internal environment, 399
Script Base, 71, 73, 85, 109, 123, 242, 244, 248, 260, 264, 266, 281, 297, 301, 306, 309, 313, 318, 320, 363, 367, 371
Script construction, 15, 71, 72, 310, 312–313, 333, 360
Script query, 71, 73
Script thwarting, 244
Secondary goal, 13–16, 20, 23, 24
Second Infant Learning Phase, 112, 113, 288
Self-actualization, 21, 401
Self-awareness, 22, 406
Semantically grounded knowledge, 37, 394
Semantically grounded system, 5
Semantic grounding, 6, 23, 28, 124, 213–215, 411

Index   431

Semantic memory, 15, 284, 349
Semantic network, 4, 14, 147, 213, 406, 407
Service function, 3, 5, 18, 36, 221, 407
Shelter, 20, 106–108, 111, 115–118, 128, 349, 350, 352–355, 359, 387, 388
Shelter building problem, 111, 115–118, 141
Shield, 128, 344, 346, 348–353, 355, 359, 384, 386–388
Sigmoidal activation functions, 131
Single instance generalization, 59, 60
SMGO problem. *See* Spatial-movement-to-goal-with-obstacle (SMGO) problem
SMG problem. *See* Spatial-movement-to-goal (SMG) problem
Smile, 81, 90
Smoking and lung cancer, 44, 45, 47
Social situation, 408
Space Invaders, 389
Spatial movement, 27, 67, 71, 72, 74, 77–80, 90–97, 104, 112, 118, 119, 149, 171, 205, 238–280, 311, 379
Spatial-movement-to-goal (SMG) problem, 61–68, 70, 71, 74–76, 79–82, 84, 85, 90, 221, 238, 241, 242, 274
Spatial-movement-to-goal-with-obstacle (SMGO) problem, 61, 74, 238–282, 319, 363, 372–375, 377–380, 383, 384, 394, 408, 411
Spatiotemporal representation, 27, 28, 54, 145, 149, 151, 155, 176, 177, 185–187, 191, 192, 216
StarCraft game environment, 26, 99, 283–343, 394
STARCRAFT-MOVEMENT-SCRIPT, 286–289, 306, 310, 311, 340
Stay concept of, 158, 159
Stress, state of, 128, 350
Striatum, 126, 127
Strict-Appear, concept of, 158–160
Strict-Disappear, concept of, 158–160
String, 26, 148, 214, 411
Substantia nigra, 126, 127
Suicide, 22
Super-Turing computable, 131
Supervised learning, 5, 129, 130, 134, 137, 412, 415
Super-visual ability, 275, 277–279
Super-visual perception, 223

Symbolic circularity, 26–28
Symbolic inference, 406
Symbolic processing, 409, 410
Symbolic representation, 28, 148, 150, 281, 409
Symptomatic perceptual condition, 30, 32
Synchronic causal condition, 52–54, 57, 243, 403
Synchronic causal factors, 51–59

T
Thalamus, 125, 126
3D model, 221, 222
3D representation, 155, 188
3D space, 146, 153, 187, 188
Thwart-handling process, 245, 248, 261
Thwarting of script, 241–243
Thunder, 50, 51
Time, representation of, 173
Tool Construction phase, 210, 211
Tool Use phase, 210, 211
Trajectory, concept of, 161, 165
Transfer learning, 4, 67
Tree, concept of, 149
Turing test, 392
2D movement operations, 162
2D object, 210, 211
2D operator, 210
2D representation, 155, 188, 210–213
2D space, 146, 167, 187, 188, 210, 224–226, 228, 244
2.5D sketch, 221, 222, 407

U
Unsupervised causal learning, 130, 136, 137, 192, 412
Unsupervised learning, 5, 41, 130, 131, 183, 184, 412
Unsupervised rapid learning, 4

V
Visual feature, 14, 29, 154, 155, 196

W
White blood cell, 414

Printed by Printforce, the Netherlands